INTERNATIONAL HARVESTER

SHOP MANUAL IH-203

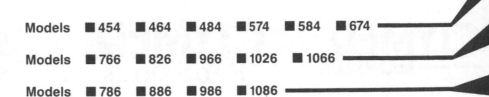

Models ■ 454 ■ 464 ■ 484 ■ 574 ■ 584 ■ 674

Models ■ 766 ■ 826 ■ 966 ■ 1026 ■ 1066

Models ■ 786 ■ 886 ■ 986 ■ 1086

I&T

SHOP MANUALS

Information and Instructions

This shop manual contains several sections each covering a specific group of wheel type tractors. The Tab Index on the preceding page can be used to locate the section pertaining to each group of tractors. Each section contains the necessary specifications and the brief but terse procedural data needed by a mechanic when repairing a tractor on which he has had no previous actual experience.

Within each section, the material is arranged in a systematic order beginning with an index which is followed immediately by a Table of Condensed Service Specifications. These specifications include dimensions, fits, clearances and timing instructions. Next in order of arrangement is the procedures paragraphs.

In the procedures paragraphs, the order of presentation starts with the front axle system and steering and proceeding toward the rear axle. The last paragraphs are devoted to the power take-off and power lift systems. Interspersed where needed are additional tabular specifications pertaining to wear limits, torquing, etc.

HOW TO USE THE INDEX

Suppose you want to know the procedure for R&R (remove and reinstall) of the engine camshaft. Your first step is to look in the index under the main heading of ENGINE until you find the entry "Camshaft." Now read to the right where under the column covering the tractor you are repairing, you will find a number which indicates the beginning paragraph pertaining to the camshaft. To locate this wanted paragraph in the manual, turn the pages until the running index appearing on the top outside corner of each page contains the number you are seeking. In this paragraph you will find the information concerning the removal of the camshaft.

More information available at haynes.com
Phone: 805-498-6703

J H Haynes & Co. Ltd.
Haynes North America, Inc.

ISBN-10: 0-87288-370-1
ISBN-13: 978-0-87288-370-3

© Haynes North America, Inc. 1990
With permission from J.H. Haynes & Co. Ltd.

Clymer is a registered trademark of Haynes North America, Inc.

Cover art by Sean Keenan

Disclaimer

There are risks associated with automotive repairs. The ability to make repairs depends on the individual's skill, experience and proper tools. Individuals should act with due care and acknowledge and assume the risk of performing automotive repairs.

The purpose of this manual is to provide comprehensive, useful and accessible automotive repair information, to help you get the best value from your vehicle. However, this manual is not a substitute for a professional certified technician or mechanic.

This repair manual is produced by a third party and is not associated with an individual vehicle manufacturer. If there is any doubt or discrepancy between this manual and the owner's manual or the factory service manual, please refer to the factory service manual or seek assistance from a professional certified technician or mechanic.

Even though we have prepared this manual with extreme care and every attempt is made to ensure that the information in this manual is correct, neither the publisher nor the author can accept responsibility for loss, damage or injury caused by any errors in, or omissions from, the information given.

INTERNATIONAL HARVESTER

Models ■454 ■464 ■484 ■574 ■584 ■674

Previously contained in I&T Shop Manual No. IH-44

SHOP MANUAL

INTERNATIONAL HARVESTER

SERIES 454-464-484-574-584-674

Engine serial number is stamped on left side of engine crankcase on all non-diesel series. Engine serial number is stamped on right side of engine crankcase on all diesel series.

Tractor serial number is stamped on name plate attached to left side of speed transmission housing on Models 454, 464, 574 and 674 or on the right rear side of front axle support on Models 484 and 584.

INDEX (By Starting Paragraph)

INDEX CONT.

CONDENSED SERVICE DATA

	454, 464 Non-Diesel	454, 464, 484 Diesel	574, 674 Non-Diesel	574, 674 Diesel	584 Diesel
GENERAL					
Engine Make	IH	IH	IH	IH	IH
Engine Model	(1)	D-179	C-200	D-239	D-206
Number of Cylinder	4	3	4	4	4
Bore-Inches	(2)	3-7/8	3-13/16	3-7/8	3-7/8
Stroke-Inches	4-25/64	5-1/16	4-25/64	5-1/16	4-3/8
Main Bearing, Number of	3	4	3	5	5
Cylinder Sleeves	...	Wet	...	Wet	Wet
Forward Speeds, Number of	8	8	8	8	8
Alternator/Starter Make			Delco-Remy and Lucas		

(1) 454-C157, 464-C175
(2) 454-3-3/8, 464-3-9/16

	454, 464 Non-Diesel	454, 464, 484 Diesel	574, 674 Non-Diesel	574, 674 Diesel	584 Diesel
TUNE-UP					
Compression Pressure	170 (1)	315-340 (1)	185 (1)	315-340 (1)	315-340 (1)
Firing order	1-3-4-2	1-3-2	1-3-4-2	1-3-4-2	1-3-4-2
Valve Tappet Gap (Hot)					
Intake	0.027	0.010	0.027	0.010	0.010
Exhaust	0.027	0.012	0.027	0.012	0.012
Valve Seat Angle (Degrees)	30	45	30	45	45
Ignition Distributor Make	(2)	...	(2)	...	...
Breaker Gap	0.020	...	0.020	...	...
Distributor Timing	See Para. 115	...	See Para. 115,117	...	...
Timing Mark Location			Crankshaft Pulley		
Spark Plug Electrode Gap	0.023	...	0.023	...	...
Carburetor Make	M-S	...	M-S or Zenith	...	...
Injection Pump Make	...	Robert Bosch	...	Robert Bosch	Robert Bosch
Injection Pump Timing	...	14°BTDC	...	14°BTDC	14°BTDC
Battery Terminal, Grounded			Negative		
Engine Low Idle RPM	625	750	625	700	750
Engine High Idle RPM, No Load	See Para. 101	See Para. 93	See Para. 101	See Para. 93	See Para. 93
Engine Full Load RPM	(3)	(3)	(3)	(3)	(3)

(1) Approximate psi, at sea level, at cranking speed.
(2) Delco-Remy or Prestolite.
(3) 454, 484, 574-2200 rpm; 464, 674-2400 rpm; 584-2300 rpm.

	454, 464 Non-Diesel	454, 464, 484 Diesel	574, 674 Non-Diesel	574, 674 Diesel	584 Diesel
SIZES-CAPACITIES-CLEARANCES					
Crankshaft Main Journal Diameter	2.748-2.749	3.1484-3.1492	2.748-2.749	3.1484-3.1492	3.1484-3.1492
Crankpin Diameter	(1)	2.5185-2.5193	2.373-2.374	2.5185-2.5193	2.5185-2.5193
Camshaft Journal Diameter					
No. 1 (Front)	2.089-2.090	2.2823-2.2835	2.089-2.090	2.2823-2.2835	2.2823-2.2835
No. 2	2.069-2.070	2.2823-2.2835	2.069-2.070	2.2823-2.2835	2.2823-2.2835
No. 3	1.499-1.500	2.2823-2.2835	1.499-1.500	2.2823-2.2835	2.2823-2.2835
No. 4	...	2.2823-2.2835	...	2.2823-2.2835	2.2823-2.2835
No. 5	...	...	...	2.2823-2.2835	2.2823-2.2835
Piston Pin Diameter	(2)	1.4172-1.4173	0.8748-0.8749	1.4172-1.4173	1.4172-1.4173
Valve Stem Diameter					
Intake	0.3715-0.3725	0.3919-0.3923	0.3715-0.3725	0.3919-0.3923	0.3919-0.3923
Exhaust	0.371-0.372	0.3911-0.3915	0.371-0.372	0.3911-0.3915	0.3911-0.3915
Main Bearing Diametral Clearance	0.0012-0.0042	0.0028-0.0055	0.0012-0.0042	0.0028-0.0055	0.0028-0.0055
Rod Bearing Diametral Clearance	(3)	0.0023-0.0048	0.0009-0.0034	0.0023-0.0048	0.0023-0.0048
Piston Skirt Diametral Clearance	0.002-0.005	0.0039-0.0047	0.002-0.005	0.0039-0.0047	0.0039-0.0047
Crankshaft End Play	0.005-0.010	0.006-0.009	0.005-0.010	0.006-0.009	0.006-0.009
Camshaft Bearing Diametral Clearance	0.0005-0.005	0.0009-0.0033	0.0005-0.005	0.0009-0.0033	0.0009-0.0033
Camshaft End Play	0.002-0.010	0.004-0.018	0.002-0.010	0.004-0.018	0.004-0.018
Cooling System Capacity-Qts.	11.5	12.5	12	14	14.5
Crankcase Oil-Qts.	7	8	7	10	9
Transmission and Differential-Gallons	9	9	9	9	9

(1) 454, 2.059-2.060; 464, 2.373-2.374
(2) 454, 0.8591-0.8593; 464, 0.8748-0.8749
(3) 454, 0.0009-0.0039; 464, 0.0009-0.0034

FRONT SYSTEM AXLE TYPE

FRONT SYSTEM AXLE TYPE

Two types of front axles are available; the swept back adjustable (Fig. 1) and the straight adjustable (Fig. 2).

AXLE MAIN MEMBER

All Models

1. To remove either the swept back or straight axle main member (10—Fig. 1 or Fig. 2), disconnect tie rods from steering arms (1 and 17). Identify and disconnect steering lines. Cap or plug openings immediately to prevent entrance of dirt or other foreign material into the system. Unbolt and remove the cylinder retaining bracket (15) and remove the cylinder rod anchor pin. Carefully lower cylinder, steering arm and tie rods from axle. Support front of tractor, remove axle adjusting bolts (12) and withdraw axle extensions (3 and 16), steering knuckles and front wheels as assemblies. Place a floor jack under axle main member (10), unbolt and remove axle pivot pin (8), then lower axle main member from front support (19).

Inspect pivot pin (8) and bushings (9, 11 and 13) and renew as required. Reinstall by reversing the removal procedure and if necessary, adjust toe-in as outlined in paragraph 2.

TIE RODS AND TOE-IN

All Models

2. The procedure for removal and disassembly of the tie rods on all models is obvious after an examination of the units. Tie rod ends are non-adjustable and faulty units will require renewal.

Adjust toe-in on all models to ¼-inch, plus or minus 1/16-inch. Adjustment is made by varying the length of the tie rods. Both tie rods should be adjusted an equal amount with not more than one turn difference when adjustment is complete.

STEERING KNUCKLES

All Models

3. To remove either steering knuckle (5—Fig. 1 or Fig. 2), jack up under axle extension and remove front wheel. Disconnect tie rod from steering arm, remove steering arm retaining bolt and

pull arm from steering knuckle. Remove the Woodruff key and lower the steering knuckle assembly from axle extension. Remove thrust bearing (4) and felt washer (7). Hub and wheel bearings can now be removed if necessary.

Inspect bushings (2) and renew as required. Install new bushings with open ends of lubrication grooves inward. Press bushings in until they are 1/16-inch below flush with top and bottom

surface of extension bore. Install thrust bearing (4) with chamfered side downward.

Pack wheel bearings using IH251H EP or equivalent No. 2 lithium grease. Tighten bearing adjusting nut while rotating the hub to 70 ft.-lbs. torque. Back the nut off and retorque to 50 ft.-lbs. Back nut off ¼-turn from this position and install cotter pin.

The balance of reassembly is the reverse of disassembly procedure.

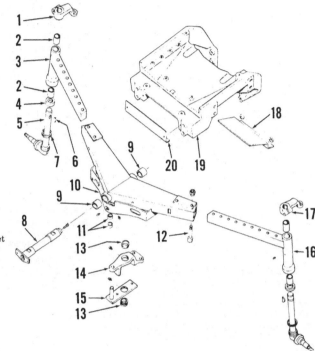

Fig. 1—Exploded view of the swept back adjustable front axle assembly.

1. Steering arm R.H.
2. Bushings
3. Axle extension R.H.
4. Thrust bearing
5. Steering knuckle
6. Key
7. Felt washer
8. Axle pivot pin
9. Bushings
10. Axle main member
11. Bushings
12. Adjusting bolt (4)
13. Bushings
14. Center steering arm
15. Cylinder retaining bracket
16. Axle extension L.H.
17. Steering arm L.H.
18. Baffle plate
19. Front support
20. Front cover

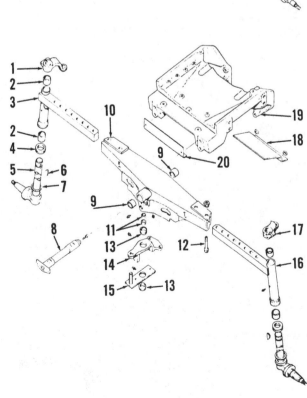

Fig. 2—Exploded view of the straight adjustable front axle assembly. Refer to Fig. 1 for legend.

HYDROSTATIC STEERING SYSTEM

NOTE: The maintenance of absolute cleanliness of all parts is of utmost importance in the operation and servicing of the hydrostatic steering system. Of equal importance is the avoidance of nicks or burrs on any of the working parts.

STEERING FLUID CIRCUITS

All Models

4. Power steering is standard equipment on all tractors and is similar except for the hand pump and steering valve assembly.

Refer to Fig. 3 for view showing the relative position of power steering components. The 12.5 gpm (Model 584) or 12 gpm (all other models) hydraulic pump is mounted on inside of multiple control valve, Fig. 4. Fluid passes from the pump to a priority type flow divider valve which supplies a controlled flow of approximately 3 gpm to the power steering. The return flow from the power steering is returned back to the multiple control valve and is controlled at a lower pressure by the pto regulating valve. If pto is engaged, the fluid flows to the pto clutch assembly with excess fluid going to oil cooler. If pto is in disengaged position, the power steering fluid returns to oil cooler, which is protected by a bypass valve.

LUBRICATION AND BLEEDING

All Models

5. The tractor rear frame serves as a common reservoir for all hydraulic and lubrication operations. The filter element, shown in Fig. 4, should be renewed at 10 hours, 100 hours, 200 hours and then every 200 hours of operation thereafter. Hydraulic fluid should be drained and new fluid added every 800 hours, or once a year, whichever occurs first.

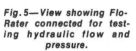

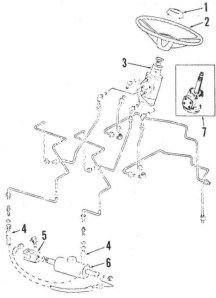

Fig. 3—Schematic view showing general layout of component parts of typical power steering system on International tractors. Ross hand pump and steering valve (3) is used on Models 454, 464, 574 and 674. Danfoss pump and steering valve (7) in inset is used on Models 484 and 584.

1. Cap
2. Steering wheel
3. Hand pump (Ross)
4. Cylinder hoses
5. Anchor clevis
6. Steering cylinder
7. Hand pump (Danfoss)

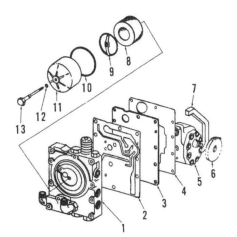

Fig. 4—Multiple control valve, hydraulic pump and filter located on left side of clutch housing.

1. Multiple control valve
2. Gasket
3. Plate
4. Gasket
5. Pump
6. Gear
7. Seal
8. Filter element
9. Filter by-pass valve
10. Gasket
11. Cover
12. "O" ring
13. Bolt

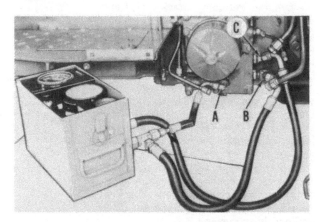

Fig. 5—View showing Flo-Rater connected for testing hydraulic flow and pressure.

A. Power steering pressure (outlet)
B. Hitch pressure (outlet)
C. Hitch return

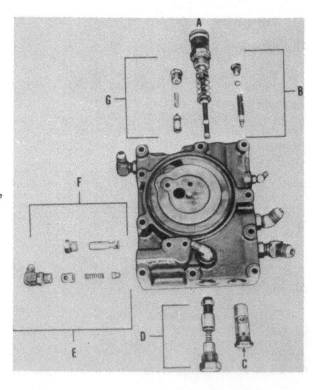

Fig. 6—Multiple control valve removed and showing control valves removed from their bores.

A. Pto valve
B. Pto pressure regulating valve
C. System relief valve (2300-2500 psi)
D. Flow divider
E. Check valve power steering return
F. Power steering relief valve (1500-1600 psi)
G. By-pass valve (oil cooler)

Only IH "Hy-Tran" fluid should be used and level should be maintained at the "FULL" mark shown on level gage (dipstick). The dipstick is located on left front side of rear frame.

Whenever power steering lines have been disconnected, or fluid changed, start engine and cycle power steering system from stop to stop several times to bleed air from system. Then, check and if necessary, add fluid.

HYDRAULIC FLOW AND PRESSURE TEST

All Models

6. A flow and pressure test can be made with a Flo-Rater or equivalent. Refer to Fig. 5 and connect inlet of Flo-Rater to the outlet of the power steering and hitch. Then, connect outlet of Flo-Rater to the return port from the hitch as shown.

7. HYDRAULIC PUMP. To check the pump free flow run engine at rated rpm (2200 on Models 454, 574 or 484, 2300 on Model 584 or 2400 on Models 464 and 674). Restrict the Flo-Rater to 1250 psi and there should be a 12.5 gpm flow on Model 584 or a 12 gpm flow on all other models.

If pump free flow is not as specified, remove and service pump as outlined in paragraphs 12 and 13.

8. STEERING RELIEF VALVE. To check the relief valve, restrict the Flo-Rater to 1500-1600 psi; at this point the valve should open. If pressure is not as specified, renew the relief valve assembly (F—Fig. 6) in the multiple control valve.

9. FLOW DIVIDER. To check the flow divider, restrict the Flo-Rater to 1500-1600 psi; approximately 2½ gpm

flow loss should occur when relief valve opens. If flow is not as specified, remove the flow valve (D—Fig. 6) in the multiple control valve and service by renewing parts.

HYDRAULIC PUMP

All Models

10. Tractors may be equipped with either Cessna (Fig. 7) or Plessey (Fig. 8) gear type pumps. Pump used on Model 584 is 12.5 gpm and pumps used on all other models are 12 gpm. A priority type flow divider takes approximately 3 gpm for the power steering and the balance is used for the draft, position and auxiliary valves in the hydraulic system.

11. REMOVE AND REINSTALL. To remove the hydraulic pump, drain all compartments of Hy-Tran and remove hydraulic filter. Disconnect all lines to Multiple Control Valve (MCV) and plug or cap openings. Disconnect pto linkage. Unbolt and remove MCV and pump. Remove cap screws securing pump to MCV.

NOTE: There are copper sealing washers under pump mounting cap screws.

12. OVERHAUL [CESSNA]. With pump removed from MCV, proceed as follows: Remove pump drive gear and key, then unbolt and remove cover (11—Fig. 7) from body (1). Balance of disassembly will be obvious after an examination of unit.

When reassembling, use new diaphragm, gaskets, back-up gasket, diaphragm seal and "O" ring. With open part of diaphragm seal (8) toward cover (11), work same into grooves of cover using a dull tool. Press protector gasket (7) and back-up gasket (6) into relief of diaphragm seal. Install check ball (12) and spring (13) in cover, then install diaphragm (5) inside the raised lip of the diaphragm seal and be sure bronze face of diaphragm is toward pump gears. Dip gear and shaft assemblies in oil and install them in cover. Position thrust plate (2) in pump body

with bronze side toward pump gears and cut-out portion toward inlet (suction) side of pump. Install pump body over gears and shafts and install retaining cap screws. Torque cap screws evenly and alternately to 40 ft.-lbs.

Check pump rotation. Pump will have a slight amount of drag but should turn evenly.

Pump gears and shafts, are available only as a pump assembly.

13. OVERHAUL [PLESSEY]. With pump removed from MCV, proceed as follows: Remove pump drive gear and key, then unbolt and remove covers (1 and 9—Fig. 8) from body. Balance of disassembly will be obvious after an examination of unit.

With pump disassembled, inspect all parts for burrs, scoring, wear or other damage.

Use all new seals, backing strip, and "O" ring, which are available only as a seal package. Any other parts are available only as a pump assembly.

Reassemble by reversing the disassembly procedure.

STEERING CONTROL VALVE

Models 454-464-574-674

14. REMOVE AND REINSTALL. To remove steering control valve, first remove cap (monogram) from steering wheel, remove nut, attach puller and remove steering wheel. Remove hood and side sheets, tunnel cover, batteries and instrument panel side covers. On diesel tractors, remove fuel shut off knob. On non-diesel tractors, remove choke knob and disconnect from back of panel. Identify and disconnect the four hydraulic lines from steering control valve. Plug and cap openings to prevent entrance of dirt in system. Remove the bolts securing the instrument panel to battery box and the front capscrews only on each side securing the steering pedestal to clutch housing and tilt pedestal rearward. Loosen the throttle indicator Allen screw and disconnect the throttle linkage inside

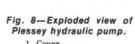

Fig. 7—Exploded view of Cessna hydraulic pump.

1. Pump body	7. Protector gasket
2. Thrust plate	(thin)
3. Driving gear	8. Diaphragm seal
4. Driven gear	9. "O" ring
5. Diaphragm	10. Seal
6. Back-up gasket	11. Cover
(thick)	12. Ball
	13. Spring

Fig. 8—Exploded view of Plessey hydraulic pump.

1. Cover
2. Backing strip
3. Sealing ring
4. Backing strip
5. Bearings
6. Gear assy.
7. Woodruff key
8. Pump body
9. Cover
10. Seal
11. Seal Retainer
12. "O" ring
13. "O" ring

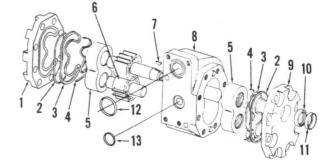

steering column. With throttle shaft between high and low speed, push the handle down to compress the spring. Clamp the shaft using vise grip pliers, then drive the two bottom roll pins out. Release the vise grip pliers and remove the remaining roll pin. Remove the four cap screws securing instrument panel to the support assembly and move the panel to the side. Remove the cap screws securing steering control valve and remove valve.

Reinstall by reversing the removal procedure; then bleed hydraulic system as outlined in paragraph 5.

15. **OVERHAUL.** To disassemble the removed steering control valve assembly, refer to Fig. 9 and proceed as follows: Install a fitting in one of the four ports in valve body (25), then clamp fitting in a vise so that input shaft (17) is pointing downward. Apply match marks across end cover and valve body for aid in reassembly. Remove cap screws (39) and end cover (38).

NOTE: Lapped surfaces of end cover (38), commutator set (33 and 34), manifold (32), stator-rotor set (31), spacer (29) and valve body (25) must be protected from scratching, burring or any other damage as sealing of these parts depends on their finish and flatness.

Remove seal retainer (35) and seal (36), then carefully remove washer (37), commutator set (33 and 34) and manifold (32). Grasp spacer (29) and lift off the spacer, drive link (30) and stator-rotor set (31) as an assembly. Separate spacer and drive link from stator-rotor set.

Remove unit from vise, then clamp fitting in vise so that input shaft is pointing upward. Remove water and dirt seal (2) and felt seal (3). Place a light mark on flange of upper cover (9) and valve body (25) for aid in reassembly. Unbolt upper cover from valve body, then grasp input shaft and remove input shaft, upper cover and valve spool assembly. Remove and dis-

card seal ring (10). Slide upper cover assembly from input shaft and remove Teflon spacer (16). Remove shims (12) from cavity in upper cover or from face of thrust washer (14) and note number of shims for aid in reassembly. Remove snap ring (4), stepped washer (5), brass washer (6), Teflon washer (7) and seal (8). Retain stepped washer (5) and snap ring (4) for reassembly. Do not remove needle bearing (11) as it is not serviced separately.

Remove snap ring (13), thrust washers (14) and thrust bearing (15) from input shaft. Drive out pin (18) and withdraw torsion bar (21) and spacer (20). Place end of valve spool on top of bench and rotate input shaft until drive ring (19) falls free, then rotate input shaft clockwise until actuator ball (23) is disengaged from helical groove in input shaft. Withdraw input shaft and remove actuator ball. Do not remove actuator ball retaining spring (24) unless renewal is required.

Remove plug (28) and recirculating ball (26) from valve body.

Thoroughly clean all parts in a suitable solvent, visually inspect parts and renew any showing excessive wear, scoring or other damage.

If needle bearing (11) is excessively worn or otherwise damaged, renew upper cover assembly (9) as bearing is not serviced separately.

Using a micrometer, measure thickness of the commutator ring (33—Fig. 9) and commutator (34). If commutator ring is 0.0015 inch or more thicker than commutator, renew the matched set.

Place the stator-rotor set (31) on the lapped surface of end cover (38). Make certain that vanes and vane springs are installed correctly in slots of the rotor. Arched back of springs must contact

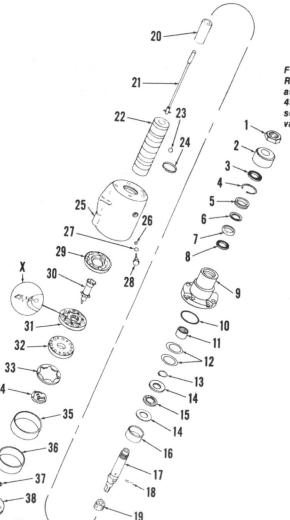

Fig. 9—Exploded view of Ross steering control valve assembly used on Models 454, 464, 574 and 674. Inset "X" shows vane and vane spring used in slot on each rotor lobe.

1. Nut
2. Water & dirt seal
3. Felt seal
4. Snap ring
5. Stepped washer
6. Brass washer
7. Teflon washer
8. Seal
9. Cover (upper)
10. Seal ring
11. Needle bearing
12. Shims
13. Snap ring
14. Thrust washers
15. Thrust bearing
16. Teflon spacer
17. Input shaft
18. Pin
19. Drive ring
20. Spacer
21. Torsion bar
22. Valve spool
23. Actuator ball
24. Retaining spring
25. Valve body
26. Recirculating ball
27. "O" ring
28. Plug
29. Spacer plate
30. Drive link
31. Stator-Rotor set
32. Manifold
33. Commutator ring
34. Commutator
35. Seal retainer
36. Seal
37. Washer
38. End cover
39. Cap screws

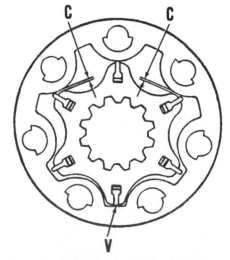

Fig. 10—With rotor positioned in stator as shown, clearances (C) must not exceed 0.006 inch. Refer to text.

vanes. See inset X—Fig. 9. Position lobe of rotor in valley of stator as shown at (V—Fig. 10). Center opposite lobe on crown of stator, then using two feeler gages, measure clearance (C) between rotor lobes and stator. If clearance is more than 0.006 inch, renew stator-rotor assembly. Using a micrometer, measure thickness of stator and rotor. If stator is 0.002 inch or more thicker than rotor, renew the assembly. Stator, rotor, vanes and vane springs are available only as an assembly.

Before reassembling, wash all parts in clean solvent and air dry. All parts, unless otherwise indicated, are installed dry. Install recirculating ball (26—Fig. 9) and plug (28) with new "O" ring (27) in valve body and tighten plug to a torque of 10-14 ft.-lbs. Clamp fitting (installed in valve body port) in a vise so that top end of valve body is facing upward. Install thrust washer (14), thrust bearing (15), second thrust washer (14) and snap ring (13) on input shaft (17). If actuator ball retaining ring (24) was removed, install new retaining ring. Place actuator ball (23) in its seat inside valve spool (22). Insert input shaft into valve spool, engaging the helix and actuator ball with a counter-clockwise motion. Use the midsection of torsion bar (21) as a gage between end of valve spool and thrust washer, then place the assembly in a vertical position with end of input shaft resting on a bench. Insert drive ring (19) into valve spool until drive ring is engaged on input shaft spline. Remove torsion bar gage. Install spacer (20) on torsion bar and insert the assembly into valve spool. Align crossholes in torsion bar and input shaft and install pin (18). Pin must be pressed into shaft until end of pin is about 1/32-inch below flush. Place spacer (16) over spool and install spool assembly into valve body. Position original shims (12) on thrust washer (14), lubricate new seal ring (10), place seal ring in upper cover (9) and install upper cover assembly. Align the match marks on cover flange and valve body and install cap screws finger tight. Tighten a worm drive type hose clamp around cover flange and valve body to align the outer diameters, then tighten cap screws to a torque of 18-22 ft.-lbs.

NOTE: If either input shaft (17) or upper cover (9) or both have been renewed, the following procedure for shimming must be used. With upper cover installed (with original shims) as outlined above, invert unit in vise so that input shaft is pointing downward. Grasp input shaft, pull downward and prevent it from rotating. Engage drive link (30) splines in valve spool and ro-

tate drive link until end of spool is flush with end of valve body. Remove drive link and check alignment of drive link slot to torsion bar pin. Install drive link until its slot engages torsion bar pin. Check relationship of spool end to body end. If end of spool is within 0.0025 inch of being flush with end of body, no additional shimming is required. If not within 0.0025 inch of being flush, remove cover and add or remove shims (12) as necessary. Reinstall cover and recheck spool to valve body position.

With drive link installed, place spacer plate (29) on valve body with plain side up. Install stator-rotor set over drive link splines and align cap screw holes. Make certain vanes and vane springs are properly installed. Install manifold (32) with circular slotted side up and align cap screw holes with stator, spacer and valve body. Install commutator ring (33) with slotted side up, then install commutator (34) over drive link end making certain that link end is engaged in the smallest elongated hole in commutator. Install seal (36) and retainer (35). Apply a few drops of hydraulic fluid on commutator. Use a small amount of grease to stick washer (37) in position over pin on end cover (38). Install end cover making sure that pin engages center hole in commutator. Align holes and install cap screws (39). Alternately and progressively tighten cap screws while rotat-

ing input shaft. Final tightening should be 18-22 ft.-lbs. torque.

Relocate the unit in vise so input shaft is up. Lubricate new seal (8) and carefully work seal over shaft and into bore with lip toward inside. Install new Teflon washer (7), brass washer (6) and stepped washer (5) with flat side up. Install snap ring (4) with rounded edge inward. Place new felt seal (3) and water and dirt seal (2) over input shaft.

Remove unit from vise and remove fitting from port. Turn unit on its side with hose ports upward. Pour clean hydraulic fluid into inlet port, rotate input shaft until fluid appears at outlet port, then plug all ports.

Models 484-584

16. **REMOVE AND REINSTALL.** To remove steering control valve, remove access panel from top of hood, lower side panels from instrument panel, battery and battery tray. Remove steering wheel cap and nut. Using a suitable puller, remove steering wheel. Identify and disconnect the four hydraulic lines from steering valve. Immediately plug or cap all openings to prevent dirt or other foreign material from entering system. Working through opening in instrument panel along side of steering column, remove steering valve mounting bolts and remove the assembly.

Reinstall by reversing the removal procedure, then bleed air from system as outlined in paragraph 5.

17. **OVERHAUL.** Clean exterior of unit and scribe match marks on end plate, rotor plate, distributor plate and valve housing as shown in Fig. 11. Unbolt and remove steering column and shaft (3) from steering valve. Clamp valve housing in a soft jaw vise in inverted position and remove end plate cap screws (25 and 26—Fig. 12).

NOTE: One cap screw (25) has a roll pin in the end to restrain the check ball (10).

Remove end plate (24), wear washer (17), rotor plate (21) and rotor (22). Remove distributor plate (19) and drive shaft (16), then remove check ball retaining threaded bushing (9) from valve housing and remove check ball (10). Push valve spool assembly from valve housing. Remove axial washer (5), needle (thrust) bearing (6), bearing race (7) and retaining ring (8) from end of spool. Remove cross pin (11) from spool, then slide spool (12) from valve sleeve (15). Remove inner and outer plate springs (13 and 14) from spool. Remove sealing ring (4), "O" ring (3) and oil seal (1) from housing (2).

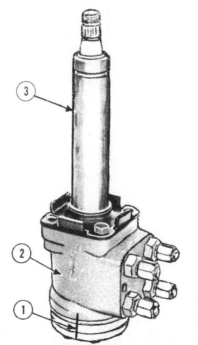

Fig. 11—View of steering valve (hand pump) removed from Model 484 or 584.

1. Match marks
2. Valve housing
3. Steering column & shaft assy.

Clean and inspect all parts and renew any showing excessive wear or other damage. If housing (2), spool (12) or sleeve (15) are not suitable for further service, renew complete assembly as these parts are not serviced separately. All "O" rings and seals are available in a seal kit. Lubricate all internal parts with Hy-Tran Fluid and coat all "O" rings with vaseline. Reassemble by reversing the disassembly procedure. Install two outer plate springs, then install four inner plate springs as shown in Fig. 13. Use cross-sectional view (Fig. 14) as a guide when reassembling. Align the previously installed match marks and tighten end plate cap screws to a torque of 260-300 in.-lbs. Reinstall steering column and upper shaft assembly.

STEERING CYLINDER

All Models

18. **R&R AND OVERHAUL.** To remove the power steering cylinder, disconnect and immediately plug the hydraulic lines. Remove cap screws from cylinder support arms (center steering arm) and separate arms to free cylinder. Remove pin retaining anchor clevis to axle main member and remove cylinder assembly from tractor.

With cylinder removed, move piston rod back and forth several times to clear oil from cylinder. Place cylinder in a vise and clamp vise only enough to prevent cylinder from turning. Using a wrench in the flats of piston rod, remove the anchor clevis on one end and

the fitting (elbow) on other end. Remove cylinder head retaining ring (11—Fig. 15) as follows: Lift end of retaining ring out of slot, then using a pin type spanner, rotate cylinder head (2) and work retainer ring out of its groove. Cylinder rod and piston assembly (10) can now be removed from tube (1). Remove remaining cylinder head in the same manner. All seals, "O" rings and back-up washers are now available for inspection and/or renewal. Clean all parts in a suitable solvent and inspect. Check cylinder tube for scoring, grooving and out-of-roundness. Light scoring can be polished out by using a fine emery cloth and oil, providing a rotary motion is used during polishing operation. A cylinder tube that is heavily scored or grooved, or is out-of-round, should be renewed. Check piston rod and piston for scoring, grooving and straightness. Polish out very light scoring with fine emery cloth and oil, using rotary motion. Renew rod and piston assembly if heavily scored or

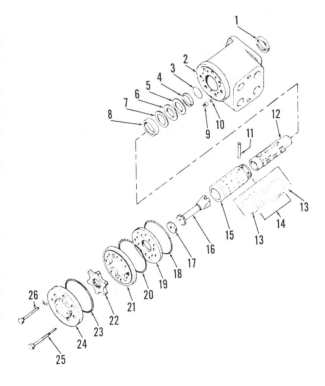

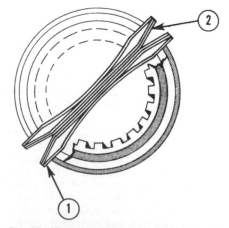

Fig. 13—View showing correct installation of the two outer plate springs (1) and four inner plate springs (2).

Fig. 12—Exploded view of steering valve used on Models 484 and 584.

1. Oil seal
2. Housing
3. "O" ring
4. Sealing ring
5. Axial washer
6. Needle bearing
7. Bearing race
8. Retaining ring
9. Threaded bushing
10. Check ball
11. Cross pin
12. Spool
13. Outer plate springs (2)
14. Inner plate springs (4)
15. Sleeve
16. Drive shaft
17. Wear washer
18. "O" ring
19. Distributor plate
20. "O" ring
21. Outer ring
22. Inner rotor
23. "O" ring
24. End plate
25. Cap screw w/roll pin
26. Cap screw (6)

Fig. 14—Cross-sectional view of Danfoss steering valve used on Models 484 and 584.

1. Plate springs
2. Cross pin
3. Valve body
4. Check ball
5. Spool
6. Sleeve
7. Drive shaft
8. Outer ring
9. Inner rotor

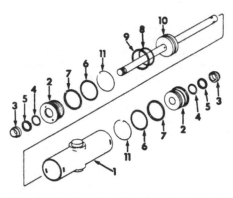

Fig. 15—Exploded view of power steering cylinder used on all models.

1. Cylinder tube
2. Cylinder head
3. Wiper seal
4. "O" ring
5. Back-up washer
6. "O" ring
7. Back-up washer
8. Piston "O" ring
9. Piston seal ring
10. Piston and rod
11. Cylinder head retaining ring

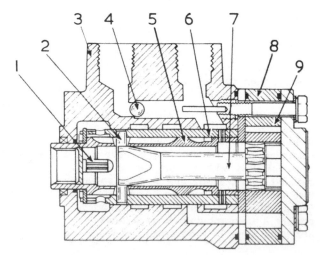

grooved, or if piston rod is bent. Inspect piston seal ring (9) for frayed edges, wear and imbedded dirt or foreign particles. Renew piston seal ring if any of the above conditions are found. Inspect balance of "O" rings, back-up washers and seals and renew as necessary. Inspect bores of cylinder heads and renew same if excessively worn or out-of-round.

Reassemble steering cylinder as follows: Install "O" ring (4), back-up washer (5), wiper seal (3), back-up washer (7) and cylinder head "O" ring (6) to cylinder head, then install cylinder head assembly over threaded end of piston rod. Lubricate piston seal ring (9) and cylinder head "O" ring (6), then using a ring compressor or a suitable

hose clamp, install piston and rod assembly into cylinder tube. Install cylinder head in cylinder tube so hole in groove will accept nib of retaining ring. Position retaining ring and pull same into its groove by rotating cylinder head. Install remaining cylinder head. Install anchor clevis and torque to 170-180 ft.-lbs.

Reinstall cylinder assembly on tractor, then fill and bleed power steering system as outlined in paragraph 5.

OIL COOLER

All Models

19. **R&R AND OVERHAUL.** The re-

turn fluid from the power steering flows through internal ports in the multiple control valve to pto control valve. Fluid not used by the pto is routed through internal ports to oil cooler by-pass valve (Fig. 6) and to oil cooler, returning back to reservoir.

Service of the oil cooler involves only removal and reinstallation, or renewal of faulty units. However, since it is hinged on one side, the oil cooler will swing out for cleaning after radiator grille is removed. Removal of the oil cooler is obvious after examination of the unit. However, outlet and inlet hoses must be identified as they are removed from oil cooler pipes so oil circuits will be kept in the proper sequence.

NON-DIESEL ENGINE AND COMPONENTS

All non-diesel models are equipped with a four cylinder engine. Model 454 has a bore and stroke of 3-3/8 x 4-25/64 inches and a displacement of 157 cubic inches. Model 464 has a bore and stroke of 3-9/16 x 4-25/64 inches and a displacement of 175 cubic inches. Models 574 and 674 have a bore and stroke of 3-13/16 x 4-25/64 inches and a displacement of 200 cubic inches.

R&R ENGINE ASSEMBLY

All Models

23. To remove the engine assembly, first remove hood and rear side panels, drain cooling system and disconnect battery ground cable. Disconnect wires to alternator and wire to headlights. Disconnect steering lines and oil cooler lines and plug openings to prevent dirt from entering the system. Disconnect air cleaner scoop and radiator hoses. Support tractor under clutch housing, attach hoist to front support, then unbolt and remove front support, radiator, axle and wheels as an assembly. Shut off fuel and disconnect fuel line at fuel pump. Disconnect electrical wires from engine, unclip harness and lay wiring harness rearward. Remove temperature sensing bulb and oil pressure tube from engine, then disconnect throttle rod, choke and tachometer cable. Attach hoist to engine, unbolt and remove engine assembly.

When reinstalling engine assembly, unbolt and remove clutch assembly and position it on transmission input shaft. Disconnect the linkage from clutch release shaft. Clutch can be bolted to flywheel after engine is installed by work-

ing through the opening at bottom of clutch housing.

NOTE: If flywheel and clutch plate balancing marks are indicated (dab of white paint), they must be aligned.

CYLINDER HEAD

All Models

24. **REMOVE AND REINSTALL.** Drain coolant, remove hood and disconnect battery cable. Unbolt and remove air cleaner. Disconnect upper radiator hose and remove coolant temperature bulb from engine. Disconnect spark plug wires, unbolt coil bracket from cylinder head and lay coil and wiring harness out of the way.

NOTE: If equipped with Magnetic Pulse ignition system, remove pulse amplifier and lay it aside.

Disconnect controls from carburetor, then unbolt and remove manifold and carburetor assembly. Remove rocker arm cover, rocker arms and shaft assembly and push rods. Remove cylinder head cap screws and lift off cylinder head.

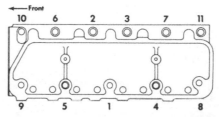

Fig. 17—Cylinder head cap screws are tightened in the sequence shown. Refer to text.

When reinstalling cylinder head, use new head gasket and make certain gasket sealing surfaces are clean. Use guide studs when installing cylinder head.

Tighten cylinder head retaining cap screws in two steps, using the sequence shown in Fig. 17. Tighten the cap screws to a torque of 65 ft.-lbs. in first step and then to 90 ft.-lbs. in final step. Adjust valve tappet gap to 0.027 inch as outlined in paragraph 25.

VALVES AND SEATS

All Models

25. Inlet and exhaust valves are not interchangeable. Inlet valves seat directly in the cylinder head; whereas, the cylinder head is fitted with renewable seat inserts for the exhaust valves. Inserts are available in standard size as well as oversize of 0.015 and 0.030 inch. Valve face and seat angle for both the inlet and exhaust is 30 degrees. Valve rotators (Rotocoils) are used on the exhaust valves and an umbrella type stem seal is used on inlet valves.

When removing valve seat inserts, use the proper puller. Do not attempt to drive chisel under insert or counterbore will be damaged. Chill new seat insert with dry ice or liquid Freon prior to installing. When new insert is properly bottomed, it should be 0.008-0.030 inch below edge of counterbore. After installation, peen the cylinder head material around the complete outer circumference of the valve seat insert. The O.D. of new standard insert is 1.5655 inches and the insert counterbore I.D. is 1.5625 inches.

Check the valves and seats against the specifications which follow:

Inlet
Face and seat angle . . . . 30°
Stem diameter . . . . 0.3715-0.3725 in.
Stem to guide diametral
   clearance . . . . 0.003-0.005 in.
Seat width . . . . 0.048-0.074 in.
Valve run-out (max.) . . . . 0.002 in.
Valve tappet gap (warm) . . . . . . 0.027 in.
Valve head margin . . . . 5/64 in.

Exhaust
Face and seat angle . . . . 30°
Stem diameter . . . . 0.371-0.372 in.
Stem to guide diametral
   Clearance . . . . 0.0035-0.0055 in.
Seat width . . . . 0.083-0.103 in.
Valve run-out (max.) . . . . 0.002 in.
Valve tappet gap (warm) . . . . . 0.027 in.
Valve head margin . . . . 5/64 in.

To adjust the valve tappet gap, refer to Fig. 18 and proceed as follows: Crank engine to position number one piston at top dead center of compression stroke. Adjust the four valves indicated on the chart. Turn engine crankshaft one revolution to position number four piston at T.D.C. (compression) and adjust four remaining valves indicated on chart.

VALVE GUIDES AND SPRINGS

All Models

26. The inlet and exhaust valve guides are interchangeable. Inlet guides should be pressed into head until top of guide is 1-1/8 inches above spring recess in head.

NOTE: Late production engines starting with C-157 S/N909, C-175 S/N1807 and C-200 S/N19520, inlet guide height above spring seat in head is 15/16-inch.

Exhaust valve guides should be pressed into head until top of guide is ¾-inch above spring recess in head. Guides are presized and, if carefully installed, should need no final sizing. Inside diameter should be 0.3755-0.3765 inch and valve stem to guide diametral clearance should be 0.003-0.005 inch for inlet valves and 0.0035-0.0055 inch for exhaust valves.

Inlet and exhaust valve springs are also interchangeable. Springs should

have a free length of 2-7/16 inches and should test 146-156 pounds when compressed to a length of 1-19/32 inches. Renew any spring which is rusted, discolored or does not meet the pressure test specifications.

VALVE TAPPETS

(Cam Followers)

All Models

27. Tappets are of the barrel type and ride directly in 0.9990-1.0005 inch diameter bores in the crankcase (cylinder block). Tappet diameter should be 0.9965-0.9970 inch. Oversize tappets are not available. Tappets can be removed from side of crankcase after removing rocker arms, push rods and side cover plate.

VALVE LEVERS

(Rocker Arms)

All Models

28. Removal of the rocker arm assembly may be accomplished after first removing the hood and valve cover. The rocker shaft hold down cap screws are also the left row of cylinder head cap screws and removal of these screws may allow the left side of the cylinder head to rise slightly and damage to the head gasket could result. Because of possible damage to the head gasket, it is recommended that the cylinder head be removed as in paragraph 24 when removing the rocker arm assembly.

The rocker arm shaft has an outside diameter of 0.748-0.749 inch. Rocker arm bushings have an inside diameter of 0.7505-0.7520 inch. Bushings are not renewable. All rocker arms are interchangeable. End rocker shaft brackets have a plug installed.

VALVE ROTATORS

All Models

29. Positive type valve rotators are used on the exhaust valves. Refer to Fig. 19.

Normal servicing of the valve rotators consists of renewing the units. It is important, however, to observe the valve action after engine is started. Valve rotator action can be considered satisfactory if valve rotates a slight amount each time the valve opens.

VALVE TIMING

All Models

30. To check valve timing, remove rocker arm cover and crank engine to position number one piston at T.D.C. of compression stroke. Adjust number one intake valve tappet gap to 0.034 inch. Place a 0.004 inch feeler gage between valve lever and valve stem of number one intake. Slowly rotate crankshaft in normal direction until valve lever becomes tight on feeler gage. At this point, number one intake valve will start to open and timing pointer should be within the range of 5 to 11 degrees before top dead center.

NOTE: One tooth "out of time" equals approximately 13 degrees.

Readjust number one intake valve tappet gap as outlined in paragraph 25.

TIMING GEAR COVER

All Models

31. To remove the timing gear cover, first remove hood, drain cooling system and disconnect battery cable. Disconnect wires to alternator and headlights. Disconnect steering lines and oil cooler lines and plug openings to prevent dirt from entering system. Disconnect air cleaner scoop and radiator hoses. Support tractor under clutch housing, attach hoist to front support, then unbolt and remove front support, radiator, axle and wheels as an assembly. Unbolt and remove alternator, alternator bracket and drive belt. Remove crankshaft nut, attach a suitable puller and remove crankshaft pulley. Puller can be attached to the tapped holes in

WITH	ADJUST VALVES (Engine Warm)							
No. 1 Piston at T.D.C. (Compression)	1	2	3		5			
No. 4 Piston at T.D.C. (Compression)				4		6	7	8

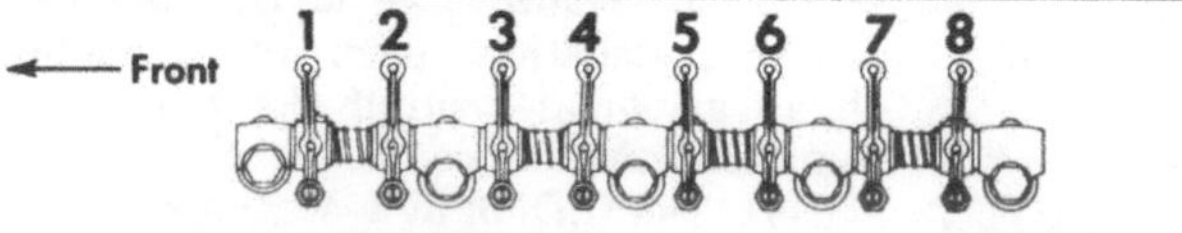

← Front   1   2   3   4   5   6   7   8

Numbering sequence of valves which correspond to chart

Fig. 18—Chart shows valve tappet gap adjusting procedure.

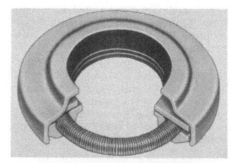

Fig. 19—Cut-away view showing construction of a "Rotocoil" valve rotator.

pulley. Unbolt timing gear cover, then pull cover forward off the dowels and remove from engine.

Reassemble by reversing the disassembly procedure. Tighten crankshaft pulley retaining nut to a torque of 95 ft.-lbs.

TIMING GEARS

All Models

32. **CRANKSHAFT GEAR.** Crankshaft gear is keyed and press fitted to the crankshaft. The gear can be removed using a suitable puller after first removing the timing gear cover as outlined in paragraph 31.

Before installing, heat gear in oil, then drift heated gear on crankshaft. Make certain timing marks on crankshaft gear and camshaft gear are aligned as shown in Fig. 20.

33. **CAMSHAFT GEAR.** Camshaft gear is keyed and press fitted on the camshaft. Backlash between camshaft gear and crankshaft gear should be 0.0032-0.0076 inch. Camshaft gear can be removed using a suitable puller after first removing timing gear cover as outlined in paragraph 31 and the gear retaining nut.

Before installing, heat gear in oil until gear will slide on shaft. Install lock and nut, then tighten nut to a torque of 110-120 ft.-lbs. Make certain timing marks on crankshaft gear and camshaft gear are aligned as shown in Fig. 20.

34. **IDLER GEAR.** To remove the idler gear, first remove the timing gear cover as outlined in paragraph 31. Idler gear shaft is attached to front of engine by a cap screw.

Idler gear shaft diameter should be 2.0610-2.0615 inches and clearance between shaft and renewable bushing in gear should be 0.0015-0.0045 inch. End clearance of gear on shaft should be 0.009-0.013 inch. Make certain that oil passage in shaft is open and clean.

When reinstalling, make certain that dowel in shaft engages hole in engine front plate. The shaft retaining cap screw should be torqued to 90 ft.-lbs.

CAMSHAFT AND BEARINGS

All Models

35. **CAMSHAFT.** To remove the camshaft, first remove timing gear cover as outlined in paragraph 31. Then, remove rocker arms assembly and push rods. Drain and remove oil pan and oil pump. Remove fuel pump. Remove engine side cover and remove

cam followers (tappets). Working through the openings in camshaft gear, remove the camshaft thrust plate retaining cap screws. Carefully withdraw camshaft from engine.

Recommended camshaft end play of 0.002-0.010 inch is controlled by the thrust plate.

Check camshaft journal diameter against the values which follow:
No. 1 (front)............2.089-2.090 in.
No. 2 ....2.069-2.070 in.
No. 3 ....1.4995-1.5005 in.

When installing the camshaft, reverse the removal procedure and make certain timing marks are aligned as shown in Fig. 20. Tighten camshaft thrust plate cap screws to a torque of 35-40 ft.-lbs.

36. **CAMSHAFT BEARINGS.** To remove the camshaft bearings, first remove the engine as outlined in paragraph 23 and camshaft as in paragraph 35. Unbolt and remove clutch, flywheel and engine rear end plate. Extract plug from behind camshaft rear bearing and remove the bearings.

Using a closely piloted arbor, install new bearings so that oil holes in bearings are in register with oil holes in crankcase. The chamfered end of bearings should be installed towards rear.

Camshaft bearings are presized and if carefully installed should need no final sizing. Camshaft bearing journals should have a diametral clearance in the bearings of 0.0005-0.005 inch.

When installing plug at rear camshaft bearing, use sealing compound on plug and bore.

ROD AND PISTON UNITS

All Models

37. Connecting rod and piston assemblies can be removed from above after removing the cylinder head as outlined in paragraph 24 and the oil pan. On all models except 454, also remove the engine balancer as in paragraph 46.

Cylinder numbers are stamped on

Fig. 20—View of gear train and timing marks.

the connecting rod and cap. Numbers on rod and cap should be in register and face toward the camshaft side of engine. The arrow stamped on the top of piston should point toward front of engine.

On Model 454, tighten connecting rod bolts to a torque of 43-49 ft.-lbs.

On all models except 454, two types of connecting rod bolts may be used. The PLACE bolt has a head that is either notched or concave and the shank and thread diameter are nearly the same. This type attains its tension by bending the bolt head and should be torqued to 50 ft.-lbs. The PITCH bolt has a standard bolt head with a washer face. The thread diameter is larger than the shank. This type attains its tension by stretching of the shank and should be torqued to 45 ft.-lbs.

PISTON AND RINGS

All Models

38. The cam ground pistons operate directly in block bores and are available in standard size as well as oversizes of 0.010, 0.020, 0.030 and 0.040 inch. With piston removed from engine, measure cylinder bores both parallel and at right angle to crankshaft centerline. If taper from top of cylinder to bottom of piston travel exceeds 0.005 inch, or if out-of-round exceeds 0.005 inch, rebore cylinder to next larger oversize.

Standard bore sizes are as follows:
454 (C-157) ....3.3750-3.3775 in.
464 (C-175) ....3.5593-3.5618 in.
574 & 674 (C-200) ....3.812-3.815 in.

When reboring, bore cylinder to within approximately 0.001 inch of desired size to allow finish honing.

Recommended clearance of piston in cylinder bore is 0.002-0.005 inch when measured between piston skirt and bore at 90 degrees to piston pin.

Pistons are fitted with two compression rings and one oil control ring. The rings should have an end gap of 0.010-0.035 inch. Side clearance of rings in piston grooves is 0.0035-0.005 inch for the top compression ring, 0.002-0.0035 inch for the second compression ring and 0.0031-0.0074 inch on all models except the 454 which is 0.002-0.0035 inch for the oil control ring.

Piston rings are available in standard sizes as well as oversizes of 0.010, 0.020, 0.030 and 0.040 inch.

PISTON PINS

All Models

39. The full floating type piston pins

are retained in the piston bosses by snap rings. Specifications are as follows:

Piston pin diameter,
    454 (C-154) . . . . 0.8591-0.8593 in.
    464 (C-175) . . . . 0.8748-0.8749 in.
    574 (C-200) . . . . 0.8748-0.8749 in.
    674 (C-200) . . . . 0.8748-0.8749 in.
Piston pin diametral clearance in piston,
    454 (C-157) . . . . 0.0000-0.0005 in.
    464 (C-175) . . . . 0.0002-0.0004 in.
    574 & 674 (C-200) . . . 0.0002-0.0004 in.
Piston pin diametral clearance in rod bushing,
    454 (C-157) . . . . 0.0002-0.0006 in.
    464 (C-175) . . . . 0.0002-0.0005 in.
    574 & 674 (C-200) . . . 0.0002-0.0005 in.
Piston pins are available in standard size as well as 0.005 inch oversize.

CONNECTING RODS AND BEARINGS

All Models

40. Connecting rod bearings are of the slip-in, precision type, renewable from below after removing oil pan and connecting rod caps. On all models except the 454, remove the engine balancer.

When installing new bearing inserts, make certain the projections on same engage slots in connecting rod and cap and that cylinder identifying numbers on rod and cap are in register and face toward camshaft side of engine. Connecting rod bearings are available in standard size as well as undersizes of 0.002, 0.010, 0.020 and 0.030 inch. Check the crankshaft crankpins and connecting rod bearings against the values which follow:

Model 454
Crankpin diameter . . . . . . 2.059-2.060 in.
Max. allowable
    out-of-round . . . . 0.0015 in.
Max. allowable taper . . . . 0.0015 in.
Diametral clearance . . . 0.0009-0.0039 in.
Rod side clearance . . . . . . 0.005-0.014 in.
Rod bolt torque . . . . 43-49 ft.-lbs.

Models 464, 574, 674
Crankpin diameter . . . . . . 2.373-2.374 in.
Max. allowable
    out-of-round . . . . 0.0015 in.
Max. allowable taper . . . . 0.0015 in.
Diametral clearance . . . 0.0009-0.0034 in.
Rod side clearance . . . . . . 0.012-0.020 in.
Rod bolt torque,
    PLACE bolt* . . . . 50 ft.-lbs.
    PITCH bolt* . . . . 45 ft.-lbs.
    *Refer to paragraph 37 for bolt identification.

CRANKSHAFT AND MAIN BEARINGS

All Models

41. The crankshaft is supported in three non-adjustable slip-in, precision type main bearings, renewable from below after removing the oil pan, engine balancer (all except 454) and main bearing caps. Crankshaft end play is controlled by the flanged center main bearing inserts. Removal of crankshaft requires R&R of engine. Check main bearings and crankshaft against the values which follow:
Crankpin diameter,
    Model 454 . . . . 2.059-2.060 in.
    Models 464,
    574, 674 . . . . 2.373-2.374 in.
Main journal diameter . . . 2.748-2.749 in.
Max. allowable taper . . . . 0.0015 in.
Max. allowable out-of-round . . 0.0015 in.
Crankshaft end play . . . . . 0.005-0.010 in.
Main bearing diametral
    clearance . . . . 0.0012-0.0042 in.
Main bearing bolt torque,
    ft.-lbs. 75-85
Main bearings are available in standard size and undersizes of 0.002, 0.010, 0.020 and 0.030 inch. Alignment dowels (IH tool FES 6-1 or equivalent) should be used when installing the rear main bearing cap.

CRANKSHAFT SEALS

All Models

42. **FRONT.** To renew the crankshaft front oil seal, first remove hood, drain

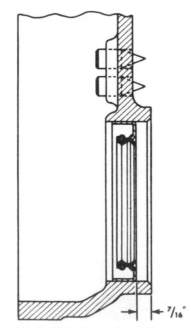

Fig. 21—Installation of the front crankshaft oil seal.

cooling system and disconnect radiator hoses. Disconnect air cleaner scoop and wire to headlights. Disconnect steering lines and oil cooler lines and plug openings to prevent dirt from entering system. Support tractor under clutch housing and attach hoist to front support. Then, unbolt and remove front support, radiator, axle and wheels as an assembly. Remove the belt from crankshaft pulley.

Remove crankshaft pulley retaining nut, attach a suitable puller and remove pulley. Remove oil seal in conventional manner. Install new oil seal in bore until front face of seal is 7/16-inch below flush with front edge of seal bore. Refer to Fig. 21. Inspect pulley seal surface and renew or recondition pulley if surface is worn or pitted. Install crankshaft pulley and tighten retaining nut to 95 ft.-lbs.

Reassemble tractor by reversing disassembly procedure.

43. **REAR.** To renew the crankshaft rear oil seal, the engine must be detached from clutch housing as follows: Disconnect battery cables and remove hood and rear side panels. Disconnect the tachometer cable and electrical wiring from engine, unclip harness and lay tachometer cable and wiring harness rearward. Shut off fuel and disconnect fuel line. Remove temperature sensing bulb from engine, then disconnect fuel line. Remove temperature sensing bulb from engine, then disconnect throttle rod, choke and oil pressure line. Disconnect and identify the wires at starter and unbolt and remove starter. Disconnect steering lines and oil cooler lines and plug openings to prevent dirt from entering system. Attach a hoist or split stand to engine and support clutch housing with a rolling floor jack. Unbolt the engine from clutch housing and roll rear section of tractor from engine. Unbolt and remove clutch assembly, then unbolt and remove flywheel.

The rear oil seal can be removed after collapsing same. Take care not to damage sealing surface of crankshaft when removing seal. Use seal installing tool and oil seal driver (IH tools FES 6-2 and 6-3) or equivalent and drive seal in until it is flush with rear of crankcase.

Coat flywheel retaining bolts with sealer, install flywheel and tighten bolts to a torque of 75 ft.-lbs.

When recoupling the tractor, position the clutch assembly on the transmission input shaft. Disconnect the linkage from clutch release shaft. Clutch can be bolted to flywheel after tractor is rejoined by working through opening at bottom of clutch housing.

NOTE: If flywheel and clutch plate balancing mark are indicated (dab of white paint), they must be aligned.

The balance of reassembly is the reverse of disassembly procedure.

FLYWHEEL

All Models

44. To remove the flywheel, refer to the procedure outlined in paragraph 43.

To install a new flywheel ring gear, heat same to approximately 500 degrees F.

ENGINE BALANCER

All Models Except 454

45. Engines are equipped with an engine balancer which is mounted on underside of engine crankcase.

Balancers are driven by a renewable gear (2—Fig. 22) on crankshaft. The balancer consists of two unbalanced gear weights which rotate in opposite directions at twice crankshaft speed. They produce forces which tend to counteract the vibration which is inherent in four cylinder engines having a single plane crankshaft (1 and 4 throws displaced 180 degrees from throws 2 and 3). It is extremely important that balancer weights are correctly timed to each other and that complete unit is timed to crankshaft.

46. R&R AND OVERHAUL. To remove the engine balancer, drain oil and remove oil pan.

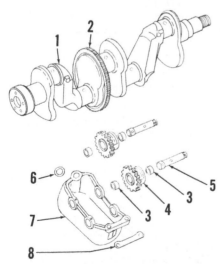

Fig. 22—Exploded view of engine balancer used on all models except 454. Balancer drive gear (2) is welded to the crankshaft on early models and a shrink fit on late models.

1. Crankshaft
2. Balancer drive gear
3. Bushing
4. Balance weight
5. Weight shaft
6. Shim L.H.
7. Housing
8. Shim R.H.

Remove Allen head screws securing balancer to crankcase, then remove balancer and shims (6 and 8—Fig. 22).

To disassemble the balancer unit, first clean with a suitable solvent, then drive out roll pins from housing (7) and shafts (5). Press or bump shafts out roll pin side of housing and lift out weight gears.

Inspect shafts, bushings, gear teeth and housing for excessive wear or other damage. Refer to the specifications which follow to determine parts renewal and operating clearances. Bushings (3) should be reamed after installation to an inside diameter of 0.752-0.753 inch. Refer to Fig. 23 for diagram of bushing installation.

Backlash, crankshaft gear to
weight gear . . . . 0.004-0.007 in.
Backlash between
weight gears . . . . 0.003-0.009 in.
Shaft bearing surface
diameter . . . . 0.7495-0.750 in.
Weight gear bushings
I.D. . . . . 0.752-0.753 in.
Weight gear operating
clearance on shaft . . . . 0.002-0.0035 in.
Weight gear end clearance in
housing . . . . 0.008-0.020 in.

When reassembling, make certain that oil passages in housing are clean. Lubricate shafts and bushings and place one weight in housing. Install the weight shaft from roll pin side of housing, through the weight gear and secure in place with a roll pin. Position the second weight assembly in housing, meshing gear teeth so that flats of weights are parallel.

NOTE: Later production balancers have timing marks stamped on gears.

Install second shaft and secure with roll pin. Using a dial indicator, check backlash between balancer weight gears. Backlash should be 0.003-0.009 inch. If backlash is excessive, renew both weight gear assemblies.

To install the balancer assembly, first rotate crankshaft to position No. 1 piston at TDC of compression stroke. Install balancer, making certain that timing mark on weight gear is in register with timing mark on balancer drive gear (2—Fig. 22). Using a dial indicator, check backlash between balancer drive gear (2) and weight gear. Backlash should be 0.004-0.007 inch. Add or remove shims (6 and 8) as necessary to obtain correct backlash. Equal thickness of shims must be used at both sides of balancer.

Install oil pan with new pan gasket. Fill crankcase to correct level with new oil.

47. RENEW BALANCER DRIVE GEAR (WELD ON TYPE). To renew balancer drive gear (2—Fig. 22), first remove crankshaft from engine. Apply Prussian Blue on crankshaft area below timing dot on ring gear. Using a square, scribe a line through the timing dot on crankshaft. Remove worn or damaged drive gear and any burrs which might be present on gear mounting surface of crankshaft.

Heat new gear to 200 degrees F., install new gear on crankshaft with timing mark aligned with scribed line. Be sure the gear is bottomed against crankshaft shoulder.

CAUTION: Do not overheat gear as this will cause distortion.

Using a low hydrogen electrode, weld gear to shaft at four places ½ to ¾-inch wide, spaced equally as shown in Fig. 24. Weld must not protrude above finished surface on crankshaft.

47A. RENEW BALANCER DRIVE GEAR (SHRINK FIT TYPE). To renew balancer drive gear (2—Fig. 22), first remove crankshaft from engine. Apply Prussian Blue on crankshaft area below timing dot on ring. Using a square, scribe a line through the timing dot on to the crankshaft. Remove worn or damaged drive gear, and any burrs which might be present on gear mounting surface of crankshaft.

Heat new gear to 550 degrees F., install new gear on crankshaft with timing mark aligned with scribed line. Be sure the gear is bottomed against crankshaft shoulder as shown in Fig. 24A.

CAUTION: Do not overheat gear as this will cause distortion. Do not weld this gear as it has a Tufftride coating.

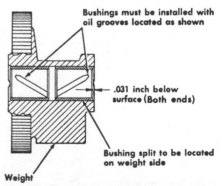

Fig. 23—Sectional view showing correct bushing installation in balancer weight. Ream bushings after installation to an inside diameter of 0.752-0.753 inch.

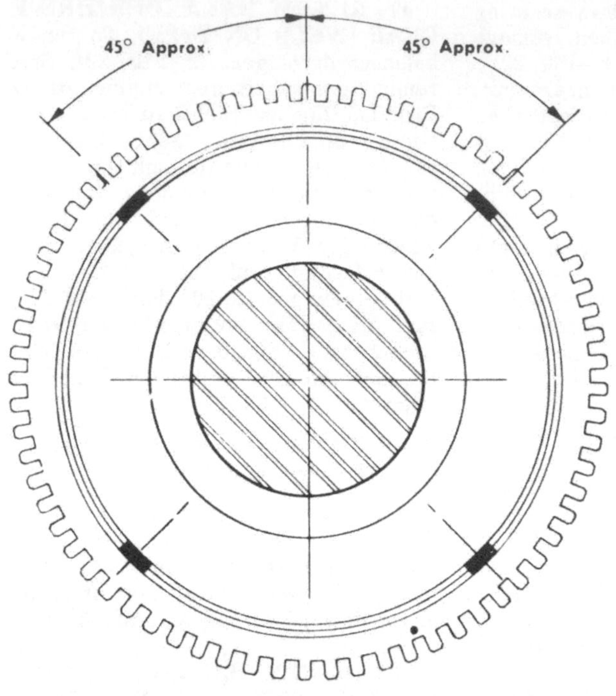

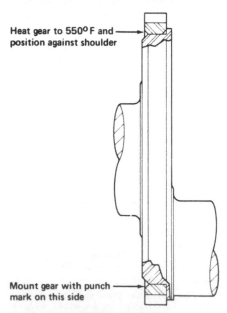

Fig. 24—Weld crankshaft balancer drive gear in four places as shown.

NOTE: When installing oil pump, crank engine until number one piston is coming up on compression stroke. Continue cranking until the TDC mark on crankshaft pulley is in register with the pointer on timing gear cover. Install oil pump so that tang on pump shaft is in the approximate position shown in Fig. 27. Retime distributor as outlined in paragraph 115 or 117.

OIL PRESSURE REGULATOR

All Models

49. The oil pressure regulator valve is located in the oil gallery at left rear side of engine. See Fig. 28. Check regulator valve and spring against the following specifications:

Valve diameter . . . . 0.743-0.745 in.
Valve clearance in bore . . 0.002-0.007 in.
Spring free length . . . . 3.3 inches
Spring test and
  length . . . . 8.4 lbs. at 1.1 in.
Oil pressure at rated rpm . . . . . . 38 psi

The filter by-pass valve is located in the spin-on type filter.

OIL PUMP

All Models

48. The gear type oil pump is gear driven from a pinion on camshaft and is accessible for removal after removing the engine oil pan. Disassembly and overhaul of pump is evident after an examination of the unit and reference to Fig. 25. Gaskets between pump cover and body can be varied to obtain the recommended 0.0025-0.0055 inch pumping (body) gear end play. Refer to the following specifications:

Pumping gears recommended
  backlash . . . . 0.003-0.006 in.
Pump drive gear recommended
  backlash . . . . 0.000-0.008 in.
Pumping gear
  end play . . . . 0.0025-0.0055 in.
Gear teeth to body radial
  clearance . . . . 0.0068-0.0108 in.
Gear shaft clearance
  in bore . . . . 0.0015-0.003 in.
Mounting bolts torque,
  ft.-lbs. 20-23

Service (replacement) pump shaft and gear assemblies (8—Fig. 25) are not drilled to accept the pump driving gear pin (1). A 1/8-inch hole must be drilled through the shaft after drive gear is installed on shaft to the dimension shown in Fig. 26.

OIL PAN

All Models

50. To remove the oil pan, place a jack under center section of tractor. Then, remove lower six bolts and loosen upper two bolts in the front bolster to engine. Remove the pan bolts; there are three pan bolts partly hidden between the pan and clutch housing. Oil pan can now be removed.

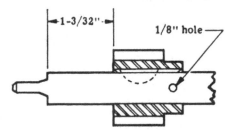

Fig. 26—New pump shaft and gear assemblies will require a 1/8-inch hole to be drilled at location shown.

Heat gear to 550° F and position against shoulder

Mount gear with punch mark on this side

Fig. 24A—Crankshaft balancer drive gear shrink fit installation.

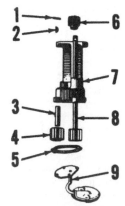

Fig. 25—Exploded view of oil pump assembly.

1. Pin
2. Woodruff key
3. Idler gear shaft
4. Idler gear
5. Gasket
6. Drive gear
7. Pump body
8. Drive shaft and gear
9. Cover and screen assembly

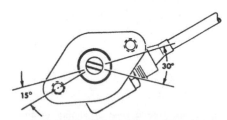

Fig. 27—Install tachometer drive housing with distributor mounting holes about 15 degrees from centerline of engine. With No. 1 piston at TDC of compression stroke, mesh oil pump drive gear so that distributor drive shaft tang is approximately 30 degrees from centerline of engine.

Fig. 28—Oil pressure regulator valve is located in oil gallery at left rear side of engine.

DIESEL ENGINE AND COMPONENTS

Models 454, 464 and 484 diesel tractors are equipped with a three cylinder engine, having a bore and stroke of 3.875 x 5.06 inches and a displacement of 179 cubic inches.

Models 574 and 674 diesel tractors are equipped with a four cylinder engine, having a bore and stroke of 3.875 x 5.06 inches and a displacement of 239 cubic inches.

Model 584 diesel tractors are equipped with a four cylinder engine, having a bore and stroke of 3.875 x 4.38 inches and a displacement of 206 cubic inches.

R&R ENGINE ASSEMBLY

All Models

51. To remove the engine assembly, first remove hood and rear side panels, drain cooling system and disconnect battery ground cable. Disconnect wires to alternator and wire to headlights. Disconnect steering lines and oil cooler lines and plug openings to prevent dirt from entering the system. Disconnect air cleaner scoop and radiator hoses. Support tractor under clutch housing, attach hoist to front support, then unbolt and remove front support, radiator, axle and wheels as an assembly. Shut off fuel and disconnect the inlet line at filter base and the return line at the tee. Disconnect the fuel shut off cable at the injection pump and the wire to the ether injector. Disconnect electrical wires from engine, unclip harness and lay rearward. Remove the breather tube assembly from right side of engine. Disconnect temperature sensing bulb and oil pressure tube from engine, then disconnect throttle rod and tachometer cable. Attach hoist to engine, then unbolt and remove engine assembly.

When reinstalling engine assembly unbolt and remove clutch assembly and position it on transmission input shaft. Disconnect the linkage from clutch release shaft. Clutch can be bolted to flywheel after engine is installed by working through the opening at bottom of clutch housing.

NOTE: If flywheel and clutch plate balancing marks are indicated (dab of white paint), they must be aligned.

CYLINDER HEAD

All Models

52. **REMOVE AND REINSTALL.** Drain coolant, remove hood and disconnect battery cable. Unbolt and remove air cleaner. Disconnect upper radiator hose and by-pass hose and remove coolant temperature bulb from engine. Disconnect injector lines from fuel injectors and injection pump. Cap or plug all fuel connections immediately after they are disconnected to prevent en-

```
→Front
  13  O   1   O   3   O   11
  O   9       5       7    O
      O       O       O

  O   8   O   6   O   10  O
  12  O   4       2       14
      O       O       O
```

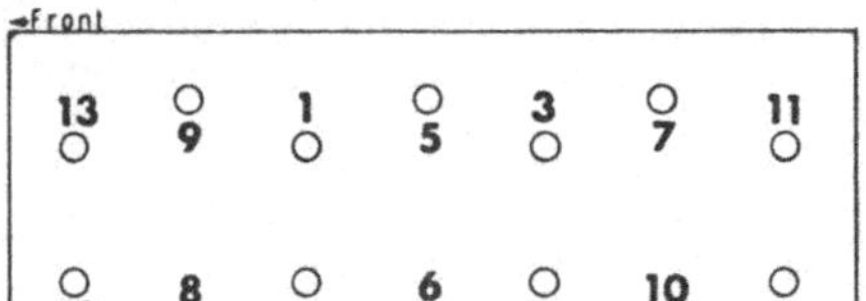

Fig. 29—Models 454, 464 and 484 cylinder head cap screws are tightened in the sequence shown. Refer to text.

```
←  Front
       14      6      4      12
  18   O  10   O   1  O  8   O  16
  O         O    O    O         O
      ( )   ( )    ( )   ( )
  O         O    O    O         O
  15   O   7   O   2  O  9   O  17
       11      3      5      13
```

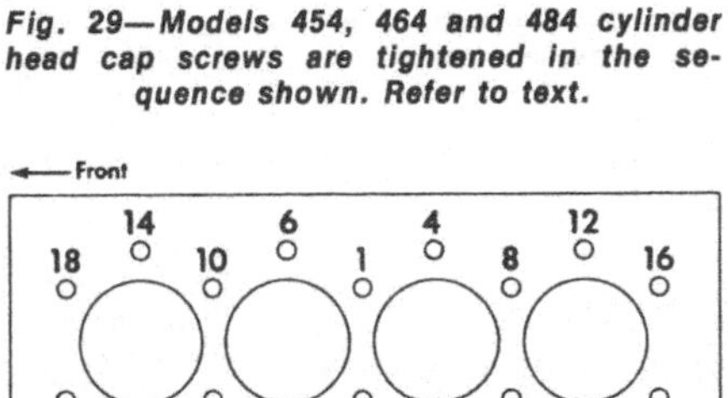

Fig. 30—Models 574, 674 and 584 cylinder head cap screws are tightened in the sequence shown. Refer to text.

trance of dirt or other foreign material. Unbolt and remove exhaust manifold and water collecting tube from right side and inlet manifold from left side of cylinder head. Remove rocker arm cover, rocker arms and shaft assembly and push rods. Remove cylinder head cap screws and lift off cylinder head.

CAUTION: The injector nozzle assemblies protrude slightly through the combustion side of cylinder head. Be extremely careful when removing or reinstalling cylinder head with injectors installed not to damage injector ends. DO NOT place cylinder head on bench with combustion side down.

When reinstalling cylinder head, use new head gasket and make certain gasket sealing surfaces are clean. Use guide studs when installing cylinder head.

Tighten cylinder head retaining cap screws in three steps using the sequence shown in Fig. 29 for the three cylinder engine and Fig. 30 for the four cylinder engine. Tighten the cap screws to a torque of 30 ft.-lbs. in first step, 60 ft.-lbs. in second step and 90 ft.-lbs. in third step. Adjust valve tappet gap to 0.010 inch (intake) and 0.012 inch (exhaust) as outlined in paragraph 54.

NOZZLE SLEEVES

All Models

53. The cylinder is fitted with brass injector nozzle sleeves which pass through the coolant passages. The nozzle sleeves are available as service items.

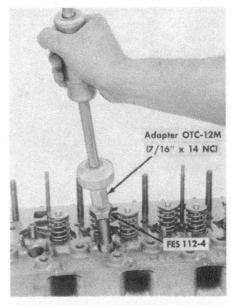

Fig. 31—Use IH tool No. FES 112-4 and slide hammer to remove injector nozzle sleeves.

To renew the nozzle sleeves, remove cylinder head as outlined in paragraph 52, then remove injectors. Use special bolt (IH tool No. FES 112-4) and turn it into sleeve. Attach a slide hammer puller and remove nozzle sleeve. See Fig. 31. Use caution during this operation not to damage the sealing areas in the cylinder head.

NOTE: Under no circumstances should screw drivers, chisels or other such tools be used in an attempt to remove injector nozzle sleeves.

When installing nozzle sleeves, be sure the sealing areas are completely clean and free of scratches. Apply a light coat of "Loctite Grade B" on sealing surfaces of nozzle sleeves, then using installing tool (IH tool No. FES 112-3) shown in Fig. 32, drive nozzle sleeves into their bores until they bottom as shown in Fig. 33.

NOTE: Injector nozzle sleeves have a slight interference fit in their bores. When installing sleeves, make certain sleeve is driven straight with its bore.

VALVES AND SEATS

All Models

54. Inlet and exhaust valves are not interchangeable. The inlet valves seat directly in the cylinder head and the exhaust valves seat on renewable seat inserts. Inserts are available in oversizes of 0.004 and 0.016 inch.

Valve face and seat angle for both the inlet and exhaust is 45 degrees. Valve rotators and valve stem seals are used on all valves.

When removing valve seat inserts, use the proper puller. Do not attempt to drive chisel under insert or counterbore will be damaged. Chill new inserts with dry ice or liquid Freon prior to installing. When new insert is properly bottomed, it should be 0.008-0.030 inch below edge of counterbore. After installation, peen the cylinder head material around the complete outer circumference of the valve seat insert.

Check the valves and seats against the following specifications:

Inlet
Face and seat angle.............45°
Stem diameter........0.3919-0.3923 in.
Stem to guide diametral
   clearance..........0.0014-0.0026 in.
Seat width ....0.076-0.080 in.
Valve run-out (max.) ....0.001 in.
Valve tappet gap (warm).....0.010 in.
Valve recession from face of
   cylinder head,
   Normal ....0.039-0.055 in.
   Maximum ....0.120 in.

Exhaust
Face and seat angle.............45°
Stem diameter........0.3911-0.3915 in.
Stem to guide diametral
   clearance.............0.0022-0.0034 in.
Seat width ....0.081-0.089 in.
Valve run-out (max.) ....0.001 in.
Valve tappet gap (warm).....0.012 in.
Valve recession from face of
   cylinder head,
   Normal ....0.047-0.063 in.
   Maximum ....0.120 in.

To adjust valve tappet gap on D179 engine, crank engine to position number one piston at top dead center of compression stroke. Adjust the four valves indicated on the chart shown in Fig. 34. Turn engine crankshaft one complete revolution to position number one piston at top dead center of exhaust stroke and adjust the remaining two valves indicated on chart.

To adjust valve tappet gap on D206 and D239 engines, crank engine to position number one piston at top dead center of compression stroke. Adjust the four valves indicated on the chart shown in Fig. 35. Turn engine crankshaft one complete revolution to position number four piston at top dead center compression stroke and adjust the remaining four valves indicated on chart.

VALVE GUIDES AND SPRINGS

All Models

55. The inlet and exhaust valve guides are interchangeable. Inlet and exhaust guides should be pressed into cylinder head until top of guides are 1.154-1.169 inches above spring recess in the head. Guides must be reamed to size after installation. Inside diameter of guides should be 0.3937-0.3945 inch and valve stem to guide diametral clearance should be 0.0014-0.0026 inch for inlet and 0.0022-0.0034 inch for exhaust valves.

Inlet and exhaust valve springs are

Fig. 33—Drive nozzle sleeves into cylinder head until they bottom.

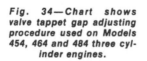

Fig. 34—Chart shows valve tappet gap adjusting procedure used on Models 454, 464 and 484 three cylinder engines.

WITH	ADJUST VALVES (Engine Warm) *					
No. 1 Piston at T.D.C. (Compression)	1	2		4	5	
No. 1 Piston at T.D.C. (Exhaust)			3			6

Fig. 35—Chart shows valve tappet gap adjusting procedure used on Models 574, 674 and 584 four cylinder engines.

WITH	ADJUST VALVES (Engine Warm)							
No. 1 Piston at T.D.C. (Compression)	1	2		4	5			
No. 4 Piston at T.D.C. (Compression)			3			6	7	8

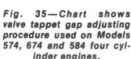

Numbering sequence of valves which correspond to chart

Fig. 32—Injector nozzle sleeve installing tool. Apply "Grade B Loctite" to sealing surface of nozzle sleeve.

also interchangeable. Springs should have a free length of 2.173-2.181 inches and should test 140-160 pounds when compressed to a length of 1.346 inches. Renew any spring which is rusted, discolored or does not meet the pressure test specifications.

VALVE TAPPETS
(CAM FOLLOWERS)

All Models

56. The 0.7862-0.7868 inch diameter mushroom type tappets operate directly in the unbushed crankcase bores. Clearance of tappets in the bores should be 0.0005-0.0024 inch. Tappets can be removed after first removing the camshaft as outlined in paragraph 65. Oversize tappets are not available.

VALVE LEVERS
(Rocker Arms)

All Models

57. Removal of the rocker arms and shaft assembly is conventional. To remove the rocker arms from the shaft, remove bracket clamp bolts and slide all parts from shaft. Outside diameter of rocker shaft is 0.8491-0.8501 inch. The renewable bushing in rocker arms should have an operating clearance of 0.0009-0.0025 inch on the rocker shaft with a maximum allowable clearance of 0.006 inch. Rocker arm adjusting screws are self-locking. If they turn with less than 12 ft.-lbs. torque, renew adjusting screw and/or rocker arm.

Reassemble rocker arms on rocker shaft keeping the following points in mind. Thrust washers are used between each spring and rocker arms, and spacer rings are used between rocker arms and brackets except between rear rocker arm and rear bracket. To insure that lubrication holes in rocker shaft are in correct position, align punch mark on front end of rocker shaft with slot in front mounting

bracket as shown in Fig. 36. End of shaft must also be flush with bracket. The rocker shaft clamp screws on brackets should be tightened to a torque of 10 ft.-lbs. Tighten the holddown nuts on bracket studs to a torque of 47 ft.-lbs.

VALVE ROTATORS

All Models

58. Positive type valve rotators are installed on both the inlet and exhaust valves. Refer to Fig. 37.

Normal servicing of the valve rotators consists of renewing the unit. It is important, however, to observe the valve action after engine is started. Valve rotator action can be considered satisfactory if valve rotates a slight amount each time the valve opens.

VALVE TIMING

All Models

59. To check valve timing, remove rocker arm cover and crank engine to position number one piston at T.D.C. of compression stroke. Adjust number one intake valve tappet gap to 0.014 inch. Place a 0.004 inch feeler gage between valve lever and valve stem of number one intake. Slowly rotate crankshaft in normal direction until valve lever becomes tight on feeler gage. At this point, number one intake valve will start to open and timing pointer should be within the range of 23 to 29 degrees before top dead center.

NOTE: One tooth "out of time" equals approximately 11 degrees.

Readjust number one intake valve tappet gap as outlined in paragraph 54.

TIMING GEAR COVER

All Models

60. To remove the timing gear cover,

first remove hood, drain cooling system and disconnect battery cable. Disconnect wires to headlights, alternator and unbolt and remove alternator. Disconnect steering lines and oil cooler lines and plug openings to prevent dirt from entering system. Disconnect air cleaner scoop and radiator hoses. Support tractor under clutch housing, attach hoist to front support, then unbolt and remove front support, radiator, axle and wheels as an assembly. Unbolt and remove fan, fan pulley, belt, cross-over fuel line and water pump. Then, unbolt and remove oil pan. Disconnect tachometer drive cable and remove tachometer drive. Do not lose the small driving tang when removing tachometer drive. Remove the three cap screws, flat washer and pressure ring from crankshaft. Tap pulley with a plastic hammer to loosen it, then slide pulley off of the wedge rings. Unbolt timing gear cover, then pull cover forward off the dowels and remove from engine.

Reassemble by reversing the disassembly procedure. When installing the crankshaft pulley, place one pressure ring on the crankshaft with thick end toward engine. Install the wedge rings in pulley bore so that slots in rings are 90 degrees apart. Slide pulley on to crankshaft and align with timing pin. Install pressure ring, flat washer and three cap screws. Tighten cap screws evenly in three steps to a torque of 22 ft.-lbs., 44 ft.-lbs. and finally to 57 ft.-lbs.

NOTE: Some late production crankshaft pulleys use different number of wedge rings. Pulley may be equipped with three sets of wedge rings, two sets of wedge rings or one clamping ring.

TIMING GEARS

All Models

61. CRANKSHAFT GEAR. Crankshaft gear is a shrink fit on crankshaft. To renew the gear, it is recommended that the crankshaft be removed from engine. Then, using a chisel and hammer, split the gear at its timing slot. The roll pin for indexing crankshaft

Fig. 36—Align punch mark on front end of rocker shaft with slot in front mounting bracket. End of shaft must be flush with bracket.

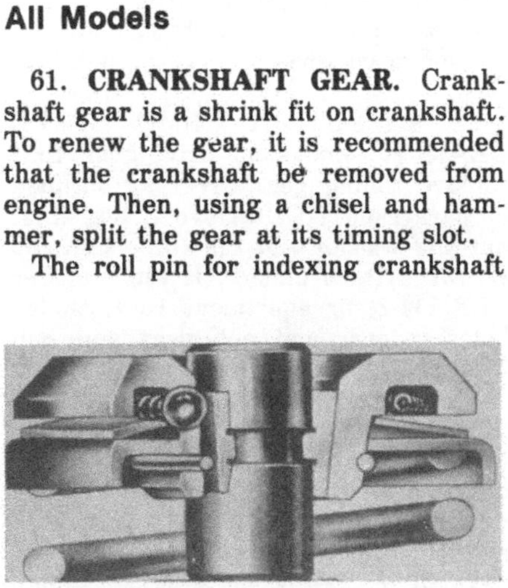

Fig. 37—Cut-away view showing installation of a "Rotocap" valve rotator.

gear on crankshaft must protrude approximately 5/64-inch. Heat new gear to 400°F. and install it against bearing journal. Make certain all timing marks are aligned as shown in Fig. 38.

62. CAMSHAFT GEAR. Camshaft gear is a shrink fit on camshaft. To renew the gear, remove camshaft as outlined in paragraph 65. Gear can now be pressed off in conventional manner, using care not to damage the tachometer drive slot in end of camshaft.

When reassembling, install thrust plate and Woodruff key. Heat gear to 400° F. and install it on camshaft.

NOTE: When sliding gear on camshaft, set thrust plate clearance at 0.004-0.018 inch.

Install camshaft assembly and make certain all timing marks are aligned as shown in Fig. 38.

63. IDLER GEAR. To remove idler gear, first remove timing gear cover as outlined in paragraph 60. Idler gear shaft is attached to front of engine by a special (left hand thread) cap screw.

Idler gear is equipped with two renewable needle bearings. A spacer is used between the bearings.

When installing idler gear, coat threads of special cap screw with "Loctite" and tighten cap screw to a torque of 67 ft.-lbs. End clearance of gear on shaft should be 0.008-0.013 inch. Align all timing marks as shown in Fig. 38.

Timing gear backlash should be as follows:

Idler to crankshaft gear,
    New gears . . . . 0.007-0.015 in.
    Used gears (max.) . . . . 0.027 in.
Idler to camshaft gear,
    New gears . . . . 0.0035-0.016 in.
    Used gears (max.) . . . . 0.0226 in.
Idler to injection pump gear,
    New gears . . . . 0.0021-0.012 in.
    Used gears (max.) . . . . 0.024 in.

64. INJECTION PUMP DRIVE GEAR. To remove the drive gear, first remove timing gear cover as outlined in paragraph 60. Remove drive shaft nut and washer and the three hub cap screws. Attach puller (IH tool number FES 111-2) or equivalent to threaded holes in gear and pull gear and hub from shaft.

When reassembling, make certain all timing marks are aligned as shown in Fig. 38. For D179 engine use timing dot next to number 3 and for D206 engine use timing dot next to number 2 and for D239 engine use timing dot next to number 4 on injection pump drive gear. Refer to paragraph 91 and retime injection pump. Tighten drive

shaft nut to a torque of 47 ft.-lbs. and three hub cap screws to a torque of 17 ft.-lbs.

CAMSHAFT AND BEARINGS

All Models

65. CAMSHAFT. To remove the camshaft, first remove timing gear cover as outlined in paragraph 60. Then, remove rocker arm assembly and push rods. Remove engine side cover and secure cam followers (tappets) in raised position with clothes pins or rubber bands. Working through openings in camshaft gear, remove camshaft thrust plate retaining cap screws. Carefully withdraw camshaft assembly.

Recommended camshaft end play is 0.004-0.018 inch. Camshaft bearing journal diameter should be 2.2823-2.2835 inches.

Install camshaft by reversing the removal procedure. Make certain timing marks are aligned as shown in Fig. 38.

66. CAMSHAFT BEARINGS. To remove the camshaft bearings, first remove the engine as outlined in paragraph 51 and camshaft as in paragraph 65. Unbolt and remove clutch, flywheel and engine rear end plate. Remove expansion plug from behind camshaft rear bearing and remove the bearings.

NOTE: Camshaft bushings are furnished semi-finished for service and must be align reamed after installation to an inside diameter of 2.2844-2.2856 inches.

Install new bearings so that oil holes in bearings are in register with oil holes in crankcase.

Normal operating clearance of camshaft journals in bearings is 0.0009-0.0033 inch. Maximum allowable clearance is 0.0045 inch.

When installing expansion plug at rear camshaft bearing, apply a light

coat of sealing compound to edge of plug and bore.

ROD AND PISTON UNITS

All Models

67. Connecting rod and piston assemblies can be removed from above after removing the cylinder head as outlined in paragraph 52 and the oil pan. On Models 574, 674 and 584 also remove the engine balancer as in paragraph 76.

Cylinder numbers are stamped on the connecting rod and cap. Stamp any new or unmarked rod and cap assemblies with correct cylinder numbers. Numbers on rod and cap should be in register and face toward the camshaft side of engine. The arrow stamped on the top of pistons should point toward front of engine.

Tighten the connecting rod nuts to a torque of 61 ft.-lbs.

PISTONS, SLEEVES AND RINGS

All Models

68. Pistons are not available as individual service parts, but only as matched units with the wet type sleeves. New pistons have a diametral clearance in new sleeves of 0.0039-0.0047 inch when measured between piston skirt and sleeve at 90 degrees to piston pin.

The wet type cylinder sleeves should be renewed when out-of-round or taper exceeds 0.006 inch. Inside diameter of new sleeves is 3.875-3.8754 inches. Cylinder sleeves can usually be removed by bumping them from the bottom with a block of wood.

Before installing new sleeves, thoroughly clean counterbore at top and seal ring groove at bottom. All sleeves should be free to rotate by hand when tried in bores without seal rings. After

Fig. 38—Gear train and timing marks. Use timing dot next to number 4 for D239 engine, timing dot next to number 2 for D206 engine and timing dot next to number 3 for D179 engine on injection pump drive gear.

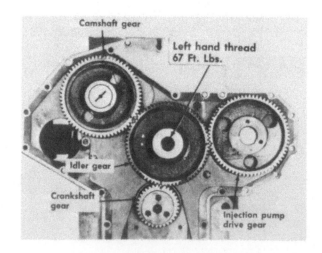

making a trial installation without seal rings, remove sleeves and install new seal rings dry into grooves in crankcase. Wet lower end of sleeve with a thick soap solution or equivalent and install sleeve.

NOTE: The cut-outs in bottom of sleeve are for connecting rod clearance and must be installed toward each side of engine. Chisel marks are provided on top edge of cylinder sleeve to aid in correct installation. Align chisel marks from front to rear of engine.

If seal ring is in place and not pinched, very little hand pressure is required to press the sleeve completely in place. Sleeve flange should extend 0.003-0.005 inch above top surface of cylinder block. If sleeve stand-out is excessive, check for foreign material under the sleeve flange. The cylinder head gasket forms the upper cylinder sleeve seal and excessive sleeve stand-out will result in coolant leakage.

Pistons are fitted with two or three compression rings and one oil control ring. Top compression ring may be conventional rectangular type or full keystone type. If full keystone ring is used proceed as follows: Prior to installing new rings, check top ring groove of piston by using Perfect Circle Piston Ring Gage No. 1 as shown in Fig. 39. If one, or both shoulders of gage touch ring land, renew the piston. It is not possible to measure ring side clearance of a full keystone ring in the groove. Check fit of top ring as follows: Place ring in its groove and push ring into groove as far as possible. Check the

distance ring is below ring land. This distance should be 0.002-0.015 inch. Refer to Fig. 40 for view showing ring fit being checked using IH tools FES 68-3 and dial indicator FES 67.

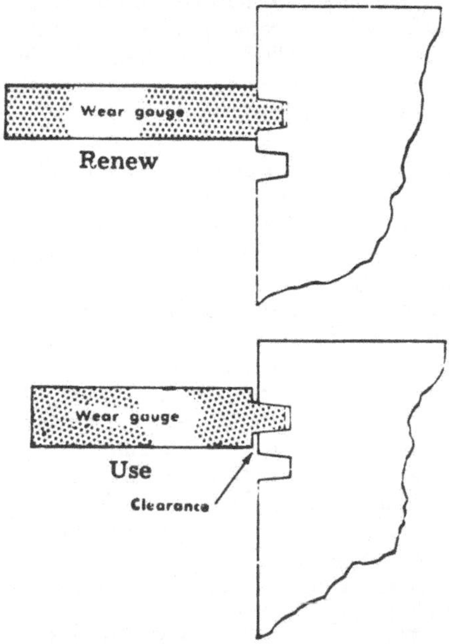

If rectangular top compression ring is used, check side clearance in conventional manner. If equipped with notch on inside diameter, install with notch to top.

The second compression ring is a taper face ring and is installed with largest outside diameter toward bottom of piston. Upper side of ring is marked TOP.

The third compression ring (if so equipped) is step cut on outside diameter and is installed with notch toward bottom of piston. Upper side of ring is marked TOP.

The oil control ring can be installed either side up, but make certain the coil spring expander is completely in its groove.

Additional piston ring information is as follows:
Ring End Gap,
  Compression rings . . . . 0.014-0.025 in.
  Oil control ring . . . . 0.010-0.019 in.
Ring Side Clearance,
  Top compression (keystone)
    ring drop . . . . 0.002-0.015 in.
  Top compression
    (rectangular) . . . . . 0.0035-0.0048 in.
  Second
    compression . . . . . . 0.0030-0.0042 in.
  Third
    compression . . . . . . 0.0024-0.0034 in.
  Oil control . . . . 0.0014-0.0024 in.

PISTON PINS

All Models

69. The full floating type piston pins are retained in the piston bosses by snap rings. Specifications are as follows:
Piston pin diameter,
  All models . . . . 1.4172-1.4173 in.
Piston pin diametral clearance in rod bushing,
  All models . . . . 0.0005-0.0010 in.
Piston pins are not available in oversize.

CONNECTING RODS AND BEARINGS

All Models

70. Connecting rod bearings are of slip-in, precision type, renewable from below after removing oil pan, engine balancer (if so equipped) and connecting rod caps.

When installing new bearing inserts, make certain that projections on same engage slots in connecting rod and cap,

and that cylinder identifying number on rod and cap are in register and face toward camshaft side of engine.

Connecting rod bearings are available in standard size as well as undersizes of 0.010, 0.020 and 0.030 inch.

Check crankshaft crankpins and connecting rod bearings against the values which follow:
Crankpin diameter . . . . 2.5185-2.5193 in.
Max. allowable
  out-of-round . . . . 0.0012 in.
Diametral clearance . . . 0.0023-0.0048 in.
Rod side clearance . . . . . . 0.006-0.010 in.
Rod bolt torque, ft.-lbs. 61

CRANKSHAFT AND MAIN BEARINGS

All Models

71. The crankshaft is supported in four main bearings in the D179 engine and five main bearings in the D206 and D239 engines. Main bearings are of the non-adjustable, slip-in, precision type, renewable from below after removing the oil pan, engine balancer (if so equipped) and main bearing caps. Crankshaft end play is controlled by the flanged rear main bearing inserts. Removal of crankshaft requires R&R of engine. Check crankshaft and main bearings against the values which follow:
Crankpin
  diameter . . . . 2.5185-2.5193 in.
Main journal
  diameter . . . . 3.1484-3.1492 in.
Max. allowable
  out-of-round . . . . 0.002 in.
Crankshaft end play . . . . . 0.006-0.009 in.

Fig. 40—Top compression (Keystone) ring should be 0.002-0.015 inch below ring land for proper fit.

Fig. 39—Use Perfect Circle Ring Gage No. 1 to check top (keystone) ring groove.

Main bearing diametral
   clearance.........0.0028-0.0055 in.
Main bearing bolt torque, ft.-lbs.
Necked down bolts,
   Marked 10.9K.................80*
   Marked 12.9K.................97*
Pitch diameter bolts,
   Marked 10.9K...............105*
   Marked 12.9K...............141*
   *With lubricated threads.

Necked down bolts have a shank diameter of 0.410 inch. Pitch diameter bolts have a shank diameter of 0.500 inch.

Main bearings are available in standard size and undersizes of 0.010, 0.020 and 0.030 inch.

CRANKSHAFT SEALS

All Models

72. **FRONT.** To renew the crankshaft front oil seal, first remove hood, drain cooling system and disconnect radiator hoses. Disconnect air cleaner scoop and wires to headlights. Disconnect steering lines and oil cooler lines and plug openings to prevent dirt from entering system. Support tractor under clutch housing, attach hoist to front support. Then, unbolt and remove front support, radiator, axle and wheels as an assembly. Remove belt from crankshaft pulley.

Remove three cap screws, flat washer and pressure ring from crankshaft. Tap crankshaft pulley with a plastic hammer to loosen it, then slide pulley off the wedge rings. Remove wedge rings and oil seal wear ring from crankshaft. Remove and renew oil seal in conventional manner. Renew "O" ring on crankshaft. Inspect wear ring and timing pin for wear or other damage and renew as necessary. Use a non-hardening sealer on timing pin. When installing wear ring, timing pin must engage slot in crankshaft gear. Install one pressure ring on crankshaft with thick end against wear ring. Place wedge rings in pulley bore so that slots in rings are 90 degrees apart. Slide pulley onto crankshaft, aligning slot in pulley with timing pin in wear ring. Install pressure ring, flat washer and three cap screws. Tighten cap screws evenly in three steps to a torque of 22 ft.-lbs., 44 ft.-lbs. and finally to 57 ft.-lbs. Refer to note in paragraph 60. Reassemble tractor by reversing disassembly procedure.

73. **REAR.** To renew crankshaft rear oil seal, the engine must be detached from clutch housing as follows: Disconnect battery cables and remove hood and rear side panels. Disconnect the tachometer cable and electrical wiring from engine, unclip harness and lay

tachometer cable and wiring harness rearward. Shut off fuel and disconnect fuel lines. Remove temperature sensing bulb from engine, then disconnect throttle rod, fuel shut-off cable and oil pressure line. Disconnect and identify the wires at starter, then unbolt and remove starter. Disconnect steering lines and oil cooler lines and plug openings to prevent dirt from entering system. Attach a hoist or split stand to engine and support clutch housing with a rolling floor jack. Unbolt the engine from clutch housing and roll rear section of tractor from engine. Unbolt and remove clutch assembly and flywheel.

Then, unbolt and remove seal retainer and seal. Check the depth that old seal is installed in seal retainer. New oil seal may be installed in retainer in any of three locations; 1/16-inch above flush with retainer (new engine original position), flush with retainer or 1/16-inch below flush with retainer. Location of seal in retainer will depend on condition of the sealing surface on crankshaft. Use new gasket and install seal retainer with new seal. Tighten retainer bolts to a torque of 14 ft.-lbs. Install flywheel and tighten bolts to a torque of 85 ft.-lbs.

When recoupling the tractor, position the clutch assembly on the transmission input shaft. Disconnect the linkage from clutch release shaft. Clutch can be bolted to flywheel after tractor is rejoined by working through opening at bottom of clutch housing.

NOTE: If flywheel and clutch plate

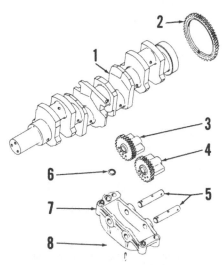

Fig. 41—Exploded view of engine balancer used on Models 574, 674 and 584. Balancer drive gear (2) is welded to the crankshaft on early models and a shrink fit on late models.

1. Crankshaft
2. Balancer drive gear
3. Idler weight assy.
4. Drive weight assy.
5. Weight shafts
6. "O" ring
7. Housing
8. Double roll pin

balancing marks are indicated (dab of white paint), they must be aligned.

The balance of reassembly is the reverse of disassembly procedure.

FLYWHEEL

All Models

74. To remove the flywheel, refer to the procedure outlined in paragraph 73.

To install a new flywheel ring gear, heat same to approximately 500 degrees F.

ENGINE BALANCER

Models 574-674-584

75. The D206 and D239 engines are equipped with an engine balancer which is mounted on underside of engine crankcase.

Balancers are driven by a renewable gear (2—Fig. 41) which is welded to crankshaft on early models and is a shrink fit to crankshaft on late models. The balancer consists of two un-balanced gear weights which rotate in opposite directions at twice crankshaft speed. They produce forces which tend to counteract the vibration which is inherent in four cylinder engines having a single plane crankshaft (1 and 4 throws displaced 180 degrees from throws 2 and 3). It is extremely important that balancer weights are correctly timed to each other and that complete unit is timed to crankshaft.

76. **R&R AND OVERHAUL.** To remove the engine balancer, drain oil and remove oil pan as outlined in paragraph 80. Unbolt and remove oil suction tube assembly and "O" ring. Remove the two remaining cap screws and remove balancer from crankcase.

To disassemble the balancer unit, first clean with a suitable solvent, then drive roll pins (8—Fig. 41) out of housing and into weight shafts (5). Press or bump shafts out roll pin side of housing (7) and lift out weight gear assemblies (3 and 4). Drive roll pins (8) out of shafts.

Inspect shafts, gear teeth, bushings and housing for excessive wear or other damage. Refer to the specifications which follow to determine parts renewal and operating clearances.

Backlash, crankshaft gear to
   weight gears.........0.010-0.016 in.
Backlash between
   weight gears.........0.007-0.009 in.
Weight gear bushing
   I.D.................0.9082-0.9084 in.
Shaft bearing surface
   diameter..........0.9051-0.9055 in.

Weight gear operating
    clearance on shaft . . . 0.0027-0.0033 in.
Weight gear end clearance in
    housing . . . . 0.008-0.016 in.

NOTE: Weight gear bushings are not catalogued separately; if bushings show excessive wear, renew weight gear assemblies.

When reassembling, make certain that oil passages in housing are clean. Lubricate shafts and weight gear bushings and place drive weight assembly in housing. Install weight shaft from roll pin side of housing, through the weight gear. Align roll pin hole in shaft with roll pin hole in housing and tap shaft into position. Install roll pins as shown in Fig. 42.

NOTE: Roll pins must be shorter than shaft diameter.

Install roll pins to a depth of 1-3/8 inches in roll pin bosses. Place idler weight in housing and with both weights down, mesh gear teeth so that timing marks are aligned. Install second shaft and secure with roll pin. Using a dial indicator, check backlash between weight gears. Backlash should be 0.007-0.009 inch. If backlash is excessive, renew weight gear assemblies.

To install the balancer assembly, first rotate crankshaft to position No. 2 piston at T.D.C. of compression stroke. Place new "O" ring (6—Fig. 41) in recess on balancer. Then, with weights toward bottom of balancer housing and weight gear timing marks aligned, install balancer so that timing mark on drive weight is aligned with timing mark on balancer drive gear (2). Install new "O" ring and oil suction tube assembly. Using a dial indicator, check backlash between balancer drive weight gear and balancer drive gear (2). Backlash should be 0.010-0.016 inch.

NOTE: When backlash between weight gears (3 and 4) is correct (0.007-0.009 inch) and backlash between balancer drive gear (2) and drive weight gear exceeds 0.016 inch, then balancer drive gear (2) is excessively worn and should be renewed. Refer to paragraph 77 or 78.

Install oil pan with new pan gasket and fill crankcase to proper level with new oil.

77. RENEW BALANCER DRIVE GEAR (WELD ON TYPE). To renew balancer drive gear (2—Fig. 41), first remove crankshaft from engine. Remove worn or damaged drive gear (gear must not be split to remove) and any burrs which might be present on gear mounting surface of crankshaft. Heat new gear to 200 degrees F., install timing roll pin in crankshaft and position the gear on crankshaft.

CAUTION: Do not overheat gear as this will cause distortion.

Make sure timing marks are facing same end of engine as the timing marks on the balancer drive weight. Using a low hydrogen electrode weld gear to shaft at four places ½ to ¾-inch wide spaced equally around gear. Weld must not protrude above finished surface on crankshaft.

78. RENEW BALANCER DRIVE GEAR (SHRINK FIT TYPE). To renew balancer drive gear (2—Fig. 41), first remove crankshaft from engine. Remove worn or damaged drive gear, by using a chisel and splitting the gear and any burrs which might be present on gear mounting surface of crankshaft. Heat new gear to 360-390 degrees F., align the single mark (below the tooth space of the drive gear) with the notch on the crankshaft flange. Make sure timing marks are facing same end of engine as the timing marks on balance drive weight.

OIL PUMP AND RELIEF VALVE

All Models

79. The internally mounted gear type oil pump is driven from the crankshaft gear, and is accessible for removal after removing the engine oil pan.

The pump is mounted to the front main bearing cap. Remove the cap and pump assembly (Fig. 43). Leave the bearing cap on pump when repairing oil pump. The cap can be clamped in the vise to support the pump. If removal of bearing cap is necessary, take off the bearing adjusting nut and tab lockwasher. Remove the idler gear to obtain access to the mounting bolts.

Removing the relief valve is obvious after an examination of unit and reference to Fig. 43. Specifications for relief valve are as follows:
Valve diameter . . . . 0.825-0.827 in.
Valve clearance in
    bore . . . . 0.003-0.007 in.
Valve spring (USED)
    Free length . . . . 2.520 in.
    Test length . . . . 1.858 in.
    Test load . . . . 18-20 lbs.
Valve spring (NEW)
    Free length . . . . 2.85 in.
    Test length . . . . 1.86 in.
    Test load . . . . 33 lbs.
To disassemble the pump, remove the snap ring from rear end of pump and the cover nuts. The cover and body have dowel pins and will have to be tapped lightly with a plastic hammer. Remove the body gears and Woodruff key from the drive gear shaft.

Inspect all parts for scoring, excessive wear or other damage. Check pump against the following specifications:
Drive shaft end play (cover
    installed) . . . . 0.000-0.0020 in.
Drive shaft running clearance
    (in bushings) . . . . 0.001-0.0032 in.
Radial clearance
    (gears to body) . . . . . . 0.007-0.0120 in.
Idler gear running clearance
    (on shaft) . . . . 0.001-0.0032 in.

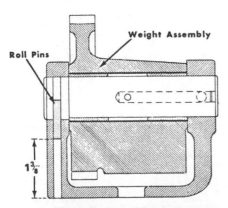

Fig. 42—Sectional view of balancer weight gear, housing and shaft used on Models 574, 674 and 584 engines showing correct installation of roll pins. Roll pins must be shorter than shaft diameter.

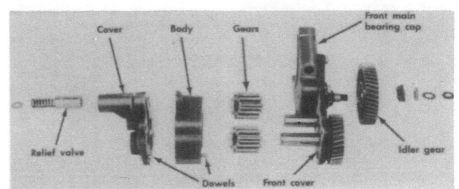

Fig. 43—Exploded view of internal oil pump used on diesel engines.

End clearance idler
   gear .... 0.002-0.0038 in.
Oil pressure at rated rpm.... 40-60 psi

When reassembling, the pump plate if removed, must be bolted to main bearing cap before idler gear is installed. Refer to Fig. 44, install idler gear and adjust the bearings to an end play of 0.000-0.002 inch. Install pump and main bearing cap on crankshaft and torque the bolts to values outlined in paragraph 71.

Install oil pan with new gasket and fill crankcase to proper level with new oil.

OIL PAN

All Models

80. To remove the oil pan, place a jack under center section of tractor. Then remove lower six bolts and loosen upper two bolts in the front bolster to engine. Remove the pan bolts, there are three pan bolts partly hidden in between the pan and clutch housing. Oil pan can now be removed.

GASOLINE FUEL SYSTEM

CARBURETOR

All Models So Equipped

83. All non-diesel tractors are equipped with Marvel-Schebler or Zenith updraft carburetors. Disassembly and overhaul procedures are obvious after an examination of carburetor and reference to Figs. 45 and 46.
Parts data follows:

Model 454 (Early)

Marvel-Schebler	IH Part No.
Model TSX-959	405873R91
Gasket package	400429R93
Overhaul package	405924R91
Main jet	388587R1
Idle jet	69929D
Idle needle	69923D
Discharge nozzle	405923R1

Venturi	405925R1
Float	381962R1
Float needle & seat	392096R91

Model 454 (Late)

Marvel-Schebler	IH Part No.
Model TSX-934	398069R92
Gasket package	400429R93
Overhaul package	405924R91
Main jet	388587R1
Idle jet	69929D
Idle needle	69923D
Discharge nozzle	405923R1
Venturi	405922R1
Float	381962R1
Float needle & seat	392096R91

Models 464 & 574

Marvel-Schebler	IH Part No.
Model TSX-959	405873R93

Gasket package	400429R93
Overhaul package	405924R91
Main jet	404156R1
Idle jet	69929D
Idle needle	69923D
Discharge nozzle	405923R1
Venturi	405925R1
Float	381962R1
Float needle & seat	392096R91

Models 574 & 674

Zenith	IH Part No.
Model 267L10	535407R92
Gasket package	68441C1
Overhaul package	537885R91
Idle jet	384863R1
Idle needle	537009R1
Main jet & Discharge nozzle	537884R1
Float	381027R1
Float needle & seat	405531R1

84. **LOW IDLE ADJUSTMENT.** Before attempting to adjust the carburetor, start engine and operate until thoroughly warmed. Then, adjust idle speed stop screw (21—Fig. 45 or 46) to obtain a low idle engine speed of 625

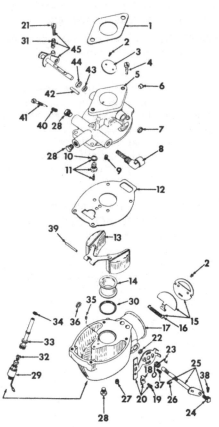

Fig. 45—Exploded view of typical Marvel-Schebler carburetor.

1. Gasket		23. Spring	
3. Throttle plate		24. Swivel	
5. Throttle body		25. Choke shaft	
6. Plug		27. Drip hole filler	
7. Cup plug		29. Fuel shut-off	
8. Strainer assy.		solenoid assy.	
9. Idle jet		30. Retainer	
10. Gasket		32. Main fuel jet	
11. Float needle & seat		33. Discharge nozzle	
12. Bowl gasket		34. Gasket	
13. Float		35. Vent	
14. Venturi		36. Cup plug	
15. Choke plate assy.		39. Float pivot shaft	
16. Spring		40. Spring	
17. Bowl assy.		41. Idle mixture screw	
18. Clamp		42. Stop pin	
20. Bracket		43. Packing	
21. Idle speed stop screw		44. Retainer	
22. Packing		45. Throttle shaft	

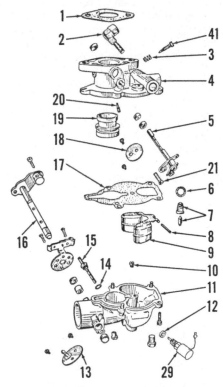

Fig. 46—Exploded view of typical Zenith carburetor.

1. Gasket		14. Gasket	
2. Strainer assy.		15. Main jet &	
3. Spring		discharge nozzle	
4. Throttle body		16. Choke shaft	
5. Throttle shaft		17. Bowl gasket	
6. Gasket		18. Throttle plate	
7. Float needle & seat		19. Venturi	
8. Float pivot shaft		20. Idle jet	
9. Float		21. Idle speed stop screw	
10. Main air bleed		29. Fuel shut-off solenoid assy.	
11. Fuel bowl		41. Idle mixture screw	
12. Gasket			
13. Choke plate			

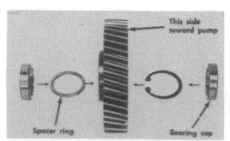

Fig. 44—View showing assembling oil pump idler gear bearing cups.

rpm. Turn idle mixture screw (41) in or out as required to obtain smoothest idle. Readjust idle speed stop screw, if necessary, to obtain correct low idle speed.

If carburetor has been disassembled, initial setting of idle mixture screw is one turn open.

85. **MAIN FUEL ADJUSTMENT.** To prevent engine "run-on" or "dieseling" after ignition is switched to "OFF" position, the carburetor is equipped with a solenoid fuel shut-off valve which stops fuel flow through the main jet.

On all models equipped with Marvel-Schebler carburetors, the main fuel adjusting screw is located in outer end of solenoid unit (29—Fig. 45). To adjust the main fuel screw, proceed as follows: With ignition switch in "OFF" position, turn adjusting screw in (clockwise) until it just contacts the solenoid plunger. Then, turn adjusting screw out 4½ turns.

On models equipped with Zenith carburetors, the solenoid unit (29—Fig. 46) is not equipped with a main fuel adjusting screw. The unit is preset and no adjustment is required.

Always use new gasket when installing either type solenoid unit.

86. **HIGH SPEED ADJUSTMENT.** The carburetor throttle lever has a stop incorporated; however, engine high idle rpm and governed rpm are controlled by the engine governor. Refer to the governor section for adjustment information.

87. **FLOAT SETTING.** To adjust the float height, invert throttle body and float assembly. On Marvel-Schebler carburetors, measure clearance between free end of float and bowl gasket. Bend float lever if necessary to obtain a clearance of ¼-inch.

On Zenith carburetors, measure distance from surface of bowl gasket to bottom of float. Bend float lever as required to obtain a measurement of 1-5/32 inches.

FUEL PUMP

All Models

88. All gasoline equipped tractors have a conventional diaphragm type fuel pump located on left side of crankcase. The fuel pump lever is actuated by an eccentric on engine camshaft. Procedure for removing pump is obvious after an examination of the unit.

DIESEL FUEL SYSTEM

The diesel fuel system on D179, D206 and D239 engines are direct injection type equipped with a Robert Bosch injection pump.

When servicing any unit of the diesel fuel system, the maintenance of absolute cleanliness is of utmost importance. Of equal importance is the avoidance of nicks or burrs on any of the working parts.

Probably the most important precaution that service personnel can impart to owners of diesel powered tractors, is to urge them to use an approved fuel that is absolutely clean and free from foreign material. Extra precaution should be taken to make certain that no water enters the fuel storage tanks.

Fig. 47—View of primary fuel filter on right side of D-179 engine. Final filter is located on left side of engine.

1. Crankcase breather
2. Vent screw
3. Primary fuel filter
4. Bowl
5. Water drain valve
6. Oil level gage
7. Oil filter

Fig. 48—View of fuel filters on right side of D-206 or D-239 engine.

1. Primary fuel filter
2. Water drain valve
3. Final fuel filter
4. Oil filter
5. Oil level gage
6. Crankcase breather
7. Vent screw

FILTERS AND BLEEDING

All Models

89. All diesel tractors are equipped with two fuel filters. Models 454, 464 and 484 have the primary filter on right side of engine, and the final filter on left side of engine. Models 574, 674 and 584 have both filters on right side of engine and the front one is the primary filter.

Model 454 has spin-on renewable filters. Models 464, 484, 574, 674 and 584 are equipped with renewable elements.

To renew filters and bleed fuel system refer to Fig. 47 and 48 and proceed as follows: Shut off fuel at tank, open vent screws at top of filters and open drain valves at bottom of filters. On Model 454 remove sediment bowl from primary filter, then unscrew and discard both filters. On Models 464, 484, 574, 674 and 584 unscrew acorn nut at top of filters, remove center bolt and sediment bowl, then discard filter elements.

On Model 454 install new filters, using new seal rings and clean and reinstall sediment bowl.

NOTE: Tighten filter only hand tight; do not use tools to tighten.

Close drain valves and vent screws. On Models 464, 484, 574, 674 and 584 install new elements, using new seal rings. Clean and reinstall the sediment bowl and secure with the center bolt and acorn nut. Tighten the nut securely but do not over tighten. Reinstall drain valves and close vent screws.

On all models, to bleed the fuel system, open fuel shut-off valve at tank and open vent screw on top of primary filter. Allow fuel to run until a solid stream with no air bubbles appears, then close vent screw. Repeat this operation on final fuel filter. Loosen

vent screw on the top of injection pump a few turns until fuel is flowing free of air bubbles, then retighten vent screw.

INJECTION PUMP

All models use a Robert Bosch Model BR or CR diesel pump which is a distributor type and is completely sealed.

Because of the special equipment needed, and skill required of servicing personnel, service of injection pumps is generally beyond the scope of that which should be attempted in the average shop. Therefore, this section will include only timing of pump to engine, removal and installation and the linkage adjustments which control the engine speeds.

If additional service is required, the pump should be turned over to an International Harvester facility which is equipped for diesel service, or to some other authorized diesel service station. Inexperienced personnel should NEVER attempt to service diesel injection pumps.

All Models

90. **REMOVE AND REINSTALL.** To remove either Model BR or CR Robert Bosch pump, first thoroughly clean injection pump, lines and side of engine. Shut off fuel and remove timing hole cover screws. Rotate cover 90 degrees and remove by prying evenly with two screwdrivers. Turn crankshaft in normal direction of rotation until timing line (TL—Fig. 49) on face cam aligns with timing pointer (TP) in pump.

NOTE: The face cam has two timing

Fig. 49—View of timing pointer (TP) and timing line (TL) on Robert Bosch Model BR injection pump. Model CR injection pump is similar.

lines. Near one of the lines, a letter "L" is etched on face cam. DO NOT use this timing line. If this line is visible with timing pointer, rotate crankshaft one complete revolution to position correct timing line in register with pointer.

At this time, timing pointer on timing gear cover should be aligned with 14 degree BTDC mark on crankshaft pulley. Disconnect throttle control rod and shut-off cable from pump. Disconnect fuel inlet and return lines from pump, then remove high pressure injection lines. Immediately cap or plug all openings. Unbolt and remove the rectangular cover from timing gear cover. Remove nut and washer from pump drive shaft, then remove three cap screws securing hub to drive gear. Remove nuts from pump mounting studs. Install a puller (IH tool FES 111-2 or equivalent) on drive gear and force pump drive shaft from hub. Remove injection pump.

CAUTION: Do not turn crankshaft while pump is removed.

When reinstalling pump, make certain that timing pointer (TP—Fig. 49) and timing line (TL) are aligned, then reinstall pump by reversing removal procedure. Align scribe mark on pump mounting flange with punch mark on engine front plate. Tighten nut on pump drive shaft to a torque of 47 ft.-lbs. Check and adjust pump timing as outlined in paragraph 91.

91. **STATIC TIMING.** To adjust static timing, first shut off fuel at tank and remove timing hole cover screws. Rotate cover 90 degrees and remove cover by prying evenly with two screwdrivers. Turn crankshaft in normal direction of rotation until number one piston is coming up on compression stroke. Continue turning crankshaft until the 14 degree BTDC mark on crankshaft pulley is aligned with timing pointer on timing gear cover. At this time, timing line (TL—Fig. 49) on face cam should be aligned with timing pointer (TP) on roller retainer ring. If not, first make certain that scribe line on pump mounting flange is aligned with punch mark on engine front plate. Then, unbolt and remove rectangular cover from front of timing gear cover. Loosen the three cap screws securing pump shaft hub to drive gear. Rotate hub as required to align timing line on face cam with the timing pointer. Tighten hub retaining cap screws. Reinstall timing cover.

92. **TIMING ADVANCE.** To check and adjust the timing advance, remove timing cover from side of pump and

install timing window (IH tool FES 111-5). With static timing line in register with timing pointer as shown in Fig. 49, note location of timing pointer in relation to the marks on timing window. Start engine and allow it to run until it reaches normal operating temperature. Check position of timing pointer with engine operating at 750 rpm for 454, 464, 484, and 584 models and at 700 rpm for 574 and 674 models. Timing pointer should be at full retard (static timing) position. While slowly increasing the engine rpm, observe the timing pointer advance. The pointer should start to advance at approximately 1050-1150 rpm and should reach full advance of 6 degrees or 2 marks at 2160 rpm on Models 454 and 484; 7 degrees or 2-1/3 marks at 2300 rpm on Model 464; 7 degrees or 2-1/3 marks at 2100 rpm on Model 574; 5.5 degrees or 1-5/6 marks at 2360 rpm on Model 674; or 5.6-6.4 degrees or 1-5/6-2-1/6 marks at 2200 rpm on Model 584.

NOTE: Each mark on the timing window is equal to 3 pump degrees.

If start of advance does not take place at specified rpm, remove and advance spring cap (A—Fig. 50) and add or remove shims between spring and cap as required. On Model CR pump, advance spring cap is on opposite side of pump. Remove shims if advance starts too late or add shims if advance starts too soon.

On early Model BR pumps, the total number of degrees advance is controlled by shims under screw on end of advance piston. Remove shims if total advance is less than specified or add shims if advance is too great. Screw is accessible after removing cap (A).

On later pumps, a shoulder in ad-

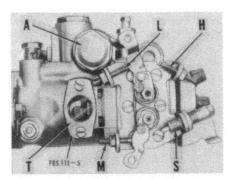

Fig. 50—View of Robert Bosch Model BR injection pump showing adjustment points. Control adjustments on Model CR pump are similar. Each mark on FES 111-5 timing window is 3 degrees. On Model CR injection pump, bolt on type advance spring cap is on opposite side of pump.

A. Advance spring cap
H. High idle stop screw
L. Low idle stop screw
M. Maximum fuel stop screw
S. Shut-off plunger
T. Timing window

vance spring cap is used to control amount of advance.

INJECTION PUMP SPEED ADJUSTMENT

All Models

93. **LOW AND HIGH IDLE.** To adjust the low idle speed, start engine and bring to operating temperature. Move speed control lever to low idle position, loosen jam nut and turn low idle speed stop screw (L—Fig. 50) as required to obtain an engine low idle speed of 750 rpm for Models 454, 464 484 and 584 and 700 rpm for Models 574 and 674. Tighten jam nut.

Move speed control lever to high idle position, loosen jam nut and turn high idle stop screw (H) as required to obtain an engine high idle no load speed as follows:

Model	Rpm
454	2420
464	2640
574	2420
674	2640
484	2420
584	2590

Tighten jam nut. With engine high idle speed properly adjusted, the rated load engine speed should be 2200 rpm for Models 454, 484 and 574, 2400 rpm for Models 464 and 674 and 2300 for Model 584.

To adjust rated load rpm, use a dynamometer and load engine to maintain rated rpm with throttle control lever in high idle position. Adjust the maximum fuel stop screw (M—Fig. 50) to obtain the following (pto) horsepower at rated rpm.

Model	Horsepower
454	40.0
464	44.4
484	42.0
574	52.5
674	61.5
584	52.0

CAUTION: Do not overfuel the engine or attempt to increase horsepower above the rated load.

94. **SHUT-OFF PLUNGER.** The shut-off plunger (S—Fig. 50) provides an excess fuel starting position for the shut-off control lever on injection pump.

To adjust the shut-off plunger on Model BR injection pump, operate engine at approximately 900 rpm. Disconnect the shut-off cable, loosen jam nut on shut-off plunger and back plunger unit out several turns. Move shut-off lever rearward until engine speed increases. Hold lever in this position, turn plunger in until end of plunger just contacts the lever, then tighten

jam nut. Move lever fully rearward to depress the plunger. Engine must shut off at this position. Connect shut-off cable to control lever.

To adjust the shut-off plunger on Model CR injection pump, disconnect the shut-off cable. Loosen the jam nut on shut-off plunger and back plunger out several turns. Operate engine at low idle rpm and move control lever on pump counter-clockwise until engine starts to surge and intermittently emit black smoke. Adjust shut-off plunger to contact control lever at this point and tighten jam nut. This is the excess fuel or starting position for the control lever. Move lever fully rearward to depress the plunger. Engine must shut off in this position at all throttle settings. Connect shut-off cable to control lever.

INJECTION NOZZLES

WARNING: Fuel leaves the injection nozzles with sufficient force to penetrate the skin. When testing, keep your person clear of the nozzle spray.

All Models

95. **TESTING AND LOCATING FAULTY NOZZLE.** If engine does not run properly and a faulty injection

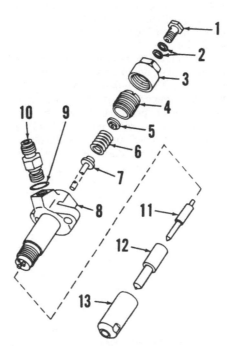

Fig. 51—Exploded view of Robert Bosch flange mounted injector nozzle assembly used on some models.

1. Hollow screw
2. Seal rings
3. Adjusting lock cap
4. Adjusting screw
5. Spring seat
6. Spring
7. Spindle
8. Body
9. Seal ring
10. Fuel inlet connector
11. Valve
12. Nozzle tip
13. Nozzle holder nut

nozzle is suspected, or if one cylinder is misfiring, locate faulty nozzle as follows: Loosen the high pressure line fitting on each nozzle holder in turn, thereby allowing fuel to escape at the union rather than enter the cylinder. As in checking spark plugs in a spark ignition engine, the faulty nozzle is the one that, when its line is loosened, least affects the running of the engine.

Remove the suspected nozzle as outlined in paragraph 96, place nozzle in a test stand and check the nozzle against the following specifications:
Opening pressure (new) ..3100-3300 psi
Opening pressure (used) ......2900 psi

Nozzle showing visible wetting on the tip after 10 seconds at 2700 psi is permissible. Maximum allowable leakage through return is 10 drops in 1 minute at 2700 psi. If nozzle requires overhauling, refer to paragraph 97.

96. **REMOVE AND REINSTALL.** To remove any injection nozzle, first remove dirt from nozzle, injection line, return line and cylinder head. Disconnect leakage return line and high pressure injection line from nozzle and immediately cap or plug all openings. Remove the injector stud nuts on flange mounted injectors or retaining bolt on clamp mounted injectors and carefully withdraw the injector assembly from cylinder head.

When reinstalling, tighten flange mounted injector stud nuts evenly and to a torque of 8 ft.-lbs. Tighten clamp mounted injector retaining bolt to a torque of 18 ft.-lbs.

NOTE: It is recommended that cooling system be drained before removing injectors. It is possible that injector tube may come out with injector and allow coolant to enter engine.

97. **OVERHAUL (FLANGE MOUNTED TYPE).** To disassemble the injector, place assembly in a vise with nozzle tip pointing upward. Remove nozzle holder nut (13—Fig. 51), then carefully remove nozzle tip (12) and valve (11). Invert the body assembly in vise and remove adjusting lock cap (3). Remove adjusting screw (4), spring seat (5), spring (6) and spindle (7), then remove fuel inlet connector (10). Thoroughly clean all parts in a suitable solvent. Clean inside the orifice end of nozzle tip with a wooden cleaning stick. The 0.012 inch diameter orifice spray holes may be cleaned by inserting a cleaning wire of proper size. Cleaning wire should be slightly smaller than spray holes. Clean the "sac" end of nozzle with a 0.047 inch drill and the fuel return passage in body with a 5/64-inch drill.

When reassembling, make certain all parts are perfectly clean and install parts while wet with clean diesel fuel. To check cleanliness and fit of valve (11) in nozzle tip (12), use a twisting motion and pull valve about ⅓ of its length out of nozzle tip. When released valve should slide back to its seat by its own weight.

NOTE: Valve and nozzle tip are mated parts and, under no circumstance, should valves and nozzle tips be interchanged.

Tighten nozzle holder nut to a torque of 50 ft.-lbs. Connect the assembled injector nozzle to a test pump and flush the valve. Adjust opening pressure by turning adjusting screw (4) in to increase or out to decrease opening pressure. Opening pressure should be adjusted to 3000 psi if old spring (6) is used or 3100-3300 psi if new spring is used. Valve should not show leakage at orifice spray holes for 10 seconds at 2700 psi. Maximum allowable leakage

through return port is 10 drops in 1 minute at 2700 psi.

98. OVERHAUL (CLAMP MOUNTED TYPE). To disassemble the injector, clamp flats of nozzle body (3—Fig. 52) in a vise with nozzle tip pointing upward. Remove nozzle holder nut (10). Remove nozzle tip (9) with valve (8) and spacer (7). Invert nozzle body (3) and remove spring seat (6), spring (5) and shims (4).

Thoroughly clean all parts in a suitable solvent. Clean inside the orifice end of nozzle tip with a wooden cleaning stick. The 0.011 inch diameter orifice spray holes may be cleaned by inserting a cleaning wire of proper size. Cleaning wire should be slightly smaller than spray holes.

When reassembling, make certain all parts are perfectly clean and install parts while wet with clean diesel fuel. To check cleanliness and fit of valve (8) in nozzle tip (9), use a twisting motion and pull valve about ⅓ of its length out of nozzle tip. When released, valve should slide back to its seat by its own weight.

NOTE: Valve and nozzle tip are mated parts and under no circumstance should valves and nozzle tips be interchanged.

Install shims (4), spring (5) and spring seat (6) in nozzle body (3). Place spacer (7) and nozzle tip (9) with valve (8) in position and install holder nut (10). Tighten holder nut to a torque of 43 ft.-lbs. Connect the assembled injector nozzle to a test pump and check opening pressure. Adjust opening pressure by varying number and thickness of shims (4). Opening pressure should be adjusted to 3000 psi if old spring (5) is used or 3100-3300 psi if new spring is installed. Valve should not show leakage at orifice spray holes for 10 seconds at 2700 psi. Maximum allowable leakage through return port is 10 drops in 1 minute at 2700 psi.

ETHER STARTING AID

All Models

99. TESTING. At temperatures below freezing, it is necessary that ether be used as a starting aid. To test the ether spray pattern, disconnect ether line at spray nozzle and remove spray nozzle from air cleaner outlet pipe. Reconnect to ether tube. Press ether injection button on dash and observe spray pattern. A good spray pattern is cone shaped. Dribbling or no spray indicates a blocked spray nozzle or lack of ether pressure. Clean spray nozzle or install new can of ether as needed.

To change the ether fluid container, turn knurled adjusting screw clockwise until container can be removed. Install new container in the bail and tighten adjusting screw (counter-clockwise) while guiding container head into position. Rotate container to be sure it is seated properly in injector body, then tighten adjusting screw to hold container firmly in position.

NOTE: In warm temperatures, ether container may be removed and a protective plug installed in injector body. DO NOT operate tractor without either the ether container or protective plug in position.

NON-DIESEL GOVERNOR

The governor used on non-diesel engines is a centrifugal flyweight type and is driven by the crankshaft gear via an idler gear. Before attempting any governor adjustments, check all linkage for binding or lost motion and correct any undesirable conditions.

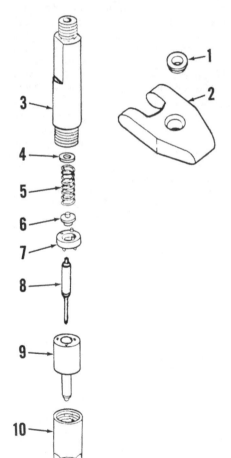

Fig. 52—Exploded view of Robert Bosch clamp mounted injector nozzle assembly used on some models.

1. Washer
2. Clamp
3. Nozzle body
4. Shim
5. Spring
6. Spring seat
7. Spacer
8. Valve
9. Nozzle tip
10. Nozzle holder nut

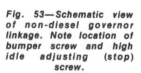

Fig. 53—Schematic view of non-diesel governor linkage. Note location of bumper screw and high idle adjusting (stop) screw.

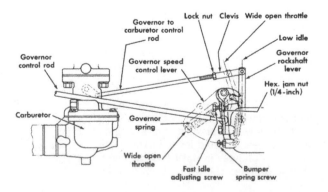

ADJUSTMENTS

All Non-Diesel Models

100. **SYNCHRONIZING GOVERNOR AND CARBURETOR.** If removal of carburetor, manifold, governor or governor linkage has been performed, or if difficulty is encountered in adjusting engine speeds, adjust governor to carburetor control rod as follows: Loosen alternator mounting bolts, remove belt from pulley and move alternator away from engine. Pull throttle lever down to put tension on governor spring, then disconnect governor to carburetor rod from governor. Hold both carburetor and governor in wide open position and adjust rod clevis until pin slides freely into clevis and rockshaft lever. Then, remove pin and lengthen rod by unscrewing clevis one full turn. Reinstall clevis pin and tighten jam nut. Install and adjust alternator belt.

101. **HIGH IDLE SPEED.** Refer to Fig. 53 and back out bumper screw ½ to ¾-inch. Start engine and bring to operating temperature. Move throttle lever to high idle position and check the engine high idle speed which should be 2420 rpm for Models 454 and 574 or 2640 rpm for Models 464 and 674. If engine high idle speed is not as stated, loosen jam nut and turn the fast idle adjusting screw as required. Tighten jam nut.

With engine high idle speed adjusted, move throttle control lever to low idle position, then quickly advance it to high idle position. Adjust the bumper screw just enough to eliminate engine surge. Repeat the operation until engine will advance from low idle to high idle without surging. Do not turn bumper screw in further than necessary as engine low idle speed could be affected.

102. **LOW IDLE SPEED.** With engine running and at operating temperature, place throttle control lever in low idle position and check engine low idle speed which should be 625 rpm for all models. If low idle speed is not as stated, adjust idle speed stop screw on carburetor as required.

If specified engine low idle speed cannot be obtained, check governor to carburetor rod adjustment as outlined in paragraph 100. It may also be necessary to vary the length of governor control rod by disconnecting and adjusting the rod ball joint located on rear end of governor rod linkage.

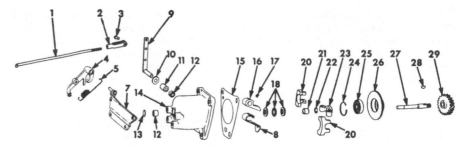

Fig. 54—Exploded view of typical governor used on non-diesel engines.

1. Governor to carburetor rod	8. Bumper spring	15. Gasket
2. Clevis	9. Rockshaft and lever	16. Fork
3. Clevis pin	10. Seal	17. Set screw
4. Governor lever	11. Bearing	18. Thrust bearing
5. Governor spring	12. Bearing	20. Weights
7. Lever bracket	13. Expansion plug	21. Thrust sleeve
	14. Housing	22. Snap ring
23. Weight carrier		
24. Snap ring		
25. Bearing		
26. Bearing carrier		
27. Governor shaft		
28. Woodruff key		
29. Drive gear		

OVERHAUL

All Non-Diesel Models

103. To remove the governor unit, remove alternator, disconnect governor control rod and governor to carburetor rod. Remove carburetor air inlet hose. Then, unbolt and remove governor assembly.

Disassembly of the governor unit will be self-evident after examination of the unit and reference to Fig. 54.

The governor weights should have an operating clearance of 0.002-0.010 inch on the pins. Bearings (18 and 25) should be free of any roughness. Rockshaft lever (9) and governor shaft (27) should operate freely in bearings (11 and 12). Make certain that seal (10) is in good condition. The lubricating holes in governor and engine front end plate must be open and clean.

A governor overhaul service package is available and consists of items 8 through 13, 15 through 21, snap ring (24) and bearing (25). A governor gear and weight assembly consisting of items 20 through 29 is also available for service.

After assembly and installation are completed, check and adjust engine speed as outlined in paragraphs 100, 101 and 102.

COOLING SYSTEM

RADIATOR

All Models

104. To remove the radiator, first drain cooling system, then remove hood and front side panels. Disconnect radiator hoses, oil cooler and power steering lines. Remove air scoop. Unbolt fan shroud and move it rearward. Unbolt and remove radiator and oil cooler as an assembly.

FAN

All Models

105. The fan is attached to the water pump shaft and one belt drives the fan, water pump and alternator.

To remove the fan, loosen alternator brace, mounting bolts and remove belt. Remove the four cap screws and lift out fan and spacer.

To adjust the drive belt, move alternator away from engine until a pressure of 25 pounds, applied midway between the water pump and crankshaft pulleys, will deflect the belt 13/16-inch.

WATER PUMP

All Non-Diesel Models

107. **R&R AND OVERHAUL.** To remove the water pump, drain cooling system and remove hood. Loosen alternator mounting bolts and remove the drive belt. Disconnect water pump inlet hose, then unbolt and remove fan and pulley. Unbolt water pump from engine and withdraw pump from left side of tractor.

To disassemble the water pump, remove plate (12—Fig. 56) and gasket (11). Remove shaft and bearing retainer (3), support pump body and press shaft assembly (4) from impeller (9) and pump body. Press shaft assembly from hub (2). Remove seal (8).

The shaft and bearing are available only as a preassembled and prelubricated unit. Water pump overhaul package (IH Part No. 398 179 R 93) is available.

When reassembling, reverse the disassembly procedure and keep the following points in mind: When installing seal (8), press only on outer diameter. Install hub with small diameter facing out. Press impeller on shaft until there is 0.031 inch clearance between machined face of pump body and face of impeller. Reinstall pump by reversing removal procedure, using a new pump mounting gasket and by-pass seal (6). Install pulley, fan and adjust the belt as outlined in paragraph 105.

All Diesel Models

108. **R&R AND OVERHAUL.** To remove the water pump, first drain the cooling system, then remove hood. Loosen alternator mounting bolts and remove the drive belt. Unbolt and remove fan (1—Fig. 57), spacer (2) and pulley (3). Remove cap screws securing pump body (7) to water pump carrier. Then, remove pump from tractor.

Disassemble the pump as follows: Remove plastic screw (13) and using a ½x2 inch NC cap screw for a jack screw, force the impeller (12) off rear end of shaft. Using two screw drivers, pry seal assembly (10) out of pump body. Support hub (4) and press out shaft. Press shaft and bearing assembly (5) out front of body. Make certain that body is supported as close to bearing as possible.

When reassembling, press shaft and

bearing assembly into body using a piece of pipe so pressure is applied only to outer race of bearing. Bearing race should be flush with front end of body. Install new "O" ring (9) and seal (10). Press only on outer diameter of seal. Clean inside of pulley hub (4) and outside of shaft with degreasing solvent. Apply a light coat of "Loctite" #601 to both surfaces. Support the shaft assembly and press hub on shaft until hub is flush with end of shaft. Install face ring (11) in impeller (12), then press impeller on shaft until there is a clearance of 0.012-0.020 inch between body and front of impeller (opposite fins). Install plastic screw.

Using a new gasket (8) reinstall pump by reversing the removal procedure. Install fan and adjust the belt as outlined in paragraph 105.

ELECTRICAL SYSTEM

ALTERNATOR AND REGULATOR

All Models

109. Delco-Remy "DELCOTRON" alternators are used on Models 454, 464, 574 and 674. Some of these models may be equipped with Delco-Remy alternator No. 1100805 and a double contact external regulator No. 1119513. Some models may be equipped with the Delco-Remy alternator No. 1100578, 1100593 or 1102920 and a solid state regulator No. 1116387 which is mounted internally and has no provision for adjustment.

Models 484 and 584 are equipped with Lucas No. 15ACR alternators. A solid state regulator is mounted in end of alternator and is non-adjustable.

CAUTION: Because certain components of the alternator can be damaged by procedures that will not affect a D.C. generator, the following precautions MUST be observed.

a. When installing batteries or connecting a booster battery, the negative post of battery must be grounded.

b. Never short across any terminal of the alternator or regulator.

c. Do not attempt to polarize the alternator.

d. Disconnect all battery ground straps before removing or installing any electrical unit.

e. Do not operate alternator on an open circuit and be sure all leads are properly connected before starting engine.

Specification data for the alternators and regulators are as follows:

Lucas Alternator 15 ACR
Field current at 80° F.,
    Amperes 3.0
    Volts 12.0
Rated output hot,
    Amperes at rpm ....28.0 at 6000

Delco-Remy Alternator 1100805
Field Current at 80° F.,
    Amperes 2.2-2.6
    Volts 12.0
Cold output at specified voltage,
    Specified volts 14.0
    Amperes at rpm ....21.0 at 2000
    Amperes at rpm ....30.0 at 5000
Rated output hot,
    Amperes 32.0

Delco-Remy Alternators 1100578 and 1100593
Field current at 80° F.,
    Amperes 4.0-4.5
    Volts 12.0
Cold output at specified voltage,
    Amperes at rpm ....32.0 at 5000
Rated output hot,
    Amperes 37.0

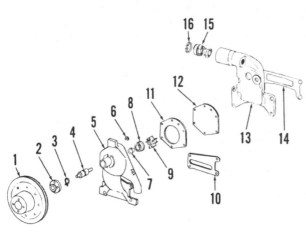

Fig. 56—Exploded view of water pump used on all non-diesel engines.

1. Pulley
2. Hub
3. Retainer
4. Pump shaft and bearing
5. Body
6. Seal (by-pass)
7. Slinger
8. Seal assembly
9. Impeller
10. Gasket
11. Gasket
12. Plate
13. Water outlet
14. Gasket
15. Thermostat
16. Snap ring

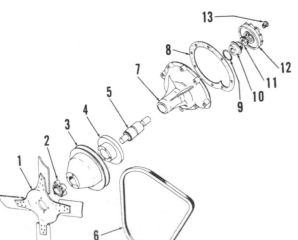

Fig. 57—Exploded view of water pump used on all diesel engines.

1. Fan
2. Fan spacer
3. Pulley
4. Hub
5. Pump shaft and bearing
6. Alternator and water pump belt
7. Body
8. Gasket
9. "O" ring
10. Seal assembly
11. Impeller face ring
12. Impeller
13. Plastic screw

Delco-Remy Alternator 1102920
Field current at 80° F.,
   Amperes....................4.0-4.5
   Volts.......................12.0
Cold output at specified voltage,
   Amperes at rpm.......22.0 at 2000
   Amperes at rpm.......38.0 at 5000
Rated output hot,
   Amperes 42

**Regulator 1119513 used with
Delco-Remy Alternator 1100805**
Ground polarity.............Negative
Field relay,
   Air gap ....0.015 in.
   Point opening...............0.030 in.
   Closing voltage range.......3.8-7.2
Voltage regulator,
   Air gap (lower points
     closed).............0.067 inch (1)
   Upper point opening (lower
     points closed)...........0.014 in.
Voltage setting,
   65° F. 13.9-15.0
   85° F. 13.8-14.8
   105° F. 13.7-14.6
   125° F. 13.5-14.4
   145° F. 13.4-14.2
   165° F. 13.2-14.0
   185° F. 13.1-13.9
(1) When bench tested, set air gap at
0.067 inch as starting point, then adjust
air gap to obtain specified difference
between voltage settings of upper and
lower contacts. Operation on lower contacts must be 0.05-0.4 volts lower than
on upper contacts. Voltage setting may
be increased up to 0.3 volt to correct
chronic battery undercharging or decreased up to 0.3 volt to correct
battery overcharging. Temperature
(ambient) is measured ¼-inch away
from regulator cover and adjustment
should be made only when regulator is
at normal operating temperature.

**110. DELCO-REMY ALTERNATOR
1100805 TESTING AND OVERHAUL.**
The only tests which can be made without removal and disassembly of alternator are the field current draw and output tests. Refer to paragraph 109 for
specifications.

To disassemble the alternator, first
scribe match marks (M—Fig. 58) on the
two frame halves (6 and 16), then
remove the four through-bolts. Pry
frame apart with a screwdriver between stator frame (11) and drive end
frame (6). Stator assembly (11) must
remain with slip ring end frame (16)
when unit is separated.

NOTE: When frames are separated,
brushes will contact rotor shaft at bearing area. Brushes MUST be cleaned of
lubricant if they are to be re-used.

Clamp the iron rotor (12) in a protected vise, only tight enough to permit
loosening of pulley nut (1). Rotor and
end frame can be separated after
pulley and fan are removed. Check
bearing surface of rotor shaft for
visible wear or scoring. Examine slip
ring surface for scoring or wear and
winding for overheating or other damage. Check rotor for grounded, shorted
or open circuits using an ohmmeter as
follows:

Refer to Fig. 59 and touch the ohmmeter probes to points (1-2) and (1-3); a
reading near zero will indicate a
ground. Touch ohmmeter probes to the
slip rings (2-3); reading should be 4.6-
5.5 ohms. A higher reading will indicate an open circuit and a lower reading will indicate a short. If windings
are satisfactory, mount rotor in a lathe
and check runout at slip rings using a
dial indicator. Runout should not exceed 0.002 inch. Slip ring surfaces can

be trued if runout is excessive or if
surfaces are scored. Finish with 400
grit or finer polishing cloth until
scratches or machine marks are removed.

Disconnect the three stator leads and
separate stator assembly (11—Fig. 58)
from slip ring end frame assembly.
Check stator windings for grounded or
open circuits as follows: Connect ohmmeter leads successively between each
pair of leads. A high reading would indicate an open circuit.

NOTE: The three stator leads have a
common connection in the center of the
windings. Connect ohmmeter leads
between each stator lead and stator
frame. A very low reading would indicate a grounded circuit. A short circuit
within the stator windings cannot be
readily determined by test because of
the low resistance of the windings.

Three negative diodes (19) are located in the slip ring end frame (16)
and three positive diodes (20) in the
heat sink (15). Diodes should test at or
near infinity in one direction when
tested with an ohmmeter, and at or
near zero when meter leads are reversed. Renew any diode with approximately equal meter readings in both
directions. Diodes must be removed
and installed using an arbor press or
vise and suitable tool which contacts
only the outer edge of the diode. Do
not attempt to drive a faulty diode out
of end frame or heat sink as shock may
cause damage to the other good diodes.
If all diodes are renewed, make certain
the positive diodes (marked with red
printing) are installed in the heat sink
and negative diodes (marked with black
printing) are installed in the end frame.

Brushes are available only in an assembly which includes brush holder
(13). Brush springs are available for
service and should be renewed if heat
damage or corrosion is evident. If
brushes are re-used, make sure all
grease is removed from surface of
brushes before unit is reassembled.

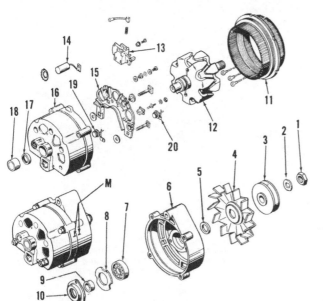

*Fig. 58—Exploded view of
"DELCOTRON" alternator
No. 1100805 used with ex-
ternal regulator. Note
match marks (M) on end
frames.*

1. Pulley nut
2. Washer
3. Drive pulley
4. Fan
5. Spacer
6. Drive end frame
7. Ball bearing
8. Gasket
9. Spacer
10. Bearing retainer
11. Stator
12. Rotor
13. Brush holder
14. Capacitor
15. Heat sink
16. Slip ring end frame
17. Felt seal and retainer
18. Needle bearing
19. Negative diode (3 used)
20. Positive diode (3 used)

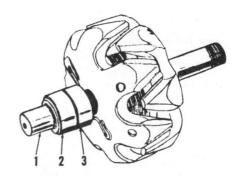

*Fig. 59—Removed rotor assembly showing
test points when checking for grounds,
shorts and opens.*

When reassembling, install brush springs and brushes in holder, push brushes up against spring pressure and insert a short piece of straight wire through hole (W—Fig. 60) and through end frame (16—Fig. 58) to outside. Withdraw the wire only after alternator is assembled.

Capacitor (14—Fig. 58) connects to the heat sink and is grounded to the end frame. Capacitor protects the diodes from voltages surges.

Remove and inspect ball bearing (7). If bearing is in satisfactory condition, fill bearing ¼-full with Delco-Remy lubricant No. 1948791 and reinstall. Inspect needle bearing (18) in slip ring end frame. This bearing should be renewed if its lubricant supply is exhausted; no attempt should be made to relubricate and re-use the bearing. Press old bearing out towards inside and press new bearing in from outside until bearing is flush with outside of end frame. Saturate felt seal with SAE 20 oil and install seal and retainer assembly.

Reassemble alternator by reversing the disassembly procedure. Tighten pulley nut to a torque of 50 ft.-lbs.

NOTE: A battery powered test light can be used instead of an ohmmeter for all electrical checks except shorts in rotor windings. However, when checking diodes, test light must not be of more than 12.0 volts.

Fig. 60—Exploded view of brush holder assembly used in Delco-Remy No. 1100805 alternator. Insert wire in hole (W) to hold brushes up. Refer to text.

111. DELCO-REMY ALTERNATOR 1100578, 1100593 OR 1102920 TESTING AND OVERHAUL. The only test which can be made without removal and disassembly of alternator is the regulator. If there is a problem with the battery not being charged, and the battery and cable connectors have been checked and are good, check the regulator as follows: Operate engine at moderate speed and turn all accessories on and check the ammeter. If ammeter reading is within 10 amperes of rated output as stamped on alternator frame (or refer to paragraph 109 for specifications) alternator is not defective. If ampere output is not within 10 amperes of rated output, ground the field winding by inserting a screwdriver into test hole (D—Fig. 62). If output is then within 10 amperes of rated output, renew the regulator.

CAUTION: When inserting screwdriver in test hole the tab is within ¾-inch of casting surface. Do not force screwdriver deeper than one inch into end frame.

If output is still not within 10 amperes of rated output, the alternator will have to be disassembled. Check the field winding, diode trio, rectifier bridge and stator as follows:

To disassemble the alternator, first scribe match marks (M—Fig. 61) on the two frame halves (4 and 16), then remove the four through-bolts. Pry frame apart with a screwdriver between stator frame (12) and drive end frame (4).

Stator assembly (12) must remain with slip ring end frame (16) when unit is separated.

NOTE: When frames are separated, brushes will contact rotor shaft at bearing area. Brushes MUST be cleaned of lubricant with a soft dry cloth if they are to be re-used.

Clamp the iron rotor (13) in a protected vise, only tight enough to permit loosening of pulley nut (1). Rotor and end frame can be separated after pulley and fan are removed. Check bearing surface of rotor shaft for visible wear or scoring. Examine slip ring surface for scoring or wear and rotor winding for overheating or other damage. Check rotor for grounded, shorted or open circuits using an ohmmeter as follows:

Refer to Fig. 59 and touch the ohmmeter probes to points (1-2) and (1-3); a reading near zero will indicate a ground. Touch ohmmeter probes to the slip rings (2-3); reading should be 5.3-5.9 ohms. A higher reading will indicate an open circuit and a lower reading will indicate a short. If windings are satisfactory, mount rotor in a lathe and check runout at slip rings using a dial indicator. Runout should not exceed 0.002 inch. Slip ring surfaces can be trued if runout is excessive or if surfaces are scored. Finish with 400 grit or finer polishing cloth until scratches or machine marks are removed.

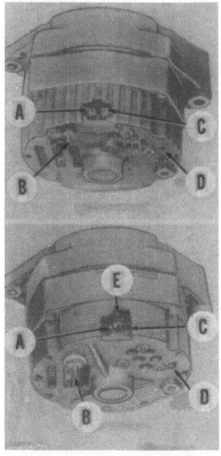

Fig. 62—View showing the terminals and test hole on Delco-Remy alternator No. 1100578 or 1100593 (Top) and alternator No. 1102920 (Bottom).

A. No. 1 terminal
B. Battery terminal
C. No. 2 terminal
D. Test hole
E. "R" terminal

Fig. 61—Exploded view of "DELCOTRON" alternator No. 1100578 or 1100593 used with internal mounted solid state regulator. Note match marks (M) on end frames.

1. Pulley nut
2. Washer
3. Spacer (outside drive end)
4. Drive end frame
5. Grease slinger
6. Ball bearing
7. Spacer (inside drive end)
8. Bearing retainer
9. Bridge rectifier
10. Diode trio
11. Capacitor
12. Stator
13. Rotor
14. Brush holder
15. Solid state regulator
16. Slip ring end frame
17. Bearing and seal assembly

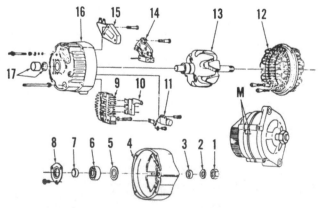

Before removing stator, brushes or diode trio, refer to Fig. 63 and check for grounds between points A to C and B to C with an ohmmeter, using the lowest range scale. Then, reverse the lead connections. If both A to C readings or both B to C readings are the same, the brushes may be grounded because of defective insulating washer and sleeve at the two screws. If the screw assembly is not damaged or grounded, the regulator is defective.

To test the diode trio, first remove the stator. Then, remove the diode trio, noting the insulator positions. Using an ohmmeter, refer to Fig. 64 and check between points A and D. Then, reverse the ohmmeter lead connections. If diode trio is good, it will give one high and one low reading. If both readings are the same, the diode trio is defective Repeat this test at points B and D and at C and D.

The rectifier bridge (Fig. 65) has a grounded heat sink (A) and an insulated heat sink (E) that is connected to the ouput terminal. Connect ohmmeter to the grounded heat sink (A) and to the flat metal strip (B). Then, reverse the ohmmeter lead connections. If both readings are the same, the rectifier bridge is defective. Repeat this test between points A and C, A and D, B and E, C and E and D and E.

Test the stator (12—Fig. 61) windings for grounded or open circuits as follows: Connect ohmmeter leads successively between each pair of leads. A high reading would indicate an open circuit.

NOTE: The three stator leads have a common connection in the center of the windings. Connect ohmmeter leads between each stator lead and stator frame. A very low reading would indicate a grounded circuit. A short circuit within the stator windings cannot be readily determined by test because of the low resistance of the windings.

Brushes and springs are available only as an assembly which includes brush holder (14—Fig. 61 or 66). If brushes are re-used, make sure all grease is removed from surface of brushes before unit is reassembled. On alternator 1102920 the brush holder is equipped with a "R" terminal for electronic tachometer (Fig. 62 and 66). When reassembling, install regulator and then brush holder, springs and brushes. Push brushes up against spring pressure and insert a short piece of straight wire through the hole and through end frame to outside. Be sure that the two screws at Points A and B (Fig. 63) have insulating washers and sleeves.

NOTE: A ground at these points will cause no output or controlled output. Withdraw the wire only after alternator is assembled.

Capacitor (11—Fig. 61) connects to the rectifier bridge and is grounded to the end frame. Capacitor protects the diodes from voltage surges.

Remove and inspect ball bearing (6— Fig. 61). If bearing is in satisfactory

Fig. 65—Bridge rectifier test points. Refer to text.

Fig. 63—Test points for brush holder. Refer to text.

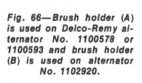

Fig. 66—Brush holder (A) is used on Delco-Remy alternator No. 1100578 or 1100593 and brush holder (B) is used on alternator No. 1102920.

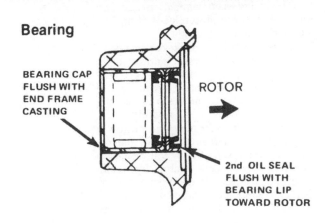

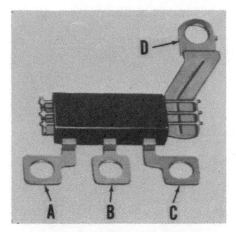

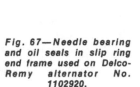

Fig. 67—Needle bearing and oil seals in slip ring end frame used on Delco-Remy alternator No. 1102920.

Fig. 64—Diode trio test points. Refer to text.

Bearing

BEARING CAP FLUSH WITH END FRAME CASTING

ROTOR

2nd OIL SEAL FLUSH WITH BEARING LIP TOWARD ROTOR

condition, fill bearing ¼ full with Delco-Remy lubricant No. 1948791 and reinstall. Inspect needle bearing (17) in slip ring end frame. This bearing should be renewed if its lubricant supply is exhausted; no attempt should be made to relubricate and re-use the bearing. Press old bearing out towards inside and press new bearing in from outside until bearing is flush with outside of end frame. Saturate felt seal with SAE 20 oil and install seal. Alternator 1102920 is equipped with two oil seals as shown in Fig. 67.

Reassemble alternator by reversing the disassembly procedure. Tighten pulley nut to a torque of 50 ft.-lbs.

112. LUCAS ALTERNATOR 15ACR TESTING AND OVERHAUL. The following component testing can be accomplished with minimum disassembly. Remove slip ring end cover. Note position of stator winding connections and unsolder connections from rectifier. Remove brush and regulator assembly. Renew brushes if overall length is less than 5/16-inch.

Check field winding continuity and resistance simultaneously by connecting a battery operated ohmmeter as shown in Fig. 68. The ohmmeter should read 4.3 ohms for 15ACR rotor with pink winding or 3.3 ohms for 15ACR rotor with purple winding.

To check rotor field winding insulation, connect a 110 volt AC 15 watt test lamp as shown in Fig. 69. The lamp should not light.

Inner stator winding short-circuiting is indicated by signs of burning of the insulation varnish covering. If this is obvious, renew the stator assembly. To check continuity of the stator windings, connect any two of the three stator winding leads in series with a 12 volt battery and 36 watt test lamp (Fig. 70). Lamp should light. Transfer one test light lead to third stator lead. Lamp should light.

Check insulation of stator windings by connecting a 110 volt AC 15 watt test lamp between lamination and any one of the three stator leads (Fig. 71). The lamp should not light.

To test rectifier diodes, connect a 12 volt battery and 1.5 watt bulb in series as shown in Fig. 72. Reverse test connections. Lamp should light during one half of test only. If one diode is unsatisfactory, renew rectifier assembly.

Continue disassembly as follows: Scribe a mark across alternator halves. Remove mounting bolts and separate alternator. Separate stator from slip ring end bracket. Remove drive pulley, fan and shaft key and separate rotor assembly from drive end bracket. Press

rotor shaft from front bearing. Unsolder field winding connections and remove slip ring. Press off rear bearing.

To reassemble, reverse disassembly

procedure. Use only "M" grade 45-55 resin-core solder to attach wires to diode pins (Fig. 74). Tighten alternator through bolts to a torque of 55 in.-lbs.

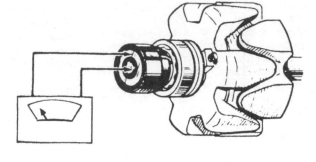

Fig. 68—Field winding continuity and resistance test on Lucas alternator rotor. Refer to text.

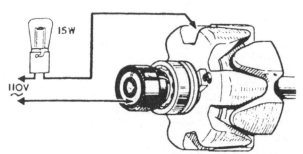

Fig. 69—Field winding insulation check. Refer to text.

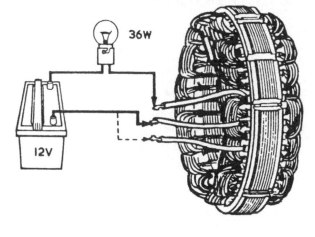

Fig. 70—Stator winding continuity test on Lucas alternator. Refer to text.

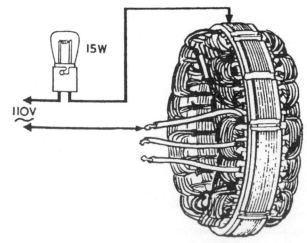

Fig. 71—Stator winding insulation check. Refer to text.

STARTING MOTORS

All Models

113. Delco-Remy and Lucas starting motors are used and specification data follows:

Delco-Remy Starter 1107867

Volts 12.0
No-load test,
  Volts .9.0
  Amperes (min). 40.0*
  Amperes (max) 140.0*
  Rpm (min) 8000
  Rpm (max). 13000
              *Includes solenoid

Delco-Remy Starter 1108394

Volts 12.0
No-load test,
  Volts .9.0
  Amperes (min). 55.0*

Amperes (max) 80.0*
Rpm (min) 3500
Rpm (max). 6000
              *Includes solenoid

Delco-Remy Starter 1113203

Volts 12.0
No-load test,
  Volts .9.0
  Amperes (min). 50.0*
  Amperes (max) 70.0*
  Rpm (min) 3500
  Rpm (max). 5500
              *Includes solenoid

Delco-Remy Starter 1113690

Volts 12.0
No-load test,
  Volts .9.0
  Amperes (min). 75.0*
  Amperes (max) 105.0*
  Rpm (min) 5000
  Rpm (max). 7000
              *Includes solenoid

Lucas Starter M-50

Volts 12.0
No-load test,
  Volts 12.0
  Amperes (max) 115.0*
  Rpm (min) 5500
  Rpm (max). 8000
              *Includes solenoid

STANDARD IGNITION

All Non-Diesel Models

114. **DISTRIBUTOR.** All non-diesel engines are equipped with Delco-Remy or Prestolite distributors and the firing order is 1-3-4-2. Refer to Figs. 75 and 76.

115. **INSTALLATION AND TIMING.** With oil pump properly installed as outlined in paragraph 48, make certain the timing pointer on timing gear cover is in register with the TDC mark on crankshaft pulley.

Install the distributor so that rotor arm is in the number one firing position and adjust the breaker contact gap to 0.020 inch on all models. Loosen distributor clamp bolts, turn distributor counter-clockwise until breaker contacts are just beginning to open. Tighten clamp bolts. Attach a timing light and with engine operating at correct high idle, no-load speed, adjust distributor to the following crankshaft pulley degree marks:

Model 454 (C-157) . . . . .17° BTDC
Model 464 (C-175) . . . . .20° BTDC
Model 574 & 674 (C-200) . . . .20° BTDC

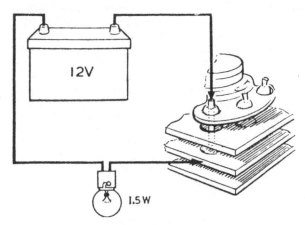

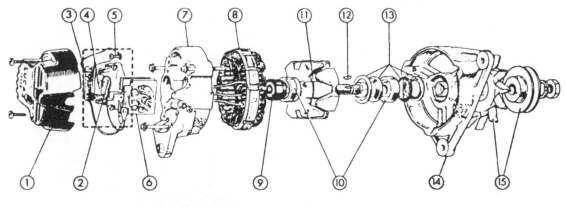

Fig. 72—Testing rectifier diodes. Refer to text.

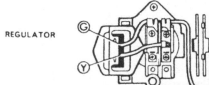

REGULATOR

Wiring Color Code:

G — Green   Y — Yellow

Fig. 73—Partially exploded view of Lucas alternator used on Models 484 and 584.

1. Moulded cover
2. Brush box assy.
3. Regulator
4. Brush & spring assy.
5. Regulator grounding & brush box screw
6. Rectifier
7. Slip ring end bracket
8. Stator winding assy.
9. Slip ring assy.
10. Ball bearings
11. Rotor & field winding
12. Woodruff key
13. Bearing assy.
14. Drive end bracket
15. Fan & Pulley

The distributor should be set to exact timing at high idle rpm. Any variance that may exist will then occur at low idle end of advance curve.

MAGNETIC PULSE IGNITION

Models 574, 674 Non-Diesel So Equipped

116. Tractors may be equipped with the "Delcotronic" transistor controlled magnetic pulse ignition system. This system consists of a special pulse distributor, pulse amplifier, resistors and special ignition coil. Refer to Fig. 77. Ignition switch, starter solenoid switch and battery are conventional.

The external appearance of the magnetic pulse distributor resembles a standard distributor; however, the internal construction is different. The timer core (5—Fig. 78) and magnetic pickup assembly (6) are used instead of the conventional breaker plate, contact point set and condenser assembly. The timer core (5) has the same number of equally spaced projections as engine cylinders. The magnetic pickup as-assembly (6) consists of a ceramic permanent magnet, pickup coil and pole piece. The flat metal pole piece has the same number of equally spaced internal projections as the timer core (one for each engine cylinder). The timer core, which is secured to advance cam (11) is made to rotate around distributor shaft (13) by advance weights (12). This provides a conventional centrifugal advance.

The pulse amplifier (3—Fig. 77) con-

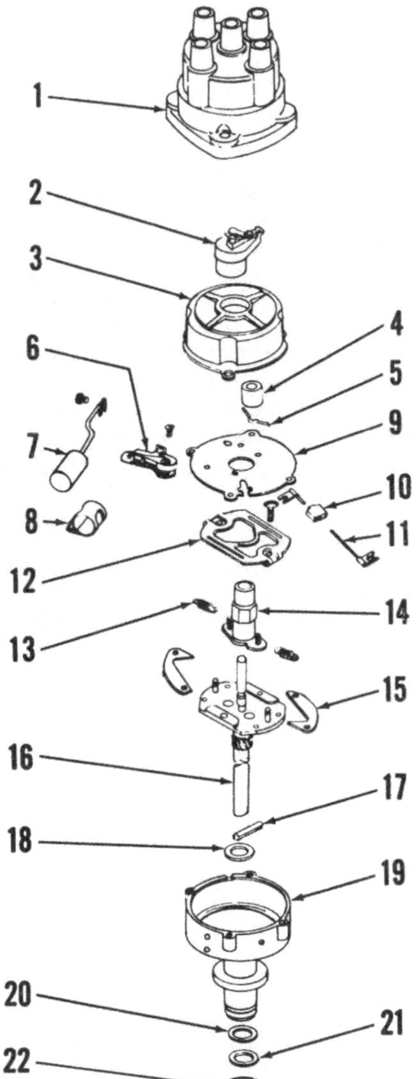

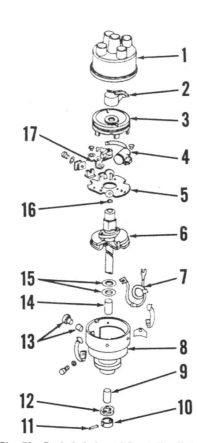

Fig. 74—Soldering stator wires to diode pins. Use long nose pliers as thermal shunt to avoid damage to diodes.

Fig. 76—Exploded view of Prestolite distributor used on some non-diesel engines.

1. Distributor cap	9. Bushing
2. Rotor	10. Collar
3. Dust cover	11. Pin
4. Condenser	12. Shim spacers
5. Breaker plate	13. Oil plug & wick
6. Shaft, weights,	14. Bushing
springs & cam assy.	15. Shim spacers
7. Primary lead	16. Retaining ring
8. Housing	17. Contact breaker set

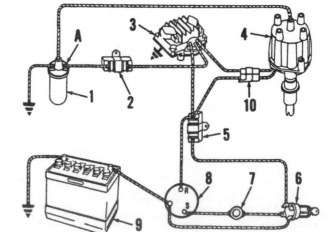

Fig. 77—Typical wiring circuit of magnetic pulse ignition system.

1. Ignition coil
2. Resistor (0.43 ohm)
3. Pulse amplifier
4. Distributor
5. Resistor (0.68 ohm)
6. Ignition switch
7. Starter switch
8. Starter solenoid switch
9. Battery
10. Connector body

Fig. 75—Exploded view of Delco-Remy distributor used on some non-diesel engines.

1. Distributor cap	12. Weight hold down plate
2. Rotor	13. Weight spring (2 used)
3. Cover	14. Breaker cam
4. Cam lubricator wick	15. Weight (2 used)
5. Retainer	16. Shaft assembly
6. Breaker contact set	17. Roll pin
7. Condenser	18. Spacer
8. Condenser bracket	19. Housing
9. Breaker plate	20. Shim
10. Grommet	21. Spacer
11. Primary wire	22. Seal (in housing)

sists primarily of transistors, resistors, capacitors and a diode mounted on a printed circuit panelboard (10—Fig. 79).

117. DISTRIBUTOR INSTALLATION. With oil pump properly installed as outlined in paragraph 48, make cer-

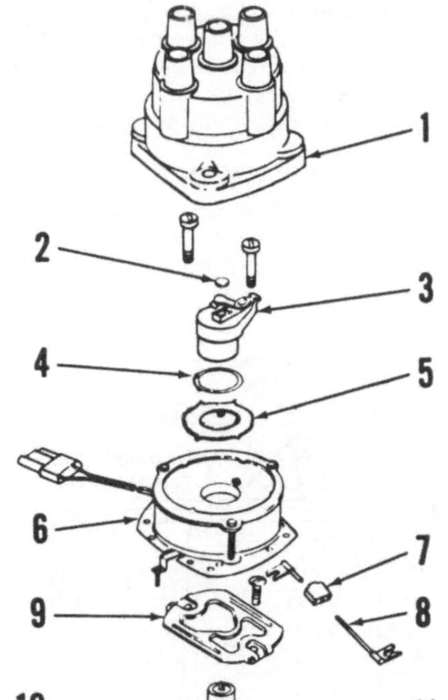

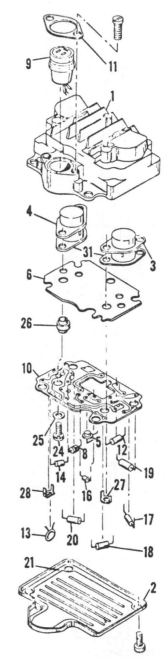

Fig. 78—Exploded view of Delco-Remy distributor used on engines equipped with magnetic pulse ignition system.

1. Distributor cap
2. Nylon button (2 used in weight base)
3. Rotor
4. Retainer
5. Timer core
6. Magnetic pickup assembly
7. Grommet
8. Primary wire
9. Weight hold down
10. Weight spring (2 used)
11. Advance cam
12. Weight (2 used)
13. Shaft assembly
14. Washer
15. Thrust washer
16. Housing
17. Roll pin
18. Washer
19. Shim
20. Shim
21. Seal (in housing)

tain the timing pointer on timing gear cover is in register with the TDC timing mark on crankshaft pulley.

Install the distributor so that rotor arm is in the number one firing position and projections on pole piece and timer core are aligned. Secure distributor with clamps and bolts.

Attach a timing light and with engine operating at correct high idle, no-load speed, adjust distributor to advanced timing of 20° BTDC.

118. TROUBLE SHOOTING. Faulty engine performance usually will be evidenced by one of the following conditions:

A. Engine miss or surge.
B. Engine will not run at all.

CAUTION: When trouble shooting the system, use extreme care to avoid accidental shorts or grounds which may cause instant damage to the amplifier. Never disconnect the high voltage lead between coil and distributor and never disconnect more than one spark plug lead unless ignition switch is in "OFF" position. To make compression checks, disconnect wiring harness plug at the pulse amplifier, then remove spark plug leads.

119. If engine misses or surges, and fuel system and governor are satisfactory, check distributor as follows: Make certain that the two distributor leads (solid white and white with green stripe) are connected to connector body as shown in Fig. 80. Disconnect the connector body halves and connect an ohmmeter (step 1) as shown in Fig. 80. Any reading above or below a range of 550-750 ohms indicates a defective pickup coil. Remove one ohmmeter lead from connector body and connect to ground (step 2—Fig. 80). Any reading less than infinite indicates a defective pickup coil. Renew magnetic pickup coil assembly (6—Fig. 78) if necessary.

A poorly grounded pulse amplifier can also cause an engine to miss or surge. To check, temporarily connect a jumper lead from amplifier housing to a good ground. If engine performance improves, amplifier is poorly grounded. Correct as required.

120. If engine will not run at all, remove one spark plug lead and hold lead about ¼-inch from engine block. Crank engine and check for spark between spark plug lead and block. If sparking occurs, the trouble most likely is not ignition. If sparking does not occur, check ignition system as follows:

With distributor connector and amplifier connector attached, connect a 12-volt test light (step 1) as shown in Fig.

81. Turn ignition switch to "ON" position. If the test light bulb does not light, there is an open between ignition switch (6—Fig. 77) and connector body (10), including leads, distributor pickup coil and resistor (5). Check distributor

Fig. 79—Exploded view of typical magnetic pulse amplifier.

1. Housing
2. Cover
3. Drive transistor (TR2)
4. Output transistor (TR1)
5. Trigger transistor (TR3)
6. Heat sink
8. Zener diode
9. Wiring connector
10. Printed circuit
11. Connector clamp
12. Capacitor
13. Capacitor
14. Resistor (10 ohm)
16. Resistor (680 ohm)
17. Resistor (1800 ohm)
18. Resistor (15000 ohm)
19. Resistor (15 ohm)
20. Resistor (150 ohm)
21. Gasket
24. Screw
25. Washer
26. Bushing insulators
27. Clip (round)
28. Clip (rectangular)
31. Insulator

pickup coil as outlined in paragraph 119. If the bulb burns at full brilliance, resistor (5) is not properly connected to ignition switch. If bulb burns at about half brilliance, resistor (5) is properly connected and circuit is satisfactory. Disconnect ignition wire between switch (6) and resistor (5). Turn ignition switch to "ON" position and press starter switch button (7) to crank engine. If bulb does not light, check for

open between solenoid switch (8) and resistor (5). If bulb lights, reconnect ignition wire and proceed with tests. Connect the 12-volt test light between connector body (step 2—Fig. 81) and coil input terminal (point A—Fig. 77). Turn ignition switch to "ON" position, press starter switch button and crank engine. If the bulb flickers, the primary circuit is operating normally. Check (in usual manner for standard ignition system) the secondary system, including spark plugs, wiring, ignition coil tower and secondary winding and the distributor cap for evidence of arc-over or leakage to ground. If bulb does not light, there is an open between test point "A" and ground. Check ground

wire and coil. If test light burns at full brilliance, check for an open between point "A" and connector body. This includes leads, connections and resistor (2). If wiring, connections and resistor are satisfactory, check for poor amplifier ground by connecting a jumper lead between amplifier housing and a good ground. If bulb now burns at half brilliance, amplifier is poorly grounded. Correct as required. If bulb remains at full brilliance, renew or repair pulse amplifier as outlined in paragraph 121.

121. **AMPLIFIER TEST AND REPAIR.** To check the pulse amplifier for defective components, remove the unit from engine and proceed as follows:

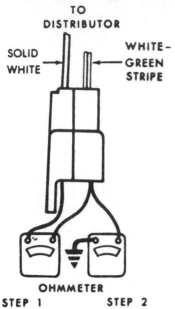

Fig. 80—View showing ohmmeter connections to distributor connector body when checking for defective pickup coil.

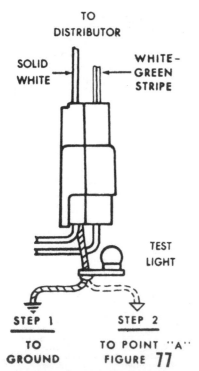

Fig. 81—View showing test light connections when checking for open circuits or defective pulse amplifier.

Fig. 82—Typical pulse amplifier with bottom cover removed showing locations of components.

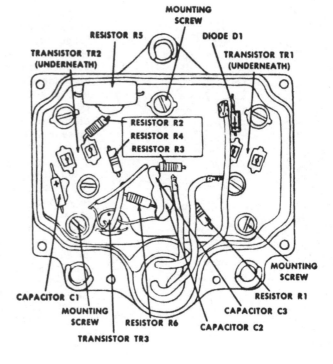

Fig. 83—View showing test points when using an ohmmeter to check the amplifier components.

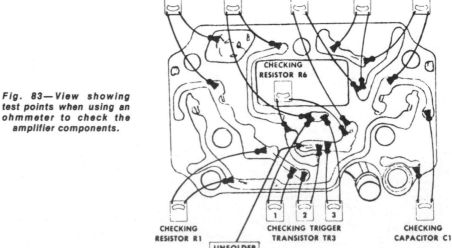
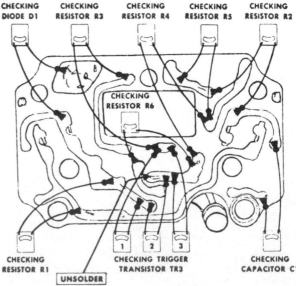

Refer to Fig. 79 and remove retaining screws, cover (2) and gasket (21). To aid in reassembly, note locations of the three lead connections to the panel board. See Fig. 82. Remove the three panelboard mounting screws and lift the assembly from amplifier housing.

CAUTION: Drive transistor TR2 (3—Fig. 79) and output transistor TR1 (4) are not interchangeable and must not be installed in reversed position. Before removing transistors, identify and mark each transistor and their respective installation on the heat sink and panelboard assembly.

Remove the screws securing TR1 and TR2 transistors to panelboard, then separate transistors and heat sink from panelboard. Note the thin insulators (31) between transistors and heat sink and the bushing insulators (26) separating the heat sink from panelboard. Visually inspect the panelboard for defects.

NOTE: To check the panelboard assembly, it is first necessary to unsolder capacitors C2 and C3 at location shown in Fig. 83. A 25 watt soldering gun is recommended and 60% tin-40% lead solder should be used when resoldering. DO NOT use acid core solder. Avoid excessive heat which may damage the panelboard. Chip away any epoxy involved and apply new epoxy (Delco-Remy part No. 1966807).

An ohmmeter having a 1½ volt cell is recommended for checking the amplifier components. The low range scale should be used in all tests except where specified otherwise. In all of the following checks, connect ohmmeter as shown in Fig. 83 and then reverse the ohmmeter leads to obtain two readings. If, during the following tests, the ohmmeter readings indicate a defective component, renew the defective part,

then continue checking the balance of the components.

Trigger Transistor TR3. If both readings in steps 1, 2 or 3 are zero or if both readings in steps 2 or 3 are infinite, renew the transistor.

Diode D1. If both readings are zero or if both readings are infinite, renew the diode.

Capacitor C1. If both readings are zero, renew the capacitor.

Capacitor C2 & C3. Connect ohmmeter across each capacitor. If both readings on either capacitor are zero, renew the capacitor.

Resistor R1. If both readings are infinite, renew the resistor.

Resistor R2. Use an ohmmeter scale on which the 1800 ohm value is within the middle third of the scale. If both readings are infinite, renew the resistor.

Resistor R3. Use an ohmmeter scale on which the 680 ohm value is within the middle third of the scale. If both readings are infinite, renew the resistor.

Resistor R4. Use an ohmmeter scale on which the 15000 ohm value is within the middle third of the scale. If either reading is infinite, renew the resistor.

Resistor R5. Use the lowest range ohmmeter scale. If either reading is infinite, renew the resistor.

Resistor R6. Use an ohmmeter scale on which the 150 ohm value is within the middle third of the scale. If both readings are infinite, renew the resistor.

Transistors TR1 & TR2. Check each transistor as shown in Fig. 84. If both readings in steps 1, 2 or 3 are zero or if both readings in steps 2 or 3 are infinite, renew the transistor.

Reassemble by reversing the disassembly procedure. When installing transistors TR1 and TR2, coat transistor side of heat sink (6—Fig. 79) and both sides of flat insulators (31) with silicone grease. The silicone grease, which is available commercially, con-

ducts heat and thereby provides better cooling.

Delco-Remy part numbers for the pulse amplifier and component parts are as follows:

Pulse amplifier assembly 1115005
Output transistor TR1 1960632
Drive transistor TR2.........1960584
Trigger transistor TR3........1960643
Zener Diode D1 1960642
Printed circuit (panelboard)....1963865
Capacitor C1 1960483
Capacitors C2 & C3 1962104
Resistor R1 (10 ohm).........1960640
Resistor R2 (1800 ohm).......1960639
Resistor R3 (680 ohm).......1960638
Resistor R4 (15000 ohms)......1960641
Resistor R5 (15 ohms)........1963873
Resistor R6 (150 ohms).......1965254

CLUTCH

122. All models are equipped with an 11 inch single plate dry disc engine clutch.

Clutch wear is compensated for by adjusting clutch linkage.

ADJUSTMENT

All Models

123. To adjust the engine clutch linkage, refer to Fig. 85. Then, adjust free height (2) of 6 inches as measured from foot platform. Loosen jam nut (B) and turn adjusting screw (A) to obtain free pedal height. Tighten jam nut (B). Then, loosen nut (C) and adjust clevis (D) until free pedal travel (1) of 1¾ inches is obtained. Tighten nut (C). This adjustment will provide a clearance of 1/8-inch between clutch release levers and release bearing.

Engine clutch linkage should be adjusted when free pedal travel has decreased to 5/8-inch.

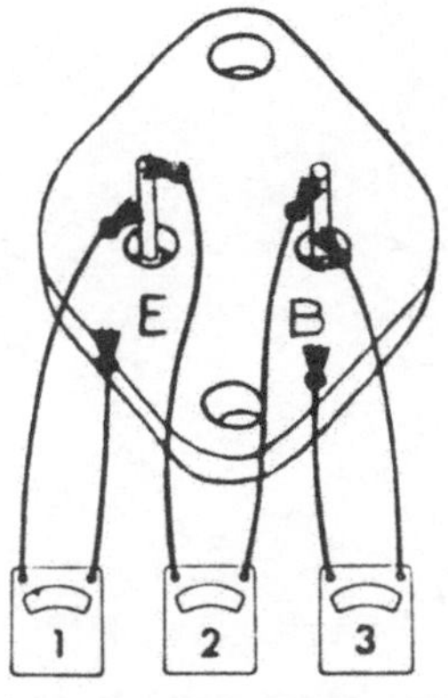

Fig. 84—Ohmmeter test points when checking TR1 and TR2 transistors. Transistors must be installed with emitter pin (E) and base pin (B) in their original positions on panelboard.

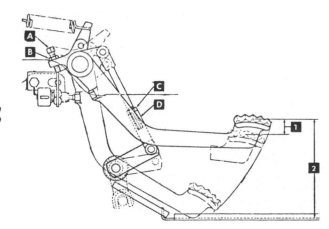

Fig. 85—View showing the adjusting points of the clutch. Refer to text.

REMOVE AND REINSTALL

Engine Clutch

124. To remove the engine clutch, it is first necessary to detach (split) engine from clutch housing as follows: Disconnect battery cables and remove hood and rear side panels. Disconnect the tachometer cable and electrical wiring from engine, unclip harness and lay tachometer cable and wiring harness rearward. Shut off fuel and disconnect fuel lines. Remove temperature sensing bulb from engine, then disconnect throttle rod, choke on non-diesel engines or fuel shut-off cable on diesel engines and oil pressure line. Disconnect and identify the wires at starter and unbolt and remove starter. Disconnect steering lines and oil cooler lines and plug openings to prevent dirt from entering system. Attach a hoist or split stand to engine and support clutch housing with a rolling floor jack. Unbolt the engine from clutch housing and roll rear section of tractor from engine. Unbolt and remove clutch assembly from flywheel.

When installing the clutch, place clutch assembly over clutch shaft. Disconnect the linkage from clutch release shaft. Clutch can be bolted to flywheel after tractor is rejoined by working through opening at bottom of clutch housing. Tighten clutch retaining cap screws to a torque of 20 ft.-lbs.

NOTE: If flywheel and clutch plate balancing mark are indicated (dab of white paint), they must be aligned.

The balance of reassembly is the reverse of disassembly procedure.

SPEED TRANSMISSION

The synchromesh speed transmission is located in the clutch housing. The speed transmission has been designed with synchronizers and therefore can be shifted with the tractor in motion by disengaging the clutch and applying a steady and continuous pressure to the speed lever until the shift is complete. There is no neutral position in the speed transmission as the neutral position is provided for in the range transmission.

LINKAGE ADJUSTMENT

All Models

125. Place the speed transmission in 1st gear (J—Fig. 97) and the speed control lever in 1st gear on quadrant, adjust clevises (K) and (L) to retain this setting.

REMOVE AND REINSTALL

All Models

126. Removal of the speed transmission requires the removal of the complete clutch housing from the tractor.

127. To remove the clutch housing, first remove the hood and rear side panels. Disconnect battery cables. Disconnect tachometer cable and electrical wiring from engine, unclip harness and lay wiring harness and tachometer cable rearward. Shut off fuel and disconnect fuel lines. Remove temperature sensing bulb from engine, then disconnect throttle rod, choke on non-diesel engines or fuel shut-off cable on diesel engines and oil pressure line. Disconnect and identify the wires at starter and unbolt and remove starter. Drain the transmission and differential. Remove right and left step plates, tunnel cover and differential lock pedal return spring. On top of speed transmission, disconnect the fuel line at coupler, two power steering lines, lube line and speed transmission shift linkage. Dis-

connect return hose at rear of brake master cylinders and brake and clutch return springs. Unbolt and remove steering and battery mount assembly. Disconnect right hand and left hand brake lines to master cylinders. Support front and rear sections of tractor. Unbolt engine from clutch housing. Move engine section away from clutch housing.

To separate clutch housing from rear frame, remove pto bearing cap retainer and then the snap ring. Using a brass drift tap the pto shaft out of the bearing as you remove the clutch housing. Also work the pto driven gear off pto shaft, working through the opening in the bottom of clutch housing.

OVERHAUL

All Models

128. With the clutch housing removed as outlined in paragraph 127, disassemble the speed transmission as follows: Disconnect clutch linkage clevis from release shaft. Remove clutch release sleeve with throw-out bearing and lube tube if so equipped, as an assembly. Late production tractors are equipped with sealed release bearings. Tap out roll pins (3—Fig. 87) and remove the shaft (4). Unbolt and remove the release sleeve carrier (2). Remove pto clutch shaft and bearing assembly (Fig. 88).

Unbolt and remove transmission top cover and shift cam assembly.

NOTE: There are two shift cam rollers on top of shift rail.

Drive roll pin from rear shift fork making sure the pin clears synchronizer, remove the other roll pin and remove shift rail to rear. Remove snap ring between 2nd and 4th gear (1—

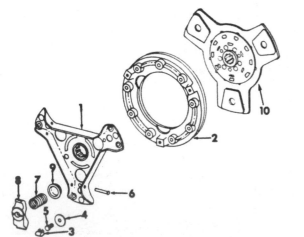

Fig. 86—Exploded view of typical engine clutch assembly. Item 10 is available as a unit only. Some diesel models are equipped with a 4 pad driven disc instead of the 3 pad disc shown.

1. Back plate
2. Pressure plate
3. Clip
4. Washer
5. Adjusting screw
6. Pin
7. Clutch spring
8. Lever
9. Spring cup
10. Driven disc

Fig. 87—View of the clutch fork and bearing release sleeve carrier.

1. Pto bearing retainer cap
2. Release bearing sleeve carrier
3. Roll pins
4. Clutch fork shaft

Fig. 89), unbolt rear bearing cage and slide main shaft out the rear. This will allow removal of the thrust washers, 2nd gear, synchronizer and 1st gear. Then, remove the other gears, synchronizer and clutch shaft. Refer to Fig. 90 for location of gears and synchronizers on the main shaft.

There is a speed shift hub (4—Fig. 91) used between 1st and 2nd gears and between 3rd and 4th gears in the synchronizers. Items 2 and 3 are available only as an assembly.

If it is necessary to renew the ball bearing (5—Fig. 92) on main shaft, then remove the two seal rings (2), then wrap shim stock around the shaft at the seal ring grooves before removing snap ring (4) to protect the ring grooves. Be sure the lube holes (1 and 3) are clean.

129. Remove the internal snap ring from rear of countershaft and snap ring (6—Fig. 93) from front of countershaft. Using a brass drift, tap the countershaft rearward and remove the gears and shaft.

At this time, all parts of the speed transmission can be inspected and parts renewed as necessary. Refer to Figs. 94 and 95 for installation dimensions and information. Use new "O" rings and seals during assembly.

130. Reassembly procedure is the reverse of disassembly, keeping the following points in mind: Lubricate all parts with Hy-Tran. Be sure when connecting main shaft to clutch shaft that there is a thrust washer on each side of snap ring. Leave the lower front pto bearing out of the case until after the clutch housing is bolted to rear frame. When installing top cover, refer to Fig. 96, position the cam in first gear and the speed transmission into the same

Fig. 91—Exploded view of the synchronizer used in the speed transmission. Items 2 and 3 are available only as an assembly.

1. Gear
2. Cups
3. Synchronizer
4. Speed shift hub
5. Gear

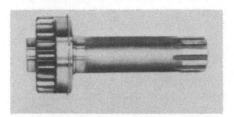

Fig. 88—View of the pto drive shaft and bearing assembly.

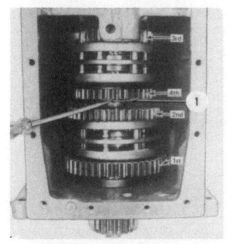

Fig. 89—View showing removing the snap ring (1) on main shaft of speed transmission. Refer to text.

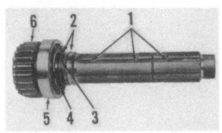

Fig. 92—View of main shaft showing lube holes and seal rings.

1. Lube holes
2. Seal rings
3. Lube supply hole
4. Snap ring
5. Bearing
6. Main shaft

Fig. 93—View of the countershaft and the gears as they are located on the shaft.

1. Bearing
2. 1st gear
3. 2nd gear
4. 4th gear
5. Constant mesh driven gear
6. Snap ring
7. Countershaft

Fig. 90—Exploded view showing the gears and synchronizers as they are on the main shaft.

1. Main shaft
2. Bearing cage
3. 1st gear
4. Synchronizer (1st & 2nd)
5. 2nd gear
6. 4th gear
7. Synchronizer (3rd & 4th)
8. 3rd gear
9. Bearing
10. Clutch shaft
11. Thrust washers
12. Snap ring
13. Spacer

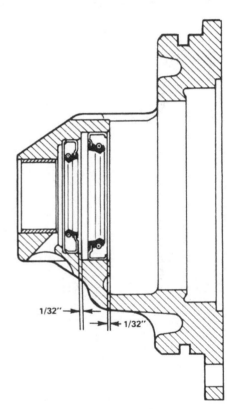

Fig. 94—Install the oil seals in the pto and clutch shaft bearing cage as shown.

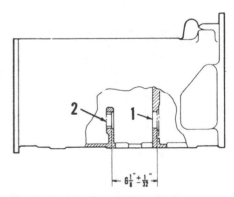

Fig. 95—Install the countershaft needle bearing (2) and the pto ball bearing (1) as shown.

gear. Measure A and B on the cam and adjust the speed transmission to correspond. When installing the lower front

bearing on pto shaft first remove the pto clutch cover on right hand side of rear frame and with a bar, hold

forward on the pump drive gear while installing the bearing and snap ring.

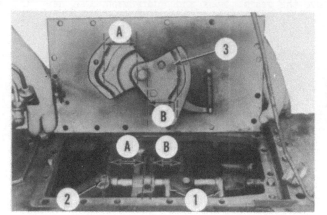

Fig. 96—View showing the points to measure for installing the cover on the speed transmission. Refer to text.

1. Shifter forks
2. Speed shift hub
3. Shift cam assembly

LUBRICATION PUMP

131. The lube pump is a 4.5 gpm capacity pump, and is located in the front of the rear section with the range transmission. The lube pump is driven by the pto and hydraulic pump drive gear assembly.

On early units with transmission serial number below 101631, the pump lubricates the speed transmission main shaft and bearings in rear frame and keeps the fluid at the proper level in rear frame. Refer to paragraph 132 for test procedures on this pump.

On later units, the speed transmission receives its lubrication from the oil cooler return line. The pump lubricates the main drive bevel gears and maintains proper fluid level in rear frame and speed transmission as required. Pump flow cannot be checked on late units.

All Models

132. TESTING. To test the lube pump on early units, first remove tunnel cover over the transmission. Then, to check pressure, remove the sending unit (A—Fig. 98) and install a pressure gage. The pressure should be 40 psi at high idle and 25 psi at low idle speed. To check lube pump output, plug the sending unit port. Disconnect the line at (B) and attach Flo-Rater inlet hose to the junction block and the return hose from Flo-Rater to the hitch housing filler plug. With the Flo-Rater restrictor valve open, run the tractor at high idle. The output flow should be 4.5 gpm.

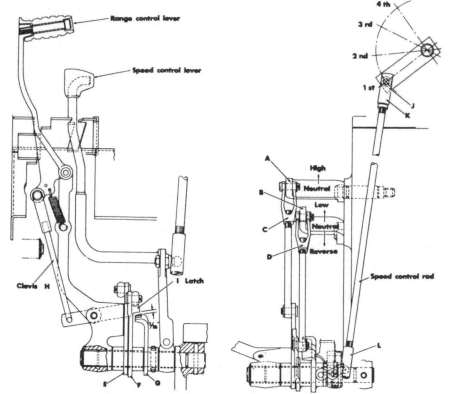

Fig. 97—View showing the adjusting points for the speed and range transmission linkage. Refer to text.

Fig. 98—View showing the test points for the lube pump. Refer to text.

Fig. 99—View of the lube pump with MCV removed.

1. Snap ring
2. Low range gear
3. Countershaft
4. Lube pump
5. Pickup screen

CAUTION: DO NOT close the restrictor valve.

If pressure or flow or both is low and the lines and the pickup screen on pump are tight and clean, renew the pump. Refer to paragraph 133 for removing and reinstalling. Pump parts are not serviced separately; renew complete pump.

133. **REMOVE AND REINSTALL.** To remove the lube pump, drain Hy-Tran fluid from all compartments. Then, remove the MCV and hydraulic pump unit. With MCV removed, disengage snap ring (1—Fig. 99) and slide gear (2) to the rear. Remove the one cap screw that holds the screen assembly (5) to the pump. Remove the tube nut on line (1—Fig. 100). Then remove the pto clutch cover on the right side of the rear frame. Remove snap ring (4), gear (3) and set screws (5) that secure pump in position. Pull the pressure tube (1) towards the MCV side enough to remove pump. Reassembly procedure is the reverse of disassembly.

RANGE TRANSMISSION

134. **The range transmission is located in the front portion of the tractor rear frame. This transmission provides four positions: HI (direct drive), Lo (underdrive), neutral and reverse. The transmission parking brake is also located in the range transmission.**

To remove the range transmission gears and shafts, the tractor rear main frame must be separated from the clutch housing and the differential assembly removed. See Fig. 106 for a view of the range transmission shafts and gears. Refer to paragraph 135 for linkage adjustment procedures and to paragraph 136 for information on overhaul of the range transmission.

LINKAGE ADJUSTMENT

All Models

135. With range lever in neutral on the quadrant and shift arms (A and B—Fig. 97) in neutral, adjust clevises (C and D) so that slots (E and F) are in alignment with stationary lever slot (G). With the range lever still in neutral, adjust clevis (H) to provide 1/16-inch clearance between latch (I) and lever (G).

R&R AND OVERHAUL

All Models

136. Before separating the rear frame from clutch housing, remove the seat, fenders and tank assembly as follows: Remove the seat. Disconnect wiring harness on right and left sides, fuel lines, and the park brake assembly from left fender. Disconnect variable valve flow control line and remove control lever knobs. Unbolt platform from rear frame and lift fenders, fuel tank and platform as a unit. Remove foot platforms.

Disconnect all necessary hydraulic lines, brake lines and wiring harness to the rear frame. Disconnect pto control valve linkage, speed shift rod and lube regulator block (early) or MCV dump orifice elbow (late). Unbolt and remove the hydraulic lift assembly. Attach split stands to clutch housing and support rear frame with a floor jack. Remove rear wheels, then unbolt and remove rear axle assemblies. Remove pto clutch cover on right side of rear frame. Refer to Fig. 101 and drive roll pin from rear of clutch assembly and 1000 rpm shaft. Remove the cover or bearing flange on rear of tractor and move the output shaft (2) rearward out

of the pto clutch assembly. Remove clutch (4) and snap ring from rear of drive gear (6), then remove gear. Separate rear frame from clutch housing. Remove the planetary drive shafts. Note that the right shaft is longer than the left shaft. Remove brake disc, install screws in the tapped holes of the brake piston (3—Fig. 102) and pry with a screwdriver equally around piston to remove. On tractors equipped with a differential lock, remove the shaft and fork.

NOTE: Shims located between the shaft and housing allow for proper disengagement.

Attach hoist to the differential, remove bolts (2—Fig. 102) and install jack screws in tapped holes (4). Remove bearing retainers, keeping the shims with the retainer for reassembly. Lift the differential from frame.

137. The range transmission can now be disassembled as follows: Refer to Fig. 103 and disconnect brake linkage at (2), remove roll pin using an open end wrench (4) to support brake band when driving pin (5) out, then remove band. Remove shifter shaft roll pin (1), then remove shaft and shifter forks being careful not to lose poppet balls and springs. Remove snap ring (4—Fig. 104) and using a hardwood block drive the brake drum (3) forward and remove the Woodruff key. Unbolt and remove main shaft bearing cage and main shaft. Keep the shims with bearing cage. Remove the countershaft front bearing retainer ring and snap ring back of low drive gear (5—Fig. 105), drive countershaft forward and remove the gears and spacer. Drive out roll pin (2) and remove reverse idler shaft and gear (1).

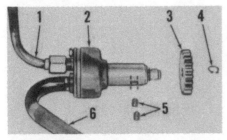

Fig. 100—Exploded view of lube pump. Internal parts of the pump are not serviceable.

1. Pressure line
2. Pump
3. Gear
4. Snap ring
5. Set screws
6. Screen

Fig. 101—View of pto clutch, drive gears, 1000 rpm shaft and lube pump gear.

1. Roll pin
2. Shaft (1000 rpm)
3. Spring and eye
4. Clutch assembly
5. Lube gear
6. Drive gear

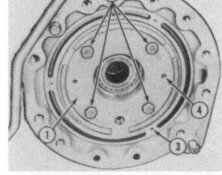

Fig. 102—View showing the holes for removing the brake piston and jack screw holes for removing carrier. Refer to text.

1. Carrier
2. Carrier cap screws
3. Brake piston holes
4. Carrier jack screw holes

138. Clean and inspect all parts and renew any which show excessive wear or other damage. Refer to Fig. 106 for location of range transmission shafts and gears. When renewing range trans-

mission pinion bearings refer to Fig. 107 for all except Series 674 row crop. Press both bearing cups (3) into bearing retainer (4), with smallest diameter toward center, until they bottom. Press rear bearing (5) on pinion shaft, with largest diameter toward gear, until it bottoms. Place bearing carrier over pinion shaft with flange toward gear (rear). Press front bearing on shaft with largest diameter toward front and as bearing cone enters bearing carrier, rotate the carrier to insure alignment of parts. Install snap ring (1).

139. On Model 674 row crop the bearing cups are part of the carrier and pinion bearing must be set-up to a rolling torque of 20 to 70 in.-lbs. as follows:

Press rear bearing cone (2—Fig. 108) against shoulder on main shaft. Position carrier (3) on shaft and rear bearing (lubricate with Hy-Tran). Then wrap a cord around O.D. of carrier and attach end of cord to a spring scale. Press the front bearing (4) into carrier until a pull of 4.5 to 9 lbs. on the spring scale is required to rotate the carrier. Install thrust washer (5). Select a snap ring that will fit into the groove without allowing a 0.002 inch feeler

gage to also go into groove. Snap rings (shims) are available from 0.066 inch to 0.093 inch in increments of 0.003 inch.

140. Reassembly sequence of the range transmission is countershaft, re-

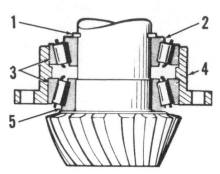

Fig. 107—View showing the bearings on bevel pinion shaft on all except Series 674 row crop.

1. Snap ring
2. Bearing cone (front)
3. Bearing cups
4. Bearing cage
5. Bearing cone (rear large I.D.)

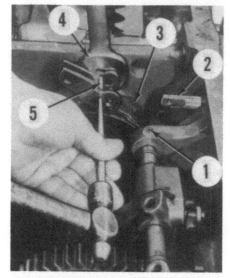

Fig. 103—View showing procedure for removing parking brake band. Refer to text.

1. Roll pin (shifter shaft)
2. Brake linkage
3. Brake band
4. Open end wrench
5. Pin

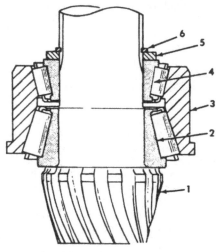

Fig. 108—View showing the bearings on bevel pinion shaft used on Model 674 row crop.

1. Pinion shaft
2. Bearing cone (rear)
3. Carrier
4. Bearing cone (front)
5. Thrust washer
6. Snap ring (shim)

Fig. 104—Bevel pinion shaft of range transmission.

1. Shifter collar (hi-range)
2. Sliding gear (forward & reverse)
3. Brake drum
4. Snap ring

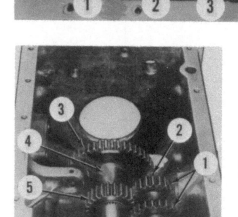

Fig. 105—Countershaft of the range transmission.

1. Reverse idler gear
2. Roll pin
3. Constant mesh gear
4. Spacer
5. Low range drive gear

Fig. 106—View showing range transmission shafts, components and location.

1. Constant mesh gear
2. Reverse idler shaft
3. Reverse idler gear
4. Countershaft
5. Pto shaft
6. Low range gear
7. Spacer
8. Shaft (speed transmission)
9. Shift collar (hi-speed)
10. Sliding gear (forward & reverse)
11. Bevel pinion shaft
12. Brake drum
13. Bearing cage

verse idler shaft and pinion shaft. To install the countershaft, install bearing on front of shaft. Start shaft in rear frame and install constant mesh gear (1—Fig. 106), spacer (7), low range drive gear (6) and snap ring securing low range gear. Install the snap ring on front of countershaft.

To install reverse idler shaft (2) and gear (3), align the hole in the shaft with the hole in frame and install roll pin.

To adjust the pinion shaft, install rear bearing assembly as outlined in paragraph 138 or 139. Note the number etched on the pinion end of shaft. Install pinion shaft (1—Fig. 109) and bearing assembly (2) without shims (3), drum (4), gear (5) and shift collar (6), then tighten bearing assembly. Install pinion gage bar IH tool No. FES 143-2 as shown (3—Fig. 110). Using a telescoping gage (2) measure the distance between gage bar and pinion (1). The reading etched on end of bevel pinion minus 2.500 inches is the specified dis-

tance at which the pinion should be set. The difference between the specified distance and the measured distance between gage bar and pinion is the correct amount of shim thickness to be installed. Shims are available in thicknesses of 0.004 and 0.007 inch.

After determining the number of shims needed, remove pinion shaft and install shims (3—Fig. 109) and reinstall pinion shaft with brake drum (4), Woodruff key, brake drum snap ring,

sliding gear (5) and shift collar (6). Install shifter forks, shaft, poppet springs and balls. Install brake band and pins.

Reinstall differential assembly in rear frame and check carrier bearing preload as outlined in paragraph 142 or 143 and backlash as outlined in paragraph 144.

Complete reassembly of tractor by reversing the disassembly procedure. Bleed the brakes as outlined in paragraph 153.

MAIN DRIVE BEVEL GEARS AND DIFFERENTIAL

The differential is carried on tapered roller bearings. The bearing on the bevel gear side of the differential is larger than that on the opposite side and therefore, it is necessary to keep the bearing cages in the proper relationship.

The differential assemblies for Models 454, 464 and 484 have two pinions with a single shaft (Fig. 111). Models 574, 674 and 584 have four pinions and four shafts (Fig. 112). The differential lock is always on the right hand side if so equipped.

ADJUSTMENT

All Models

141. **CARRIER BEARING PRE-LOAD.** The carrier bearings can be adjusted by either of two methods; however, in either case the platform, fenders, fuel tank and hydraulic lift assembly should be removed as outlined in paragraph 136 and the final drives removed as outlined in paragraph 148.

142. To use the direct measurement

Fig. 109—View showing component parts on the bevel pinion shaft.

1. Pinion shaft	4. Brake drum
2. Bearing cage	5. Gear
3. Shim	6. Shift collar

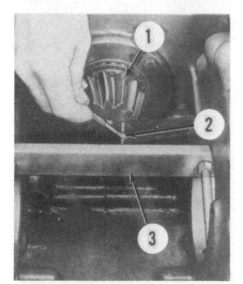

Fig. 110—Use IH tool No. FES 143-2 and telescoping gage when checking pinion shaft setting.

1. Pinion shaft
2. Telescoping gage
3. IH tool No. FES 143-2

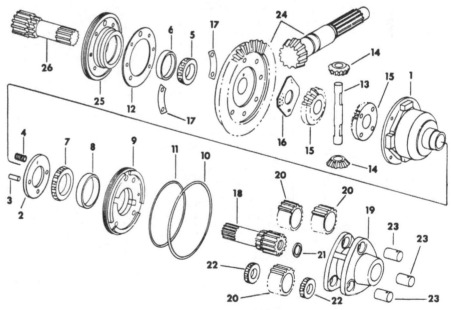

Fig. 111—Exploded view of differential, planetary unit and associated parts used on Models 454, 464 and 484.

1. Differential case	9. Differential bearing cage (RH)	15. Bevel gears	22. Bearing cone
2. Differential lock ring	10. "O" ring	16. Thrust washer	23. Pins
3. Pin	11. "O" ring	17. Lock plates	24. Ring and pinion gear
4. Spring	12. Bearing cage shims	18. Planetary drive shaft (RH)	25. Differential bearing cage (LH)
5. Bearing cone (LH)	13. Differential pinion shaft	19. Carrier	26. Planetary drive shaft (LH)
6. Bearing cup (LH)	14. Pinion gears	20. Planetary gears	
7. Bearing cone (RH)		21. Shim	
8. Bearing cup (RH)			

method, proceed as follows: Install differential in rear frame with no "O" rings or shims behind bearing cages and tighten right bearing cage cap screws to 85 ft.-lbs. torque. Rotate the differential and tighten the left hand bearing cage cap screws in steps of 25, 50, 75 and 100 in.-lbs. torque. Now loosen the left hand bearing cage cap screws, rotate differential and retighten cap screws to 20 in.-lbs. Then, without rotating differential, tighten bearing cage cap screws to 30 in.-lbs. torque. Insert a depth gage through puller bolt holes and measure between the surface of rear main frame and machined surface (Fig. 113) of bearing cage and average the readings. Remove the left bearing cage and measure the thickness as shown in Fig. 114. Subtract second reading from first reading which will give the required thickness of shim pack.

If bearing cage has feeler gage slots (Fig. 115) measure with feeler gage for the right thickness of shim pack. Shim pack on either type of bearing cage

must be within 0.002 inch of the determined shim pack thickness. Shims are available in thicknesses of 0.003, 0.007, 0.012 and 0.0299 inch. Shims can be divided between the two bearing cages to provide the proper backlash as outlined in paragraph 144.

143. To use the rolling torque method, use the original shim packs behind bearing cages and rotate the differential as bearing cage cap screws are tightened. Be sure there is some backlash maintained between bevel gear and pinion. Place the range transmission in neutral, then loosen the left bearing cage so there is no preload on bearing (cap screws finger tight). Wrap a cord around differential, attach to a spring scale and note the pounds pull required to keep differential and drive pinion in motion. Tighten bearing cage cap screws evenly to a torque of 85 ft.-lbs. and recheck the rolling torque. Now vary the shims until 1 to 3 pounds more pull is required to keep differential and transmission in motion than

when no preload was applied to bearings. With bearing preload determined, refer to paragraph 144 to set backlash between bevel gear and pinion.

144. BACKLASH ADJUSTMENT. With the differential carrier bearing preload determined as outlined in paragraph 142 or 143, the backlash between bevel gear and drive pinion should be checked and adjusted as follows: Mount a dial indicator and while holding drive pinion forward, check backlash in at least three places during a revolution of the differential. Correct backlash is 0.006-0.011 inch. If the backlash is not as stated, shift bearing cage shims from one side to the other as required. Do not add or remove shims as the previously determined bearing preload

Fig. 113—Using a depth micrometer to determine shim pack required on early production tractors. Refer to text.

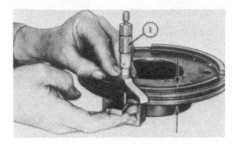

Fig. 114—Measuring the thickness of left hand carrier. Refer to text.

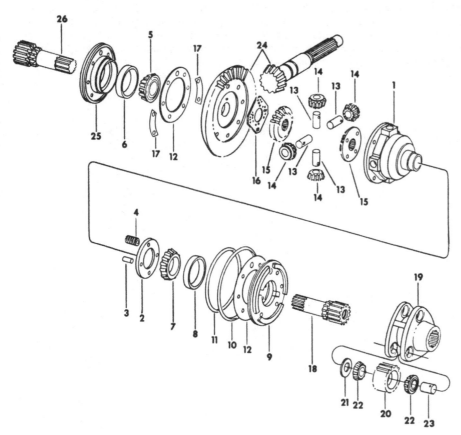

Fig. 112—Exploded view of differential, planetary unit and associated parts used on Models 574, 674, and 584.

1. Differential case	9. Differential bearing cage (RH)	15. Bevel gears	22. Bearing cone
2. Differential lock ring	10. "O" ring	16. Thrust washer	23. Pin
3. Pin	11. "O" ring	17. Lock plates	24. Ring and pinion gear
4. Spring	12. Bearing cage shims	18. Planetary drive shaft (RH)	25. Differential bearing cage (LH)
5. Bearing cone (LH)	13. Differential pinion shafts	19. Carrier	26. Planetary drive shaft (LH)
6. Bearing cup (LH)	14. Pinion gears	20. Planetary gears	
7. Bearing cone (RH)		21. Shim	
8. Bearing cup (RH)			

Fig. 115—On late production tractors with feeler gage slots use feeler gage to determine shim pack required. Refer to text.

will be changed. Shifting 0.010 inch shim thickness from one side to the other will change backlash approximately 0.0075 inch.

R&R BEVEL GEARS

All Models

145. The main drive bevel pinion is also the range transmission mainshaft. The procedure for removing, reinstalling and adjusting pinion setting is outlined in the range transmission section paragraphs 134 through 140.

To remove the bevel ring gear, follow the procedure outlined in paragraph 146 for R&R of differential. The ring gear is secured by the differential case type 8 cap screws which should be tightened to a torque of 115 to 130 ft.-lbs. Use "Loctite" #262 on the cap screw threads.

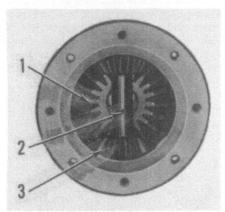

Fig. 116—View of the differential bevel gear and pinion gears used in Models 454, 464 and 484.

1. Bevel gear
2. Pinion shaft
3. Pinion gear (2 used)

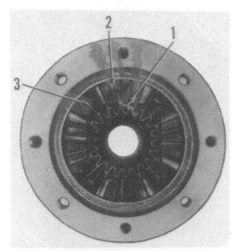

Fig. 117—View of the differential bevel gear and pinion gears used in Models 574, 674 and 584.

1. Pinion shaft (4 used)
2. Pinion gear (4 used)
3. Bevel gear

R&R DIFFERENTIAL

All Models

146. Tractor Models 454, 464 and 484 are equipped with a two pinion differential (Fig. 116). Models 574, 674 and 584 are equipped with a four pinion differential (Fig. 117). To remove the differential, remove final drives as outlined in paragraph 148 and the hydraulic lift assembly as outlined in paragraph 169.

With final drives and hydraulic lift assembly removed, attach a hoist to the differential assembly, remove both differential bearing cages and carefully lift differential from rear frame.

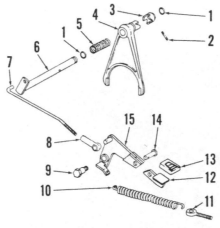

Fig. 118—Exploded view of differential lock linkage on tractors so equipped. Pedal (12 through 15) is used on early Models 454 and 574. All other models use a one-piece pedal.

1. "O" ring
2. Roll pin
3. Cam
4. Fork
5. Spring
6. Shaft
7. Rod
8. End
9. Shoulder bolt
10. Return spring
11. Eye bolt
12. Pedal (front)
13. Pad
14. Pin
15. Pedal (rear)

To disassemble differential, use a puller and remove the carrier bearings and note that bearing on bevel gear side of differential is a larger bearing. If equipped with a differential lock it will be removed with the carrier bearing on the right side. Remove the cap screws securing the ring gear to differential. Any further disassembly is obvious.

When reassembling differential, line holes in pinion shaft or shafts with holes in differential case. Some early units were equipped with type 5 cap screws and lock plates. Discard lock plates and type 5 cap screws and reassemble with type 8 cap screws. Use "Loctite" #262 on screw threads and tighten the cap screws to a torque of 115 to 130 ft.-lbs. Refer to paragraph 141 through 144 for carrier bearing preload and backlash adjustment.

DIFFERENTIAL LOCK

All Models So Equipped

Tractors may be equipped with a differential lock which operates on the right hand side of the differential with a dowel pin-type coupling locking the differential gears. It is operated with a pedal at the rear of foot plate on right hand side of tractor.

Refer to Fig. 118 for an exploded view of differential lock linkage and to items 2, 3, 4 and 15—Fig. 111 or 112 for differential lock.

147. **R&R AND OVERHAUL.** Removal and overhaul of the differential lock requires removal and separation of the differential assembly as outlined in paragraph 146.

Overhaul of the differential lock is obvious after an examination of the unit.

When reassembling, adjust the linkage rod clevis (A—Fig. 119) so that

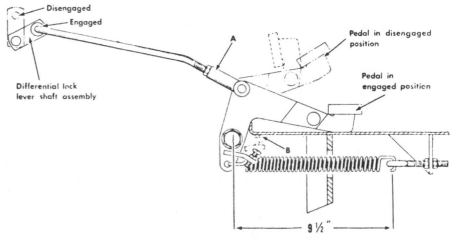

Fig. 119—Differential lock linkage adjustment. Refer to text.

fork (4—Fig. 118) and cam (3) are bottomed out in fully engaged position, with the differential lock pedal hitting the platform (Fig. 119). Adjust set screw (B) so that arm on differential lock lever shaft is vertical when disengaged. Adjust the return spring to 9½ inches while in the disengaged position.

FINAL DRIVE

The final drive assemblies consist of the rear axle, planetary unit and outer brake disc ring and can be removed from the tractor as a unit.

All Models

148. **REMOVE AND REINSTALL.** To remove either final drive, remove drain plug and drain housing. Remove the seat. Disconnect wiring harness on right and left sides, fuel lines and the park brake assembly from left fender. Disconnect variable valve flow control line and remove control lever knobs. Unbolt platform from rear frame and lift fenders, fuel tank and platform as a unit. When removing final drive on left side, remove the hydraulic supply line to the hitch and auxiliary valves from the MCV. Disconnect 3-point hitch lower link from axle housing. Support rear frame with a floor jack and remove rear wheel. Unbolt and remove the final drive assembly as shown in Fig. 120.

Reinstall by reversing the removal procedure. Use new gaskets and tighten axle housing retaining cap screws to a torque of 170 ft.-lbs.

149. **OVERHAUL PLANET CARRIER.** With final drive assembly removed as outlined in paragraph 148 proceed as follows: Remove outer brake disc (20—Fig. 121) and cap screw securing the planetary unit to axle shaft.

NOTE: When removing the planetary unit there are shims (10) between retainer (14) and axle (1). Keep the shims together.

With planet carrier removed from final drive assembly, drive roll pins into the planet pins (12) from outside of carrier (13). Then, drive or press planet pins out of carrier and remove each planet gear, bearings and retainer (14). Remove roll pins from planet pins for reassembly.

Inspect gears and bearings for damage or wear and renew as necessary.

When reassembling it will be necessary to determine shims needed for preload of planet gear bearings as follows: Using two flat washers (2 inch O.D. x 17/32-inch I.D. x 1/8-inch thick), cut on dotted line as shown in Fig. 122. Then, using a ½-inch NF x 3 inch cap screw and nut, secure the cap screw in vise with threaded end pointing upward. Lubricate the bearing in Hy-Tran and position over the cap screw with

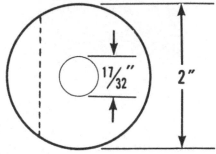

Fig. 122—Dimension for making washer to determine the shims needed for planet gear preload. Refer to text.

Fig. 120—View showing final drive assembly being removed.

1. Sling	4. Planetary carrier	7. Aligning dowels
2. Rear axle	5. Brake outer disc	8. Brake center disc
3. Axle carrier	6. Drive shaft	9. Cap screw

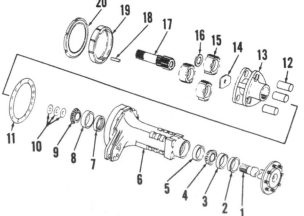

Fig. 121—Exploded view of planetary and rear axle components.

1. Axle
2. Sleeve (wear)
3. Seal
4. Bearing cone
5. Bearing cup
6. Housing
7. Seal
8. Bearing cup
9. Bearing cone
10. Shims
11. Gasket
12. Pin
13. Carrier
14. Retainer
15. Gear (planet)
16. Shim
17. Shaft (planetary)
18. Dowel pin
19. Ring gear
20. Brake disc

Fig. 123—Using a micrometer to take a reading on outer faces of bearing cone where washers are cut away. Refer to text.

bearing and washer on each side of gear, with washer cuts in alignment. Using an inch pound torque wrench, tighten the nut to 10 in.-lbs. while rotating the gear. Using a micrometer and referring to Fig. 123, take a reading on outer face of bearing cones where the washers are cut away. With the preceding reading, refer to chart Fig. 124 for the required shims to obtain bearing preload. Keep each shim pack and gear together after correct shims have been selected.

When reassembling, install one planet gear and its shim pack into carrier and secure in place with planet gear pin and roll pin. Then, place retainer (1—Fig. 125) in carrier and install the other two planet gears and their selected shim packs.

150. OVERHAUL REAR AXLE AND RING GEAR. If it is necessary to remove the ring gear (19—Fig. 121), remove planet carrier as outlined in paragraph 149. Using a pusher arrangement push on inner end of axle to remove axle and bearings. Then with suitable puller remove ring gear from dowels.

Inspect bearings and seals for damage or wear and renew as necessary.

To reassemble, install axle bearing cups and inner oil seal in housing if removed. Install outer oil seal and outer bearing cone on axle shaft. Pack the inner and outer bearing cones with a multi-purpose grease. Install the axle into the housing, drive the inner bearing cone on axle until the planetary carrier assembly can be installed. Tighten the cap screw to draw the bearing cone in until there is 0.001 to 0.010 inch end play, using a dial indicator to check. Using an inch pound torque wrench on the planet carrier cap screw, turn at a slow rotation and record the reading. Continue to draw the bearing in until rolling torque is 20 to 30 in.-lbs. OVER the first reading. Then remove cap screw and planet carrier from axle housing.

With carrier removed measure the distance between end of axle and inner bearing face as shown in Fig. 126 and

record the reading. Refer to Fig. 127 and measure the distance from face of hub to retainer using the planetary drive shaft (3) to hold retainer (2) in place. Select shims within 0.002 inch of the difference between first and last reading. This will be the correct shim pack to be installed. Shims of 0.007, 0.012 and 0.0299 inch thicknesses are available in a rear axle shim package.

Install ring gear with dowels, if removed, by using a wood block and tapping ring gear equally all the way around. Then, coat selected shim pack with grease and install on end of axle. Install planet carrier on axle being careful not to drop the shims. Install cap screw and torque to 250 ft.-lbs. Install outer brake disc.

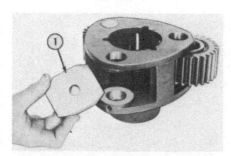

Fig. 125—Install retainer before the last two planet gears are installed.

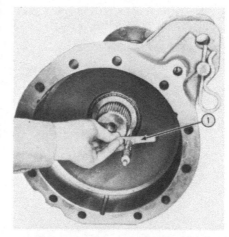

Fig. 126—Using a depth micrometer to measure distance from end of axle to inner bearing cone. Refer to text.

BRAKES

Brakes on all models are self equalizing hydraulic actuated wet type single disc brakes. Brakes are located on the differential output shafts and are accessible after removing final drive units as outlined in paragraph 148. Return Hy-Tran fluid from the oil cooler maintains a full master cylinder for both cylinders. This is referred to as keep fill system. Brake operation can be accomplished with engine inoperative because of the keep fill line which keeps the master cylinders filled. Also in the brake system is an equalizer valve (3—Fig. 128).

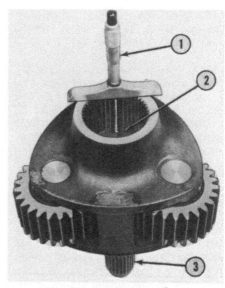

Fig. 127—With planetary drive shaft supporting the retainer use a depth micrometer to obtain hub to retainer dimension as shown. Refer to text.

1. Depth micrometer
2. Retainer
3. Planetary drive shaft

Fig. 128—View of brake pedal adjusting bolts and equalizer. Refer to text.

1. L.H. brake line
2. R.H. brake line
3. Equalizer
4. Equalizer plate
5. Eccentric bolts
6. Lock nuts

Tractors	Figure Obtained (inches)	Shims Required (inch)
454, 464 & 484	1.706-1.712	0.042
	1.699-1.705	0.049
	1.692-1.698	0.056
574, 674 & 584	2.123-2.127	0.042
	2.118-2.122	0.049
	2.113-2.117	0.056

Fig. 124—Chart to determine the shims needed for planet gear bearings. Refer to text.

Tho valve permits equal flow to both brake pistons when both brake pedals are engaged. If one brake is applied, the equalizer does not function.

Service (foot) brakes MUST NOT be used for parking or any other stationary job which requires the tractor to be held in position. Even a small amount of fluid seepage would result in brakes loosening and severe damage to equipment or injury to personnel could result. USE PARK BRAKE when parking tractor.

BRAKE ADJUSTMENT

All Models

151. The only external adjustment that can be made on brakes is the brake pedal maximum travel.

152. To make the brake pedal maximum travel adjustment, refer to Fig. 128 and proceed as follows: Loosen nuts (6). Rotate the eccentric bolts (5) to obtain 6 inch pedal free height from the top of the brake pedal to the platform. Retighten nuts (6).

BLEED BRAKES

All Models

153. To bleed the brakes, start the tractor and let it run to insure the brake lines are filled. Keep engine running when bleeding brakes. Clamp brake return hose to the transmission. Attach two plastic hoses ¼-inch I.D. by 30 inches long over brake bleeder fittings. Run the hoses into the hydraulic filler hole.

Bleeder fittings are located in hydraulic housing above final drives. Depress brake pedal and while holding in this position open bleeder valve and when a solid flow of oil appears, close valve. Repeat operation on opposite

brake. Check brake pedal feel. If brake pedal operation feels spongy rather than having a solid feel, repeat the bleeding operation.

BRAKE ASSEMBLIES

All Models

154. **R&R AND OVERHAUL.** Removal of either brake is accomplished by removing final drive as outlined in paragraph 148. Refer to Fig. 120 showing final drive removed from tractor. Remove the planetary drive shaft. Remove brake disc (1—Fig. 129), inner brake ring (2) and brake piston (3) with "O" ring (4). Refer to Fig. 130 for removal procedure of brake piston.

With brake disassembled, clean and

Fig. 130—Install screws (1) and pry brake piston (2) from housing.

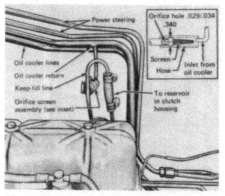

Fig. 131—View showing the location of orifice and screen on early models. Later models are similar.

inspect all parts. Renew any parts showing excessive wear or other damage.

When reassembling install new "O" ring (4—Fig. 129) on piston and install piston in rear frame being careful not to damage the "O" ring.

Reinstall by reversing the removal procedure. When assembly is completed, check external adjustment as outlined in paragraph 152, start engine and bleed brakes as in paragraph 153.

BRAKE MASTER CYLINDERS AND EQUALIZER VALVE

All Models

155. **REMOVE AND REINSTALL.** Remove brake master cylinder and equalizer valve as follows: Remove hood and rear side panels. Some mechanics prefer to remove the battery and the lower plate of battery box. Disconnect all necessary lines from the master cylinders. Remove the snap ring securing the equalizer plate to equalizer and remove pedal adjusting eccentric bolts. Unbolt and remove master cylinders separately or unbolt bracket and remove as an assembly.

Between the oil cooler return and the keep fill line there is an orifice screen located in the hose. Refer to Fig. 131. When this orifice screen is installed the screen should be toward the oil cooler return side. This orifice provides the proper amount of oil to the master cylinders. There also is a filter screen located in the lines going into each master cylinder.

Reinstall by reversing the removal procedure. Adjust the brake pedal travel as outlined in paragraph 152 and bleed the brakes as outlined in paragraph 153.

156. **OVERHAUL MASTER CYLINDER.** With master cylinder removed refer to Fig. 132 and remove boot (5) and snap ring (3). Remove plunger (7), seal (8) and all other internal parts (9 through 14) as a unit.

Clean and inspect all parts. Pay particular attention to cylinder bore for

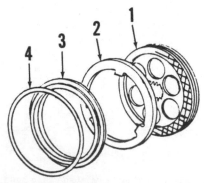

Fig. 129—Exploded view of brake piston and disc.

1. Brake disc	3. Piston
2. Ring	4. "O" ring

Fig. 132—Exploded view of brake master cylinder.

1. Body
2. Retainer
3. Snap ring
4. Push rod
5. Boot
6. Sleeve
7. Plunger
8. Seal
9. Retainer
10. Spring
11. Spacer (valve)
12. Spring washer
13. Valve stem
14. Seal

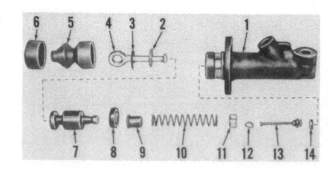

scoring or ridges. Use all new seals and reassemble by reversing disassembly procedure.

When installing valve seal (14), the smallest diameter is to be installed first on the head of valve stem (13). Also refer to Fig. 133 for installing valve seal.

157. **OVERHAUL EQUALIZER VALVE.** With equalizer valve removed refer to Fig. 134 and remove snap ring (12) and equalizing plate (11). Tap the spool (3) from snap ring end to knock plug (1) out the opposite end, then remove spool and spring. Remove the fittings (10), springs (8), 5/16-inch balls (7) and 3/16-inch balls (6).

Clean and inspect all parts. Renew all "O" rings and any other parts that are worn or damaged.

Reassemble by reversing the disassembly procedure.

POWER TAKE-OFF

The power take-off used on all models is an independent type. The pto is available as a dual speed unit having both 540 and 1000 rpm rear output shafts on some early models prior to tractor serial number U100001, or as a 540 rpm rear unit with an optional 1000 rpm side mount unit on later models. Oil for the pto unit on all models is furnished from the hydraulic pump through the MCV. From the MCV the oil is supplied through a cored passage in the rear frame to the 1000 rpm output shaft (early models) or lower pto drive shaft (later models) to actuate the pto clutch. Operation of the pto unit is controlled by a spool type valve located in the MCV housing.

PTO LINKAGE ADJUSTMENT

All Models So Equipped

158. Loosen the pto control handle bolts (Fig. 135). With the bolts loose adjust linkage (A) so that pto valve movement is 1-3/16 inches ± 1/32-inch from disengaged to engaged or overcenter position. Move the control handle to disengaged position and adjust the lever to center of notch. Tighten bolts in control handle. Then, with engine running move the control lever to the "feathering" first notch position. When control lever is released it should return freely to the disengaged position notch.

OPERATING PRESSURE

All Models So Equipped

159. To check operating pressure, run tractor until hydraulic fluid is approximately 100 degrees F. Refer to Fig. 136 and remove pressure line (1) from tee on left side of tractor. Attach a test gage capable of registering at least 300 psi. Engage pto and operate engine at 2200 rpm. If pressure reading is below 220 psi remove plug (4) on MCV and install shims as required under the valve plug. When the pressure is set at 220 psi to 250 psi remove the gage and connect the pressure line back to the tee. Then, remove the cap from the tee and connect pressure gage as shown in Fig. 137. Operate the engine at 2200 rpm and with the pto engaged there should be a gage read-

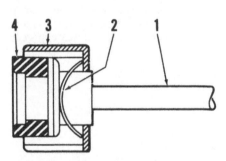

Fig. 133—View showing the seal installed on valve stem.

1. Valve stem	3. Spacer (valve)
2. Spring washer	4. Valve seal

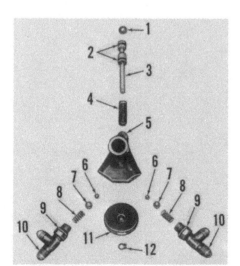

Fig. 134—Exploded view of equalizer valve.

1. Plug	7. Check balls (5/16)
2. "O" rings	8. Check springs
3. Spool	9. "O" ring
4. Spring	10. Tee
5. Body	11. Equalizer plate
6. Check balls (3/16)	12. Snap ring

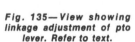

Fig. 135—View showing linkage adjustment of pto lever. Refer to text.

Fig. 136—View showing gage connection for testing pto pressure. Refer to text.

1. Pressure line	3. Gage (0-300 psi)
2. Connector	4. Pto regulating valve

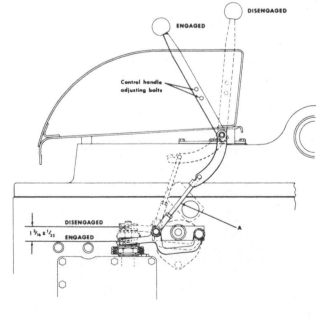

ing of 220-250 psi if the sealing rings in the pto clutch piston and the sealing rings and bushing on the 1000 rpm output shaft (early models) or lower pto drive shaft (later models), are sealing properly. If pressure is not as stated remove and overhaul pto unit as outlined in paragraph 160.

R&R AND OVERHAUL

All Models Equipped With Rear Mount

160. To remove pto clutch, drain the rear main frame, then remove the pto side cover. Refer to Fig. 138 and drive roll pin (1) from the clutch and 1000 rpm output shaft (early models) or lower pto drive shaft (later models). Remove the lower bearing flange on

rear of rear frame. Refer to Fig. 139 and remove shaft (10). If tractor is not equipped with 1000 rpm shaft that protrudes, there will be a lower pto drive shaft located in same position as the 1000 rpm shaft. Lower drive shaft will not protrude and will have a bearing cap. After the 1000 rpm shaft or lower drive shaft is pulled rearward the pto clutch can be removed. Refer to Fig. 140 and with a pair of "C" clamps compress the piston return springs and remove snap ring (2). Then remove clutch backing plate (8—Fig. 141), return springs (7), driven discs (6) and drive plates (5). There are two driven plates between each driven disc.

NOTE: Early Models 454 and 574 prior to tractor serial number U100001 have six driven plates, four driven discs and four return springs. Later Models 454 and 574 and all Model 464, 484, 674 and 584 tractors have eight driven plates, five driven discs and six return springs.

Remove piston return plate (4—Fig. 141) and brake ring (3). Remove snap ring (12), thrust washer (11), clutch hub (10), then place the clutch cup gear

Fig. 137—Testing for leaks in pto clutch or sealing rings. Refer to text.

1. Pressure line
2. Connector
3. Gage (0-300 psi)

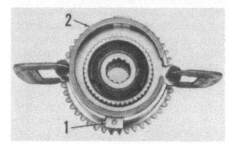

Fig. 140—View showing "C" clamps used to compress the backing plate and springs for removal of snap ring.

1. Backing plate
2. Snap ring

Fig. 138—Roll pin to be removed in pto clutch to pull the 1000 rpm shaft (early models) or lower pto drive shaft (later models).

1. Roll pin
2. Shaft
3. Spring and eye
4. Clutch assembly
5. Lube gear
6. Drive gear

Fig. 141—Exploded view of early pto clutch and component parts. Later pto clutch is similar. Refer to text.

1. Clutch cup gear
2. Piston
3. Brake ring
4. Piston return plate
5. Driven plate
6. Driven disc
7. Return springs
8. Backing plate
9. Snap ring
10. Hub assembly
11. Thrust washer
12. Snap ring

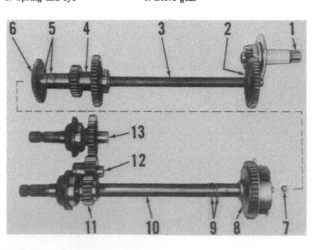

Fig. 139—Exploded view of typical pto drive train.

1. Drive shaft
2. Driven gear
3. Driven shaft
4. Countershaft drive
5. Countershaft bearing
6. Pto clutch and hydraulic pump drive gear
7. Pilot bearing
8. Clutch assembly (pto)
9. Sealing rings (two)
10. Output shaft (1000 rpm)
11. Drive gear
12. Idler gear
13. Gear and shaft (540 rpm)

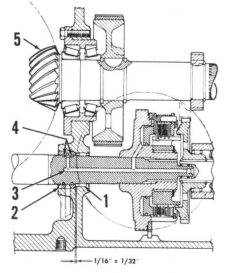

Fig. 142—Sectional view of pto clutch. Refer to text.

1. Needle bearing
2. Bushing
3. Pressure oil passage
4. Sealing rings
5. Pinion

1/16" ± 1/32"

(1) with gear up on the bench. Apply air pressure in the port hole of the clutch cup gear to remove piston (2).

161. With unit disassembled, clean and inspect all parts and renew if necessary. Pay particular attention to the clutch discs which should be free of scoring or warpage. Use all new "O" rings, seals and gaskets during reassembly. The bushing (2—Fig. 142) for the 1000 rpm output shaft or lower drive shaft is in the front of differential compartment. This bushing should be replaced if a new output shaft is required or if the sealing rings have grooved the bushing. When installing bushing (2), bottom it in bore. Position the needle bearing within 1/16-inch ± 1/32-inch of the front face of bushing. Refer to Fig. 142.

Reassembly is the reverse of disassembly. After unit is reassembled and before installing in tractor, the operation of the clutch piston can be checked by using air pressure in the oil port of the clutch cup gear. This will make sure that the sealing rings on the piston were not cut when installed.

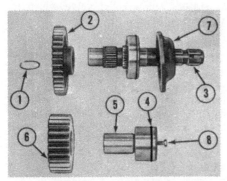

Fig. 143—View showing the 540 rpm pto output shaft.

1. Snap ring	5. Idler shaft
2. Output gear	6. Idler gear
3. Output shaft (540 rpm)	7. Bearing & seal retainer
4. "O" ring	8. Cap screw

To remove the 540 rpm output shaft (3—Fig. 143), gear (2), idler shaft (5) and idler gear (6), first remove the hydraulic lift assembly as outlined in paragraph 169. Remove snap ring (1) and unbolt bearing and seal retainer (7). Withdraw output shaft assembly and remove gear (2) through top opening. Install a cap screw (8) in rear of idler shaft (5), remove idler shaft from rear frame and remove idler gear (6) from above.

Use new "O" ring (4), oil seal and gasket and reinstall by reversing the removal procedure. Refill rear frame to proper level with IH Hy-Tran fluid.

All Models Equipped With Side Mount

162. To remove the 1000 rpm side mount pto, drain the rear main frame. Disconnect the linkage and unbolt and remove unit. Drive out roll pin and remove idler shaft (12—Fig. 144) and gear (13). Remove bearing retainer (1). Pull output shaft (8) with bearings.

NOTE: When shaft is removed from gear the two detent balls (17) and spring (16) will fall out.

Remove pipe plug from housing and drive roll pin from the selector fork shaft and remove shaft. Remove gear and fork from housing. If output shaft needle bearing is to be renewed, the new bearing must be pressed in to 3-13/16 inches ± 1/32-inch as shown in Fig. 145. If idler gear needle bearing is to be renewed press in to dimension shown in Fig. 146.

163. With unit disassembled, clean and inspect all parts and renew if necessary. Use new seal and gaskets during reassembly.

While reassembly is the reverse of disassembly, the following points are to be considered during reassembly.

When installing output shaft, install the poppet spring and two detent balls. Hold the balls in place with a hose clamp as the shaft is pushed through the gear. Remove clamp after gear is positioned on shaft. Install the idler shaft and gear with the shoulder of gear toward the idler gear support.

HYDRAULIC LIFT SYSTEM

The hydraulic lift system provides load (draft) and position control in conjunction with the 3-point hitch.

The load (draft) control is taken from the third (upper) link and transferred through a bell crank on the torsion bar and linkage to the draft control valve located in the hydraulic lift housing. The externally mounted torsion bar is located on the rear side of lift unit.

The hydraulic lift housing, which also serves as the cover for the differential portion of the tractor rear main frame, contains the work cylinder, rockshaft, valves and the necessary linkage. Auxiliary valves (either one or two), on trac-

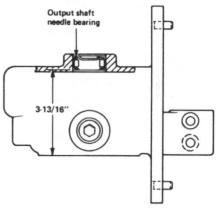

Fig. 145—Output shaft needle bearing is pressed in housing to the dimension shown.

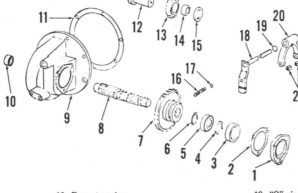

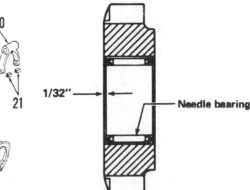

Fig. 144—Exploded view of side mount 1000 rpm pto.

1. Bearing retainer
2. Gasket
3. Seal
4. Snap ring
5. Bearing
6. Snap ring
7. Gear
8. Shaft (output)
9. Housing
10. Needle bearing
11. Gasket
12. Idler shaft
13. Idler gear
14. Needle bearing
15. Spacer
16. Poppet spring
17. Poppet ball (2 used)
18. Shaft
19. "O" ring
20. Fork
21. Blocks

Fig. 146—Bearing installation in idler gear of pto side mount.

tor so oquipped, are mounted to the right side of the lift housing. The pump which supplies the hydraulic system is attached to the MCV which is mounted on left side of rear frame. Pump is driven by a gear on the rear of pto driven shaft. The oil used in the hydraulic system is drawn from the rear main frame and through a full-flow filter which is a part of MCV on left side of rear frame.

TROUBLE SHOOTING

All Models

164. The following are symptoms which may occur during the operation of the hydraulic lift system. By using this information in conjunction with the Test and Adjust information, no trouble should be encountered in servicing the hydraulic lift system.

1. Hitch will not lift. Could be caused by:
   a. Faulty main relief valve.
   b. Faulty cushion relief valve.
   c. Internal linkage disconnected.
2. Hitch lifts when auxiliary valve is actuated. Could be caused by:
   a. Unloading valve orifice plugged.
   b. Unloading valve piston sticking.
   c. Unloading valve body assembly not seating or body assembly loose.
3. Hitch lifts load very slowly. Could be caused by:
   a. Unloading valve leaking.
   b. Excessive load.
   c. Faulty main relief valve.
   d. Faulty cushion relief valve.
   e. Work cylinder or piston scored or "O" ring faulty.
   f. Flow control valve stuck in slow position.
   g. Inefficient pump.

4. Hitch will not lower. Could be caused by:
   a. Main control valve spool sticking or spring faulty.
   b. Drop poppet "O" ring is damaged or drop poppet is sticking.
5. Hitch lowers very slowly. Could be caused by:
   a. Action control valve spool or piston sticking.
   b. Action control valve linkage maladjusted.
   c. Drop poppet valve "O" ring damaged.
6. Hitch lowers too fast in action control zone. Could be caused by:
   a. Improper adjustment of action control valve.
7. Hitch will not maintain position. Could be caused by:
   a. Work cylinder or piston scored or piston "O" ring damaged.
   b. Cylinder cushion relief valve leaking.
   c. Drop poppet check valve ball not seated.
   d. Drop poppet ball seat binding in the drop poppet.
8. Hydraulic system stays on high pressure. Could be caused by:
   a. Linkage maladjusted, broken or disconnected.
   b. Auxiliary valve not in neutral.
   c. Mechanical interference.
9. System stays on high pressure after lifting load, but returns to low pressure after slight movement of position control lever toward "Lower" position. Caused by:
   a. Leak in piston side of unloading valve circuit.
10. Hitch senses with load control in "OFF" position. Could be caused by:
    a. Improper adjustment of load sensing linkage.

TEST AND ADJUST

All Models

Before proceeding with any testing or adjusting, be sure the hydraulic pump is operating satisfactorily, hydraulic fluid level is correct and filter is in good condition. All tests should be conducted with hydraulic fluid at operating temperature which is normally 120-180 degrees F. Cycle system if necessary to insure that system is completely free of air.

165. **RELIEF VALVE.** To check the hydraulic lift system relief valve, disconnect the pressure line (P—Fig. 147) and return line (R). Attach the Flo-Rater inlet hose to (P) and the outlet hose to (R). Start and run the engine at 2200 rpm for Models 454, 484 and 574, 2300 rpm for Model 584 and 2400 rpm for Models 464 and 674. Restrict the Flo-rater to 1250 psi and the hydraulic pump should deliver 12.5 gpm on Model 584 or 12 gpm on all other models. Continue to restrict the Flo-Rater to 2300 to 2500 psi and there should be no flow recorded on the gage. If pressure is not as stated, renew relief valve which is available only as a unit.

166. **QUADRANT LEVERS FRICTION ADJUSTMENT.** To test and adjust system, first check quadrant levers (draft, position and raise response); refer to Fig. 148. Move position control to forward position, then with draft control lever in rear position it should require 4-6 pounds of force, applied at knob to move lever. If adjustment is necessary, adjust by turning nut (A). Now reverse the position of the levers and check the position control lever,

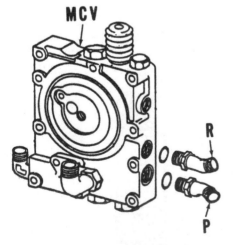

Fig. 147—View showing pressure and return ports from MCV to the hydraulic lift.

P. Pressure port
R. Return port

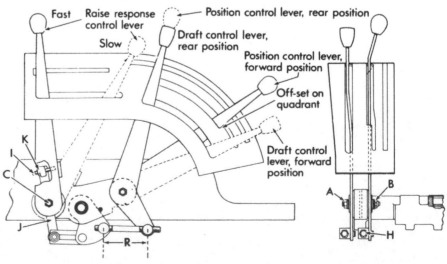

Fig. 148—View showing adjusting points on hydraulic lift controls. Nominal setting at point (R) is: Position control 3-5/16 inches and draft control 3-13/32 inches. Refer to text.

which should require 2-4 pounds of force to move lever. If necessary to adjust, turn nut (B). With the response control lever in fast position, it should require 2-4 pounds of force to move lever. If necessary to adjust turn nut (C).

167. **DROP POPPET.** To adjust the drop poppet, first place the draft control lever in full forward (Off) position. With hitch weighted, raise hitch to top of its travel with position control lever, then lower to its mid-point. The system should come off high pressure without "hiccups". If the unit does not respond properly proceed as follows: To make the adjustment first make a locating bolt. Refer to Fig. 149. Using a ¼-inch bolt 6 inches long, bend head at 90° and thread cap screw to length of 4½ inches. Drill hole for a 1/16-inch roll pin ¾-inch long parallel with face of bolt head approximately 1/8-inch from end. Remove linkage access cover and replace variable valve flow control tube with hose as shown in Fig. 150. With hitch weighted, cycle the hitch a few times and position the hitch so that it is approximately 6 to 10 inches off the floor. Attach the locating bolt to the main valve actuator assembly and mount the dial indicator as shown. Now move the position control lever to raise the hitch to obtain the peak movement of valve.

This will require only about ¼ to 3/16-inch of movement of the position lever. Record the reading that was obtained on the dial indicator. Next move the position lever down very slowly and determine the point at which the dial indicator maintains a set position. The difference between readings should be 0.019 to 0.021 inch. If reading is over 0.021 inch, turn adjusting screw in and if less than 0.019 inch turn screw out as shown in inset Fig. 150.

168. **CONTROL LINKAGE.** To adjust the control linkage, start the

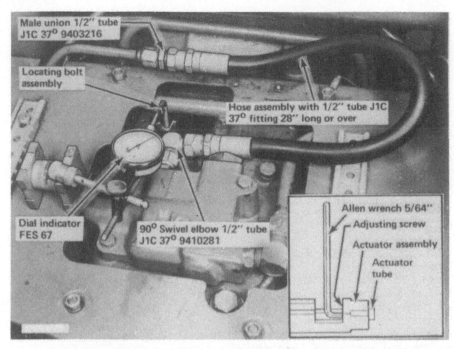

Fig. 150—Showing the dial indicator and locating bolt being used. Refer to text.

tractor and cycle the hitch with the position control lever at least five times to free the system of air. To adjust hitch maximum raise, use a straightedge attached to torsion bar mounting face as shown in Fig. 151. Measurement at (1) as shown should be 11/32 to 17/32-inch for early Model 454 with serial number prior to 7101 and 1-3/32

to 1-9/32 inches for later Model 454 and all Models 464, 484, 574, 674 and 584. If measurements are not as stated adjust the control valve spool by loosening locknut (D—Fig. 152) and turning the spool (E) in or out until the rockshaft arms are within tolerance. System should not be on high pressure. Then, tighten locknut (D). Rotation of one flat of the main valve actuator tube assembly will change the measurement approximately 5/16-inch.

NOTE: If hiccupping develops readjust drop poppet as in paragraph 167.

To adjust the drop control, move position control lever slowly forward (down) until rockshaft arms JUST reach fully lowered position. Refer to Fig. 152 and adjust screw (F) until it just contacts drop control valve arm and tighten locknut (G).

With an implement or weight attached to the hitch, move the position control lever forward (down) until it just contacts drop control valve arm. Then adjust linkage at (H—Fig. 148) so that position control lever is at off-set on quadrant.

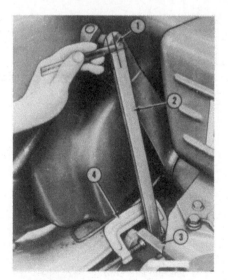

Fig. 151—View showing how to measure for maximum raise. Refer to text.

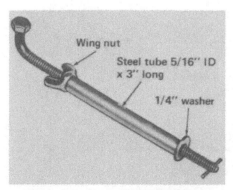

Fig. 149—View showing locating bolt for adjusting the drop poppet valve.

Fig. 152—View showing adjusting points. Refer to text.

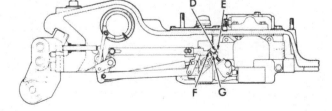

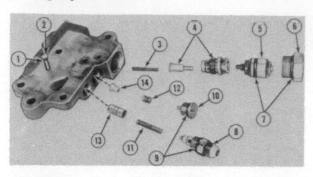

Fig. 153—Exploded view
of unloading and flow control valve.

1. "O" ring
2. Locating pin
3. Unloading valve spring
4. Unloading valve assembly
5. Unloading valve piston
6. Plug
7. "O" rings
8. Fitting
9. "O" rings
10. Plug
11. Spring
12. Spring
13. Unloading and flow control valve
14. Check poppet valve

HYDRAULIC LIFT HOUSING

All Models

169. REMOVE AND REINSTALL. To remove the hydraulic lift unit from tractor, first remove the seat assembly. Disconnect wiring harness on right and left sides, fuel lines and the park brake assembly from left fender. Disconnect variable valve flow control line and remove control lever knobs. Unbolt platform from rear frame and lift fenders, fuel tank and platform as a unit. Disconnect and remove the draft control pressure tube and auxiliary valve return tube. Remove all other necessary hydraulic lines.

Remove the hitch lift links from rockshaft arms. Unbolt lift housing, attach hoist and lift assembly from housing.

Reinstall by reversing the removal procedure.

UNLOADING AND FLOW CONTROL VALVE

All Models

170. R&R AND OVERHAUL. Remove the unloading and flow control valve from top of the lift housing.

NOTE: If only the valve is to be serviced, it will not be necessary to remove lift housing.

Remove plug (6—Fig. 153), piston (5) and unscrew valve body assembly (4). Remove plug (10), spring (12) and check poppet valve (14). Remove fitting (8), spring (11) and unloading flow control valve (13).

NOTE: Valve (13) is not serviced separately.

Remove retaining shaft (2). On later models this is a roll pin. The retaining shaft is used to position the valve in housing.

Clean and inspect all parts and renew any showing excessive wear or other damage. Using all new "O" rings reassemble by reversing the disassembly procedure.

CYLINDER AND VALVE UNIT

All Models

171. R&R AND OVERHAUL. To remove the work cylinder and valve assembly, first remove the lift unit from tractor as outlined in paragraph 169. Turn upside down on a work bench. Disconnect control linkage from main valve and other linkage. Disconnect sensing spring, then unbolt and remove cylinder and valve assembly from housing as shown in Fig. 154.

Remove the four cap screws and sep-

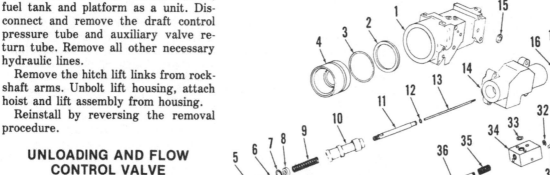

Fig. 155—Exploded view of the lift cylinder, draft control valve and action control valve assemblies.

1. Cylinder	11. Actuator tube	19. "O" ring	29. "O" ring
2. "O" ring	12. Snap ring	20. Drop poppet valve	30. Spring
3. Back-up washer	13. Drop valve actuating rod	21. "O" ring	31. Piston
4. Piston	14. Draft control valve body	22. Ball	32. "C" ring
5. Actuator link	15. "O" ring	23. Pilot valve spring	33. "O" ring
6. Drop valve adjusting screw	16. "O" ring	24. Poppet valve spring	34. Action control valve body
7. Snap ring	17. Pilot valve seat	25. "O" ring	35. Spring
8. Spring retainer	18. Back-up washers	26. Plug	36. Variable orifice spool
9. Spring		27. Snap ring	
10. Valve spool		28. Plug	

Fig. 154—View showing lift cylinder and valve assembly being removed from lift housing.

Fig. 156—View showing control linkage and variable flow valve.

1. Connecting rods
2. Inner retaining plate
3. Snap ring
4. Spacer
5. Roll pin
6. Load control shaft
7. Draft control lever
8. Slow motion actuating lever

arate the action control valve and pipe from the draft control valve. Remove switch lever bracket and eccentric plate assembly from cylinder housing. Remove three cap screws and retarding arm shoulder bolt, then separate draft control valve from cylinder housing.

To remove piston (4—Fig. 155) from cylinder (1) bump open end of cylinder on a wooden block. Inspect the cylinder and piston for scoring or wear. Small defects may be removed using crocus cloth. Install new back-up ring (3) and "O" ring (2) on piston, lubricate parts with IH "Hy-Tran" fluid and reinstall piston in cylinder.

To disassemble the draft control valve, remove snap ring (7) and withdraw spool assembly from valve body (14).

NOTE: Be careful not to drop or otherwise damage the actuating rod (13).

Loosen the locknut on actuator tube (11), then count and record the number of threads, while removing actuator link (5). Remove locknut, spring retainer and spring and after first removing snap ring (12), withdraw actuator tube from spool (10).

Remove snap ring (27) and plug (26), then remove drop poppet valve (20) and pilot valve seat (17) with their spring and ball.

Inspect all parts for nicks, burrs, scoring and undue wear and renew as necessary. Valve body (14) and spool (10) are not available separately.

Use all new "O" rings and back-up rings, lubricate all parts with IH "Hy-Tran" fluid and reassemble parts in valve body by reversing disassembly procedure.

To disassemble the action control valve, remove plug (28), spring (30) and pressure actuated piston (31). Hold the variable orifice spool (36) in against its spring, remove "C" ring (32), then withdraw the spool and spring from valve body. Check the pressure actuated piston and variable orifice spool and their bores in valve body for scoring or undue wear. Valve is available only as an assembly except for return spring (35) and "C" ring (32).

Lubricate all parts with IH "Hy-Tran" fluid and reassemble by reversing disassembly procedure.

After assembly, mount the draft control valve on the cylinder housing and the action control valve on the draft control valve. Tighten the cap screws. Place cylinder and valve assembly in housing and install the hold down bolts. Complete the balance of reassembly and adjust as outlined in paragraphs 166, 167 and 168.

CONTROL LEVERS, LINKAGE AND VARIABLE FLOW VALVE

All Models

172. **REMOVE AND REINSTALL.** With the hydraulic lift removed as outlined in paragraph 169 and lift cylinder removed as outlined in paragraph 171, refer to Fig. 156 and remove the inner and outer retaining plates (2). Drive out roll pin (5), remove draft sensitivity lever (7) and spacer (4). Remove snap ring (3) and slow motion actuating lever (8).

NOTE: There are timing marks on the spline for installation.

Remove control rods (1) and load control shaft (46—Fig. 157) with bearings. Remove three cap screws holding control lever support to lift housing. Disconnect linkage from flow control arm (27) and remove position control shaft (31). Remove flow control lever (29), friction disc (28), drive out roll pin and remove pivot bolt. Remove flow control arm (27) with return spring. Remove flow control valve plug and spool (26).

Inspect the flow control spool and housing for nicks, burrs or scoring and renew as necessary. Using new "O" rings and back-up washers, reassemble in reverse of disassembly.

To remove the control linkage, remove snap ring securing follow-up walking beam (14) and remove the follow-up walking beam with links. Remove snap rings securing position control walking link (12) to rockshaft actuating hub (9) and position control eccentric (13). Remove the right rockshaft arm and move rockshaft to the left to allow clearance for removing the position control walking link (12). Remove snap ring on outside of lift housing and remove position control eccentric (13). Draft control rod (10) and oil seal can be removed after removing draft sensing assembly (4) from housing.

Reinstall by reversing the removal procedure and adjust the system as outlined in paragraphs 166, 167 and 168.

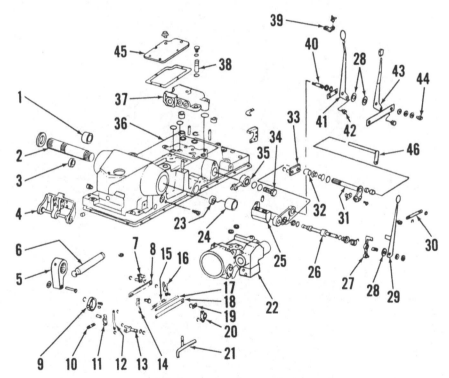

Fig. 157—Exploded view of hydraulic lift control linkage.

1. Bushing	13. Position control	24. Bushing
2. Rockshaft	eccentric	25. Support assy.
3. Seal	14. Walking beam	26. Variable flow spool
4. Draft sensing	15. Sensing spring	27. Arm
assembly	16. Draft control lever	28. Friction disc
5. Rockshaft bellcrank	17. Sensing bar	29. Flow control lever
6. Piston rod	18. Walking beam bar	30. Rod
7. Eccentric	19. Link	31. Position control
8. Link	20. Lever	shaft
9. Actuating hub	21. Tube	32. Spacer
10. Draft rod	22. Cylinder and valve	33. Lever
11. Link	assy.	34. Plug
12. Link	23. Seal	35. Connector

36. Lift housing
37. Unloading and flow
control valve
38. Cushion valve
39. Stop
40. Bolt
41. Position control
lever
42. Pin
43. Draft control lever
44. Pin
45. Cover
46. Load control shaft

ROCKSHAFT

All Models

173. **REMOVE AND REINSTALL.** To remove the rockshaft (2—Fig. 157), first remove lift housing as outlined in paragraph 169 and lift cylinder and valve as outlined in paragraph 171. Remove set screws from actuating hub (9 —Fig. 157) and rockshaft bellcrank (5). Remove the right lift arm and slide rockshaft from right to left out of housing, bellcrank and actuating hub. Remove actuating hub key as soon as it is exposed. If actuating hub sticks on rockshaft, pry against it with a heavy screwdriver so linkage will not be damaged.

Always renew the oil seals whenever rockshaft is removed. Rockshaft bushing can be removed and reinstalled using a proper sized bushing driver.

When reinstalling the rockshaft, align master splines of rockshaft and bellcrank and position bellcrank until set screw seat in rockshaft is aligned with set screw hole in bellcrank, then install set screw. Install actuating hub key, slide the actuating hub over the key until set screw hole in hub is centered on key, then install set screw.

Complete reassembly by reversing disassembly procedure.

HYDRAULIC PUMP

All Models

174. Tractors may be equipped with either Cessna or Plessey gear type pumps. For removing and reinstalling refer to paragraph 11 and for overhaul to paragraphs 12 or 13.

AUXILIARY CONTROL VALVE

All Models

175. **R&R AND OVERHAUL.** To remove the auxiliary valve or valves, unbolt and remove side panel from right hand console. Disconnect hydraulic lines, valve handles and remove the valves.

To disassemble, refer to Fig. 158 and proceed as follows: Remove end cap (1), then unscrew the actuator (9) and remove the actuator and detent assembly. Remove sleeve (17) and pull balance of parts from body. Check valve assembly can be removed at any time after removing snap ring (29). Detent (3, 4, 5 and 6) can be disassembled after removing plug (2). Push unlatching piston (8) out of actuator (9) with a long thin punch. Using a carburetor jet tool remove actuator valve plug (12), "O" ring (11) and back-up washer (10).

Inspect all parts for nicks, burrs, scoring and undue wear and renew parts as necessary. Spool (22) and body (20) are not available separately.

Use all new "O" rings and reassemble by reversing the disassembly procedure. Be sure the bolt holes in the retainer (26) are lined up with bolt holes in body. When installing actuator plug (1—Fig. 159), "O" ring (2) and back-up washer (3) should be installed to distance of 0.171 to 0.181 inches at (4). IH tool for installing the plug is part number FES 143-7. Detent unlatching pressure is adjusted by plug (2—Fig. 158). Unlatching pressure is 2000 psi to 2250 psi. Be sure filter in end cap (1) is clean (no paint) and in satisfactory condition.

When reinstalling auxiliary valves, tighten the mounting cap screws to 20-25 ft.-lbs. torque in 5 ft.-lbs. increments.

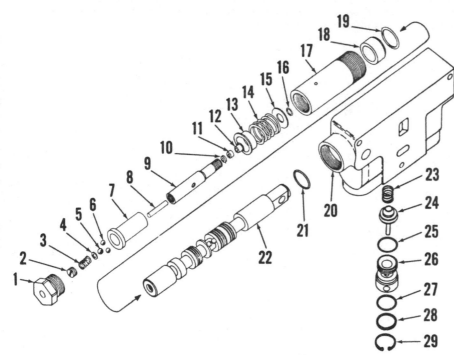

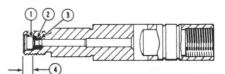

Fig. 158—Exploded view of auxiliary control valve.

1. Cap
2. Plug
3. Detent spring
4. Washer
5. Actuating ball
6. Detent ball
7. Sleeve
8. Unlatching piston
9. Actuator
10. Back-up washer
11. "O" ring
12. Plug
13. Spring retainer
14. Centering spring
15. Washer
16. "O" ring
17. Sleeve
18. "O" ring retainer
19. "O" ring
20. Valve body
21. "O" ring
22. Spool
23. Poppet spring
24. Poppet
25. "O" ring
26. Poppet retainer
27. "O" ring
28. Back-up washer
29. Snap ring

Fig. 159—Sectional view of actuator in the auxiliary valve. Refer to text.

1. Actuator valve plug
2. "O" ring
3. Back-up washer
4. Distance for installing plug (0.171-0.181 inch)

INTERNATIONAL HARVESTER

Models ■766 ■826 ■966 ■1026 ■1066

Previously contained in I&T Shop Manual No. IH-37

SHOP MANUAL
INTERNATIONAL HARVESTER
SERIES
766-826-966-1026-1066

Engine serial number is stamped on left side of enginge crankcase on all series except series 826 equipped with the D358 diesel engine. Engine serial number is stamped on right side of engine crankcase on the D358 engine. Engine serial number will be preceded by engine model number.

Tractor serial number is stamped on name plate attached to left side of clutch housing or hydrostatic drive housing.

INDEX (By Starting Paragraph)

CONDENSED SERVICE DATA

GENERAL	766 Non-Diesel	766 Diesel	826 Non-Diesel	826 Diesel	966 Diesel	1026 Turbo-Diesel	1066 Turbo-Diesel
Engine Make	IH	IH	IH	IH	IH	IH	IH
Engine Model	C-291	D-360	C-301	D-358	D-414	DT-407	DT-414
Number of Cylinders	6	6	6	6	6	6	6
Bore-Inches	3.750	3.875	3.813	3.875	4.300	4.321	4.300
Stroke-Inches	4.390	5.085	4.390	5.060	4.750	4.625	4.750
Displacement-Cubic Inches	291	360	301	358	414	407	414
Main Bearing, Number of	4	7	4	7	7	7	7
Cylinder Sleeves	Dry	Wet	None	Wet	Wet	Dry	Wet
Forward Speed without T.A.	8	8	8	8	8	....	8
Forward Speed with T.A.	16	16	16	16	16	....	16
Forward Speed with Hydrostatic Drive	....	....	Infinite	Infinite	Infinite	Infinite	Infinite
Alternator and Starter Make	Delco-Remy	Delco-Remy	Delco-Remy	Delco-Remy	Delco-Remy	Delco-Remy	Delco-Remy

TUNE-UP							
Compression Pressure	190 (1)	385-415 (1)	185 (1)	315-340 (1)	335-365 (1)	350-425 (1)	335-365 (1)
Firing Order	1-5-3-6-2-4	1-5-3-6-2-4	1-5-3-6-2-4	1-5-3-6-2-4	1-5-3-6-2-4	1-5-3-6-2-4	1-5-3-6-2-4
Valve Tappet Gap (Hot)							
Intake	0.027	0.012	0.027	0.010	0.020	0.013	0.020
Exhaust	0.027	0.021	0.027	0.012	0.025	0.025	0.025
Valve Seat Angle (Degrees)							
Intake	30	30	30	45	30	45	30
Exhaust	30	45	30	45	45	45	45
Ignition Distributor Make	IH	....	IH (2)	....	....	....	....
Breaker Gap	0.020	....	0.020	....	....	....	....
Distributor Timing Retard	1° BTDC	....	TDC	....	....	....	....
Distributor Timing Advanced	18° BTDC	....	22° BTDC	....	....	....	....
Timing Mark Location	Pulley or Flywheel	Flywheel	Crankshaft	Pulley or Flywheel	Flywheel	Flywheel	Flywheel
Spark Plug Electrode Gap	0.023	....	0.023	....	....	....	....
Carburetor Make	IH	....	IH	....	....	....	....
Injection Pump Make	....	American Bosch	....	Robt. Bosch	American Bosch	Roosa-Master	American Bosch
Injection Pump Timing	....	18° BTDC	....	16° BTDC	18° BTDC	6° BTDC	18° BTDC
Battery Terminal Grounded	Negative	Negative	Negative	Negative	Negative	Negative	Negative
Engine Low Idle Rpm	425	700	425	600	700	650	700
Engine High Idle Rpm, No Load	2640	See Para. 192	2650	2650	See Para. 192	2615	See Para. 192
Engine Full Load Rpm	2400	See Para. 192	2400	2400	See Para. 192	2400	See Para. 192

(1) Approximate psi, at sea level, at cranking speed
(2) Delco-Remy, if equipped with Magnetic Pulse ignition

SIZES-CAPACITIES-CLEARANCES

	766 Non-Diesel	766 Diesel	826 Non-Diesel	826 Diesel	966 Diesel	1026 Turbo-Diesel	1066 Turbo-Diesel
Crankshaft Main Journal Diameter	2.7480-2.7490	3.3742-3.3755	2.7480-2.7490	3.1484-3.1492	3.3742-3.3755	3.3742-3.3755	3.3742-3.3755
Crankpin Diameter	2.3730-2.3740	2.9977-2.9990	2.3730-2.3740	2.5185-2.5193	2.9977-2.9990	2.9980-2.9990	2.9977-2.9990
Camshaft Journal Diameter							
No. 1 (Front)	2.1090-2.1100	2.2824-2.2835	2.1090-2.1100	(3)	2.2824-2.2835	2.4290-2.4300	2.2824-2.2835
No. 2	2.0890-2.0900	2.2824-2.2835	2.0890-2.0900	(3)	2.2824-2.2835	2.0890-2.0900	2.2824-2.2835
No. 3	2.0690-2.0700	2.2824-2.2835	2.0690-2.0700	(3)	2.2824-2.2835	2.0690-2.0700	2.2824-2.2835
No. 4	1.4995-1.5005	2.2824-2.2835	1.4995-1.5005	(3)	2.2824-2.2835	1.4990-1.5000	2.2824-2.2835
Piston Pin Diameter	0.8748-0.8749	1.6248-1.6250	0.8748-0.8749	1.4172-1.4173	1.6248-1.6250	1.6248-1.6250	1.6248-1.6250
Valve Stem Diameter							
Intake	0.3715-0.3725	0.3718-3.3725	0.3715-0.3725	0.3919-0.3923	0.3718-0.3725	0.4348-0.4355	0.3718-0.3725
Exhaust	0.3710-0.3720	0.3718-0.3725	0.3710-0.3720	3.3911-0.3915	0.3718-0.3725	0.4348-0.4355	0.3718-0.3725
Main Bearing Diametral Clearance	0.0012-0.0042	0.0018-0.0051	0.0012-0.0042	0.0029-0.0055	0.0018-0.0051	0.0018-0.0051	0.0018-0.0051
Rod Bearing Diametral Clearance	0.0009-0.0034	0.0018-0.0051	0.0009-0.0034	0.0023-0.0048	0.0018-0.0051	0.0018-0.0051	0.0018-0.0051
Piston Skirt Diametral Clearance	0.0010-0.0045	0.0045-0.0065	0.0010-0.0045	0.0039-0.0047	0.0045-0.0065	0.0049-0.0069	0.0045-0.0065
Crankshaft End Play	0.0050-0.0130	0.0060-0.0120	0.0050-0.0130	0.0060-0.0090	0.0060-0.0120	0.0070-0.0185	0.0060-0.0120
Camshaft Bearing Diametral Clearance	0.0005-0.0050	0.0018-0.0051	0.0005-0.0050	0.0009-0.0033	0.0018-0.0051	0.0010-0.0055	0.0018-0.0051
Camshaft End Play	0.0020-0.0100	0.0020-0.0100	0.0020-0.0100	0.0040-0.0180	0.0050-0.0130	0.0020-0.0100	0.0050-0.0130
Cooling System Capacity, Qts.	20½	24	19	25	26	22	26
Crankcase Oil, Qts	9	11	9	12	12	13	18
Transmission and Differential Gallons (approximate)							
Standard Transmission	23	23	22½	22½	24	....	24
Hydrostatic Drive	....	....	28	28	27	28	27
Front Differential Housing (All Wheel Drive Tractor) Qts.	....	....	10	10	10	10	10

(3) All seven journals are 2.2823-2.2835

FRONT SYSTEM TRICYCLE TYPE

Farmall tractors are available with either of the two single front wheels shown in Fig. 1, or may also be equipped with a dual wheel tricycle type shown in Fig. 2.

SINGLE WHEEL
Farmall Models

1. The single front wheel is mounted in a fork which is bolted to the steering pivot shaft and depending on the tire size, two wheel types are used. On tractors with 9.00×10 or 11.00×10 tires, the male and female wheel halves (18 and 19—Fig. 1) are used. On tractors with 7.50×16 or 7.50×20 tires, a solid disc type wheel (12) is used in conjunction with hub (11). In all cases taper roller wheel bearings are used.

Wheel bearings are adjusted to a slight preload with adjusting nut (14). Lock adjusting nut with jam nut (16) after adjustment is complete.

DUAL WHEELS
Farmall Models

2. The pedestal for the dual tricycle wheels is bolted to the steering pivot shaft. The pedestal is available as a pre-riveted assembly (17—Fig. 2), or the pedestal and axle (18) are available as separate repair parts. Wheel bearings are adjusted to a slight preload.

FRONT SYSTEM AXLE TYPE

AXLE MAIN MEMBER
Farmall Models

3. For Farmall tractors equipped with an adjustable wide tread front axle, refer to Fig. 3, 4 and 5. The axle main member (24) pivots on pin (26)

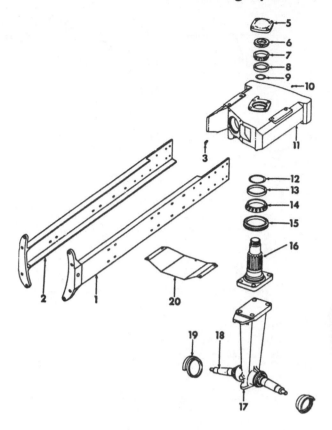

Fig. 2–Exploded view of front support, tricycle type front axle and components used on Farmall tractors.

1. Front frame R.H.
2. Front frame L.H.
3. Grease fitting
5. Cover
6. Shaft nut
7. Bearing cone
8. Bearing cup
9. "O" ring
10. Plug
11. Front support
12. "O" ring
13. Bearing cup
14. Bearing cone
15. Oil seal
16. Steering pivot shaft
17. Pedestal
18. Axle
19. Dust shield
20. Pulley shield

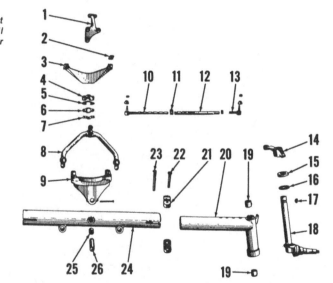

Fig. 3–Adjustable wide front axle available for Farmall 826 and 1026 tractors. Refer also to Fig. 4 and 5.

1. Steering arm (center)
2. Washer (4 used)
3. Stay rod support
4. Ball socket
5. Shim
6. Socket cap
7. Lock plate
8. Stay rod assembly
9. Axle support
10. Tie rod extension
11. Clamp
12. Tube
13. Tie rod end
14. Steering arm
15. Thrust bearing
16. Felt washer
17. Woodruff key
18. Steering knuckle
19. Bushing
20. Axle extension
21. Axle clamp
22. Clamp bolt
23. Clamp pin
24. Axle main member
25. Pivot bushing
26. Pivot pin

which is pinned in the axle support (9). The two pivot pin bushings (25) are pressed into the axle main member and should be reamed after installation, if necessary, to provide a free fit for the pivot pin.

To remove the axle main member, disconnect tie rods from steering arms, then remove axle clamps (21—Fig. 3 or 4) and on Fig. 5 remove the pin and loosen clamp bolts and withdraw the axle extension, knuckle and wheel assemblies. Remove cotter pin and pivot pin retaining pin, then disconnect stay rod ball from its support. Save the

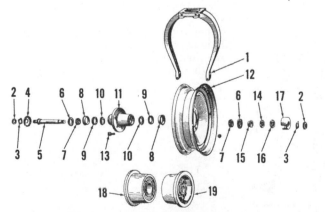

1. Fork
2. Jam nut
3. Lock washer
4. Shield (short)
5. Axle
6. Oil seal
7. Bearing cone
8. Seal retainer
9. Bearing cup
10. Grease retainer
11. Hub
12. Wheel
13. Wheel bolt
14. Adjusting nut
15. Spacer
16. Jam nut
17. Shield (long)
18. Wheel half (male)
19. Wheel half (female)

Fig. 1–Exploded view showing both types of fork mounted single front wheels. Axle (5) is fitted with taper roller bearings and is used with both types of wheels.

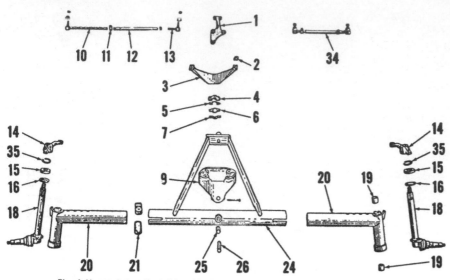

1. Steering arm (center)	14. Steering arm
2. Washer (4 used)	15. Thrust bearing
3. Stay rod support	16. Felt washer
4. Ball socket	18. Steering knuckle
5. Shim	19. Bushing
6. Socket cap	20. Axle extension
7. Lock plate	21. Axle clamp
9. Axle support	24. Axle main member
10. Tie rod extension	25. Pivot bushing
11. Clamp	26. Pivot pin
12. Tube	34. Tie rod assembly
13. Tie rod end	35. Retaining ring

Fig. 4–Heavy duty adjustable wide front axle available for all Farmall tractors.

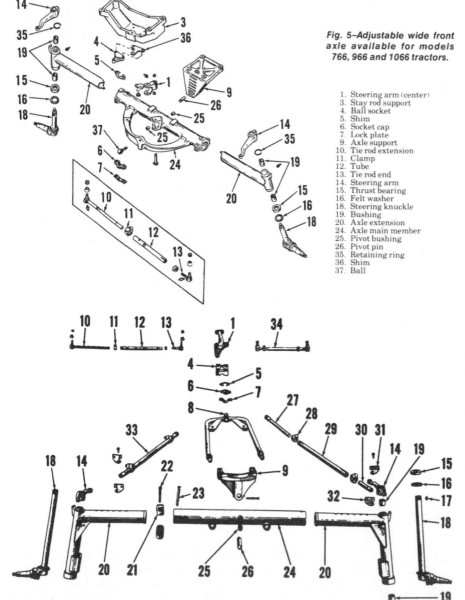

Fig. 5–Adjustable wide front axle available for models 766, 966 and 1066 tractors.

1. Steering arm (center)	
3. Stay rod support	
4. Ball socket	
5. Shim	
6. Socket cap	
7. Lock plate	
9. Axle support	
10. Tie rod extension	
11. Clamp	
12. Tube	
13. Tie rod end	
14. Steering arm	
15. Thrust bearing	
16. Felt washer	
18. Steering knuckle	
19. Bushing	
20. Axle extension	
24. Axle main member	
25. Pivot bushing	
26. Pivot pin	
35. Retaining ring	
36. Shim	
37. Ball	

Fig. 6–Exploded view of adjustable wide front axle used on Farmall "Hi-Clear" models. Note the auxiliary stay rod assemblies.

1. Steering arm (center)	19. Bushing
4. Ball socket	20. Axle extension
5. Shim	21. Axle clamp
6. Socket cap	22. Clamp bolt
7. Lock plate	23. Clamp pin
8. Stay rod	24. Axle main member
9. Axle support	25. Pivot bushing
10. Tie rod extension	26. Pivot pin
11. Clamp	27. Adjustable end
12. Tube	28. Clamp
13. Tie rod end	29. Tube
14. Steering arm	30. Threaded end
15. Thrust bearing	31. Upper bracket
16. Felt washer	32. Lower bracket
17. Woodruff key	33. Auxiliary stay rod
18. Steering knuckle	assembly

shims (5) located between socket (4) and socket cap (6). On models with the external cylinder disconnect and cap the hoses. Drive pivot pin out of axle support and axle main member and remove axle main member. On all models except Farmall equipped with heavy duty front axle (Fig. 4 or 5), stay rod (8) can now be removed from axle main member if necessary.

Reinstall by reversing the removal procedure and if necessary, adjust the front wheel toe-in as outlined in paragraph 9.

Farmall Hi-Clear

4. Farmall high clearance tractors are equipped with a front axle as shown in Fig. 6.

Removal procedure for this axle is the same as given in paragraph 3 except that the auxiliary stay rods (33) should be disconnected when removing the axle extension and wheel assemblies. Refer to paragraph 9 for toe-in adjustment.

International Models

5. International tractors are available with a non-adjustable as well as an adjustable front axle as shown in Figs. 7 and 8. Stay rod, stay rod ball and axle main member of both axles are available separately for service. The axle, or axle main member, pivots on pin (14—Fig. 7 or 8) which is retained in the axle support with a pin and a bolt and nut. The two pivot pin bushings are pressed into the axle main member, and should be reamed after installation, if necessary, to provide a free fit for the pivot pin.

6. To remove the non-adjustable axle proceed as follows: Remove nut and washer from top of steering knuckles, place correlation marks on steering arms and knuckles and with tie rods attached, remove steering arms from

knuckles. Remove retaining rings (7—Fig. 7), then raise front of tractor and withdraw steering knuckles and wheels as assemblies. Remove pivot pin retaining bolt and pin from axle support, disconnect stay rod ball and save the shims (20) located between socket (19) and cap (21). Axle main member and stay rod can now be removed after driving out pivot pin. Stay rod can be separated from axle, if necessary.

Reinstall by reversing removal procedure and refer to paragraph 9 if necessary to adjust toe-in.

7. To remove the adjustable axle main member (9—Fig. 8), proceed as follows: Remove nut and washer from top of steering arms, place correlation marks on steering arms and knuckles, then with tie rods attached, remove steering arms from knuckles. Remove axle clamps (24), raise front of tractor and remove the axle extensions and wheels as assemblies. Remove pivot pin retaining bolt and pin from axle support, disconnect stay rod ball and save shims (20) located between socket (19) and cap (21). Axle main member and stay rod (18) can be removed after driving out pivot pin (14). Stay rod can be separated from axle main member if necessary.

Reinstall by reversing removal procedure and refer to paragraph 9 if necessary to adjust toe-in.

STAY ROD AND BALL
All Models

8. The stay rod and stay rod ball are available as individual parts, except on heavy duty adjustable wide front axle (Fig. 4). On axle shown in (Fig. 5), the

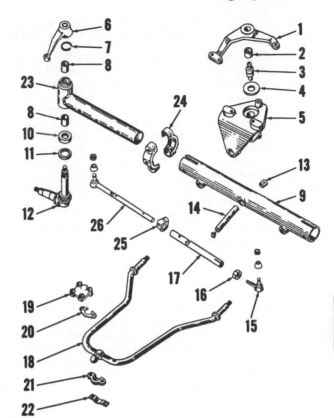

Fig. 8–Adjustable front axle and component parts available for 826 and 1026 International tractors.

1. Steering arm (center)
2. Bushing
3. Pivot pin
4. Shim (0.010 and 0.020)
5. Axle support
6. Steering arm
7. Retaining ring
8. Bushing
9. Axle main member
10. Thrust bearing
11. Felt washer
12. Steering knuckle
13. Pivot bushing
14. Pivot pin
15. Tie rod end
16. Jam nut
17. Tube
18. Stay rod
19. Ball socket
20. Shim (0.010 and 0.020)
21. Socket cap
22. Lock plate
23. Axle extension
24. Axle clamp

stay rod ball is available as individual parts but not stay rod. Clearance between stay rod ball and socket can be adjusted by adding or subtracting shims which are located between socket and cap. The socket cap screws are tightened to 85-100 ft.-lbs. torque and locked by bending tabs of lock plate against flats of screw heads. Farmall models equipped with adjustable wide tread front axle have a socket support bolted to side rails and any service required on support is obvious.

TIE RODS AND TOE-IN
All Models

9. The procedure for removal and disassembly of the tie rods on all models so equipped is obvious after an examination of the units. Tie rod ends are non-adjustable and faulty units will require renewal.

Adjust the toe-in on all models to ¼-inch, plus or minus 1/16-inch. Adjustment is made by varying the length of the tie rods. Both tie rods should be adjusted an equal amount with not more than one turn difference when adjustment is complete.

STEERING KNUCKLES
All Models

10. Removal of steering knuckles from axle extensions, or axle main member is obvious after an examination of the unit and reference to Figs. 3, 4, 5, 6, 7 and 8. Note that steering knuckles used on International model tractors (Fig. 7 and 8) and Farmall heavy duty axle (Fig. 4 and 5) have a retaining ring located at top end of knuckle pivot shaft.

When renewing steering knuckle bushings be sure to align oil hole in bushing with oil hole in axle or axle extension. Install bushing so outer ends are flush with bore. Ream bushing after installation, if necessary, to provide an operating clearance of 0.001-0.006.

Fig. 7–Non-adjustable front axle and component parts available for 826 and 1026 International tractors. Refer also to Fig. 8.

1. Steering arm (center)
2. Bushing
3. Pivot pin
4. Shim (0.010 and 0.020)
5. Axle support
6. Steering arm
7. Retaining ring
8. Bushing
9. Axle main member
10. Thrust bearing
11. Felt washer
12. Steering knuckle
13. Pivot bushing
14. Pivot pin
15. Tie rod end
16. Jam nut
17. Tube
18. Stay rod
19. Ball socket
20. Shim
21. Socket cap
22. Lock plate

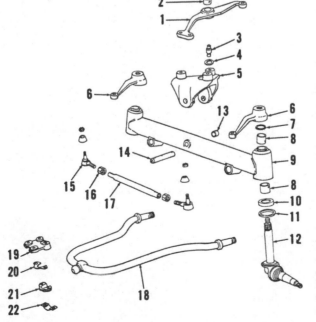

FRONT SYSTEM ALL-WHEEL DRIVE

11. Farmall and International tractors are available as 4-wheel drive (All Wheel Drive) units. The front axle assembly consists of a one-piece center housing having flanged ends to which stub axle ends are bolted. Wheel spindles and wheel hubs are carried on taper roller bearings. The axle center housing incorporates a straddle mounted pinion and a four pinion differential gear unit. Full floating axles extend outward from the differential and attach to Cardan type universal joints located inside wheel hubs.

Power for the front axle assembly is taken from a gear reduction unit (transfer case) which is bolted to the left side of tractor rear frame and driven by an idler gear on the reverse idler shaft of the range transmission. The reduction unit and the front axle are connected by a drive shaft fitted with two conventional universal joints of which the rear has a slip joint that compensates for oscillation of the front axle.

The gear reduction unit (transfer case) has a shifting mechanism which permits shifting to neutral, thus disconnecting power to the front axle. Gear sets with various ratios are available for the reduction unit to match the front and rear tire size combinations. Therefore, it is essential that the correct front and rear tire sizes be used with the correct gear set. If a change in tire sizes or gear sets is contemplated, contact the International Harvester Company for information concerning the proper combinations.

WHEEL AND PIVOT BEARINGS
Models So Equipped

12. **WHEEL BEARINGS.** Wheel bearings should be removed, cleaned and repacked annually. Removal of inner wheel bearing requires removal of hub assembly.

To remove both wheel bearings, refer to Fig. 9 and proceed as follows: Support axle assembly, then remove wheel cover (51) and the wheel and tire assembly. Clip and remove lock wire from axle retaining cap screws, then remove cap screws and tapered bush-

ings (47) and remove axle by pulling straight outward as shown in Fig. 10. NOTE: Use caution when removing axle shaft not to damage the oil seal (21—Fig. 9) located in center housing as shown in Fig. 11. Swing power yoke (41—Fig. 10) aside, straighten tabs of retainer plates (25—Fig. 12), then unbolt and remove bearing plate (34) and shims (24). Outer wheel bearing, power yoke, compensating ring and hub can now be removed from spindle. See Fig. 13.

To remove inner wheel bearing (33—Fig. 9) from spindle (22), loosen jam nut and adjusting screw (23), then position a small punch against outer end of adjusting wedge (30—Fig. 13) and bump wedge inward until clamp ring (31) is free and remove clamp ring. Insert a pin punch through knock-out holes provided in the spindle and bump on inner side of bearing inner race until bearing is about ½-inch from inner flange of spindle, then attach puller, if necessary, and complete removal of bearing. Be sure to keep bearing straight while removing or damage (scoring) to spindle could result.

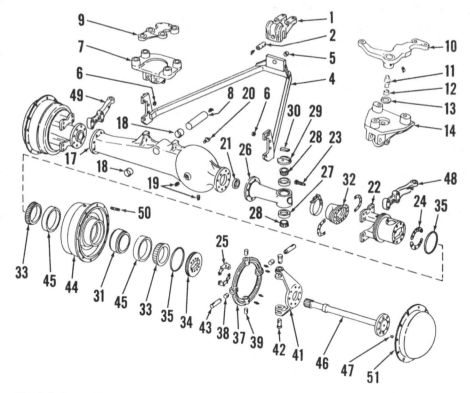

Fig. 9—Exploded view of the front drive axle used when tractors are equipped with "All Wheel Drive". Note items 10 through 14 which are used on International tractors and differ from items 7 and 9 which are used on Farmall tractors.

1. Stay bar support
2. Rear pivot pin
4. Stay rod
5. Bushing
6. Gusset washer
7. Axle support (Farmall)
8. Pivot pin
9. Center steering arm (Farmall)
10. Center steering arm (International)
11. Pivot pin
12. Bushing
13. Shim
14. Axle support (International)
17. Housing
18. Bushing
19. Plugs
20. Vent
21. Oil seal
22. Spindle
23. Wedge adjusting screw
24. Shims
25. Lock plates
26. Axle stub
27. Bearing cup
28. Pivot bearing
29. Bearing cap
30. Adjusting wedge
31. Clamp ring
32. Boot
33. Wheel bearing
34. Bearing plate
35. Seal
37. Compensating ring
38. Bushing
39. Bushing
41. Power yoke
42. Yoke pin
43. Hub pin
44. Hub
45. Bearing cup
46. Axle shaft
47. Tapered bushing
48. Steering arm L.H.
49. Steering arm R.H.
50. Wheel stud
51. Wheel cover

Fig. 10—View showing axle (46) being removed. Item (41) is power yoke. Refer also to Fig. 11.

Fig. 11—When removing axle use caution not to damage the oil seal (21) shown. Differential and carrier have been removed for clarity.

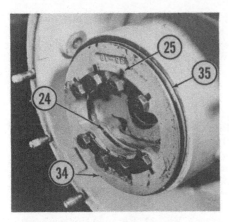

Fig. 12–View showing bearing adjusting plate and shims. Power yoke and compensating ring are removed for clarity.

24. Shims	34. Bearing plate
25. Lock plates	35. Seal

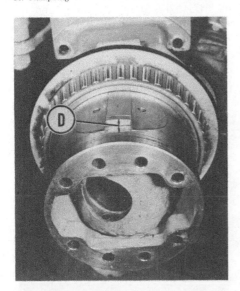

Fig. 13–Spindle assembly with outer wheel bearing and hub removed. Note outer end of pivot bearing adjusting wedge (30).

22. Spindle	33. Inner bearing
30. Adjusting wedge	35. Seal
31. Clamp ring	

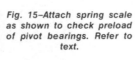

Fig. 14–A distance (D) of 1⅜-inches between end of adjusting wedge and end of slot in pivot bearing cap should be maintained. Refer to text.

NOTE: In some cases, the top pivot bearing cap may come out of bore in spindle when clamp ring and inner wheel bearing are removed and the spindle may drop and rest on the stub axle. If this occurs, proceed as follows when reassembling. Place a wood block under spindle, then with a hydraulic jack under block, raise spindle into its proper position and place upper pivot bearing and cap in its bore. Start inner wheel bearing on spindle and over upper bearing cap. This will hold parts in position. Start clamp ring over spindle and complete installation of both bearing and clamp ring.

13. Clean and inspect all parts for wear, excessive scoring or other damage and renew as necessary. Repack the wheel bearings with a good grade of multi-purpose lithium grease.

14. Reassemble and adjust wheel bearings and pivot bearings as follows: Install inner wheel bearing on spindle with largest diameter toward inside. Place clamp ring over spindle and against inner wheel bearing and make sure it does not contact the cage of inner wheel bearing. Tighten adjusting wedge until it feels solid, then measure distance from end of wedge to end of slot in pivot bearing cap. This distance (D—Fig. 14) should be at least 1⅜ inches. If measured distance is less than 1⅜ inches, remove spindle and add shims under bottom pivot bearing cone. Shims can be made from shim stock and each 0.012 shim will change the measured distance about ⅛-inch.

NOTE: This operation is to insure that a satisfactory adjustment can be made for the pivot bearings.

With clamp ring installed as outlined above and a new seal on spindle, install the hub, compensating ring, power yoke assembly and outer wheel bearing. With new seal on bearing plate, install original shims, bearing plate and lock plates. Tighten cap screws securely and check rotation of wheel hub. Hub should rotate with a slight drag. Add or subtract shims under bearing plate to obtain correct bearing adjustment. Shims are avail-

able in thicknesses of 0.002, 0.010 and 0.030. Lock cap screws with lock plates when adjustment is complete.

Disconnect tie rod from steering arm, then tighten wedge adjusting screw until a slight drag is felt when moving spindle through its full range of travel. Move spindle back and forth several times, then using a spring scale attached to tie rod hole of steering arm, check the pounds pull required to keep spindle in motion. See Fig. 15. Pull should not exceed 12 pounds and must be read while spindle is in motion. Tighten adjusting screw jam nut when adjustment is complete. Reinstall axle using caution not to damage oil seal located in center housing. Install tapered bushings and the axle retaining cap screws and tighten cap screws securely. Lock wire the axle cap screws. Reinstall wheel and tire and wheel cover and mate holes of wheel cover with grease fittings of compensating ring if tractor is equipped with this type wheel cover.

15. **PIVOT BEARINGS.** To remove the pivot bearings, the wheel bearings must be removed as outlined in paragraph 12.

With wheel bearings removed, disconnect tie rod from steering arm, then remove pivot bearing cups from their recesses in the stub axle by driving on a punch inserted in the knockout holes (H—Fig. 16) provided in outer end of

Fig. 16–Knock-out holes (H) are provided for removal of pivot bearing cups.

Fig. 15–Attach spring scale as shown to check preload of pivot bearings. Refer to text.

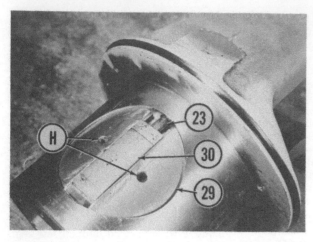

Fig. 17—View of pivot bearing top cap (29) showing location of knock-out holes (H). Note also adjusting wedge (30) and wedge adjusting screw (23).

removing the driving axles, which is involved in work of either operation, overhaul of either portion of the "All-Wheel Drive" axle can be accomplished without disturbing the other section.

stub axle. Use caution during this operation not to let punch slip past cups and damage bearings. With bearing cups free of their recesses, tilt inner end of spindle upward and remove from stub axle. Do not use force to remove spindle. If spindle does not come off readily, it is probable that bearing cups are not completely free of their recesses. The lower pivot bearing can be removed from spindle by driving on a punch inserted through the knock-out holes provided in spindle. Upper pivot bearing can be removed from upper bearing cap in the same manner. See (H—Fig. 17).

16. Clean and inspect all parts. Check bearings for roughness, damage or undue wear. Renew parts if any doubt exists as to their condition.

17. To reinstall pivot bearings, first lubricate lower bearing, drive it into position in spindle and place lower bearing cup over lower bearing. Place spindle over stub axle and support it with a jack and wood block as shown in Fig. 18. Align bearing cup with its recess in stub axle, raise jack and press lower bearing cup into its recess. Note: If necessary, bearing cup can be seated by driving on top side of stub axle while spindle is supported. Drive upper bearing cup into its recess on top side of stub axle. Install upper pivot bearing on top cap and install assembly in spindle with shallowest end of ad-

justing wedge slot toward outside. Be sure adjusting wedge is installed with rounded surface upward.

Wheel bearings can now be installed and pivot bearings adjusted as outlined in paragraph 14.

FRONT DRIVE AXLE
Models So Equipped

18. **R & R AXLE ASSEMBLY.** To remove the "All-Wheel Drive" front axle and wheels as an assembly, first disconnect tie rods from steering arms. Remove "U" bolts from drive shaft front universal joint and separate universal joint. Disconnect stay rod from bracket under clutch housing and position a rolling floor jack under stay rod to support assembly. Support front of tractor, drive out pivot pin, then raise front of tractor and roll the axle and wheels assembly forward away from tractor.

19. **OVERHAUL.** Overhaul of the axle assembly can be considered as two operations and both operations can be accomplished without removing the axle assembly from the tractor. One operation concerns the stub axle along with the hub and the parts which make up the outer end of the axle. The other operation concerns the differential and carrier which is carried in the axle center housing. With the exception of

20. **AXLE OUTER END.** Outer end of axle can be disassembled as follows: Remove wheel cover (51—Fig. 9) and the wheel and tire assembly. Clip lock wire and remove axle retaining cap screws and tapered bushings (47) and pull axle from housing. Use caution when removing axle not to damage oil seal (21) located in axle center housing. Also see Fig. 11. Remove the two plugs and pins (43—Fig. 9) and remove power yoke (41) and compensating ring (37) assembly from hub. Note: Pins (43) are tapped so cap screws can be used to aid in removal as shown in Fig. 19. Remove roll pins and yoke pins (42—Fig. 9) and separate power yoke and compensating ring. Straighten lock plate tabs and remove bearing adjustment plate (34), shims (24), outer wheel bearing (33) and hub (44). Note: Identify and keep removed shims in their original relationship. Loosen jam nut and wedge adjusting screw (23), then with a small punch positioned in slot of upper pivot bearing cap (29), bump adjusting wedge (30) inward until clamp ring (31) is free. See Fig. 13. Use punch in the knock-out holes provided in spindle and bump inner wheel bearing about ½-inch toward outer end of spindle, then attach a puller, if necessary, and complete removal of bearing. Note: Be sure to keep bearing straight during removal as damage (scoring) to spindle could result. Use punch in knockout holes (H—Fig. 16) provided in stub axle and bump pivot bearing cups from recesses in stub axle, then tilt inner end of spindle upward and pull assembly from stub axle.

NOTE: Do not use force to remove spindle from stub axle. If spindle cannot be removed freely it is probable that one or the other pivot bearing cups

Fig. 18—Support spindle on a wood block during installation of pivot bearings.

Fig. 19—Use cap screws as shown to pull hub to compensating ring pins.

are not completely free of stub axle.

If necessary, stub axle (26—Fig. 9) can now be removed from center housing (17).

21. Clean and inspect all parts. Pay particular attention to wheel bearings and pivot bearings in regard to roughness, damage or wear and renew any which are in any way doubtful. Wheel bearing cups can be bumped from hub, if necessary. Compensating ring bushings are available for service and renewal procedure is obvious. Be sure also that wear on driving pins (42 and 43) is not excessive. Check dowel pins in hub and roll pins in yoke pins to see that they are straight and not unduly worn.

Reassemble by reversing the disassembly procedure and adjust pivot bearings and wheel bearings as outlined in paragraph 14.

22. DIFFERENTIAL UNIT. To remove and overhaul the differential unit, first drain differential housing and disconnect left tie rod from left steering arm. Remove "U" bolts from front universal joint and separate universal joint. Remove both front wheel covers, then remove both front axles. Unbolt carrier (1—Fig. 20) and pull

carrier and differential assembly from axle center housing. See Fig. 21.

With unit removed, remove bearing caps (2—Fig. 20), then remove differential, bearing cups (24) and bearing adjusters (4) from carrier. Differential carrier bearings (25) can be removed from differential at this time, if necessary. Remove cap screws retaining oil seal retainer (17) and bearing cage (10) to carrier and remove pinion and bearing assembly. Identify and save bearing cage shims (13, 14 and 15). Remove nut (23), washer (22), yoke (21) and seal retainer (17) from pinion (7), then press pinion from outer bearing (12) and bearing cage (10). Spacer (16), inner bearing (12) and pilot bearing (6) can now be removed from pinion although pilot bearing (6) should be unstaked from pinion prior to removal. Bearing cups (11) can be driven from bearing cage (10) after pinion shaft is out. Match mark the differential case halves (26 and 27), remove case retaining bolts, then separate case halves

and remove spider (34), pinions (32), thrust washers (33), side gears (30) and side gear thrust washers (31). If not already done, bearings (25) can now be removed from differential case halves. If necessary. lubricator (29) can also be removed. If bevel drive gear (8) is to be renewed, remove rivets (9) by drilling and punching.

Clean all parts and inspect for undue wear or scoring, chipped teeth or other damage and renew parts as necessary.

Reassembly is the reverse of disassembly, however, consider the following information during assembly. Shims (13 and 14) are available in 0.003 thickness. Shims (15) are available in thicknesses of 0.005, 0.010 and 0.030. Spacer (16) is available in widths of 0.506 through 0.526 in increments of 0.001. Pinion and bevel ring gear are available only as a matched set. Note also that axle oil seals (21—Fig. 9) can be renewed when carrier and differential assembly is out.

23. To reassemble differential unit, proceed as follows: Place inner bearing cone (12) over pinion shaft (7) with largest diameter toward gear and press bearing on shaft until it bottoms. Invert pinion shaft and press pilot bearing (6) on end of pinion shaft and stake in at least four positions. Install bearing cups (11) in bearing cage (10) with smallest inside diameters toward center. Insert pinion shaft and inner bearing in bearing cage and install spacer (16), outer bearing cone (12), seal assembly retainer (17), yoke (21), washer (22) and nut (23). Attach a holding fixture to yoke and while turning bearing cage (10), tighten nut (23) to a torque of 255 ft.-lbs. With nut tightened, check the rotation of the pinion shaft which should require a rolling torque of 8-15 in.-lbs. If shaft preload is not as stated, change the spacer (16) as required. Spacers are available in thicknesses of 0.506

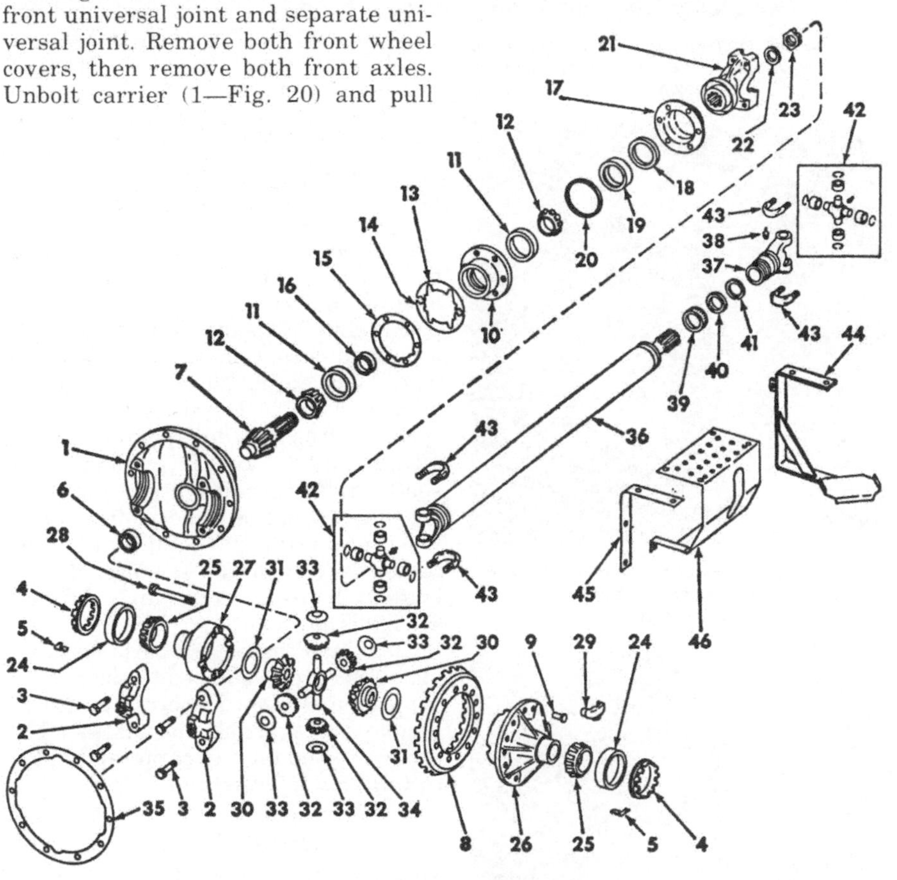

Fig. 20–Exploded view of "All Wheel Drive" front axle differential and carrier assembly showing component parts and their relative positions.

1. Carrier	14. Shim (lower)	25. Bearing	37. Sleeve yoke
2. Bearing cap	15. Shim	26. Case half (flanged)	38. Grease fitting
4. Bearing adjuster	16. Spacer	27. Case half (plain)	39. Dust cap
5. Adjuster lock	17. Seal retainer	29. Lubricator	40. Cork washer
6. Pilot bearing	18. Felt washer	30. Side gear	41. Steel washer
7. Pinion	19. Oil seal	31. Thrust washer	42. "U" joint package
8. Bevel gear	20. Cork washer	32. Pinion gear	43. "U" bolt
9. Rivet	21. Yoke	33. Thrust washer	44. Support (Farmall)
10. Bearing cage	22. Washer	34. Spider	45. Shield support (Farmall)
11. Bearing cup	23. Nut	35. Gasket	46. Shield (Farmall)
12. Bearing	24. Bearing cup	36. Stub (drive) shaft	
13. Shim (upper)			

Fig. 21–View showing differential carrier and differential unit removed from center housing.

through 0.526 in increments of 0.001. Use original shims (13, 14 and 15) as a starting point and install pinion shaft assembly in carrier.

Reassemble the differential assembly by reversing the disassembly procedure. Place differential in carrier, install bearing caps and bearing adjusters and tighten bearing adjusters until differential bearings have zero end play. Differential assembly can be moved left or right as required by loosening one bearing adjuster and tightening the opposite. Move the differential toward the pinion shaft until backlash is approximately 0.006-0.012. Paint ten or twelve teeth of bevel ring gear with red lead or prussian blue and turn pinion in direction of normal rotation. Tooth contact pattern should be located approximately midway of both length and width (depth) of bevel ring gear teeth. If pattern is too far toward toe of bevel gear teeth, turn adjusters as required to move gear away from pinion. If pattern is too far toward heel, turn adjusters (4) as required to move gear toward pinion. If pattern is too far toward root of bevel ring gear teeth, vary shims (13 and 14) to move pinion away from bevel drive gear. If pattern is too near top of bevel ring gear teeth, vary shims (13 and 14) to move pinion toward bevel drive gear. Continue this operation until tooth pattern is centered on the bevel ring gear teeth and backlash is 0.006-0.012 between pinion and bevel gear.

When a satisfactory pattern is obtained, slightly preload differential carrier bearings by tightening each bearing adjuster one notch. Lock both bearing adjusters.

Complete reassembly by reversing the disassembly procedure.

DRIVE HOUSING ASSEMBLY (TRANSFER CASE)
Models So Equipped

The drive housing which transmits power to the front axle is mounted on the left front side of the tractor rear frame and is driven by an idler gear on the reverse idler shaft of the range transmission.

24. REMOVE AND REINSTALL. To remove drive housing, remove shield, then remove "U" bolts of universal joint located closest to drive housing and separate the universal joint. Use tape or some other suitable means, to retain universal bearing cups on spider. Disconnect shifter rod from shifter shaft lever, drain rear frame, then unbolt and remove drive housing from tractor rear frame.

Before installing the unit on tractor, first determine the number and thickness of shim gaskets to be used between drive housing and rear frame as follows: Place a punch mark on "U" joint yoke 2 inches from center line of output shaft as shown in Fig. 22. Using a dial indicator as shown, record the backlash between the small spool gear and output gear. Next, engage the transmission park lock and measure

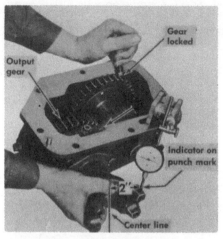

Fig. 22—Install punch mark on yoke 2 inches from center line of output shaft. Check backlash as shown.

Fig. 23—With park lock engaged check backlash between park lock, reverse driven gear and reverse drive gear.

Fig. 24—Install unit using one thin shim gasket and check total backlash. Refer to text.

the backlash between park lock, reverse driven gear and reverse drive gear as shown in Fig. 23. Record the backlash. Then, with park lock engaged, install drive housing assembly using one thin shim gasket. With dial indicator tip on the previously installed punch mark on yoke as shown in Fig. 24, check and record the backlash. This backlash reading should be 0.003-0.007 greater than the sum of the two previous backlash readings. If the backlash increase is less than 0.003, install one thick gasket or a combination of shim gaskets to increase the backlash. Gaskets are available in two thicknesses, (thin) 0.011-0.019 and (thick) 0.0016-0.024.

Installation of drive housing unit to tractor rear main frame will be facilitated if two ½-inch guide studs are used.

25. OVERHAUL. With unit removed as outlined in paragraph 24, wedge gears and remove the esna nut which retains the universal joint yoke (28—Fig. 25) to output shaft and remove yoke from shaft. Loosen jam nut and lock screw (24) in shifter fork and remove shifter shaft (25), spacer (27) and fork (23). Position housing with open side up, remove the roll (spring) pin (8) which retains spool gear shaft in housing, then remove spool gear shaft (5), spool gear (9) and thrust washers (12). Needle bearings in spool gear can be renewed at this time. Remove output shaft oil seal cage (20) and gasket. Lift rear snap ring (16) from its groove in output shaft and slide rearward. Slide output shaft gear (17) rearward and catch detent balls (14) and spring (15) as they emerge from bore of output shaft. Complete removal of shaft and lift gear from housing. Ball bearing (18) can be removed from shaft after removing snap ring (19) and needle bearing (3) can be removed from bore in housing. Any further disassembly is obvious.

Clean and inspect all parts for scoring, undue wear, chipped teeth or other damage and renew as necessary. Reassemble by reversing the disassembly procedure and refer to paragraph 24 for installation information.

DRIVE SHAFT
Models So Equipped

The drive shaft between drive housing and the front axle differential is conventional and can be removed and serviced as outlined in paragraph 26.

26. R&R AND OVERHAUL. To remove drive shaft from tractor, first remove shield, then remove "U" bolts from front and rear universal joint

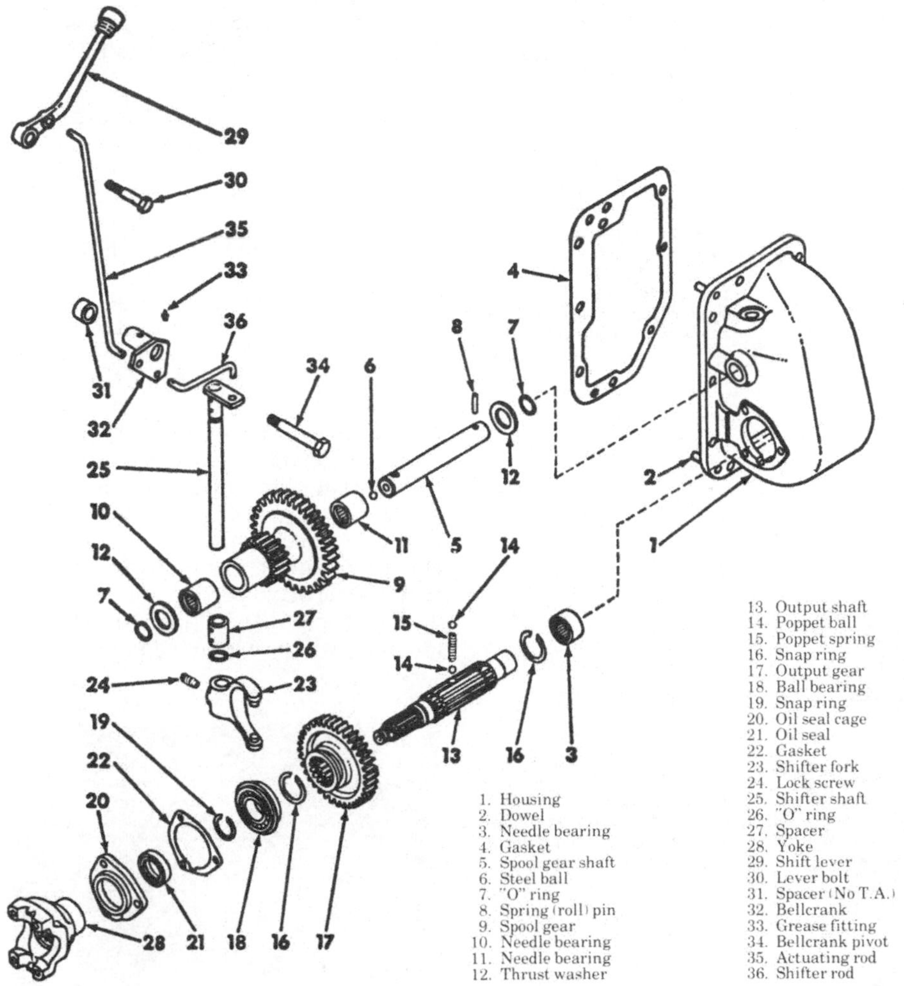

POWER STEERING SYSTEM

Note: The maintenance of absolute cleanliness of all parts is of utmost importance in the operation and servicing of the hydraulic power steering system. Of equal importance is the avoidance of nicks or burrs on any working parts.

OPERATION
All Gear Drive Models

27. Power steering is standard equipment on all tractors and except for steering cylinders and front support which differ between Farmall and International tractors, components for all tractors are similar. Refer to Figs. 26, 27, 28 and 29 for views showing the general lay-out of component parts.

The pressurized oil used for power steering, power brakes and direct drive or torque amplifier clutches is furnished by a 9 gallon per minute pump, located on left side of clutch housing and mounted on the inner side of multiple control valve. Of the 9 gpm supplied by the pump, a priority of 3 gpm is taken by a flow divider located in the multiple control valve and is utilized by the power steering, brakes and

1. Housing
2. Dowel
3. Needle bearing
4. Gasket
5. Spool gear shaft
6. Steel ball
7. "O" ring
8. Spring (roll) pin
9. Spool gear
10. Needle bearing
11. Needle bearing
12. Thrust washer
13. Output shaft
14. Poppet ball
15. Poppet spring
16. Snap ring
17. Output gear
18. Ball bearing
19. Snap ring
20. Oil seal cage
21. Oil seal
22. Gasket
23. Shifter fork
24. Lock screw
25. Shifter shaft
26. "O" ring
27. Spacer
28. Yoke
29. Shift lever
30. Lever bolt
31. Spacer (No T.A.)
32. Bellcrank
33. Grease fitting
34. Bellcrank pivot
35. Actuating rod
36. Shifter rod

Fig. 25—Exploded view of drive housing (transfer case) used when tractors are equipped with "All Wheel Drive".

yokes and lift shaft from tractor. The four exposed bearing cups can now be removed from the universal joint spiders. Unscrew dust cap from sleeve yoke, pull sleeve yoke from drive shaft and remove dust cap, steel washer and cork washer. Remove bearing cup retaining snap rings from yokes, then remove bearing cups by driving spider first one way then the other.

Individual parts available for service are sleeve yoke, drive shaft, dust cap, steel washer and cork washer. Universal joint bearing cups and spider are available only as a package.

Reassembly is the reverse of disassembly, however, be certain that bearing cups are packed with grease (either by hand during assembly or through grease fitting after assembly). Bearing cups can usually be pressed into yokes simultaneously by using a bench vise. Small spacers can be used to complete installation of cups and allow installation of snap rings. Be sure ends of spider do not catch ends of bearing needles when pressing in the bearing cups.

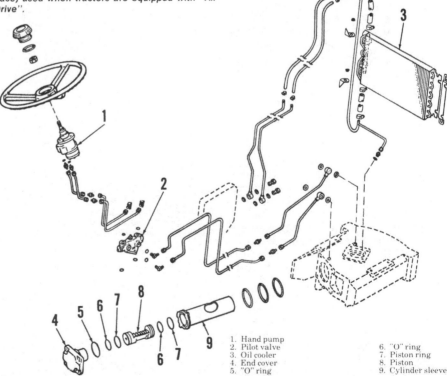

1. Hand pump
2. Pilot valve
3. Oil cooler
4. End cover
5. "O" ring
6. "O" ring
7. Piston ring
8. Piston
9. Cylinder sleeve

Fig. 26—Schematic view showing general lay-out of component parts on models 826 and 1026 comprising power steering system on Farmall tractors. Oil cooler (3) is similar on hydrostatic drive tractors.

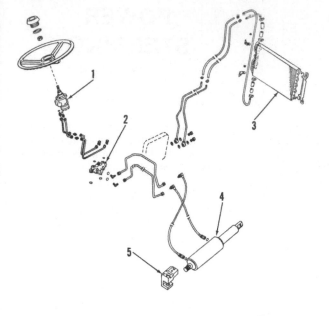

Fig. 27–Schematic view showing general lay-out of component parts on models 826 and 1026 comprising power steering system on International tractors. Oil cooler (3) is similar on hydrostatic drive tractors.

1. Hand pump
2. Pilot valve
3. Oil cooler
4. Steering cylinder
5. Cylinder mounting clevis

drive (torque amplifier) clutches and the transmission before it is returned to the reservoir.

All Hydrostatic Drive Models

28. The power steering is similar on hydrostatic drive to gear drive, (refer to Figs. 26, 27, 28 and 29) but differ in the pumps. The hydrostatic drive has 2 pumps, the #1 pump is a 12 gallon per minute pump located on right side of clutch housing and mounted on the inner side of the multiple control valve. The #1 pump delivers oil to the orifice and flow control valve which supplies 3 gpm to the power steering. The return flow from power steering is directed to the back side of the power steering relief valve. The return from power steering goes to the hydrostatic charging circuit through a charge circuit filter and to the multi-valve. The hydrostatic drive portion and pump #2 which is mounted "piggy back" on pump #1 will be explained later.

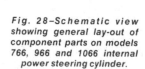

Fig. 28–Schematic view showing general lay-out of component parts on models 766, 966 and 1066 internal power steering cylinder.

1. Hand pump
4. End cover
5. "O" ring
6. "O" ring
7. Piston ring
8. Piston
9. Cylinder sleeve
10. "O" ring
11. Tube (RH)
12. Tube (LH)
13. Tube (return)
14. Tube (pressure)

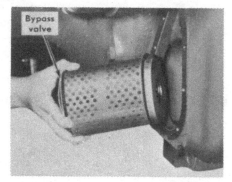

Fig. 30–Hydraulic filter is located on right side of tractor rear frame. Refer also to Figs. 28 and 29.

torque amplifier clutches, while the remaining 6 gallons are sent through the oil cooler and returned to lubricate the differential assembly and cool oil in the reservoir.

The priority 3 gpm which was diverted to the power steering has the return oil flow controlled by a pressure regulator (in multiple control valve) and a second priority flow of 1 gpm is taken to operate the power brakes. The remaining 2 gpm of the original 3 gpm is diverted to the torque amplifier clutches, and as either the direct drive clutch or torque amplifier clutch is always engaged during tractor operation, the 2 gpm is available for other work so it is again controlled by a lubrication regulator (in multiple control valve) and utilized to lubricate the

Fig. 29–Schematic view showing general lay-out of component parts on models 766, 966 and 1066 external power steering cylinder.

1. Hand pump
4. Steering cylinder
5. Hose (RH)
6. Hose (LH)
7. Tube (return)
8. Tube (pressure)

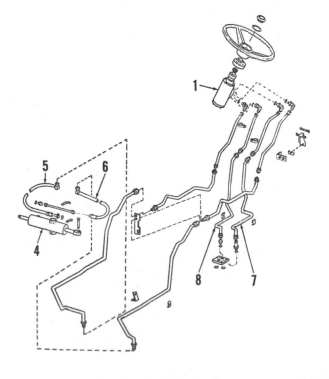

LUBRICATION AND BLEEDING
All Models

29. The tractor rear frame serves as a common reservoir for all hydraulic and lubrication operations. The filter, shown in Fig. 31 or 32, should be renewed at 10 hours, 100 hours, 250 hours, and then every 250 hours thereafter. The tractor rear frame should be drained and new fluid added every 1000 hours, or once a year, whichever occurs first.

Only IH "Hy-Tran" fluid should be used and level should be maintained at the "FULL" mark shown on level gage (dip stick). The dip stick is located on the top right front of rear frame on all series.

Whenever power steering lines have been disconnected, or fluid drained, start engine and cycle power steering system from stop to stop several times to bleed air from system, then if necessary, check and add fluid to reservoir.

TROUBLE SHOOTING
All Models

30. The following table lists some of the troubles which may occur in the operation of the power steering system. When the following information is used in conjunction with the information in the Power Steering Operational Tests section (paragraphs 33 through 42), no trouble should be encountered in locating system malfunctions.

1. No power steering or steers slowly.
   a. Binding mechanical linkage.
   b. Excessive load on front wheels and/or air pressure low in front tires.
   c. Steering cylinder piston seal faulty or cylinder damaged.
   d. Faulty power steering supply pump.
   e. Faulty commutator in hand pump.
   f. Flow divider valve spool sticking or leaking excessively.
   g. Control (pilot) valve spool sticking or leaking excessively. (Models 826 and 1026).
   h. Circulating check ball not seating.
   i. Flow control valve orifice plugged.

2. Will not steer manually.
   a. Binding mechanical linkage.
   b. Excessive load on front wheels and/or air pressure low in front tires.
   c. Pumping element in hand pump faulty.
   d. Faulty seal on steering cylinder or cylinder damaged.
   e. Pressure check valve leaking.
   f. Control (pilot) valve spool binding or centering spring broken. (Models 826 and 1026).
   g. Check valve in clutch housing inlet tube stuck in closed position.

3. Hard steering through complete cycle.
   a. Low pressure from supply pump.
   b. Internal or external leakage.
   c. Line between hand pump and control (pilot) valve obstructed. (Models 826 and 1026).
   d. Faulty steering cylinder.
   e. Binding mechanical linkage.
   f. Excessive load on front wheels and/or air pressure low in front tires.
   g. Cold hydraulic fluid.

4. Momentary hard or lumpy steering.
   a. Air in power steering circuit.
   b. Control (pilot) valve sticking. (Models 826 and 1026).

5. Shimmy.
   a. Control (pilot) valve centering spring weak or broken. (Models 826 and 1026).
   b. Control (pilot) valve centering spring washers bent, worn or broken. (Models 826 and 1026).

OPERATING PRESSURE AND RELIEF VALVE
All Gear Drive Models

31. System operating pressure and relief valve operation can be checked as follows: Remove the small orifice chamber plug (13—Fig. 33) which is the bottom plug located on rear side of the multiple control valve and install a gage capable of registering at least 3000 psi, then start engine and operate until hydraulic fluid is warm. Run engine at high idle rpm, turn front wheels in either direction until they reach stop, then continue to apply steering effort to steering wheel so system is pressurized and note reading on the gage. Gage should read approximately 1800-1900 psi for 826. For models 766, 966 and 1066 the gage should read, 1600 psi for internal steering cylinder and 1800 psi for external steering cylinder. If pressure is not as specified, renew the safety relief valve (6—Fig. 34) located on bottom side of the multiple control valve. Relief valve is available as a unit only.

All Hydrostatic Drive Models

32. Steering pressure and flow can be checked as follows: Remove the power steering line from multiple control valve, and connect Flo-Rater or equivalent (Fig. 35). Start engine and check flow which should be approximately 3 gpm with engine speed at 1500 rpm. With speed ratio lever in neutral, restrict the Flo-Rater; pressure should read approximately 1800-1900 psi for 826 and 1026. For models 966 and 1066 the gage should read, 1850 psi for internal steering cylinder and 2150 psi for external steering cylinder. If pressure is not as specified, renew safety relief valve (8—Fig. 36) located on rear side of the multiple control valve. Relief valve is available as a unit only.

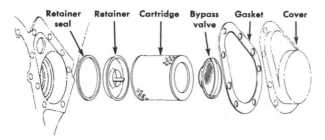

Retainer seal · Retainer · Cartridge · Bypass valve · Gasket · Cover

Fig. 31—Exploded view of hydraulic filter assembly used on 826 gear drive.

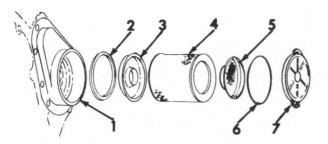

Fig. 32—Exploded view of hydraulic filter assembly used on 826 and 1026 hydrostatic drive and all 766, 966 and 1066 models.

1. Filter frame
2. Retainer seal
3. Retainer
4. Cartridge
5. Bypass valve
6. "O" ring
7. Filter frame cover

Fig. 33—Orifice chamber plug (13) is the bottom plug on rear side of all gear drive multiple control valve. Also see Fig. 34.

OPERATIONAL TESTS
All Models

33. The following tests are valid only when the power steering system is completely void of any air. If necessary, bleed system as outlined in paragraph 29 before performing any operational tests.

34. MANUAL PUMP. With transmission pump inoperative (engine not running), attempt to steer manually in both directions. NOTE: Manual steering with transmission pump not running will require high steering effort. If manual steering can be accomplished with transmission pump inoperative, it can be assumed that the manual pump will operate satisfactorily with the transmission pump operating.

Refer also to paragraphs 36 and 37 for information regarding steering wheel (manual pump) slip.

35. CONTROL (PILOT) VALVE. On models 826 and 1026, attempt to steer manually (engine not running). Manual steering will require high

Fig. 35–View showing the Flo-Rater connected for checking pressure and flow.

steering effort but if steering can be accomplished, control (pilot) valve is working.

No steering can be accomplished if control valve is stuck on center. A control valve stuck off center will allow steering in one direction only.

36. STEERING WHEEL SLIP (CIRCUIT TEST). Steering wheel slip is the term used to describe the inability of the steering wheel to hold a given position without further steering movement. Wheel slip is generally due to leakage, either internal or external, or a faulty hand pump, steering cylinder or control (pilot) valve on models 826 and 1026. Some steering wheel slip, with hydraulic fluid at operating temperature, is normal and permissible. On models 826 and 1026 a maximum of four revolutions per minute and on models 766, 966 and 1066 a maximum of one revolution per minute is acceptable. By using the steering wheel slip test and a process of elimination, a faulty unit in the power steering system can be located.

However, before making a steering wheel slip test to locate faulty components, it is imperative that the complete power steering system be completely free of air before any testing is attempted.

To check for steering wheel slip (circuit test), proceed as follows: Check reservoir (rear frame) and fill to correct level, if necessary. Bleed power steering system, if necessary. Bring power steering fluid to operating temperature, cycle steering system until all components are approximately the same temperature and be sure this temperature is maintained throughout the tests. Remove steering wheel cap (monogram), then turn front wheels until they are against stop. Attach a torque wrench to steering wheel nut. NOTE: Either an inch-pound, or a foot-pound wrench may be used. Advance hand throttle until engine reaches rated rpm, then apply 72 inch-pounds (6 foot-pounds) to torque wrench in the same direction as the front wheels are positioned against the stop. Keep this pressure (torque) applied for a period of one minute and count the revolutions of the steering wheel. Use same proce-

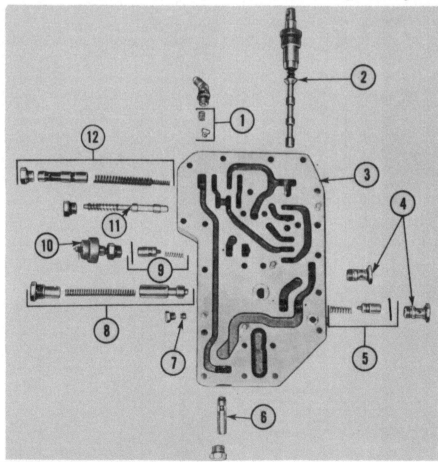

Fig. 34–All gear drive models, multiple control valve with front and rear plates and pump removed and showing control valves removed from their bores.

1. Brake check valve
2. Drive selector valve
3. Multiple control valve body
4. Oil cooler unions
5. Oil cooler by-pass valve
6. Safety relief valve
7. Flow control orifice
8. Flow control valve
9. Lubrication regulator valve
10. Pressure switch
11. Clutch dump valve
12. Pressure regulator valve

dure and check the steering wheel slip in the opposite direction. A maximum of four revolutions for 826 and 1026 or one revolution for 766, 966 and 1066 per minute in either direction is acceptable and system can be considered as operating satisfactorily. If, however, the steering wheel revolutions per minute exceed the maximum, record the total rpm for use in checking the steering cylinder or hand pump.

NOTE: While four revolutions per minute of steering wheel slip for 826 and 1026 is acceptable, it is generally considerably less in normal operation.

37. MANUAL (HAND) PUMP TEST. If steering wheel slip is more than four rpm, on models 826 or 1026 disconnect one line between the hand pump and the control (pilot) valve. Plug the openings securely. With engine running at rated rpm apply 72 inch-pounds (6 foot-pounds) with torque wrench on steering wheel nut in direction to pressurize the plugged line. Check the steering wheel revolutions for one minute. A maximum of two rpm is acceptable and hand pump is good. If the slip is more than two rpm, repair or renew the hand pump.

38. STEERING CYLINDER TEST. If steering wheel slip, as checked in paragraph 36, exceeds the maximum of four revolutions for 826 and 1026 or one revolution for 766, 966 and 1066 per minute, proceed as follows: Be sure operating temperature is being main-

tained, then disconnect and plug the steering cylinder lines. Repeat the steering wheel slip test, in both directions, as described in paragraph 36. If steering wheel slip is ½ rpm or more, **less** than that recorded in paragraph 36, overhaul or renew the steering cylinder.

39. FLOW DIVIDER GEAR DRIVE. When checking flow divider operation, also check orifice (7—Fig. 34) to see that it is open and clean as this orifice is the unit that actually meters the 3 gpm used to operate the power steering.

To check operation of flow divider and orifice, proceed as follows: Disconnect one of the steering cylinder lines, and connect Flo-Rater or equivalent to the supply side of disconnected line. Start engine and run at rated rpm and keep control (pilot) valve open by applying steering effort to steering wheel in the direction of disconnected line. Fluid flow from line should be approximately 3 gallons per minute.

If fluid flow is not as specified, service flow divider as outlined in paragraph 56. Bleed power steering system after reconnecting steering cylinder line, if necessary.

40. FLOW DIVIDER HYDROSTATIC DRIVE. When checking flow divider operation, also check orifice (6 —Fig. 36) to see that it is open and clean as this orifice is the unit that actually meters the 3 gpm used to operate the power steering.

To check operation of flow divider and orifice, proceed as follows: Remove power steering line from multiple control valve and connect Flo-Rater or equivalent (Fig. 35). Start engine and check flow which should be approximately 3 gallons per minute at 1500 rpm.

If fluid flow is not as specified, service flow divider as outlined in para-

graph 57. Bleed power steering system after test is made, if necessary.

41. HYDRAULIC PUMP. To check the hydraulic pump operating pressure, refer to paragraph 31 or 32. To check the hydraulic pump free flow on gear drive models, proceed as outlined in paragraph 42.

42. To check pump free flow on gear drive models, first remove the orifice plug (13—Fig. 33) and remove orifice (7 —Fig. 34).

NOTE: Removal of orifice will be simplified if a small wire is attached to the screw driver and extended about two inches beyond end of bit. Insert wire through orifice hole and wire will support orifice during removal from multiple control valve.

Disconnect the lower oil cooler line from front side of multiple control valve and place a container under opening to catch fluid. If available, connect a Flo-Rater to the flow divider port and secure discharge end in the filler hole of main frame. If no flow rating equipment is available, use a short piece of hose attached to flow divider port and place discharge end in a suitable container. Run engine at rated speed and check output (free flow) of pump which should be approximately 9 gallons per minute, with a slight leakage coming from the open oil cooler port.

If pump free flow is not as specified, remove and service pump as outlined in paragraphs 43, 45 and 47.

PUMP

All Models

43. REMOVE AND REINSTALL. To remove the power steering pump, first remove bottom plug and drain clutch housing or hydrostatic drive housing. On gear drive models, disconnect the "Torque-Amplifier" control rod (A—Fig. 37) from operating bell-

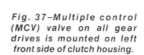

Fig. 36—Exploded view of multiple control valve used on all hydrostatic drives, showing control valves removed from their bores.

1. Multiple control valve body	13. Stud
2. Spring	14. Servo regulator valve seat
3. Plug	15. Servo regulator poppet valve
4. "O" ring	16. Guide
5. Flow divider valve	17. Spring
6. Flow divider orifice	18. Seal washer
7. "O" ring	19. "O" ring
8. Safety (relief) valve	20. Union (power steering)
9. "O" ring	21. Plug
10. Tee (power steering return)	22. Valve (brake check)
11. Cap	
12. Lock nut	

Fig. 37—Multiple control (MCV) valve on all gear drives is mounted on left front side of clutch housing.

A. Torque-Amplifier control rod
B. Bracket
C. Clutch rod
O. Oil cooler lines
P. Pressure switch
S. Brake line
T. Turnbuckle
V. Safety (relief) valve

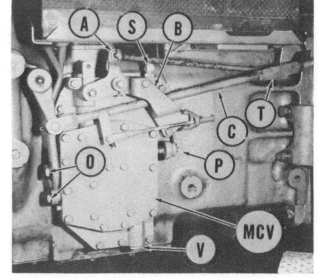

Fig. 38–Multiple control (MCV) valve on all hydrostatic drives is mounted on right front side of hydrostatic housing. Also showing the four cap screws at (B) retaining hydraulic pump to MCV.

heavy (0.016-0.024) thicknesses. In addition to sealing, the gasket also controls backlash between pump drive gear and the driving gear. When reinstalling pump and valve assembly, use new gasket of same thickness as original and reinstall assembly by reversing removal procedure.

44. Overhaul procedure for the 9 gpm pump on gear drive tractors and the 12 gpm or No. 1 pump on hydrostatic drive tractors will be covered in the following paragraphs. The 4.5 or No. 2 pump which is "piggy back" on No. 1 pump (Fig. 43A), which is used only on hydrostatic drive tractors, will be explained in paragraph 270.

45. **OVERHAUL (CESSNA 9 GPM).** With pump removed as outlined in paragraph 43, remove the esna nut which retains drive gear and remove gear and the square drive key. Remove cover (3—Fig. 44) from body. All parts are now available for inspection and/or renewal. Items 4, 5, 6, 7 and 10 are not available separately and must be ordered in a package which includes all the "O" rings and seals.

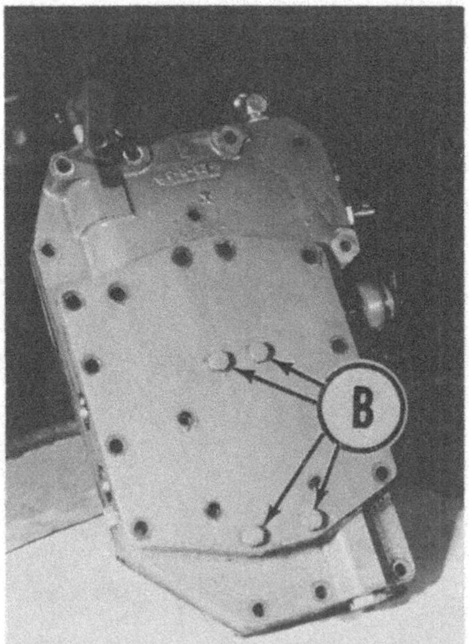

crank, unbolt bracket (B). Disconnect clutch operating rod (C) at both ends and remove clutch rod and bracket (B). Disconnect brake supply line (S), both oil cooler lines (O) and wire from oil pressure switch (P).

On hydrostatic drive models, disconnect power steering lines and brake line (Fig. 38). On all models remove all the multiple control valve mounting cap screws, except the pump retaining cap screws shown in (B—Fig. 38 or 39) and remove multiple control valve and pump assembly. See Fig. 40 or 41.

NOTE: During removal of pump assembly be very careful not to lose or damage the small check valve and spring which is located in clutch housing as shown in Fig. 42 or 43.

This check valve allows fluid to be

drawn into power steering circuit when steering with engine inoperative.

Pump can now be removed from multiple control valve by removing the four cap screws shown in (B—Fig. 38 or 39).

Before reinstalling pump, measure the gasket located between multiple control valve and clutch housing with a micrometer, to determine which thickness of gasket is used. These gaskets are available in light (0.011-0.019) and

Fig. 40–Multiple control valve removed from gear drive showing the power steering pump.

Fig. 42–View showing gear drive with multiple control valve off. Use caution not to lose or damage the small check valve and spring shown at (CV). Pump is driven by the pto driven gear (G).

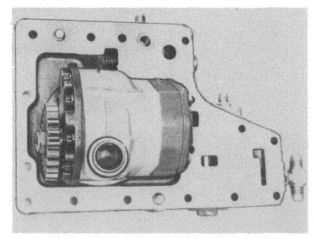

Fig. 41–Multiple control valve removed from hydrostatic drive showing the power steering pump.

Fig. 39–Multiple control valve on all gear drives, showing the four cap screws at (B) retaining hydraulic pump to inner face.

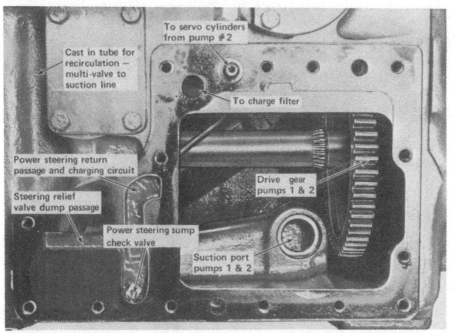

Fig. 43–View showing hydrostatic drive with multiple control valve removed. Use caution not to lose or damage the power steering sump check valve. Note pump is driven by the pto driven gear.

When reassembling install seal for drive gear shaft with lip toward inside of pump. Use new "O" rings in mounting surface. Install seal and protector in cover in same position as original. Bronze surfaces of diaphragms (7 and 10) face pump gears. Tighten body bolts to a torque of 25 ft.-lbs. and fill pump with oil prior to mounting on multiple control valve.

Refer to the following table for dimensional information.

Cessna 9 GPM Pump

Gear width (new) 0.444
Gear width (min. allowable) 0.441
Gear width variation (max. allowable) 0.001
Bearing I.D. (new) 0.6886
Bearing I.D. (max. allowable) . 0.691
Gear shaft diameter (new) 0.6875
Gear shaft diameter (min. allowable) 0.685
Body bore I.D. (new) 1.716
Body bore I.D. (max. allowable) . 1.719
Gear to body clearance (max.) .. 0.005

Fig. 43A–View showing the hydraulic pumps used on hydrostatic drive tractors in the multiple control valve. Pump (1) is a 12 gpm and pump (2) is 4.5 gpm. Pumps shown are Thompson. Cessna is also used and they are similar. Pump 2 mounts "piggy back" on pump 1.

46. OVERHAUL (CESSNA 12 GPM). With pumps removed as outlined in paragraph 43, remove the esna nut which retains drive gear and remove gear and the square drive key. Separate No. 2 pump from No. 1 pump (Fig. 43A). Remove cover (2—Fig. 45). All parts are now available for inspection and/or renewal. Items 5, 6, 7, 8 and 15 are not available separately and must be ordered in a package which includes all the "O" rings and seals.

When reassembling, use new diaphragms, gaskets, back-up washers,

Fig. 44–Exploded view of Cessna power steering pump used on gear drive tractors.

2. Seal
3. Cover
4. Diaphragm seal
5. Protector gasket
6. Back-up gasket
7. Diaphragm
8. Gears and shafts
9. Key
10. Body diaphragm
11. Pump body

Fig. 45–Exploded view of Cessna 12 gpm power steering pump used on hydrostatic drive tractors.

1. Seal
2. Cover
3. Check ball
4. Check spring
5. Diaphragm seal
6. Protector gasket
7. Back-up gasket
8. Pressure diaphragm
9. Key
10. Gears and shafts
11. "O" ring
12. Body
15. Wear plate

diaphragm seal and "O" rings. With open part of diaphragm seal (5) towards cover (2), work same into grooves of cover using a dull tool. Press protector gasket (6) and back-up gasket (7) into the relief of diaphragm seal. Install check ball (3) and spring (4) in cover, then install diaphragm (8) inside the raised lip of the diaphragm seal and be sure bronze face of diaphragm is toward pump gears. Dip gear and shaft assemblies in oil and install them in cover. Position wear plate (15) in pump body with bronze side toward pump gears and cut-out portion toward inlet (suction) side of pump. Install pump body over gears and shafts and install retaining cap screw. Torque cap screw to 45 ft.-lbs.

Check pump rotation. Pump will have a slight amount of drag but should turn evenly.

Refer to the following table for dimensional information.

Cessna 12 GPM Pump

O.D. of shafts at bushings (min.) 0.810
I.D. of bushing in body and cover (max.) 0.816
Thickness (width) of gears (min.) 0.572
I.D. of gear pockets (max.) 2.002
Max. allowable shaft to bushing clearance 0.006

47. OVERHAUL (THOMPSON 9 GPM). With pump removed as outlined in paragraph 43, remove the esna nut which retains drive gear and remove gear and the square drive key.

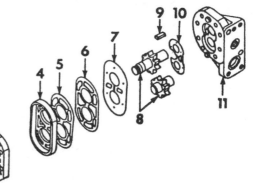

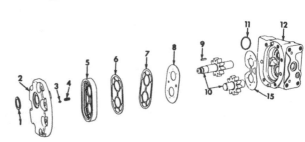

Remove cover (2—Fig. 46) from body. All parts are now available for inspection and/or renewal. Note position of pressure plate (6) as it is removed. Items 3, 4, 5, 6 and 9 are not available separately but must be ordered in a package which includes all the "O" rings and seals.

When reassembling, install oil seal for drive gear shaft with lip toward inside of pump. Use new "O" rings in mounting surface. Install sealing web, web backings, pressure plate and wear plate in same position as original. Tighten body bolts to a torque of 30 ft.-lbs. and fill pump with oil prior to installation on multiple control valve.

Refer to the following table for dimensional information.

Thompson 9 GPM Pump

Gear width (new)	0.540
Gear width (min. allowable)	0.5395
Gear width variation (max. allowable)	0.0005
Bearing I.D. (new)	0.8150
Bearing I.D. (max. allowable)	0.8155
Gear shaft diameter (new)	0.8125
Gear shaft diameter (min. allowable)	0.8120
Body bore I.D. (new)	1.7710
Body bore I.D. (max. allowable)	1.7715
Gear to body clearance (max.)	0.0029

48. OVERHAUL (THOMPSON 12 GPM). With pump removed as outlined in paragraph 43, remove the esna nut which retains drive gear and remove gear and the square drive key. Separate the No. 2 pump from No. 1 pump (Fig. 43A). Remove cover (2—Fig. 47). Bearings (7), pressure plate spring (6), "O" ring retainers (5), "O" rings (4), back-up washers (3) and oil seal (1) can now be removed from cover. Note location of bearings (7) so they can be reinstalled in the same position. Remove "O" rings (11 and 12), wear plate (16) and the pump gears and shafts (9) from pump body. Wear plate (16) is installed with reliefs towards pressure side of pump. Items 6, 12 and 16 are available separately. "O" rings and gaskets are available in a package.

Lubricate all parts during assembly, use all new gaskets and seals and be sure bearings in cover are reinstalled in their original positions. Tighten cover to body cap screws to a torque of 30 ft.-lbs.

Check pump rotation. Pump will have a slight amount of drag but should turn evenly.

Refer to the following table for dimensional information.

Thompson 12 GPM Pump

O.D. of shafts at bearings (min.)	0.812
I.D. of bearings in body and cover (max.)	0.816
Thickness (width) of gears (min.)	0.7765
I.D. of gear pockets (max.)	1.772
Max. allowable shaft to bearing clearance	0.004

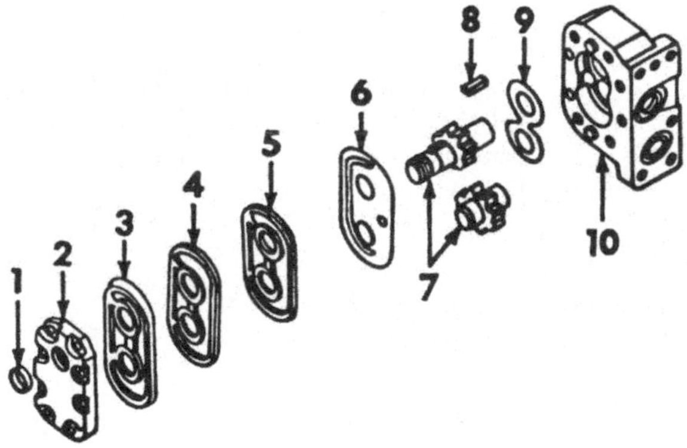

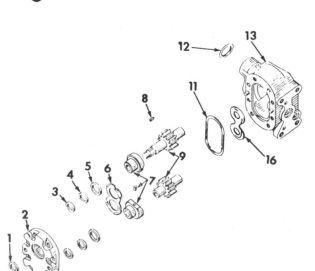

Fig. 46—Exploded view of Thompson power steering pump used on gear drive tractors. Note position of pressure plate (6).

1. Seal
2. Cover
3. Sealing web
4. Backing web (paper)
5. Backing web
6. Pressure plate
7. Pump gears and shafts
8. Drive key
9. Wear plate
10. Pump body

Fig. 47—Exploded view of Thompson power steering pump used on hydrostatic drive tractors.

1. Seal
2. Cover
3. Back-up washer (2 used)
4. "O" ring (2 used)
5. Retainer (2 used)
6. Pressure plate spring
7. Bearings
8. Key
9. Gears and shafts
11. "O" ring
12. "O" ring
13. Pump body
16. Wear plate

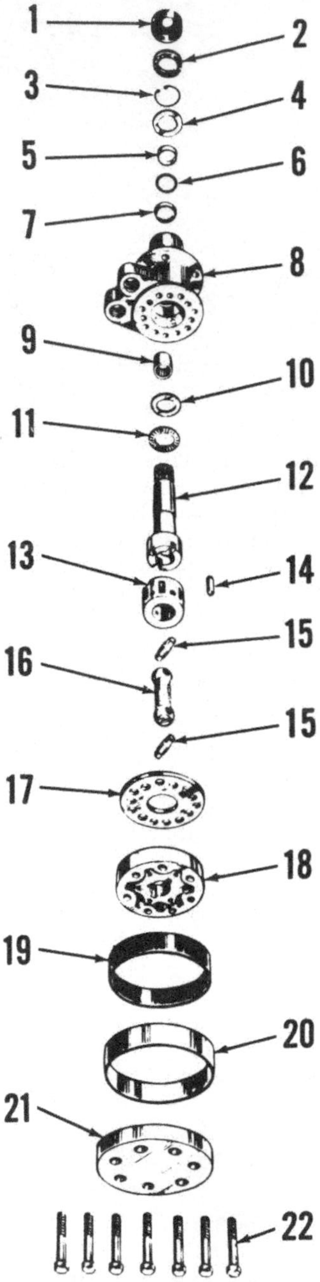

Fig. 48—Exploded view of the power steering hand pump used on 826 and 1026 models.

1. Dirt shield
2. Felt seal
3. Retaining ring
4. Washer
5. Nylon spacer
6. Back-up ring
7. Seal
8. Body
9. Needle bearing
10. Bearing race
11. Thrust bearing
12. Shaft
13. Commutator
14. Commutator pin
15. Link pins
16. Rotor drive link
17. Spacer plate
18. Stator-Rotor set
19. Seal
20. Seal retainer
21. End plate
22. Cap screws

HAND PUMP
Models 826 and 1026

49. REMOVE AND REINSTALL. To remove the power steering hand pump, first remove cap (monogram) from steering wheel, remove nut, attach puller and remove steering wheel. NOTE: DO NOT drive on steering wheel or shaft to remove steering wheel or damage to hand pump could occur. Remove the small plate just below steering wheel on rear side of steering wheel support and disconnect lines from hand pump. Remove button plug from hand throttle lever, then remove nut, washer and hand throttle lever from throttle control shaft. Unbolt support cover and lift cover and hand pump from support. Separate hand pump from support cover.

50. OVERHAUL STEERING MOTOR. To disassemble the removed steering motor, refer to Fig. 48 and clamp unit in a soft jawed vise with end plate (21) on top side. Remove end plate cap screws (22) and lift off end plate.

NOTE: Lapped surfaces of end plate (21), stator-rotor set (18), spacer plate (17) and body (8) must be protected from scratching, burring or any other damage as sealing of these parts depends only on their finish and flatness.

Remove seal retainer (20), seal (19) stator-rotor assembly (18) and spacer (17) from body. Remove commutator (13) and drive link (16), with link pins (15) and commutator pin (14). Remove body from vise, remove dirt shield (1) and felt seal (2), smooth any burrs which may be present on input shaft (12), then withdraw input shaft from body. Remove bearing race (10) and thrust bearing (11) from input shaft. Remove retaining ring (3), washer (4), nylon spacer (5), back-up ring (6) and seal (7). Do not remove needle bearing (9) unless renewal is required. If it should be necessary to renew needle bearing, press same out of commutator end of body.

Clean all parts in a suitable solvent and if necessary, remove paint from outer edges of body, spacer and end plate by passing these parts lightly over crocus cloth placed on a perfectly flat surface. Do not attempt to dress out any scratches or other defects since these sealing surfaces are lapped to within 0.0002 of being flat.

Inspect commutator and body for scoring and undue wear. Bear in mind that burnish marks may show, or discolorations from oil residue may be present on commutator after unit has been in service for some time. These can be ignored providing they do not interfere with free rotation of commutator in body.

Check fit of commutator pin (14) in the commutator. Pin should be a snug fit and if bent, or worn until diameter at contacting points is less than 0.2485, renew pin.

Measure inside diameter of input shaft bore in body and outside diameter of input shaft. If body bore is 0.006 or more larger than shaft diameter, renew shaft and/or body and commutator. Body and commutator are not available separately.

Check thrust bearing and race (10 and 11) for excessive wear or other damage and renew if necessary.

Inspect stator, rotor, vanes and vane springs for scoring, excessive wear or other damage. Place stator on lapped surface of end plate and place rotor in stator. Install vanes (V—Fig. 49) and vane springs (S) in rotor. NOTE: Arched back of springs must contact vanes. Position lobe of rotor in valley of stator as shown at (X—Fig. 50). Center opposite lobe on crown of stator, then using two feeler gages, measure clearance (C) between rotor lobes and stator. If clearance is more than 0.006, renew stator-rotor assembly. Use a micrometer and measure width (thickness) of stator and rotor. If stator is 0.002 or more wider (thicker) than rotor, renew the assembly. Stator and rotor are

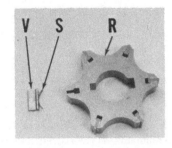

Fig. 49–When installing vanes (V) and springs (S) in rotor (R), arched back of spring contacts vane.

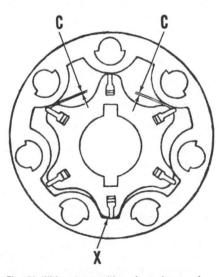

Fig. 50–With rotor positioned as shown, clearance (C) must not exceed 0.006. Refer to text.

available only as an assembly. Springs and vanes are available in a set.

Check end plate (21—Fig. 48) and spacer (17) for wear, scoring and flatness. Do not confuse the polish pattern on end plate and spacer with wear. This pattern, which results from rotor rotation, is normal.

When reassembling, use all new seals and back-up ring. All parts, except those noted below, are installed dry. Reassemble as follows: If needle bearing (9) was removed, lubricate with power steering fluid and install from commutator end of body. Press bearing into bore until inside end is 3 13/16 to 3⅞ inches from face of body as shown in Fig. 51. Lubricate thrust bearing and race and install on input shaft, then install shaft and bearing assembly in body. Install a link pin in one end of drive link, then install drive link in input shaft by engaging flats on link pin with slots in input shaft. Use a small amount of grease to hold commutator pin in commutator, then install commutator and pin in body while engaging pin in one of the long slots in input shaft. Commutator is correctly installed when edge of commutator is slightly below sealing surface of body.

Clamp body in a soft jawed vise with input shaft pointing downward. Place spacer on body and align screw holes with those in body. Place link pin in exposed end of drive link and install rotor while engaging flats of link pin with slots in rotor. Position stator over rotor and align screw holes in stator with those in spacer and body. Install vanes and vane springs in slots in rotor. Install seal (19—Fig. 48) in seal retainer (20), then install seal and retainer over stator. Install end plate, align screw holes and install cap screws. Tighten cap screws evenly to a torque of 18-22 ft.-lbs.

NOTE: If input shaft does not turn

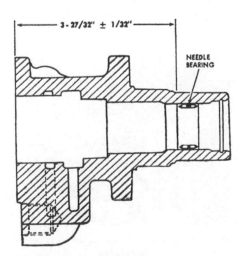

Fig. 51–When installing needle bearing in body, install same to dimension shown.

evenly after cap screws are tightened, loosen screws and retighten them again. However, bear in mind that unit was assembled dry and some drag is normal. If stickiness or binding cannot be eliminated, disassemble unit and check for foreign material or burrs which could be causing interference.

Lubricate input shaft seal (7) and install seal, back-up ring (6), nylon spacer (5), washer (4) and retaining ring (3). Install felt seal (2) and dirt shield (1). Items 2, 5, 6 and 7 are a service kit.

Turn unit on side with ports upward. Fill unit with power steering fluid and rotate input shaft slowly until interior is thoroughly lubricated. Drain excess fluid and plug ports.

Models 766, 966 and 1066

51. **REMOVE AND REINSTALL.** To remove the power steering hand pump, first remove cap (monogram) from steering wheel, remove nut, attach puller and remove steering wheel. NOTE: DO NOT drive on steering wheel or shaft to remove steering wheel or damage to hand pump could occur. Remove rear hood channel, cap screw in instrument panel and move panel forward. Drive roll pin out of throttle control collar and remove throttle control. Remove plate that contains ether starting button and disconnect hydraulic lines (2—Fig. 52). Remove allen head screws (1) and remove hand pump from tractor. Remove rubber boot and support cover.

Reassemble by reversing disassembly procedure.

52. **OVERHAUL STEERING MOTOR.** To disassemble the removed steering motor and control valve assembly, refer to Fig. 53 and proceed as follows: Install a fitting in one of the four ports in valve body (25), then clamp fitting in a vise so that input shaft (17) is pointing downward. Remove cap screws (39) and remove end cover (38).

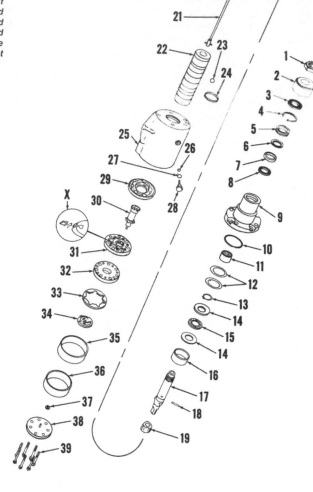

Fig. 53-Exploded view of Ross steering motor and control valve assembly used on models 766, 966 and 1066. Inset "X" shows vane and vane spring used in slot on each rotor lobe.

1. Nut
2. Water & dirt seal
3. Felt seal
4. Snap ring
5. Stepped washer
6. Brass washer
7. Teflon washer
8. Seal
9. Cover (upper)
10. Seal ring
11. Needle bearing
12. Shims
13. Snap ring
14. Thrust washers
15. Thrust bearing
16. Teflon spacer
17. Input shaft
18. Pin
19. Drive ring
20. Spacer
21. Torsion bar
22. Valve spool
23. Actuator ball
24. Retaining spring
25. Valve body
26. Recirculating ball
27. "O" ring
28. Plug
29. Spacer plate
30. Drive link
31. Stator-Rotor set
32. Manifold
33. Commutator ring
34. Commutator
35. Seal retainer
36. Seal
37. Washer
38. End cover
39. Cap screws

NOTE: Lapped surfaces on end cover (38), commutator set (33 and 34), manifold (32), stator-rotor set (31), spacer (29) and valve body (25) must be protected from scratching, burring or any other damage as sealing of these parts depends on their finish and flatness.

Remove seal retainer (35) and seal (36), then carefully remove wear washer (37), commutator set (33 and 34) and manifold (32). Grasp spacer (29) and lift off the spacer, drive link (30) and stator-rotor set (31) as an assembly. Separate spacer and drive link from stator-rotor set.

Remove unit from vise, then clamp fitting in vise so that input shaft is pointing upward. Remove water and dirt seal (2) and felt seal (3). Place a light mark on flange of upper cover (9) and valve body (25) for aid in reassembly. Unbolt upper cover from valve body, then grasp input shaft and remove input shaft, upper cover and valve spool assembly. Remove and discard seal ring (10). Slide upper cover assembly from input shaft and remove

teflon spacer (16). Remove shims (12) from cavity in upper cover or from face of thrust washer (14) and note number of shims for aid in reassembly. Remove snap ring (4), stepped washer (5), brass washer (6), teflon washer (7) and seal (8). Retain stepped washer (5) and snap ring (4) for reassembly. Do not remove needle bearing (11) unless renewal is required.

Remove snap ring (13), thrust washers (14) and thrust bearing (15) from input shaft. Drive out pin (18) and withdraw torsion bar (21) and spacer (20). Place end of valve spool on top of bench and rotate input shaft until drive ring (19) falls free, then rotate input shaft clockwise until actuator ball (23) is disengaged from helical groove in input shaft. Withdraw input shaft and remove actuator ball. Do not remove actuator ball retaining spring (24) unless renewal is required.

Remove plug (28) and recirculating ball (26) from valve body.

Thoroughly clean all parts in a suitable solvent, visually inspect parts and

Fig. 52-View showing removing of hand pump. Refer to text.

1. Allen head screws
2. Hydraulic lines

renew any showing excessive wear, scoring or other damage.

If needle bearing (11) must be renewed, press same out toward flanged end of cover. Press new bearing in from flanged end of cover to the dimension shown in Fig. 54. press only on numbered end of bearing, using a piloted mandrel.

Using a micrometer, measure thickness of the commutator ring (33—Fig. 53) and commutator (34). If commutator ring is 0.0015 or more thicker than commutator, renew the matched set.

Place the stator-rotor set (31) on the lapped surface of end cover (38). Make certain that vanes and vane springs are installed correctly in slots of the rotor.

NOTE: Arched back of springs must contact vanes. (See inset X—Fig. 53). Position lobe of rotor in valley of stator as shown at (V—Fig. 55). Center opposite lobe on crown of stator, then using two feeler gages, measure clearance (C) between rotor lobes and stator.

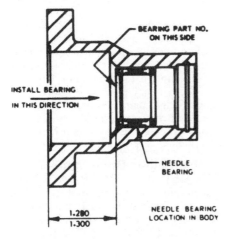

Fig. 54–When installing needle bearing in upper cover, press bearing in to dimension shown.

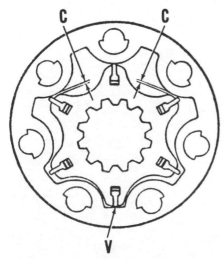

Fig. 55–With rotor positioned in stator as shown, clearances (C) must not exceed 0.006. Refer to text.

If clearance is more than 0.006, renew stator-rotor assembly. Using a micrometer, measure thickness of stator and rotor. If stator is 0.002 or more thicker than rotor, renew the assembly. Stator, rotor, vanes and vane springs are available only as an assembly.

Before reassembling, wash all parts in clean solvent and air dry. All parts, unless otherwise indicated, are installed dry. Install recirculating ball (26—Fig. 53) and plug (28) with new "O" ring (27) in valve body and tighten plug to a torque of 10-14 ft. lbs. Clamp fitting (installed in valve body port) in a vise so that top end of valve body is facing upward. Install thrust washer (14), thrust bearing (15), second thrust washer (14) and snap ring (13) on input shaft (17). If actuator ball retaining ring (24) was removed install new retaining ring. Place actuator ball (23) in its seat inside valve spool (22). Insert input shaft into valve spool, engaging the helix and actuator ball with a counter-clockwise motion. Use the mid-section of torsion bar (21) as a gage between end of valve spool and thrust washer, then place the assembly in a vertical position with end of input shaft resting on a bench. Insert drive ring (19) into valve spool until drive ring is engaged on input shaft spline. Remove torsion bar gage. Install spacer (20) on torsion bar and insert the assembly into valve spool. Align cross-holes in torsion bar and input shaft and install pin (18). Pin must be pressed into shaft until end of pin is about 1/32-inch below flush. Place spacer (16) over spool and install spool assembly into valve body. Position original shims (12) on thrust washer (14), lubricate new seal ring (10), place seal ring in upper cover (9) and install upper cover assembly. Align the match marks on cover flange and valve body and install cap screws finger tight. Tighten a worm drive type hose clamp around cover flange and valve body to align the outer diameters, then tighten cap screws to a torque of 18-22 ft. lbs.

NOTE: If either input shaft (17) or upper cover (9) or both have been renewed, the following procedure for shimming must be used. With upper cover installed (with original shims) as outlined above, invert unit in vise so that input shaft is pointing downward. Grasp input shaft, pull downward and prevent it from rotating. Engage drive link (30) splines in valve spool and rotate drive link until end of spool is flush with end of valve body. Remove drive link and check alignment of drive link slot to torsion bar pin. Install drive link until its slot engages torsion bar pin. Check relationship of spool end to body end. If end of spool is within 0.0025 of being flush with end of body, no additional shimming is required. If not within 0.0025 of being flush, remove cover and add or remove shims (12) as necessary. Reinstall cover and recheck spool to valve body position.

With drive link installed, place spacer plate (29) on valve body with plain side up. Install stator-rotor set over drive link splines and align cap screw holes. Make certain vanes and vane springs are properly installed. Install manifold (32) with circular slotted side up and align cap screw holes with stator, spacer and valve body. Install commutator ring (33) with slotted side up, then install commutator (34) over drive link end making certain that link end is engaged in the smallest elongated hole in commutator. Install seal (36) and retainer (35). Apply a few drops of hydraulic fluid on commutator. Use a small amount of grease to stick wear washer (37) in position over pin on end cover (38). Install end cover making sure that pin engages center hole in commutator. Align holes and install cap screws (39). Alternately and progressively tighten cap screws while rotating input shaft. Final tightening should be 18-22 ft. lbs. torque.

Relocate the unit in vise so input shaft is up. Lubricate new seal (8) and carefully work seal over shaft and into bore with lip toward inside. Install new teflon washer (7), brass washer (6) and stepped washer (5) with flat side up. Install snap ring (4) with rounded edge inward. Place new felt seal (3) and water and dirt seal (2) over input shaft.

Remove unit from vise and remove fitting from port. Turn unit on its side with hose ports upward. Pour clean hydraulic fluid into inlet port, rotate

Fig. 56–Installed view of power steering control (pilot) valve, on 826 gear drive.

input shaft until fluid appears at outlet port, then plug all ports.

PILOT VALVE
Model 826 Gear Drive

53. **REMOVE AND REINSTALL.** To remove the power steering control (pilot) valve, which is located on top side of clutch housing, it will be necessary to remove left battery and battery tray on diesel models. See Fig. 56. Disconnect hand pump lines and steering cylinder lines, then remove the three mounting cap screws and lift valve from clutch housing.

When reinstalling, renew "O" rings located between pilot valve and clutch housing. Bleed power steering system as outlined in paragraph 29.

Model 826 and 1026 Hydrostatic Drive

54. **REMOVE AND REINSTALL.** To remove the power steering control (pilot) valve, which is located on top side of hydrostatic drive housing, it will be necessary to remove right battery and battery tray on diesel and non-diesel models. See Fig. 57. Disconnect hand pump lines and steering cylinder lines, then remove the three mounting cap screws and lift valve from clutch housing.

When reinstalling, renew "O" rings located between pilot valve and hydrostatic housing. Bleed power steering system as outlined in paragraph 29.

All 826 and 1026 Models

55. **OVERHAUL.** With valve assembly removed, disassemble as follows: Refer to Fig. 58 and remove end caps (1) with "O" rings (2). Pull spool and centering assembly from valve body (9). Place a punch or small rod in hole of centering spring screw (3) and remove screw, centering spring (5) and centering spring washers (4). Remove plug (11), "O" ring (12) and circulating check ball (10). Remove retainer (16), seat (15), pressure check valve (14) and spring (13).

Wash all parts in a suitable solvent and inspect. Valve spool and spool bore in body should be free of scratches, scoring or excessive wear. Spool should fit its bore with a snug fit and yet move freely with no visible side play. If spool or spool bore is defective, renew complete valve assembly as spool (6) and valve body (9) are not available separately.

Inspect pressure check valve and seat. Renew parts if grooved or scored.

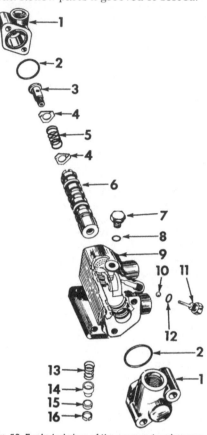

Fig. 58—Exploded view of the power steering control (pilot) valve.

1. End cap	8. "O" ring
2. "O" ring	9. Valve body
3. Centering spring screw	10. Steel ball
	11. Plug assembly
4. Centering spring washer	12. "O" ring
5. Centering spring	13. Spring
6. Spool	14. Check valve
7. Plug	15. Seat
	16. Retainer

Reassembly is the reverse of disassembly and the following points should be observed. Coat all parts with Hy-Tran fluid, or its equivalent, prior to installation. Install spring end of spool to opposite end from circulating check ball.

Measure distance between gasket surface of circulating check ball plug and end of roll pin as shown in Fig. 59. This distance should be 15/16-inch and if necessary, obtain this measure by adjusting roll pin in or out. Tighten end cap retaining cap screws to a torque of 186 in.-lbs.

Reinstall valve by reversing removal procedure and bleed power steering system as outlined in paragraph 29.

FLOW DIVIDER
All Gear Drive Models

56. **R&R AND OVERHAUL.** The flow divider valve is located in rear side of multiple control valve assembly and is under the second from bottom plug. See Fig. 34. Valve (spool) can be removed after removing plug and spring.

Service of flow divider valve consists of renewing parts. Carefully inspect spool and bore for scratches, grooves or nicks. Spool should fit bore snugly, yet be free enough to slide easily in bore with both spool and bore lubricated. Flow divider spring (8) has free length of 4.08 inches and should test 33.66 pounds when compressed to a length of 2.50 inches.

All Hydrostatic Drive Models

57. **R&R AND OVERHAUL.** The flow divider valve is located in bottom side of multiple control valve assembly and is the only plug in the bottom. See Fig. 36. Valve (spool) can be removed after removing plug and spring.

Service of flow divider valve consists of renewing parts. Carefully, inspect spool and bore for scratches, grooves or

Fig. 57—Installed view of power steering control (pilot) valve, on 826 and 1026 hydrostatic drive.

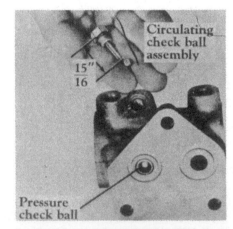

Fig. 59—Distance between gasket surface of circulating check valve plug and inner end of roll pin should be 15/16-inch as shown.

nicks. Spool should fit bore snugly, yet be free enough to slide easily in bore with both spool and bore lubricated. Flow divider spring (2) has free length of 4.08 inches and should test 33.50 pounds when compressed to a length of 2.50 inches.

STEERING CYLINDER
Farmall Models with Internal Cylinder

58. REMOVE AND REINSTALL. On Farmall models, the power steering cylinder and piston is incorporated into

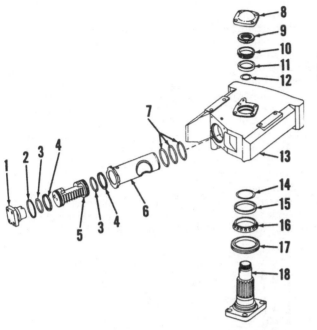

Fig. 60–Exploded view of steering cylinder, pivot shaft and components used on Farmall models.

1. End cover
2. "O" ring
3. "O" ring
4. Piston seal ring
5. Piston
6. Cylinder sleeve
7. "O" rings
8. Upper cover
9. Nut
10. Bearing cone
11. Bearing cup
12. "O" ring
13. Front support
14. "O" ring
15. Bearing cup
16. Bearing cone
17. Oil seal
18. Pivot shaft

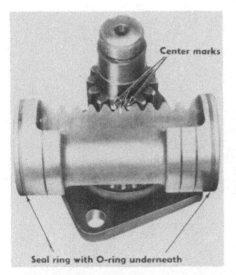

Fig. 61–Pivot shaft and piston are correctly meshed when timing marks align as shown.

Center marks

Seal ring with O-ring underneath

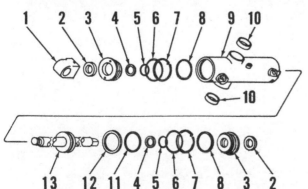

Fig. 62–Exploded view of external cylinder used on some models of 766, 966 and 1066 tractors.

1. Rod end
2. Seal
3. Cylinder head
4. Back-up washer
5. "O" ring
6. "O" ring
7. Retaining ring
8. Back-up washer
9. Cylinder
10. Bushing
11. "O" ring
12. Piston seal
13. Piston rod

the front support and removal of cylinder requires removal of the front support.

To remove front support, first remove hood, then remove radiator as outlined in paragraph 213. Support tractor and on tricycle models, remove wheels and pedestal from steering pivot shaft. On adjustable wide axle models, unbolt axle support from front support, center steering arm from pivot shaft and stay rod ball from stay rod support, then roll complete assembly away from tractor. On all models, disconnect the power steering lines from front support, then attach hoist to front support, unbolt from side rails and remove from tractor. Note: If necessary, loosen engine front mounting bolts to provide additional removal clearance for front support.

Reinstall by reversing removal pro-

cedure and bleed power steering system as outlined in paragraph 29.

59. OVERHAUL. Before disassembling power steering cylinder, match mark the cylinder flange and front support. The cylinder sleeve (6—Fig. 60) is bored off-center and the cam action resulting from rotating the cylinder will regulate the backlash between the teeth of pivot shaft (18) and teeth of piston (5). If the backlash prior to disassembly is satisfactory, the marks installed will provide the correct backlash during reassembly. If backlash is not correct, a point of reference will be established to simplify the backlash adjustment. Backlash should be adjusted to as near zero as possible without binding.

To remove the power steering cylinder and piston, first remove cap from top of pivot shaft, then straighten lock on nut, remove nut from top end of pivot shaft and pull pivot shaft from bottom of front support. Remove cylinder end cover and pull cylinder and piston from upper bolster. Piston can now be removed from cylinder. Any further disassembly required is obvious.

Reassembly is the reverse of disassembly. Renew all "O" rings and seals. When installing pivot shaft be sure to position the marked center tooth of piston rack between the two punch marked teeth of the pivot shaft as shown in Fig. 61. Tighten pivot shaft nut to provide a slight pre-load on pivot shaft and secure by bending lock flange on nut into notch in pivot shaft. Lobes are provided on flange of cylinder so mesh position of pivot shaft and piston can be adjusted to zero backlash (at straight ahead position) by using a small punch and hammer.

Farmall Models with External Cylinder

60. R&R AND OVERHAUL. To remove the external power steering cylinder, first disconnect lines from cylinder and plug lines to prevent oil drainage. Disconnect steering cylinder from center steering arm and cylinder rod anchor pin and remove from tractor.

61. With cylinder removed, move piston rod back and forth several times to clear oil from cylinder. Refer to Fig. 62 and proceed as follows: Place barrel of cylinder in a vise and clamp vise only enough to prevent cylinder from turning. Using a spanner wrench, turn cylinder head (3) until free end of retaining ring (7) appears in slot of barrel (9). Lift end of ring to outside of barrel and continue turning until all of ring is outside of barrel, lift nib from hole and remove the ring. Pull the piston rod (13) out of anchor end of cylinder. Loosen the set screw in the anchor (1) and remove anchor. Remove head from piston rod. Remove head from other end of barrel using the same procedure as before. All seals, "O" rings and backup washers are now available for inspection and/or renewal.

Clean all parts in a suitable solvent and inspect. Check cylinder for scoring, grooving and out-of-roundness. Light scoring can be polished out by using a fine emery cloth and oil providing a rotary motion is used during the polishing operation. A cylinder that is heavily scored or grooved, or that is out-of-round, should be renewed. Check piston rod and cylinder for scoring, grooving and straightness. Polish out very light scoring with fine emery cloth and oil, using a rotary motion. Renew rod and piston assembly if heavily scored or grooved, or if piston rod is bent. Inspect piston seal for frayed edges, wear and imbedded dirt or foreign particles. Renew seal if any of the above conditions are found. NOTE: Do not remove the "O" ring (11) located under the piston seal unless renewal is indicated. Inspect balance of "O" rings, back-up washers and seals and renew same if excessively worn.

Reassemble steering cylinder as follows: Lubricate piston seal and cylinder head "O" rings. Using a ring compressor, or a suitable hose clamp, install piston and rod assembly into cylinder. Install cylinder heads in cylinder so hole in cylinder heads will accept nib of retaining rings and pull same into its groove by rotating cylinder heads with a spanner wrench. Complete balance of reassembly by reversing the disassembly procedure.

Reinstall unit on tractor, then fill and bleed the power steering system as outlined in paragraph 29.

International Models

62. **R&R AND OVERHAUL.** To remove the power steering cylinder on International models, first disconnect lines from cylinder and plug lines to prevent oil drainage. Disconnect steering cylinder from center steering arm and rear mounting clevis and remove from tractor.

63. With cylinder removed, move piston rod back and forth several times to clear oil from cylinder. Refer to Fig. 63 and proceed as follows: Place barrel of cylinder in a vise and clamp vise only enough to prevent cylinder from turning. Turn cylinder head (11) until free end of retaining ring (10) appears in slot of barrel (4). Lift end of ring to outside of barrel and continue turning head until all of ring is outside of bar-

rel, lift nib from hole and remove the ring. Remove cylinder head assembly from piston rod, then pull piston rod and piston (7) from cylinder. All seals, "O" rings and back-up washers are now available for inspection and/or renewal.

Clean all parts in a suitable solvent and inspect. Check cylinder for scoring, grooving and out-of-roundness. Light scoring can be polished out by using a fine emery cloth and oil providing a rotary motion is used during the polishing operation. A cylinder that is heavily scored or grooved, or that is out-of-round, should be renewed. Check piston rod and cylinder for scoring, grooving and straightness. Polish out very light scoring with fine emery cloth and oil, using a rotary motion. Renew rod and piston assembly if heavily scored or grooved, or if piston rod is bent. Inspect piston seal for frayed edges, wear and imbedded dirt or foreign particles. Renew seal if any

of the above conditions are found. NOTE: Do not remove the "O" ring (6) located under the piston seal unless renewal is indicated. Inspect balance of "O" rings, back-up washers and seals and renew same if excessively worn. Be sure air bleed holes in the cylinder head assembly are open and clean.

Reassemble steering cylinder as follows: Lubricate piston seal and cylinder head "O" rings and using a ring compressor, or a suitable hose clamp, install piston and rod assembly into cylinder. Install cylinder head in cylinder so hole in cylinder head will accept nib of retaining ring and pull same into its groove by rotating cylinder head. Complete balance of reassembly by reversing the disassembly procedure.

Reinstall unit on tractor, then fill and bleed the power steering system as outlined in paragraph 29.

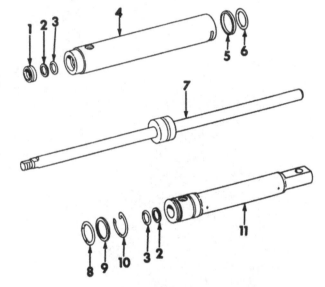

Fig. 63–Exploded view of the power steering cylinder used on International model tractors.

1. Seal
2. Back-up washer
3. "O" ring
4. Cylinder barrel
5. Piston ring
6. "O" ring
7. Piston rod
8. "O" ring
9. Back-up washer
10. End retainer
11. Cylinder head

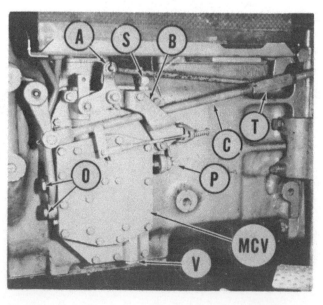

Fig. 64–Multiple control valve (MCV) on gear drive is mounted on left front side of clutch housing.

A. Torque-Amplifier control rod
B. Bracket
C. Clutch rod
O. Oil cooler lines
P. Pressure switch
S. Brake line
T. Turnbuckle
V. Safety (relief) valve

MULTIPLE CONTROL VALVE
All Gear Drive Models

64. **R&R AND OVERHAUL.** The multiple control valve is mounted on left front of clutch housing as shown in Fig. 64. The multiple control valve has a 9 gpm pump mounted on inner side which furnishes pressurized oil for operation of power steering, brakes and torque amplifier clutches as well as providing lubrication for the torque amplifier assembly and speed transmission assembly. The multiple control valve also contains the spools, valves and passages necessary to control these operations.

When servicing the multiple control valve, cleanliness is of the utmost importance as well as the avoidance of any nicks or burrs. When reinstalling the multiple control valve and pump assembly, be sure to use the same thickness gasket (5 or 6—Fig. 66) as original. Use a micrometer to measure gasket. Gaskets are available in light (0.011-0.019) and heavy (0.016-0.024) thicknesses.

To remove and reinstall the multiple control valve, refer to paragraph 43.

With unit removed, all spools, plungers and springs can be removed and the procedure for doing so is obvious. However, note that before both by-pass valves (26—Fig. 66) can be removed, the small retaining pins must be removed. Pins are retained in position by the inner and outer plates.

Refer to the following table for spring test data and to Fig. 66 for spring location. Spring call-out numbers are in parenthesis.

Brake Check Valve (18)

Free length—inches 0.500
Test length lbs. @ in . . . . . 0.5 @ 0.300

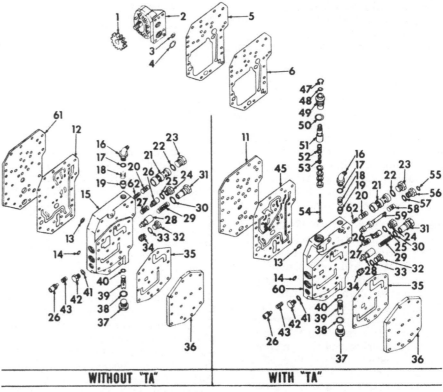

Fig. 66–Exploded view of the multiple control valve used on gear drive models.

1. Pump drive gear	20. Spring, outer	33. "O" ring	49. Valve retaining
2. Pump	21. Pressure regulator	34. Orifice	screw
3. "O" ring	valve	35. Gasket	50. "O" ring
4. "O" ring	22. "O" ring	36. Outer cover (plate)	51. Valve stem
5. Gasket (heavy)	23. Plug	37. Plug	52. Spring, outer
6. Gasket (light)	24. Plug	38. "O" ring	53. Spring, inner
11. Inner plate (with TA)	25. "O" ring	39. Safety valve	54. Valve guide
12. Inner gasket	26. By-pass valve	40. "O" ring	55. "O" ring
13. "O" ring	27. Lubrication regulator	41. "O" ring	56. Plug
14. "O" ring	spring	42. Plug	57. "O" ring
15. Body	28. Flow control valve	43. Oil cooler by-pass	58. Spring
16. Elbow	29. Spring	spring	59. Clutch dump valve
17. "O" ring	30. "O" ring	45. Inner gasket	60. Body
18. Spring	31. Plug	47. Retaining ring	61. Inner plate (no TA)
19. Brake check valve	32. Plug	48. "O" ring	62. Spring, inner

Pressure Regulator Valve (20 & 62)

Outer spring free length—
inches 3.508
Test length lbs. @ in. . . . . 64.5 @ 2.170
Inner spring free length—
inches 2.910
Test length lbs. @ in. . . . . 19.5 @ 1.940

Lubrication Regulator Valve (27)

Free length—inches 1.072
Test length lbs. @ in. . . . . . 5.2 @ 0.787

Flow Divider (29)

Free length—inches 4.08
Test length lbs. @ in. . . . . 33.7 @ 2.50

Oil Cooler By-Pass Valve (43)

Free length—inches 1.244
Test length lbs. @ in. . . . . . 26 @ 0.712

Drive Selector Valve (52 & 53)

Outer spring free length—
inches 1.578
Test length lbs. @ in. . . . . . 28 @ 1.281
Inner spring free length—
inches 1.880
Test length lbs. @ in. . . . . . 18 @ 1.254

Clutch Dump Valve (58)

Free length—inches 1.720
Test length lbs. @ in. . . . . . . 5 @ 1.150

Reassembly is the reverse of disassembly. Use all new "O" rings and refer

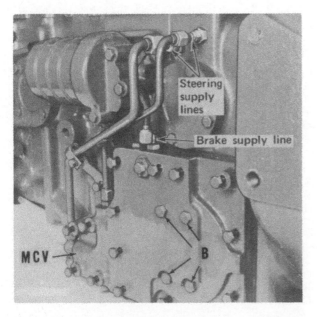

Fig. 65–Multiple control (MCV) valve on hydrostatic drive is mounted on right front side of hydrostatic housing.

to paragraph 248 when adjusting "Torque-Amplifier" drive selector and to paragraph 240 or 241 when adjusting the "Torque-Amplifier" dump valve.

Safety relief valve (39—Fig. 66) is heavily staked and cannot be disassembled. Valve should relieve at approximately 1800-1900 psi for 826. For models 766, 966 and 1066 the valve should relieve at approximately 1600 psi for internal steering cylinder and 1800 psi for external steering cylinder and faulty valves are renewed as a unit. Refer to paragraph 31 for information on checking valve.

All Hydrostatic Drive Models

65. R&R AND OVERHAUL. The multiple control valve is mounted on right front of hydrostatic housing as shown in Fig. 65. The multiple control valve has a 12 gpm pump with a 4.5 gpm pump mounted "piggy back" on the 12 gpm pump. The 12 gpm pump furnishes pressurized oil for operation of power steering and charge circuit of the hydrostatic drive. The 4.5 gpm pump furnishes pressurized oil for operation of power brakes, seat, servo cylinders and charge circuit of the hydrostatic drive. The excess oil of both pumps is used for lubrication of ring gear and pinion and transmission mainshaft rear bearing. The lubrication is controlled by the multi-valve. See Fig. 68 and refer to paragraph 67. The multiple control valve also contains the spools, valves and passages necessary to control these operations.

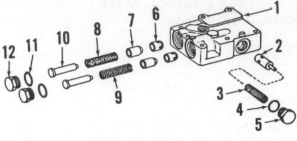

Fig. 66—Exploded view of multi-valve which is located on left top side of hydrostatic drive housing.
1. Multi-valve body
2. By-pass valve (oil cooler)
3. Spring
4. "O" ring
5. Plug
6. Bushing (poppet valve)
7. Poppet valve
8. Spring (light)
9. Spring (heavy)
10. Guide
11. "O" ring
12. Plug

When servicing the multiple control valve, cleanliness is of utmost importance as well as the avoidance of any nicks or burrs. When reinstalling the multiple control valve and pump assembly be sure to use the same thickness gasket as original. Use a micrometer to measure gasket. Gaskets are available in light (0.011-0.019) and heavy (0.016-0.024) thicknesses.

To remove and reinstall the multiple control valve, refer to paragraph 43.

With unit removed, all valves, plungers and springs can be removed and the procedure for doing so is obvious. The spring test on flow divider spring (3—Fig. 67) is as follows: Free length is 4.08 inches and test load is 33.5 pounds at 2.50 inches. To reset the servo regulator, connect gauge to servo test port which is right below brake check valve (18—Fig. 67). With engine operating at 1700 rpm and higher and S-R lever in forward or reverse, servo pressure should be approximately 400 psi. Servo pressure is adjustable by turning adjusting screw (20) "IN" to raise the pressure and "OUT" to lower it. Allow system a few seconds to stabilize before making additional adjustments.

Refer to paragraph 32 for information on checking safety relief valve (8—Fig. 36 or 6—Fig. 67). Valve should relieve at approximately 1800-1900 psi for 826 and 1026. For models 966 and 1066 the valve should relieve at approximately 1850 psi for internal steering cylinder and 2150 psi for external steering cylinder. If valve is faulty it is replaced as a unit.

Using all new "O" rings, reassemble by reversing the disassembly procedure.

OIL COOLER
All Gear Drive Models

66. R&R AND OVERHAUL. An oil cooler, shown in Fig. 26 or 27, is incorporated into the power steering system. Of the 9 gpm supplied by the power steering pump, only 3 gpm is used to operate the power steering and tractor controls. The remaining 6 gpm is directed, via the flow divider valve, to the oil cooler where it is cooled and returned to lubricate the tractor differen-

tial and provide cooling for oil in the reservoir. Pressure regulation of the oil to oil cooler is controlled by the oil cooler by-pass valve (5—Fig. 34) in the multiple control valve.

Service of the oil cooler involves only removal and reinstallation, or renewal of faulty units. Removal of the oil cooler is obvious after removal of the radiator grille and hood and an examination of the unit. However, outlet and inlet hoses must be identified as they are removed from oil cooler pipes so oil circuits will be kept in the proper sequence.

MULTI-VALVE AND OIL COOLER
All Hydrostatic Drive Models

67. R&R AND OVERHAUL. The oil coolers are similar to the ones shown in Fig. 26 or 27, except hydrostatic drive models have a multi-valve to control the oil flow. NOTE: Do not confuse with the multiple control valve. The main flow to the multi-valve is from the 12 gpm pump which is mounted in the rear frame on the left side of tractor. This pump supplies oil to the hitch and auxiliary valves. The excess oil from the other two pumps mounted on the multiple control valve, also is routed to the multi-valve (Fig. 68). The multi-valve controls pressure and flow to the oil cooler, lubrication

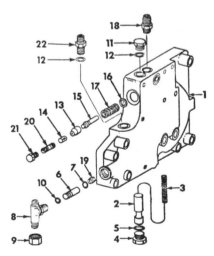

Fig. 67—Exploded view of the multiple control valve used on hydrostatic drive models.

1. Multiple control valve body	13. Servo regulator poppet valve
2. Flow divider valve	14. Servo regulator valve seat
3. Spring	15. Guide
4. Plug	16. Seal washer
5. "O" ring	17. Spring
6. Safety (relief) valve	18. Valve (brake check)
7. "O" ring	19. Flow divider orifice
8. Tee (power steering return)	20. Stud
9. Cap	21. Lock nut
10. "O" ring	22. Union (power steering)
11. Plug	
12. "O" ring	

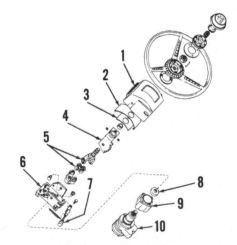

Fig. 69—Exploded view of tilt steering wheel assembly available for all models.

1. Cover	6. Support
2. Shield	7. Release lever
3. Back-up shim	8. Dust seal
4. Position quadrant	9. Spacer hub
5. Universal joint	10. Steering hand pump

circuit, center section of hydrostatic drive and recirculation line. Refer to Fig. 68. The by-pass valve (2) relieves at a 105 psi. Spring test on the by-pass valve spring (3) is as follows: Free length is 3.03 inches and test load is 35.4 pounds at 1.68 inches.

Removal of oil cooler is obvious after removal of radiator grille and hood and an examination of the unit. However, outlet and inlet hoses must be identified as they are removed from oil cooler pipes so oil circuits will be kept in the proper sequence.

TILT STEERING WHEEL
All Models

68. The tilt steering wheel assembly (Fig. 69) is available for all models. To adjust steering wheel position, move release lever (7) rearward, tilt steering wheel to desired position, then allow the spring to engage release lever in notch on quadrant (4). Steering wheel can be tilted to five different positions. Disassembly and repair of the unit will be obvious after an examination of the assembly and reference to Fig. 69.

ENGINE AND COMPONENTS (NON-DIESEL)

Series 766 non-diesel tractor is equipped with an engine having a bore and stroke of 3.750 x 4.390 inches and a displacement of 291 cubic inches.

Engine is a six cylinder design, and is fitted with dry type cylinder sleeves and is equipped with only gasoline fuel system.

Series 826 non-diesel tractor is equipped with an engine having a bore and stroke of 3.813 x 4.390 inches and a displacement of 301 cubic inches.

Engine is a six cylinder design, of the sleeveless type and may be equipped with either gasoline or L-P gas fuel system. Dry type air filters are used in the air induction system.

R&R ENGINE ASSEMBLY
All Non-Diesel Models

69. Removal of engine is best accomplished by removing front support, radiator, front axle and wheels assembly and the fuel tank assembly, then unbolting and removing the engine and side rails from clutch housing or hydrostatic drive housing as follows:

Drain cooling system, disconnect battery cable and remove hood. Disconnect radiator hoses and air cleaner inlet hose. Identify and disconnect power steering lines and hydraulic oil cooler lines. Remove operator assist handles and the torque amplifier lever, if so equipped. Notice that TA lever and shaft have match marks affixed to insure correct assembly. If equipped with hydrostatic drive, remove RANGE lever. Remove center hood and steering support cover. Disconnect all interfering lines and wiring from fuel tank and support. Attach hoist to tank, then unbolt and lift fuel tank and support from tractor.

Disconnect remaining engine controls and electrical wiring. Support tractor under clutch housing or hydrostatic drive housing, attach hoist to front support, unbolt front support from side rails, disconnect stay rod ball, if so equipped, and roll complete front assembly forward from tractor.

NOTE: If additional clearance is needed for removal of front support from side rails, loosen the engine front mounting cap screws.

Attach hoist to engine, then unbolt and remove engine and side rails from clutch housing or hydrostatic drive housing.

When reinstalling engine assembly on standard (gear) transmission, unbolt and remove clutch assembly from flywheel. Place pressure plate assembly and drive disc on shaft in clutch housing. Clutch can be bolted to flywheel after engine is installed by working through the opening at bottom of clutch housing.

CYLINDER HEAD
All Non-Diesel Models

70. **REMOVE AND REINSTALL.** To remove the cylinder head, first remove fuel tank assembly as outlined in paragraph 69. Drain cooling system and remove the coolant temperature bulb. Remove air cleaner and bracket assembly. Disconnect spark plug wires, unbolt coil bracket from cylinder head and lay coil and wiring harness out of the way. NOTE: If equipped with Magnetic Pulse ignition system, remove pulse amplifier and lay it aside. Unbolt and remove fan assembly. Disconnect upper radiator hose and by-pass hose. Disconnect controls from carburetor, then unbolt and remove manifold and carburetor assembly. Remove rocker arm cover, rocker arms and shaft assembly and push rods. Remove cylinder head cap screws and lift off cylinder head.

Use guide studs when reinstalling cylinder head, use new head gasket and make certain gasket sealing surfaces are clean. Tighten cylinder head retaining cap screws to a torque of 90 ft.-lbs. using the tightening sequence shown in Fig. 70. Manifold retaining cap screws should be tightened to a torque of 45 ft.-lbs. Adjust valve tappet gap to 0.027 as outlined in paragraph 71.

VALVES AND SEATS
All Non-Diesel Models

71. Inlet and exhaust valves are not interchangeable. Inlet valves seat directly in the cylinder head; whereas, the cylinder head is fitted with renewable seat inserts for the exhaust valves. Inserts are available in standard size as well as oversizes of 0.015 and 0.030. Valve face and seat angle for both the inlet and the exhaust is 30 degrees. Valve rotators (Rotocoils) are used on exhaust valves and umbrella type stem seals are used on inlet valves.

When removing valve seat inserts, use the proper puller. Do not attempt to drive chisel under insert or counterbore will be damaged. Chill new seat insert with dry ice or liquid Freon prior to installing. When new insert is properly bottomed, it should be 0.008-0.030 below edge of counterbore. After installation, peen the cylinder head material around the complete outer circumference of the valve seat insert. The O.D. of new standard insert is 1.5655 and the insert counterbore I.D. is 1.5625.

Check the valves and seats against the specifications which follow:

Inlet

Face and seat angle	30°
Stem diameter	0.3715-0.3725
Stem to guide diametral clearance	0.003-0.005
Seat width	0.048-0.074
Valve run-out (max.)	0.002
Valve tappet gap (warm)	0.027
Valve head margin	5/64

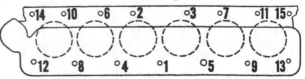

Fig. 70—Tightening sequence for cylinder head cap screws.

Exhaust

Face and seat angle .... 30°
Stem diameter 0.371-0.372
Stem to guide diametral
    clearance 0.0035-0.0055
Seat width 0.083-0.103
Valve run-out (max.) 0.002
Valve tappet gap (warm) 0.027
Valve head margin 5/64

To adjust valve tappet gap, crank engine to position number one piston at top dead center of compression stroke. Adjust the six valves indicated on the chart in Fig. 71. Turn engine one revolution to position number six piston at TDC (compression) and adjust the remaining six valves indicated on chart.

VALVE GUIDES AND SPRINGS

All Non-Diesel Models

72. The inlet and exhaust valve guides are interchangeable. Inlet guides are pressed into cylinder head until top of guide is 1 3/16 inches above the spring recess of the cylinder head. The exhaust valve guides are pressed into cylinder head until top of guide is ¾-inch above the spring recess of head. Guides are pre-sized and if carefully installed, should require no final sizing. Inside diameter of valve guides is 0.3755-0.3765 and valve stem to guide diametral clearance is 0.003-0.005 for inlet valves and 0.0035-0.0055 for exhaust valves.

Inlet and exhaust valve springs are interchangeable. Springs should have a free length of 2 7/16 inches and should test 146-156 pounds when compressed to a length of 1 19/32 inches. Renew any spring which is rusted, discolored or does not meet the pressure test specifications.

VALVE TAPPETS
(CAM FOLLOWERS)

All Non-Diesel Models

73. Tappets are of the barrel type and operate in unbushed bores in the crankcase. Tappet diameter should be 0.9965-0.9970 and should operate with a clearance of 0.002-0.004 in the 0.9990-1.0005 crankcase bores. Tappets are available in standard size only and can be removed from side of crank-

case after removing rocker arm cover, rocker arms and shaft assembly, push rods and engine side cover.

VALVE ROCKER ARM COVER

All Non-Diesel Models

74. Removal of the rocker arm cover is obvious, after removal of front hood. When reinstalling cover, use new gasket to insure an oil tight seal.

VALVE TAPPET LEVERS
(ROCKER ARMS)

All Non-Diesel Models

75. **REMOVE AND REINSTALL.** Removal of the rocker arms assembly may be accomplished in some cases without removing the fuel tank. The rocker shaft bracket hold down screws are also the left row of cylinder head cap screws and removal of these screws may allow the left side of cylinder head to raise slightly and damage to the head gasket could result. Because of possible damage to the head gasket, it is recommended that the cylinder head be removed as in paragraph 70 when removing the rocker arm assembly.

76. **OVERHAUL.** The two-piece rocker arm shaft has an outside diameter of 0.748-0.749. Rocker arm bushings have an inside diameter of 0.7505-0.7520. Bushings are not renewable. All rocker arms are interchangeable. A shaft indexing roll pin (RP—Fig. 72) and a bracket plug are installed in the front and rear rocker shaft brackets. Numbers 2, 4 and 6 rocker shaft brackets are equipped with dowel sleeves for alignment.

VALVE ROTATORS

All Non-Diesel Models

77. Positive type valve rotators (Rotocoils) are installed on the exhaust valves of all non-diesel engines. Normal servicing of the valve rotators consists of renewing the units. It is important to observe the valve action after engine is started. Valve rotator action can be considered satisfactory if valve rotates a slight amount each time the valve opens. See Fig. 73.

VALVE TIMING

All Non-Diesel Models

78. To check valve timing, remove rocker arm cover and crank engine to position number one piston at TDC of compression stroke. Adjust number one intake valve tappet gap to 0.034. Place a 0.004 feeler gage between valve lever and valve stem of number one intake. Slowly rotate crankshaft in normal direction until valve lever becomes tight on feeler gage. At this point, number one intake valve will start to open and timing pointer should be within the range of 5 to 11 degrees before top dead center.

NOTE: One tooth "out of time" equals approximately 13 degrees.

Readjust number one intake valve tappet gap as outlined in paragraph 71.

TIMING GEAR COVER

All Non-Diesel Models

79. To remove the timing gear cover, drain cooling system, remove hood and disconnect battery cables. On Farmall models, identify and disconnect power steering cylinder lines. On International models, disconnect forward end of power steering cylinder from center steering arm. Then identify and disconnect the hydraulic oil cooler lines. Disconnect air cleaner inlet hose and radiator hoses. Support tractor under clutch housing. On some Farmall

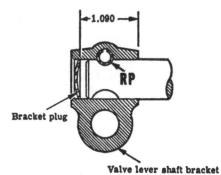

Fig. 72–Roll pin (RP) is installed with the slot away from the rocker shaft and should engage the smaller notch in the shaft. Note installation dimension of valve lever shaft bracket plug.

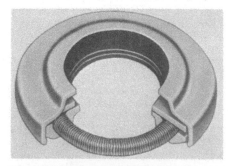

Fig. 73–Cut-away view showing construction of a "Rotocoil" valve rotator.

ADJUST VALVES (ENGINE WARM)											
With No. 1 Piston at T.D.C. (Compression)	1	2	3	5		7		9			
With No. 6 Piston at T.D.C. (Compression)					4		6		8	10	11	12

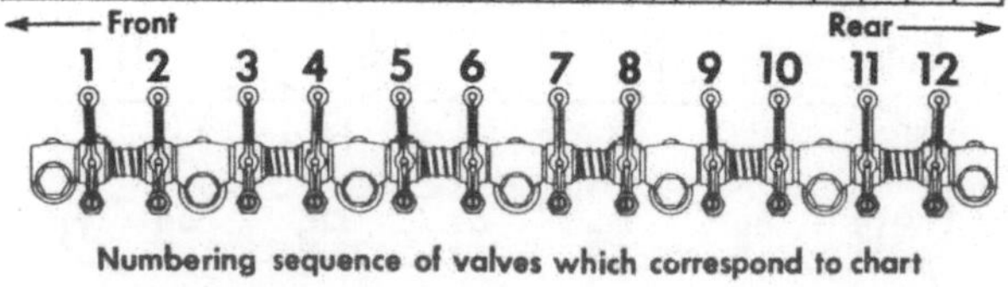

← Front   Rear →

1 2 3 4 5 6 7 8 9 10 11 12

Numbering sequence of valves which correspond to chart

Fig. 71–Charts show the valve tappet gap adjusting procedure.

models equipped with wide front axle and all International models, disconnect stay rod ball. On all models, attach hoist to front support, unbolt front support from side rails and roll complete front assembly forward from tractor.

Unbolt and remove fan, alternator and drive belts. Remove crankshaft nut, attach a suitable puller and remove crankshaft pulley. Remove cap screws retaining oil pan to timing gear cover and loosen the remaining oil pan cap screws. Unbolt timing gear cover, then pull cover forward off the dowels and remove from engine. Use care not to damage oil pan gasket.

Reassemble by reversing the disassembly procedure. Tighten crankshaft pulley retaining nut to a torque of 95 ft.-lbs.

TIMING GEARS

All Non-Diesel Models

80. **CRANKSHAFT GEAR.** Crankshaft gear is keyed and press fitted on the crankshaft. The gear can be removed using a suitable puller after first removing the timing gear cover as outlined in paragraph 79.

Before installing, heat gear in oil, then drift heated gear on crankshaft. Make certain timing marks are aligned as shown in Fig. 74.

81. **CAMSHAFT GEAR.** Camshaft gear is keyed and press fitted on the camshaft. Backlash between camshaft gear and crankshaft gear should be 0.0032-0.0076. Camshaft gear can be removed using a suitable puller after first removing the timing gear cover as outlined in paragraph 79 and the gear retaining nut.

Before installing, heat gear in oil until gear will slide on shaft. Install lock and nut, then tighten nut to a torque of 110-120 ft.-lbs. Make certain timing marks are aligned as shown in Fig. 74.

82. **IDLER GEAR.** To remove the idler gear, first remove the timing gear cover as outlined in paragraph 79. Idler gear shaft is attached to front of engine by a cap screw.

Idler gear shaft diameter should be 2.0610-2.0615 and clearance between shaft and renewable bushing in gear should be 0.0015-0.0045. End clearance of gear on shaft should be 0.009-0.013. Make certain that oil passage in shaft is open and clean.

When reinstalling, make certain that dowel on shaft engages hole in engine front plate. The shaft retaining cap screw should be torqued to 85-90 ft.-lbs.

CAMSHAFT AND BEARINGS

All Non-Diesel Models

83. **CAMSHAFT.** To remove the camshaft, first remove timing gear cover as outlined in paragraph 79. Remove rocker arms assembly and push rods. Drain and remove oil pan and oil pump. Remove engine side cover and lift out cam followers (tappets). Working through openings in camshaft gear, remove the camshaft thrust plate retaining cap screws. Carefully withdraw camshaft from engine.

Recommended camshaft end play of 0.002-0.010 is controlled by the thrust plate.

Check camshaft journal diameter against the values which follow:

No. 1 (front) 2.109-2.110
No. 2 ....2.089-2.090
No. 3 ....2.069-2.070
No. 4 .... 1.4995-1.5005

When installing the camshaft, reverse the removal procedure and make certain timing marks are aligned as shown in Fig. 74. Tighten camshaft thrust plate cap screws to a torque of 38 ft.-lbs. Refer to paragraph 94 for oil pump installation.

84. **CAMSHAFT BEARINGS.** To remove the camshaft bearings, first remove the engine as in paragraph 69 and camshaft as in paragraph 83. Unbolt and remove clutch or flex drive plate, flywheel and the engine rear end plate. Remove expansion plug from behind camshaft rear bearing and remove the bearings.

Using a closely piloted arbor, install the bearings so that oil holes in bearings are in register with oil holes in crankcase. The chamfered end of the bearings should be installed towards the rear.

Camshaft bearings are pre-sized and if carefully installed should need no final sizing. Camshaft bearing journals should have a diametral clearance in the bearings of 0.0005-0.005.

Fig. 74–Gear train and timing marks on non-diesel engine.

CA. Camshaft gear   G. Governor gear
CR. Crankshaft gear   I. Idler gear

When installing plug at rear camshaft bearing, use sealing compound on plug and bore.

ROD AND PISTON UNITS

All Non-Diesel Models

85. Connecting rod and piston assemblies can be removed from above after removing the cylinder head as outlined in paragraph 70 and the oil pan.

Cylinder numbers are stamped on the connecting rod and cap. Numbers on rod and cap should be in register and face towards the camshaft side of engine. The arrow or front marking on top of piston should be toward front of tractor.

Two types of connecting rod bolts may be used. The PLACE bolt has a head that is either notched or concave and the shank and thread diameter are nearly the same. This type attains its tension by bending the bolt head and should be torqued to 50 ft.-lbs. The PITCH bolt has a standard bolt head with a washer face. The thread diameter is larger than the shank. This type attains its tension by stretching of the shank and should be torqued to 45 ft.-lbs.

PISTONS AND RINGS

Model 826 Non-Diesel

86. The cam ground pistons operate directly in block bores and are available in standard size as well as oversizes of 0.010, 0.020, 0.030 and 0.040.

Tractors are equipped with pistons having two compression rings and one oil control ring.

Check pistons and rings against the values which follow:
Ring End Gap
    Compression 0.010-0.020
    Oil Control 0.010-0.018
Ring Side Clearance
    Top compression 0.0025-0.0040
    Second compression . 0.0020-0.0035
    Oil control 0.0025-0.0040
    Standard cylinder bore is 3.8125-3.8150. Pistons should have a diametral clearance in cylinder bores of 0.001-0.0045 when measured at bottom of skirt and 90° to piston pin.

PISTONS, SLEEVES AND RINGS

Model 766 Non-Diesel

87. The cam ground aluminum pistons are fitted with two compression rings and one oil control ring and are available in standard size only. New pistons have a diametral clearance of 0.001-0.0045 when measured between piston skirt and installed in sleeve at 90° to piston pin.

The dry type cylinder sleeves should be renewed when out-of-round or taper exceeds 0.008. Inside diameter of new sleeve is 3.7500-3.7525. With piston and connecting rod assemblies removed from block, use a hydraulic sleeve puller to remove cylinder sleeves.

Crankcase bore classifications (1, 2 or 3) are stamped in consecutive order on or near the top of the oil filter mounting pad. Sleeves are available in two outside diameter classes; Class 1 & 2 or Class 2 & 3. A Class 1 & 2 sleeve fits bore classifications 1 or 2 and Class 2 & 3 sleeve fits bore classifications 2 or 3.

Sleeve outside diameters and cylinder block bores are as follows:

Sleeve Type	Sleeve O.D.
Class 1 & 2	3.8765-3.8770
Class 2 & 3	3.8770-3.8775
Cylinder Bore	**Bore I.D.**
Class 1	3.8750-3.8755
Class 2	3.8755-3.8760
Class 3	3.8760-3.8765

Clean the sleeve and cylinder block bore with solvent and dry with compressed air. Lubricate sleeve and bore with clean diesel fuel. Using a hydraulic sleeve installing tool, press sleeve into cylinder block until sleeve flange is 0.00-0.005 above top surface of cylinder block. The force required to press this sleeve into position will vary from a minimum of 600 pounds to a maximum of 3000 pounds. To translate pounds of force into gage pressure for use with hydraulic ram, the effective area of the ram cylinders must be calculated by the formula $d^2 \times 0.784$. With the ram cylinder effective area calculated, the result can be divided into the stated pounds of force and gage pressure psi range determined. Using an OTC twin piston ram with each piston diameter of 1½ inches the following example is given: $(d^2 + d^2) \times 0.784$ equals $(2.25 + 2.25) \times 0.784$ equals 3.52 sq. in. effective area of both rams. When 3.52 is divided into 600 (lbs. of force), a gage pressure of 171 psi is obtained. When 3.52 is divided into 3000 (lbs. of force), a gage pressure of 852 psi is obtained. Therefore, sleeve installation must occur between 171 and 852 psi gage pressure.

The top compression ring is a chrome barrel face ring and is installed with counterbore up.

The second compression ring is a taper face ring and is installed with largest outside diameter toward bottom of piston. Upper side of ring is marked TOP.

The chrome slotted oil control ring can be installed either side up. A coil spring expander is used with this ring.

Additional piston ring information is as follows:

Ring End Gap
  Compression rings 0.010-0.020
  Oil control ring 0.010-0.018
Ring Side Clearance
  Top compression 0.0025-0.0040
  Second compression . 0.0025-0.0035
  Oil control 0.0025-0.0040

PISTON PINS

All Non-Diesel Models

88. The full floating type piston pins are retained in the piston bosses by snap rings. Specifications are as follows:

Piston pin diameter ... 0.8748-0.8749
Diametral clearance
  in piston 0.0002-0.0004
Diametral clearance
  in rod bushing 0.0002-0.0005
  Piston pins are available in 0.005 oversize.

CONNECTING RODS AND BEARINGS

All Non-Diesel Models

89. Connecting rod bearings are of the slip-in, precision type, renewable from below after removing oil pan and rod caps. When installing new bearing inserts, make certain the projections on same engage slots in connecting rod and cap and that cylinder identifying numbers on rod and cap are in register and face toward camshaft side of engine. Connecting rod bearings are available in standard size and undersizes of 0.002, 0.010, 0.020 and 0.030. Check the crankshaft crankpins and connecting rod bearings against the values which follow:

Crankpin diameter 2.373-2.374
Max. allowable out-of-round .. 0.0015
Max. allowable taper 0.0015
Rod bearing diametral
  clearance 0.0009-0.0034
Rod side clearance 0.007-0.013
Rod bolt torque
  PLACE bolt* .... 50 ft.-lbs.
  PITCH bolt* .... 45 ft.-lbs.
  *Refer to paragraph 85 for bolt identification.

CRANKSHAFT AND MAIN BEARINGS

All Non-Diesel Models

90. Crankshaft is supported in four main bearings and thrust is taken by the third (rear intermediate) bearing. Main bearings are of the shimless, non-adjustable, slip-in precision type, renewable from below after removing the oil pan and main bearing caps. Removal of crankshaft requires R&R of engine. Check crankshaft and main

bearings against the values which follow:

Crankpin diameter 2.373-2.374
Main journal diameter 2.748-2.749
Max. allowable out-of-round .. 0.0015
Max. allowable taper 0.0015
Crankshaft end play 0.005-0.013
Main bearing diametral
  clearance 0.0012-0.0042
Main bearing bolt torque ... 80 ft.-lbs.

Main bearings are available in standard size and undersizes of 0.002, 0.010, 0.020 and 0.030. Alignment dowels (IH tool FES 6-1 or equivalent) should be used when installing the rear main bearing cap.

CRANKSHAFT SEALS

All Non-Diesel Models

91. **FRONT.** To renew the crankshaft front oil seal, first remove hood, drain cooling system and disconnect radiator hoses. On Farmall models, identify and disconnect power steering lines. On International models, disconnect forward end of power steering cylinder from center steering arm. Then, on all models, identify and disconnect the hydraulic oil cooler lines. Support tractor under clutch housing or hydrostatic drive housing. On Farmall models equipped with wide front axle and all International models, disconnect stay rod ball. On all models, attach hoist to front support, unbolt front support from side rails and roll complete front assembly from tractor. Remove fan and alternator drive belts. Remove crankshaft pulley retaining nut, attach a suitable puller and remove crankshaft pulley. Remove and renew oil seal in conventional manner. Drive new seal in until it is seated against shoulder in timing gear cover. Install crankshaft pulley and tighten retaining nut to a torque of 95 ft.-lbs. Reassemble tractor by reversing disassembly procedure.

92. **REAR.** To renew the crankshaft rear oil seal, the engine must be detached from clutch housing or hydrostatic drive housing as outlined in paragraphs 244, 245 or 267. Then, unbolt and remove clutch or flex drive plate and flywheel. The lip type seal can be removed after collapsing same. Take care not to damage sealing surface of crankshaft when removing seal. Use seal installing tool and oil seal driver (IH tools FES6-2 and FES6-3) or equivalent and drive seal in until it is flush with rear of crankcase. Lip of seal must be toward front of engine.

FLYWHEEL

All Non-Diesel Models

93. To remove the flywheel, first split

tractor as outlined in paragraphs 244, 245 or 267, then unbolt and remove flex drive plate or clutch assembly. Remove six cap screws and lift flywheel from crankshaft. When installing flywheel, coat cap screws with sealer and tighten cap screws to a torque of 75 ft.-lbs.

To install a new flywheel ring gear, heat same to approximately 500 degrees F.

OIL PUMP

All Non-Diesel Models

94. The gear type oil pump is gear driven from a pinion on camshaft and is accessible for removal after removing the engine oil pan. Disassembly and overhaul of pump is obvious after an examination of the unit

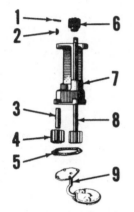

and reference to Fig. 75. Gaskets (5) between pump cover and body can be varied to obtain the recommended 0.0025-0.0055 pumping (body) gear end play.

Refer to the following specifications:
Pumping gears recommended
  backlash 0.003-0.006
Pump drive gear recommended
  backlash 0.000-0.008
Pumping gear end play   0.0025-0.0055
Gear teeth to body radial
  clearance 0.0068-0.0108
Drive shaft clearance .. 0.0015-0.0030
Mounting bolt torque ...... 22 ft.-lbs.

Service (replacement) pump shaft and gear assemblies are not drilled to accept the pump driving gear pin. A ⅛-inch hole must be drilled through the shaft after the gear is installed on the shaft to the dimension shown in Fig. 76.

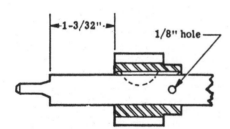

Fig. 75–Exploded view of oil pump used on all non-diesel engines.

1. Pin
2. Woodruff key
3. Idler gear shaft
4. Idler gear
5. Gasket
6. Drive gear
7. Pump body
8. Drive shaft and gear
9. Cover and screen

Fig. 76–New pump shaft and gear assemblies will require a ⅛-inch hole to be drilled at location shown.

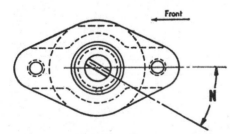

Fig. 77–On non-diesel engine, position No. 1 piston at TDC on compression stroke and mesh oil pump drive gear so angle (N) of drive shaft tang is approximately 30 degrees to centerline of engine.

NOTE: When installing the oil pump on engine, time the pump as follows: Crank engine until number one piston is coming up on compression stroke. Continue cranking until the TDC mark on crankshaft pulley or flywheel is in register with the timing pointer. Install oil pump so that tang on oil pump shaft is in the approximate position shown in Fig. 77. Retime distributor as outlined in paragraph 227 or 229.

OIL PRESSURE RELIEF VALVE

All Non-Diesel Models

95. Non-diesel tractors are equipped with the die cast base oil filter assembly shown in Fig. 78 or 79. The pressure regulator (relief) valve, bypass valve and check valve are located in the oil filter base (12—Fig. 77). Tractors equipped with spin-on-filter the pressure regulator (relief) valve is also located in the oil filter base (2—Fig. 79). All valves and their springs are retained in position by snap rings and removal is obvious after an examination of the unit and reference to Fig. 78 or 79. Specifications for filter in Fig. 78 are as follows:
Check valve,
  Valve diameter 0.770
  Spring free length .... 1 5/64 in.
By-pass valve,
  Valve diameter 0.770
  Spring free length .... 2⅞ in.
  Spring test and
    length .... 4.5 lbs. @ 1 5/64 in.
Oil pressure at 1800 RPM ... 30-40 psi

OIL PAN

All Non-Diesel Models

96. Removal of oil pan is conventional and on tricycle model tractors, can be accomplished with no other disassembly.

On Farmall models with adjustable

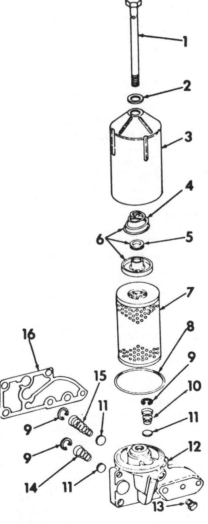

Fig. 78–Exploded view of die cast base oil filter assembly.

1. Center tube
2. Gasket
3. Case
4. Hold-down spring
5. Grommet
6. Retainer
7. Element
8. Gasket
9. Snap ring
10. Check valve spring
11. Valve
12. Filter base
13. Drain plug
14. Relief valve spring
15. By-pass spring
16. Gasket

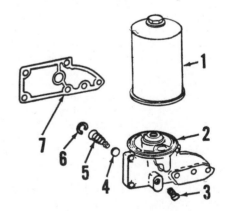

Fig. 79–Exploded view of die cast base spin-on oil filter assembly used on models 766 and some 826.

1. Element (spin-on)
2. Filter base
3. Plug
4. Regulating valve
5. Regulating spring
6. Snap ring
7. Gasket

wide front axle, the stay rod bracket must be removed from side rails before oil pan can be removed.

On International and "All Wheel Drive" models, disconnect stay rod (or bracket) from clutch housing and front axle from front support and raise front of tractor to provide clearance for oil pan removal.

ENGINE AND COMPONENTS
(DIESEL)

Model 766 diesel tractor is equipped with engine having a bore and stroke of 3.875x5.085 inches and a displacement of 360 cubic inches.

Model 826 diesel tractor is equipped with engine having a bore and stroke of 3.875x5.060 inches and a displacement of 358 cubic inches.

Models 966 and 1066 diesel tractors are equipped with engine having a bore and stroke of 4.300x4.750 and a displacement of 414 cubic inches.

Model 1026 diesel tractor is equipped with engine having a bore and stroke of 4.321x4.625 and a displacement of 407 cubic inches.

All engines are a six cylinder design. All engines except 407 cubic inch are equipped with wet type sleeves and the 407 cubic inch is fitted with dry type sleeves. Dry type air filters are used in the air induction systems of all engines. Models 1026 and 1066 diesel engines are equipped with turbochargers.

R&R ENGINE ASSEMBLY
All Diesel Models

97. Removal of engine is best accomplished by removing front support, radiator, front axle and wheel assembly and fuel tank assembly; then unbolting and removing engine and side rails from clutch housing or hydrostatic drive housing as follows:

Drain cooling system, disconnect battery cables and remove hood. Disconnect radiator hoses and air cleaner inlet hose. Identify and disconnect power steering lines and hydraulic oil cooler lines. Remove operator assist handles and the torque amplifier lever if so equipped; notice that TA lever and shaft have match marks affixed to insure correct assembly. If equipped with hydrostatic drive, remove RANGE lever. Remove center hood and steering support cover. Disconnect all interfering lines and wiring from fuel tank and support. Attach hoist to tank, then unbolt and lift fuel tank and support from tractor.

Disconnect remaining engine controls and electrical wiring. Support tractor under clutch housing or hydrostatic drive housing, attach hoist to front support, unbolt front support from side rails, disconnect stay rod

ball, if so equipped and roll complete front assembly forward from tractor.

NOTE: If additional clearance is needed for removal of front support from side rails, loosen the engine front mounting cap screws.

Attach hoist to engine, then unbolt and remove engine and side rails from clutch housing.

When reinstalling engine assembly on standard (gear) transmission, unbolt and remove clutch assembly from flywheel. Place pressure plate assembly and drive disc on shaft in clutch housing. Clutch can be bolted to flywheel after engine is installed by working through the opening at bottom of clutch housing.

CYLINDER HEAD
All Diesel Models

98. To remove the cylinder head, first remove fuel tank assembly as outlined in paragraph 97. Drain cooling system and remove coolant temperature bulb. Unbolt and remove air cleaner assembly and on models 1026 and 1066 remove turbocharger as outlined in paragraphs 201 and 202.

On **model 826** disconnect injection lines from fuel injectors and injection pump and remove excess fuel return lines from injectors. Cap all fuel connections immediately. CAUTION: Injector nozzle assemblies protrude slightly through combustion side of cylinder head. It is recommended that injector nozzles be removed before removing cylinder head. Disconnect engine oil cooler lines, unbolt and remove exhaust manifold and water collecting tube (manifold) from right side and inlet manifold from left side of cylinder head. Remove rocker arm cover, rocker arms and shaft assembly and push rods. Remove cylinder head retaining nuts and lift off cylinder head.

On **model 1026** remove breather outlet tube from left front of cylinder head. Disconnect external oil line from left side of cylinder head. Disconnect elbows of the oil filter tubes from engine oil cooler and pressure regulator block, then unbolt and remove filters from cylinder head. Loosen by-pass hose clamps and remove thermostat housing. Unbolt and remove exhaust manifold. Allow fuel filters to rest on frame channel. Disconnect injection lines from fuel injectors and injection pump. Remove excess fuel return lines from injectors. Cap all fuel connections immediately. CAUTION: Injector nozzle assemblies protrude slightly (0.087-0.120) through the combustion side of cylinder head. It is recommended that injector nozzles be removed before removing cylinder head. Remove rocker arm cover, rocker arms and shaft assembly and push rods. Remove cylinder head retaining cap screws and remove cylinder head.

On **models 766, 966 and 1066** disconnect and remove injection lines from fuel injectors and injection pump and remove excess fuel return line from injectors. Cap all fuel connections immediately. Remove injector nozzles and place nozzle assemblies where they will not be damaged. Unbolt and remove exhaust and intake manifolds. Remove rocker arm cover, rocker arms and shaft assembly and push rods. Remove cylinder head retaining bolts and lift off cylinder head using the lifting eyes.

On all models, check cylinder head for warpage as follows: Place a straight edge across the machined (combustion) side of cylinder head and measure between cylinder and straight edge. If a 0.003 feeler gage can be inserted between straight edge and cylinder head within any six-inch distance, the cyl-

Fig. 80—Series 826 diesel engine, cylinder head retaining nuts tightening sequence.

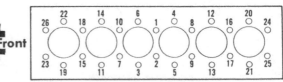

1st Torque	50 FT.-LBS.
2nd Torque	70 FT.-LBS.
3rd Torque	90 FT.-LBS.

inder head should be refaced with a surface grinder. PROVIDING, not more than 0.010 inch of material is removed.

The standard distance from rocker arm cover surface to combustion surface of cylinder head is as follows:

D358 .... 3.890-3.910
DT407 .... 5.098-5.102
D360, D414 and DT414 ... 4.198-4.202

NOTE: After cylinder head is resurfaced, check the valve head recession or protrusion specifications outlined in paragraphs 100, 101 and 102. Correct as necessary.

When reinstalling cylinder head on all models, use new head gasket and make certain that gasket sealing surfaces are clean and dry. DO NOT use sealants or lubricants on head gasket, cylinder head or block.

CAUTION: Because of the minimum amount of clearance that exists between valves and piston tops, loosen rocker arm adjusting screws before installing the rocker arms and shaft assembly. Refer to paragraphs 100, 101 and 102 for information concerning valve adjustment.

On model 826 tighten cylinder head retaining nuts in three steps using the sequence shown in Fig. 80. Tighten the nuts to a torque of 50 ft.-lbs. during the first step, 70 ft.-lbs. during the second step and 90 ft.-lbs. during the third step.

On model 1026 tighten cylinder head cap screws in three steps using the sequence shown in Fig. 81. Tighten cap screws to a torque of 65 ft.-lbs. during the first step, 110 ft.-lbs. during the second step and 135 ft.-lbs. during the third step.

On models 766, 966 and 1066 tighten cylinder head cap screws in two steps

using the sequence shown in Fig. 82. Tighten cap screws to a torque of 110 ft.-lbs. during the first step and 165 ft.-lbs. during the second step.

COOLING TUBES, WATER DIRECTORS AND NOZZLE SLEEVES
All Diesel Models

99. The cylinder head is fitted with brass injector nozzle sleeves which pass through the coolant passages. In addition, models 826 and 1026 used cooling jet tubes (one for each cylinder) and models 766, 966 and 1066 used water directors (two for each cylinder) to direct a portion of the coolant to the valve seat and nozzle sleeve area. Both the nozzle sleeves and cooling tubes or water directors are available as service items.

To renew the nozzle sleeves, remove injectors and cylinder head as outlined in paragraph 98. On D358 engines, use special bolt (IH tool No. FES 112-4) and turn it into sleeve. Attach a slide hammer puller and remove sleeve. See Fig. 83. On DT407 engines, use adapter (IH tool No. FES 25-9) and expanding screw (IH tool No. FES 25-10) along with a slide hammer and pull nozzle sleeve as shown in Fig. 84. On D360, D414 and DT414 engines, use adapter (IH tool No. FES 25-13) and turn it into sleeve. Attach a slide hammer puller and remove sleeve. See Fig. 85.

NOTE: Use caution during sleeve removal not to damage sealing areas in cylinder head. Under no circumstances should screwdrivers, chisels or other such tools be used in an attempt to remove injector nozzle sleeves.

When installing nozzle sleeves, be sure sealing areas are completely clean and free of scratches. Apply a light coat

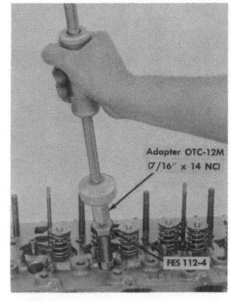

Fig. 83—On D 358 engines use IH tool No. FES 112-4 and slide hammer to remove injector nozzle sleeves.

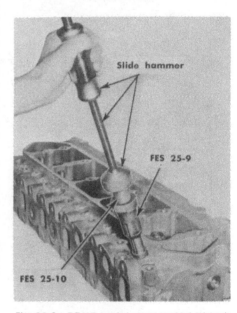

Fig. 84—On DT407 engines use special IH tools shown to remove injector nozzle sleeves.

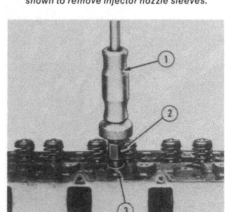

Fig. 85—On D360, D414 and DT414 engines use IH tool No. FES 25-13 and slide hammer to remove injector nozzle sleeves.

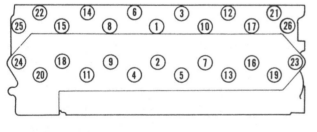
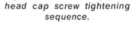

Fig. 81—Series 1026 cylinder head cap screw tightening sequence.

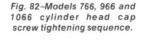

Fig. 82—Models 766, 966 and 1066 cylinder head cap screw tightening sequence.

of "Grade B Loctite" on sealing surfaces of nozzle sleeve. On D358 engine use installing tool (IH tool No. FES 112-3) shown in Fig. 86 and drive injector nozzle sleeves into their bores until they bottom as shown in Fig. 87. On DT407 engine use installing tool

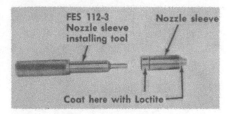

Fig. 86–Injector nozzle sleeve installing tool used on D358 engine. Apply "Grade B Loctite" to sealing surface of nozzle sleeve.

(IH tool No. FES 68-5) shown in Fig. 88 and install nozzle using same procedure. On D360, D414 and DT414 engines use installing tool (IH tool No. FES 148-3) shown in Fig. 89 and install nozzle using same procedure.

NOTE: Injector sleeves have an interference fit in their bores. When installing sleeves, be sure sleeve is driven straight with its bore and is completely bottomed.

To remove the cooling jet tubes, on D358 and DT407, thread inside diameter and install a cap screw to assist in removal. Install new cooling tubes so coolant is directed between the valves of each cylinder. Tubes must be installed flush with cylinder head surface.

To remove the water directors on

Fig. 87–Drive nozzle sleeves into D358 cylinder head until they bottom.

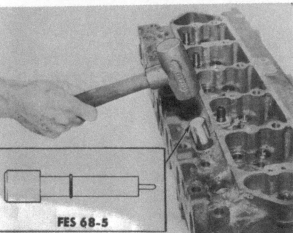

Fig. 88–On DT407 engine, use installing tool shown and drive nozzle sleeves in cylinder head bores until they bottom.

D360, D414 and DT414, use a slide hammer with a small jaw that will hook under water director. Install new water directors by tapping in place with small hammer. The directors must be recessed below the cylinder head surface and the opening aimed as shown in Fig. 90.

VALVES AND SEATS
Model 826 Diesel

100. Inlet and exhaust valves are not interchangeable. The inlet valves seat directly in the cylinder head and the exhaust valves seat on renewable seat inserts. Inserts are available in oversizes of 0.004 and 0.016. Valve rotators (Rotocaps) and valve stem seals are used on all valves.

When removing valve seat inserts, use the proper puller. Do not attempt to drive chisel under insert as counterbore will be damaged. Chill new insert with dry ice or liquid Freon prior to installing. When new insert is properly bottomed it should be 0.008-0.030 below edge of counterbore. After installation, peen the cylinder head material around the complete outer circumference of the valve seat insert.

Check the valves and seats against the following specifications:
Inlet
Face and seat angle .... 45°
Stem diameter 0.3919-0.3923
Stem to guide diametral clearance,
  Normal 0.0014-0.0026
  Maximum allowable 0.006
Seat width 0.076-0.080
Valve run-out (max.) 0.001
Valve tappet gap (warm) 0.010
Valve recession from face of cylinder head,
  Normal 0.039-0.055
  Maximum allowable 0.120

Fig. 89–On D360, D414 and DT414 engines use installing tool IH No. FES 148-3 and drive nozzle sleeves in cylinder head bore until they bottom.

Fig. 90–View showing the way that the water directors should be installed.

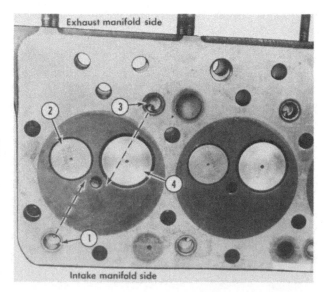

Exhaust

Face and seat angle .... 45°
Stem diameter 0.3911-0.3915
Stem to guide diametral clearance,
   Normal 0.0022-0.0034
   Maximum allowable 0.006
Seat width 0.081-0.089
Valve run-out (max.) 0.001
Valve tappet gap (warm) 0.012
Valve recession from face of cylinder
head,
   Normal 0.047-0.063
   Maximum allowable 0.120

CAUTION: Due to close clearance between valves and pistons, severe damage can result from inserting feeler gage between valve stem and valve lever (rocker arm) with engine running. DO NOT attempt to adjust valve tappet gap with engine running.

To adjust valve tappet gap, crank engine to position number one piston at top dead center of compression stroke. Adjust the six valves indicated on the chart shown in Fig. 91.

NOTE: The valve arrangement is exhaust-intake-exhaust-intake and so on, starting from front of cylinder head.

Turn engine crankshaft one complete revolution to position number six piston at TDC (compression) and adjust the remaining six valves indicated on chart.

Model 1026 Diesel

101. Inlet and exhaust valves are not interchangeable. The inlet valves seat directly in the cylinder head and the exhaust valves seat on renewable seat inserts. However, inlet valve seat inserts are available for service. Both inlet and exhaust valve seat inserts are available in standard size and oversizes of 0.005 and 0.015. Valve rotators (Rotocoils) are used on the exhaust valves.

When removing valve seats, use the proper puller. Do not attempt to drive chisel under insert as counterbore will be damaged. Chill new insert with dry ice or liquid Freon prior to installing. When new insert is properly bottomed, it should be 0.008-0.030 below edge of counterbore. After installation, peen the cylinder head material around the complete outer circumference of the valve seat insert.

VALVE SEAT INSERT CHART (OVERSIZE)

Engine	Oversize Insert	Diameter of Cylinder Head Counterbore	
		Intake	Exhaust
D-360	.002"	None	1.534-1.535
	.015"	None	1.547-1.548
D and DT-414	.002"	1.998-1.999	1.626-1.627
	.015"	2.011-2.012	1.639-1.640

Fig. 92–Chart showing the diameter of cylinder head counterbore for oversize valve seat insert.

Check the valves and seats against the specifications which follow:

Inlet and Exhaust

Face and seat angle .... 45°
Stem diameter 0.4348-0.4355
Stem to guide diametral clearance,
   Normal 0.0015-0.0032
   Maximum allowable 0.008
Seat width,
   Inlet 5/64-inch
   Exhaust 3/32-inch
Valve run-out (max.) 0.001
Valve tappet gap (warm),
   Inlet 0.013
   Exhaust 0.025

With valves installed in cylinder head, measure valve head position in relation to cylinder head surface. The inlet valve head should be flush plus or minus 0.0065 with cylinder head surface. Maximum allowable recession is 0.0225. The exhaust valve head should protrude 0.0815-0.0945 above cylinder head surface. Minimum allowable protrusion is 0.0655.

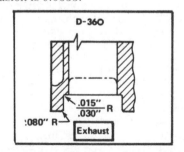

Fig. 93–Chart showing the correct radii for valve seat installation of D360 engine.

CAUTION: Due to the close valve to piston clearance, severe damage can result from inserting feeler gage between valve stem and valve lever (rocker arm) with engine running. DO NOT attempt to adjust valve tappet gap with engine running.

To adjust valve tappet gap, proceed as follows: Remove timing hole cover on right side of flywheel housing. Using a pry bar on flywheel teeth, turn flywheel in normal direction of rotation to position number one piston at TDC of compression stroke. Adjust the six valves indicated on the chart shown in Fig. 91.

NOTE: The valve arrangement is exhaust-intake-exhaust-intake and so on, starting from front of cylinder head.

Turn flywheel one complete revolution in normal direction of rotation to position number six piston at TDC (compression) and adjust the remaining six valves indicated on chart.

Models 766, 966 and 1066 Diesel

102. Inlet and exhaust valves are not interchangeable. On model 766, the inlet valves seat directly in the cylinder head and the exhaust valves seat on renewable seat inserts. Models 966 and 1066, inlet and exhaust valves seat on renewable seat inserts. Inserts are available in oversize of 0.002 and 0.015. Refer to Figs. 92, 93 and 94 for dimension to install oversize inserts. On all models valve rotators (Rotocoils) are used on all valves.

When removing valve seat inserts,

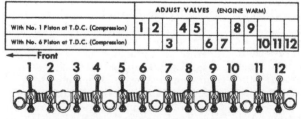

Fig. 91–Chart shows valve tappet gap adjusting procedure used on D358 and DT407 engines. Refer to text.

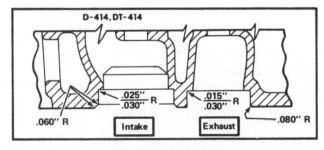

Fig. 94–Chart showing the correct radii for valve seats installation of D414 and DT414 engines.

use the proper puller. Do not attempt to drive chisel under insert as counterbore will be damaged. Chill new insert with dry ice or liquid Freon prior to installing. When new insert is properly bottomed, it should be 0.087-0.093 for exhaust valves and 0.120-0.127 for inlet valves below edge of counterbore.

Check the valves and seats against the following specifications:

Inlet

Face and seat angle	30°
Stem diameter	0.3718-0.3725

Stem to guide diametral clearance,

Normal	0.0015-0.0032
Maximum allowable	0.006
Seat width	0.075-0.085
Valve run-out (max)	0.0015

Valve tappet gap (warm)

D-360	0.012
D-414 & DT-414	0.020

Valve recession from face of cylinder head,

Normal	0.000-0.014

Exhaust

Face and seat angle	45°
Stem diameter	0.3718-0.3725

Stem to guide diametral clearance,

Normal	0.0015-0.0032
Maximum allowable	0.006
Seat width	0.075-0.085
Valve run-out (max)	0.0015

Valve tappet gap (warm)

D-360	0.021
D-414 & DT-414	0.025

Valve recession from face of cylinder head,

Normal	0.000-0.014

CAUTION: Due to close clearance between valves and pistons, severe damage can result from inserting feeler gage between valve stem and valve lever (rocker arm) with engine running. DO NOT attempt to adjust valve tappet gap with engine running.

To adjust valve tappet gap, crank engine to position number one piston at top dead center of compression stroke. Adjust the six valves indicated on the chart shown in Fig. 95.

NOTE: The valve arrangement is intake-exhaust-intake-exhaust and so on starting from front of cylinder head.

Turn engine crankshaft one complete revolution to position number six piston at TDC (compression) and adjust the remaining six valves indicated on chart.

VALVE GUIDES AND SPRINGS
Model 826 Diesel

103. The inlet and exhaust valve guides should be pressed into cylinder head until top of guides is 1 5/32 inches above spring recess in head. After installation, guides must be reamed to an inside diameter of 0.3940-0.3945. Valve stem to guide diametral clear-

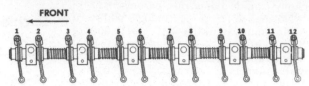

Fig. 95–Chart shows valve tappet gap adjusting procedure used on D360, D414 and DT414 engines. Refer to text.

ance should be 0.0014-0.0026 for inlet valves and 0.0022-0.0034 for exhaust valves. Maximum allowable stem clearance in all guides is 0.006.

Inlet and exhaust valve springs are also interchangeable. Springs should have a free length of 2.173 inches and should test 146-160 pounds when compressed to a length of 1 11/32 inches. Renew any spring which is rusted, discolored or does not meet the pressure test specifications.

Model 1026 Diesel

104. The inlet and exhaust valve guides are interchangeable in the DT407 engine. Inlet and exhaust valve guides should be pressed into cylinder head until top of guides is 1 1/16 inches above spring recess in head. The inside diameter of valve guides is knurled for oil control. Guides are pre-sized; however, since they are a press fit in cylinder head, it is necessary to ream them to remove any burrs or slight distortion caused by the pressing operation. Inside diameter should be 0.437-0.438. Valve stem to guide diametral clearance should be 0.0015-0.0032 with maximum allowable clearance of 0.008.

Each valve is equipped with two (inner and outer) valve springs and springs are not interchangeable between inlet and exhaust valves. Valve spring specifications are as follows:

Inlet valve springs

Free length,

Inner	2 11/32 in.
Outer	2 9/16 in.

Test load and length,

Inner	85 lbs. @ 1 1/2 in.
Outer	136 lbs. @ 1 45/64 in.

Exhaust valve springs

Free length,

Inner	1 63/64 in.
Outer	2 21/64 in.

Test load and length,

Inner	81 lbs. @ 1 7/32 in.
Outer	135 lbs. @ 1 7/16 in.

Valve rotors (Rotocoil) are used under exhaust valve springs and spring spacers are used under inlet valve springs. Install springs with dampener coils towards cylinder head. Renew any spring which is rusted, discolored or does not meet test specifications.

Models 766, 966 and 1066 Diesel

105. The inlet and exhaust valve guides are interchangeable. Valve guides should be pressed into cylinder head until the top of guides is as follows above the spring recess in head:

D-360

Intake	1.002
Exhaust	1.282

D-414—DT-414

Intake	1.217
Exhaust	1.297

Guides are pre-sized; however, since they are a press fit in cylinder head, it is necessary to ream them to remove any burrs or slight distortion caused by the pressing operation. Inside diameter should be 0.3740-0.3750. Valve stem to guide diametral clearance should be 0.0015-0.0032 with maximum allowable clearance of 0.006.

Inlet and exhaust valve springs are also interchangeable. Springs should have a free length of 2.340 inches and should test 156-164 pounds when compressed to a length of 1.552 inches. Renew any spring which is rusted, discolored or does not meet the pressure test specifications.

VALVE TAPPETS
(CAM FOLLOWERS)
Model 826 Diesel

106. The 0.7862-0.7868 diameter mushroom type tappets operate directly in the unbushed crankcase bores. Clearance of tappets in the bores should be 0.0005-0.0024. Tappets can be removed after removing the oil pan and the camshaft as outlined in paragraph 136. Oversize tappets are not available.

Model 1026 Diesel

107. Valve tappets used in DT407 engines are the mushroom type and operate in unbushed bores in crankcase. Tappet diameter is 0.623-0.624 and should operate with a clearance of 0.001-0.004 in the 0.625-0.627 crankcase bores. Tappets are available in standard size only and removal requires the removal of camshaft as outlined in paragraph 138 and engine oil pan.

Models 766, 966 and 1066 Diesel

108. Valve tappets used in D-360, D-414 and DT-414 engines are the barrel type and operate in unbushed bores in crankcase. Tappet diameter is 0.9965-0.9970 and length is 2.4370 and should operate with a clearance of 0.0025-0.0040 in the crankcase bore. Tappets are available in standard size only and removal requires the removal of cylinder head as outlined in paragraph 98.

VALVE ROCKER ARM COVER
All Diesel Models

109. Removal of the rocker arm cover is obvious after removal of front hood. Disconnect air inlet from inlet manifold and on models so equipped, remove fuel tank left front support.

When reinstalling cover, use new gasket to insure an oil tight seal.

VALVE TAPPET LEVERS (ROCKER ARMS)
Model 826 Diesel

110. **REMOVE AND REINSTALL.** Removal of the rocker arms and shaft assembly is conventional after removal of rocker arm cover. When reinstalling, tighten the hold-down nuts on bracket studs to a torque of 47 ft.-lbs.

111. **OVERHAUL.** To remove the rocker arms from the one-piece shaft, remove bracket clamp bolts and slide all parts from shaft. Outside diameter of rocker shaft is 0.8491-0.8501. The renewable bushings in rocker arms should have an operating clearance of 0.0009-0.0025 on rocker shaft with a maximum allowable clearance of 0.008. Rocker arms may be refaced providing the original contour is maintained. Rocker arm adjusting screws are self-locking. If they turn with less than 12 ft.-lbs. torque, renew adjusting screw and/or rocker arm.

Reassemble rocker arms on rocker shaft keeping the following points in mind. Thrust washers are used between each spring and rocker arms and spacer rings are used between rocker

arms and brackets except between rear rocker arm and rear bracket. To insure that lubrication holes in rocker shaft are in correct position, align punch mark on front end of shaft with slot in front mounting bracket as shown in Fig. 96. End of shaft must also be flush with bracket. Rocker shaft clamp screws on brackets should be tightened to a torque of 10 ft.-lbs.

Model 1026 Diesel

112. **REMOVE AND REINSTALL.** Removal of rocker arms and shaft assembly is accomplished as follows: Remove front hood, center hood and steering support cover. Remove air inlet elbow and hose from cylinder head inlet manifold. Disconnect fuel tank supports from engine and completely remove left front tank support. Remove rear tank support bolts except the top one, then unclip and/or disconnect all interfering wires, cables and piping. Pivot front of fuel tank upward and block in position. Rocker arm cover and the rocker arms and shaft assembly can now be removed.

113. **OVERHAUL.** The rocker shaft is a one-piece unit with an outside diameter of 0.872-0.873. Inside diameter of rocker arm bushings is 0.8745-0.8760 which provides an operating clearance of 0.0015-0.004. Maximum allowable clearance is 0.007. All rocker arms are interchangeable and bushings are renewable in rocker arms. Rocker arms may be refaced providing the original contour is maintained. Rocker shaft ends are fitted with plugs and numbers 1 and 5 shaft brackets have locating pins. Rocker arms, springs and brackets can be removed from shaft after removing retainers from ends of shaft.

Models 766 and 966 Diesel

114. **REMOVE AND REINSTALL.** Removal of rocker arms and shaft assembly is accomplished as follows: Remove all hoodsheets, heat baffle and heat shield assemblies. Remove fuel lines and fuel tank. Disconnect air cleaner assembly and any interfering wires or cables. When reinstalling be sure to use new packing rings under the valve cover bolt washers. Torque the valve cover bolts to 25 ft.-lbs.

115. **OVERHAUL.** To remove the rocker arms from the one-piece shaft, remove snap rings from end of shaft and slide all parts from shaft. Refer to Fig. 97. Outside diameter of rocker shaft is 0.8491-0.8501. The rocker arms should have an operating clearance of 0.0009-0.0039 on rocker shaft with a maximum allowable clearance of 0.007. All rocker arms are interchangeable and may be refaced providing the original contour is maintained. If necessary to replace rocker shaft (2—Fig. 97) press new plugs (1) in end of shafts.

When reassembling, lay the shaft marked "TOP" up and the bolt grooves toward the assembler and refer to Fig. 97.

Model 1066 Diesel

116. **REMOVE AND REINSTALL.** Removal of rocker arms and shaft assembly is accomplished as follows: Remove all hoodsheets, heat shield, air cleaner assembly and turbocharger air outlet pipe. Rocker arm cover and the rocker arms and shaft assembly can now be removed. When reinstalling be sure to use new packing rings under the valve cover bolt washers. Torque the valve cover bolts to 25 ft.-lbs.

117. **OVERHAUL.** To remove the rocker arms from the one-piece shaft, remove snap rings from end of shaft and slide all parts from shaft. Refer to Fig. 97. Outside diameter of rocker shaft is 0.8491-0.8501. The rocker arms should have an operating clearance of 0.0009-0.0039 on rocker shaft with a maximum allowable clearance of 0.007. All rocker arms are interchangeable and may be refaced providing the original contour is maintained. If necessary to renew rocker shaft (2—Fig. 97), press new plugs (1) in end of shafts.

When reassembling lay the shaft marked "TOP" up and the bolt grooves toward the assembler and refer to Fig. 97.

Fig. 97–View showing the valve lever and shaft assembly.

1. Plug (2 used)
2. Shaft
3. Spring
4. Support bracket
5. Rocker arm
6. Snap ring (2 used)

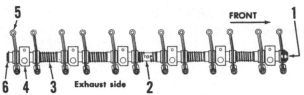

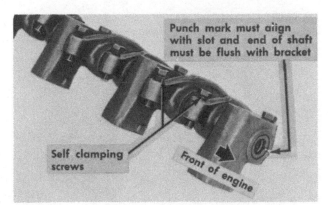

Fig. 96–On series 826, align punch marks on front end of rocker shaft with slot in front mounting bracket. End of shaft must be flush with bracket.

VALVE ROTATORS
All Diesel Models

118. Two types of positive valve rotators are used (Rotocaps and Rotocoils). Rotocaps are used on the D358 engine and Rotocoils are used on the D-360, DT407, D-414 and DT-414 engines.

Normal servicing of the valve rotators consists of renewing the units. It is important to observe the valve action after engine is started. Valve rotator action can be considered satisfactory if the valve rotates a slight amount each time the valve opens.

The Rotocoil rotates the valve at a slower speed than the Rotocap. Rotocaps should not be installed on engines where Rotocoil rotators are specified.

VALVE TIMING

Valve timing is correct when the timing (punch) marks on the timing gear train are properly aligned as shown in Figs. 98, 99 or 100. To check valve timing on an assembled engine, follow the procedure outlined in the following paragraphs.

Model 826 Diesel

119. To check valve timing, remove rocker arm cover and crank engine to position number one piston at TDC of compression stroke. Adjust number one cylinder intake valve tappet gap to 0.014. Place a 0.004 feeler gage between valve lever and valve stem of number one intake valve. Slowly rotate crankshaft in normal direction until valve lever becomes tight on feeler gage. At this point, number one intake valve will start to open and timing pointer should be within the range of 23 to 29 degrees BTDC on the flywheel.

NOTE: One tooth "out of time" equals approximately 11 degrees.

Readjust number one intake valve tappet gap as outlined in paragraph 100.

Model 1026 Diesel

120. To check valve timing, remove rocker arm cover and crank engine to position number one piston at TDC of compression stroke. Adjust number one cylinder intake valve tappet gap to 0.017. Place a 0.004 feeler gage between valve lever and valve stem of number one intake valve. Slowly rotate crankshaft in normal direction until valve lever becomes tight on feeler gage. At this point, number one intake valve will start to open and timing pointer should be within the range of 22 to 28 degrees BTDC on the flywheel.

NOTE: One tooth "out of time" equals approximately 12 degrees.

Readjust number one intake valve tappet gap as outlined in paragraph 101.

Model 766 Diesel

121. To check valve timing, remove rocker arm cover and crank engine to position number one piston at TDC of compression stroke. Adjust number one cylinder intake valve tappet gap to 0.016. Place a 0.004 feeler gage between valve lever and valve stem of number one intake valve. Slowly rotate crankshaft in normal direction until valve lever becomes tight on feeler gage. At this point number one intake valve will start to open and timing pointer should be 30 degrees ±3 degrees BTDC on the flywheel.

NOTE: One tooth "out of time" equals approximately 11 flywheel degrees.

Readjust number one intake valve tappet gap as outlined in paragraph 102.

Models 966 and 1066 Diesel

122. To check valve timing, remove rocker arm cover and crank engine to position number one piston at TDC of compression stroke. Adjust number one cylinder intake valve tappet gap to 0.024. Place a 0.004 feeler gage between valve lever and valve stem of number one intake valve. Slowly rotate crankshaft in normal direction until valve lever becomes tight on feeler gage. At this point number one intake valve will start to open and timing pointer should be 24 degrees ±3 degrees BTDC on the flywheel.

NOTE: One tooth "out of time" equals approximately 11 flywheel degrees.

Readjust number one intake valve tappet gap as outlined in paragraph 102.

TIMING GEAR COVER
All Diesel Models

123. To remove the timing gear cover, first remove hood, drain cooling system and disconnect battery cables. Disconnect air cleaner inlet hose and radiator hoses. On International models, disconnect forward end of power steering cylinder from center steering arm. On Farmall models, identify and disconnect power steering lines. On all models, identify and disconnect hydraulic oil cooler lines. Plug or cap openings to prevent dirt or other foreign material from entering hydraulic system. Support tractor under clutch housing or hydrostatic drive housing. On Farmall models equipped with wide front axle and all International models, disconnect stay rod ball. Attach hoist to front support, unbolt front support from side rails and roll complete front assembly forward from tractor. Unbolt and remove fan, alternator and drive belts.

On **model 826**, unbolt and remove air cleaner assembly. Remove pump pulley, then unbolt and remove water pump and carrier assembly. Disconnect tachometer drive cable and remove tachometer drive unit from front cover. Do not lose the small driving tang when removing tachometer drive. Remove the three cap screws, flat washer and pressure ring from crankshaft. Tap crankshaft pulley with a plastic hammer to loosen pulley, then slide pulley off of the wedge rings. Remove cap screws retaining oil pan to timing gear cover and loosen the remaining oil pan cap screws. Unbolt and remove timing gear cover. Reassemble by reversing the disassembly procedure. When installing the crankshaft pulley, place one pressure ring on crankshaft with thick end towards engine. Install the wedge rings in pulley bore so that slots in rings are 90 degrees apart. Slide pulley onto crankshaft and align with timing pin. Install pressure ring, flat pressure washer and three cap screws. Tighten the cap screws evenly to a torque of 55 ft.-lbs.

On **model 1026**, remove crankshaft pulley retaining nut, attach a suitable puller and remove crankshaft pulley.

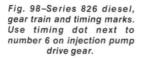
Fig. 98—Series 826 diesel, gear train and timing marks. Use timing dot next to number 6 on injection pump drive gear.

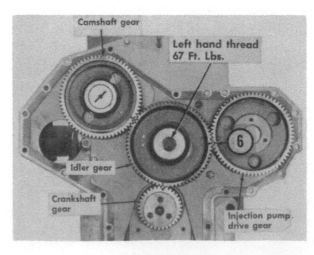

NOTE: Attach puller to tapped holes in pulley. Remove cap screws retaining oil pan to timing gear cover and loosen the remaining oil pan cap screws. Unbolt timing gear cover from engine and side rails. Pull cover forward off the dowels and remove from engine. Use care not to damage oil pan gasket. Reinstall timing gear cover by reversing the removal procedure. Tighten crankshaft pulley retaining nut to a torque of 205 ft.-lbs.

On **models 766, 966 and 1066,** remove alternator and fan assembly. Drain engine oil and remove oil pan. Remove three cap screws and retaining plate from end of crankshaft. Attach suitable puller (IH FES 10-17 or equivalent) and remove crankshaft pulley.

NOTE: Attach the puller to tapped holes in pulley ONLY. The front pulley is also the vibration dampener and any other type puller can damage the elastic member.

Unbolt and remove oil pump from front of cover. The front oil seal is in the oil pump housing. Unbolt and remove timing gear cover. Reassemble by reversing the disassembly procedure. When reinstalling the crankshaft pulley, heat in boiling water for a period not to exceed ONE hour. Then install on crankshaft and secure with retaining plate and cap screws torqued to a 125 ft.-lbs.

NOTE: Do not install pulley cold, and when heating do not heat more than one hour as the elastic member can be damaged.

TIMING GEARS
Model 826 Diesel

124. **CRANKSHAFT GEAR.** Crankshaft gear is a shrink fit on crankshaft. To renew the gear, it is recommended that the crankshaft be removed from engine. Then, using a chisel and hammer, split the gear at its timing slot.

The roll pin for indexing crankshaft gear on crankshaft must protrude approximately 5/64-inch. Heat new gear to 400° F. and install it against bearing journal.

When reassembling, make certain all timing marks are aligned as shown in Fig. 98.

125. **CAMSHAFT GEAR.** Camshaft gear is a shrink fit on camshaft. To renew the gear, remove camshaft as outlined in paragraph 136. Gear can now be pressed off in conventional manner, using care not to damage the tachometer drive slot in end of camshaft. When reassembling, install thrust plate and Woodruff key. Heat gear to 400° F. and install it on camshaft. NOTE: When sliding gear on camshaft, set thrust plate clearance at

0.004-0.017. Install camshaft assembly and make certain all timing marks are aligned as shown in Fig. 98.

126. **IDLER GEAR.** To remove the idler gear, first remove timing gear cover as outlined in paragraph 123. Idler gear shaft is attached to front of engine by a special (left hand thread) cap screw. Idler gear is equipped with two renewable needle bearings. A spacer is used between the bearings.

When installing idler gear, align all timing marks as shown in Fig. 98. Coat threads of special cap screw with "Grade B Loctite" and tighten cap screw to a torque of 67 ft.-lbs. End clearance of gear on shaft should be 0.008-0.013.

Timing gear backlash should be as follows:

Idler to crankshaft gear,
  New gears 0.007-0.015
  Used gears (max.) 0.0295
Idler to camshaft gear,
  New gears 0.0035-0.0107
  Used gears (max.) 0.0215
Idler to injection pump gear,
  New gears 0.0021-0.012
  Used gears (max.) 0.024

127. **INJECTION PUMP DRIVE GEAR.** To remove the pump drive gear, first remove timing gear cover as outlined in paragraph 123. Remove pump drive shaft nut and washer and the three hub cap screws. Attach puller (IH tool No. FES 111-2 or equivalent) to threaded holes in gear and pull gear and hub from shaft.

When reassembling, make certain all timing marks are aligned as shown in Fig. 98. Use timing dot next to number 6 on injection pump drive gear. Refer to paragraph 186 and retime injection pump. Tighten pump drive shaft nut to a torque of 47 ft.-lbs. and the three hub cap screws to a torque of 17 ft.-lbs.

Model 1026 Diesel

128. **CRANKSHAFT GEAR.** The crankshaft gear is keyed and press

fitted to the crankshaft. The gear can be removed using a suitable puller after first removing the timing gear cover as outlined in paragraph 123.

Before installing, heat gear in oil, then drift heated gear on crankshaft. Make certain timing marks are aligned as shown in Fig. 99.

129. **CAMSHAFT GEAR.** The camshaft gear is keyed and press fitted to camshaft. The camshaft gear can be removed using a suitable puller after timing gear cover is removed as outlined in paragraph 123, and the retaining nut removed.

Backlash between the camshaft gear and crankshaft gear should be 0.003-0.012 with a maximum allowable backlash of 0.016.

Before installing, heat gear in oil until gear will slide on shaft. Install and tighten gear retaining nut to a torque of 55 ft.-lbs. Make certain timing marks are aligned as shown in Fig. 99.

130. **IDLER GEAR.** To remove the idler gear, first remove timing gear cover as outlined in paragraph 123.

The idler gear rotates on two taper roller bearings and idler gear shaft is attached to front of cylinder block by a **left hand thread**, twelve point head cap screw. Removal of idler gear, bearings, bearing spacer and shaft is obvious.

The idler gear and idler gear shaft are available separately but the two taper bearings, two bearing cups and the bearing spacer must be renewed as an assembly.

When reinstalling, make certain timing marks are aligned as shown in Fig. 99 and tighten the idler gear shaft cap screw to a torque of 88 ft.-lbs.

131. **INJECTION PUMP DRIVE GEAR.** To remove the injection pump drive gear, first remove the timing gear cover as outlined in paragraph 123.

Fig. 99–Series 1026 diesel gear train and timing marks.

CA. Camshaft gear
CR. Crankshaft gear
ID. Idler gear
IN. Injector pump drive gear

Position engine in number 1 cylinder firing position as outlined in paragraph 187. Then remove nut and washer, attach a suitable puller and remove the gear.

Reassemble by reversing the disassembly procedure and align timing marks as shown in Fig. 99.

Models 766, 966 and 1066 Diesel
132. **CRANKSHAFT GEAR.** Crankshaft gear is a shrink fit on crankshaft. To renew the gear, it is recommended that the crankshaft be removed from engine. When renewing crankshaft gear, the oil pump drive spline (2—Fig. 101) will also have to be renewed. With crankshaft removed, use a chisel and hammer and split the oil pump drive spline and then split crankshaft gear at the roll pin slot.

Install new roll pin in crankshaft if it was removed. Heat new gear and drive spline to 370°-395°F. Install crankshaft gear against shoulder and over roll pin. Then, slide the drive spline against crankshaft gear.

When reassembling, make certain all timing marks are aligned as shown in Fig. 100.

133. **CAMSHAFT GEAR.** The camshaft gear is keyed and pressed fitted to camshaft. The camshaft gear can be removed using a suitable puller after camshaft is removed as outlined in paragraph 140.

Backlash between the camshaft gear and idler gear should be 0.003-0.007 with a maximum allowable backlash of 0.016.

When installing, heat gear to 370°-395°F. and press gear against shoulder of shaft. Make certain timing marks are aligned as shown in Fig. 100.

134. **IDLER GEAR.** To remove the idler gear, first remove timing gear cover as outlined in paragraph 123.

The idler gear rotates on two taper roller bearings and idler gear shaft is attached to front of cylinder block by a twelve point head cap screw. Removal of idler gear, bearings, bearing spacer and shaft is obvious.

The idler gear and idler gear shaft are available separately but the two taper bearings, two bearing cups and the bearing spacer must be renewed as an assembly.

When reinstalling, make certain timing marks are aligned as shown in Fig. 100 and tighten the idler gear shaft cap screw to a torque of 85 ft.-lbs.

135. **INJECTION PUMP DRIVE GEAR.** To remove the injection pump drive gear, first remove the pump access cover from front of timing cover.

Turn engine until pointer (3—Fig. 102) is aligned with mark on pump hub, then remove three cap screws while holding pump shaft as shown.

Reassemble by reversing the disassembly procedure and making sure that the timing marks are aligned.

CAMSHAFT AND BEARINGS

Model 826 Diesel
136. **CAMSHAFT.** To remove the camshaft, first remove timing gear cover as outlined in paragraph 123. Remove rocker arm cover, rocker arms assembly and push rods. Remove engine side cover and secure cam followers (tappets) in raised position with clothes pins or rubber bands. Working through openings in camshaft gear, remove camshaft thrust plate retaining cap screws. Carefully withdraw camshaft assembly.

Recommended camshaft end play is 0.004-0.017. Camshaft bearing journal diameter should be 2.2823-2.2835 for all journals.

Install camshaft by reversing the removal procedure. Make certain timing marks are aligned as shown in Fig. 98.

137. **CAMSHAFT BEARINGS.** To remove the camshaft bearings, first remove the engine as outlined in paragraph 97 and camshaft as in paragraph 136. Unbolt and remove clutch or flex drive plate, flywheel and engine rear support plate. Remove expansion plug

from behind camshaft rear bearing and remove the bearings.

NOTE: Camshaft bearings are furnished semi-finished and must be align reamed after installation to an inside diameter of 2.2844-2.2856.

Install new bearings so that oil holes in bearings are in register with oil holes in crankcase.

Normal operating clearance of camshaft journals in bearings is 0.0009-0.0033. Maximum allowable clearance is 0.006.

When installing expansion plug at rear camshaft bearing, apply a light coat of sealing compound to edge of plug and bore.

Model 1026 Diesel
138. **CAMSHAFT.** To remove the camshaft, first remove timing gear cover as outlined in paragraph 123. Remove rocker arm cover, rocker arms assembly and push rods. Unbolt and remove the externally mounted oil

Fig. 101–View showing crankshaft gear and oil pump drive spline.

1. Crankshaft
2. Drive spline
3. Crankshaft gear

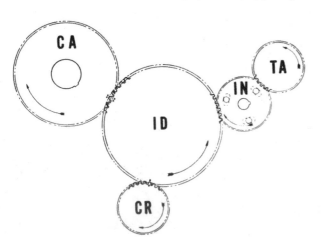

Fig. 100–Series 766, 966 and 1066 diesel gear train and timing marks.

CA. Camshaft gear
CR. Crankshaft gear
ID. Idler gear
IN. Injector pump drive gear
TA. Tachometer drive gear

Fig. 102–View showing timing marks for injection pump.

1. Idler gear
2. Pump drive gear
3. Timing pointer

pump and the engine side covers. Raise and secure cam followers (tappets) with clothes pins or magnets. Working through openings in camshaft gear, remove camshaft thrust plate retaining cap screws. Carefully withdraw camshaft assembly.

Recommended camshaft end play of 0.002-0.010 is controlled by the thrust plate.

Check the camshaft against the values which follow:

Journal Diameter

No. 1 (front) 2.429-2.430
No. 2 .... 2.089-2.090
No. 3 .... 2.069-2.070
No. 4 .... 1.499-1.500

When installing camshaft, reverse the removal procedure and make certain timing marks are aligned as shown in Fig. 99. Tighten camshaft thrust plate retaining cap screws to a torque of 33 ft.-lbs.

139. CAMSHAFT BEARINGS. To remove the camshaft bearings, first remove the engine as outlined in paragraph 97 and camshaft as in paragraph 138. Unbolt and remove flex drive plate, flywheel and engine rear support plate. Bearings can now be removed.

Using a closely piloted arbor, install new camshaft bearings so that oil holes in bearings are in register with oil holes in crankcase.

Camshaft bearings are pre-sized and should need no final sizing if carefully installed. Camshaft journals should have a diametral clearance of 0.001-0.0055 in the camshaft bearings. Maximum allowable clearance is 0.008.

Models 766, 966 and 1066 Diesel

140. CAMSHAFT. To remove the camshaft, first remove timing gear cover as outlined in paragraph 123. Remove idler gear as outlined in paragraph 134. Remove cylinder head as outlined in paragraph 98. Remove camfollowers from above. Working through openings in camshaft gear, remove camshaft thrust plate retaining cap screws. Carefully withdraw camshaft assembly.

Recommended camshaft end play is 0.005-0.017; if excessive, renew thrust plate. Camshaft bearing journal diameter should be 2.2824-2.2835 for all journals.

Install camshaft by reversing the removal procedure. Make certain timing marks are aligned as shown in Fig. 100.

141. CAMSHAFT BEARING. To remove the camshaft bearings, first remove the engine as outlined in paragraph 97 and camshaft as in paragraph 140. Unbolt and remove clutch or flex drive plate, flywheel and engine rear support plate. Remove expansion plug

from behind camshaft rear bearing and remove the bearings.

Install new bearings so that oil holes in bearings are in register with oil holes in crankcase.

Camshaft bearings are pre-sized and should need no final sizing if carefully installed. Camshaft journals should have a diametral clearance of 0.0010-0.0056 in the camshaft bearings.

When installing expansion plug at rear camshaft bearing, apply a light coat of sealing compound to edge of plug and bore.

PISTON AND ROD UNITS
All Diesel Models

142. Connecting rod and piston assemblies can be removed from above after removing cylinder head as outlined in paragraph 98 and oil pan as in paragraph 166.

Models 826 and 1026 cylinder numbers are stamped on connecting rod and cap. Numbers on rod and cap should be in register and face towards camshaft side of engine. The arrow or FRONT marking stamped on the tops of pistons should be towards front of engine.

Models 766, 966 and 1066 numbers are stamped on connecting rod and cap. NOTE: Be sure to note and record the numbers on the rod and the cylinder from which it came.

Be sure the numbers on the rods face AWAY from the camshaft, and the marking on top of the pistons is TOWARD camshaft side of engine.

On series 826 equipped with D358 engine, tighten connecting rod nuts to a torque of 45 ft.-lbs.

On series 1026 equipped with DT407 engine, tighten connecting rod bolts to a torque of 105 ft.-lbs.

On series 766, 966 and 1066 equipped with D360, D414 and DT414 engines, tighten connecting rod cap screws in two steps. Tighten cap screw to a torque of 60 ft.-lbs. in first step and 130 ft.-lbs. in second step.

PISTONS, SLEEVES AND RINGS
Model 826 Diesel

143. Pistons are not available as individual service parts, but only as matched units with the wet type sleeves. New pistons have a diametral clearance in new sleeves of 0.0039-0.0047 when measured between piston skirt and sleeve at 90 degrees to piston pin.

The wet type cylinder sleeves should be renewed when out-of-round or taper exceeds 0.006. Inside diameter of new sleeve is 3.8750-3.8754. Cylinder sleeves can usually be removed by bumping them from the bottom, with a block of wood.

Before installing new sleeves, thoroughly clean counterbore at top of block and seal ring groove at bottom. All sleeves should enter crankcase bores full depth and should be free to rotate by hand when tried in bores without seal rings. After making a trial installation without seal rings, remove sleeves and install new seal rings, dry, in grooves in crankcase. Wet lower end of sleeve with a soap solution or equivalent and install sleeve.

NOTE: The cut-outs in bottom of sleeve are for connecting rod clearance and must be installed toward each side of engine. Chisel marks are provided on top edge of cylinder sleeves to aid in correct installation. Align chisel marks from front to rear of engine.

If seal ring is in place and not pinched, very little hand pressure is required to press the sleeve completely in place. Sleeve flange should extend 0.003-0.005 above top surface of cylinder block. If sleeve stand-out is excessive, check for foreign material under the sleeve flange. The cylinder head gasket forms the upper cylinder sleeve seal and excessive sleeve standout could result in coolant leakage.

Pistons are fitted with three compression rings and one oil control ring. Prior to installing new rings, check top ring groove of piston by using Perfect Circle Piston Ring Gage No. 1 as shown in Fig 103. If one, or both, shoulders of gage touch ring land, renew the piston.

The top compression ring is a full keystone ring and it is not possible to measure ring side clearance in the groove with a feeler gage. Check fit of top ring as follows: Place ring in its

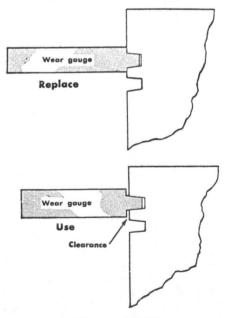

Fig. 103–Perfect Circle Ring Gage No. 1 to check top ring groove of pistons used in D358 and DT407 engines.

groove and push ring into groove as far as possible. Measure the distance ring is below ring land. This distance should be 0.002-0.015. Refer to Fig. 104 for view showing ring fit being checked using IH tools FES 68-3 and dial indicator FES 67.

The second compression ring is a taper face ring and is installed with largest outside diameter towards bottom of piston. Upper side of ring is marked TOP.

The oil control ring can be installed either side up, but make certain the coil spring expander is completely in its groove.

Additional piston ring information is as follows:

Ring End Gap
Compression rings 0.014-0.022
Oil control ring 0.010-0.016
Ring Side Clearance
Top compression
(ring drop) 0.002-0.015
Second compression . 0.0030-0.0042
Third compression . . 0.0024-0.0034
Oil control 0.0014-0.0024

Model 1026 Diesel

144. The cam ground pistons are fitted with two compression rings and one oil control ring. Pistons and rings are available for service in standard size only. Prior to installing new rings, check top ring groove of piston by using a Perfect Circle Piston Ring Gage No. 1 as shown in Fig. 103. If one or both, shoulders of gage touch ring land, renew the piston.

The top compression ring is a full keystone ring and it is not possible to measure ring side clearance in the groove with a feeler gage. Check fit of top ring as follows: Install ring in its groove and push ring into groove as far as possible. Measure the distance ring is above or below ring land. This distance must be 0.019 below to 0.013 above ring land on DT407 engine. See Fig. 104 for view showing ring fit being checked using IH tools FES 68-3 and dial indicator FES 67.

The second compression ring is a taper face ring and is installed with largest outside diameter toward bottom of piston. Upper side of ring is marked TOP.

The oil control ring is a one-piece slotted ring and can be installed either side up.

Additional piston ring information is as follows:

Ring End Gap
Top compression ring . . . 0.013-0.023
Second compression ring 0.013-0.023
Oil control ring 0.013-0.028
Ring Side Clearance
Top compression ring
(ring drop) . . . . . −0.019 to +0.013

Second compression
ring 0.0030-0.0048
Oil control ring 0.0020-0.0035

Cylinder sleeves are removed from top of cylinder block after rod and piston assemblies are out. These dry type sleeves are a tight fit in the cylinder block and must be removed and installed with a hydraulic ram. Cylinder sleeves should be renewed when out-of-round or taper exceeds 0.006. Inside diameter of new sleeves is 4.3209-4.3219 for DT407 engines. New pistons should have a diametral clearance in new sleeves of 0.0049-0.0069 on DT407 engine when measured between piston skirt and installed sleeve at 90 degrees to piston pin.

Three sizes of sleeves available for service are as follows: Standard outside diameter, 0.002 oversize outside diameter and 0.010 oversize outside diameter.

The fit of sleeves in cylinder block should be so that installation can be accomplished between a minimum of 750 pounds of force and a maximum of 2000 pounds of force.

To translate pounds of force into pump gage pressure for use with hydraulic ram, the effective area of the ram cylinders must be calculated by the formula $d^2 \times 0.784$. With ram cylinder effective area calculated, the result can be divided into stated pounds of force and pump gage pressure determined. Using an OTC twin piston ram with each piston diameter of 1½ inches, the following example is given: $(d^2 + d^2) \times 0.784$ equals $(2.25 + 2.25) \times 0.784$ equals 3.52 sq. in. effective area of both ram pistons. When 3.52 is divided into the stated pounds of force,

the following pump gage pressures are obtained:

Pounds of Force	Pump Gage Pressure
750	213 psi
2000	568 psi

Therefore, sleeve installation must occur between 213 and 568 psi pump gage pressure on DT407 engines.

Before installing new sleeves, clean sleeve and cylinder block bore with solvent and dry with compressed air. Lubricate sleeve and bore with clean diesel fuel. Press sleeve into cylinder block until sleeve flange projects 0.040-0.045 above face of block.

When installing the 0.002 oversize or 0.010 oversize sleeve, cylinder block bore will have to be bored and/or honed to the correct oversize. Diameter of standard cylinder block bore is 4.4680-4.4694.

Models 766, 966 and 1066 Diesel

145. Pistons are not available as individual service parts, but only as matched units with the wet type sleeves. New pistons have a diametral clearance in new sleeves of 0.0045-0.0065 when measured between piston skirt and sleeve at 90 degrees to piston pin.

The wet type cylinder sleeves should be renewed when out-of-round or taper exceeds 0.004. Inside diameter of new sleeve is as follows:
D360 . . . . 3.875-3.876
D414 . . . . 4.300-4.301
DT414 . . . . 4.300-4.301

Cylinder sleeves can usually be removed by bumping them from the bottom, with a block of wood.

Before installing new sleeves, thoroughly clean counterbore of block using a wire brush or steel wool. All sleeves should enter crankcase bores full depth and should be free to rotate by hand when tried in bores without "O" rings. With sleeves in bores without "O" rings, sleeve flange should extend 0.002-0.005 above top surface of cylinder block. If sleeves do not extend above top surface of cylinder block, shims are available and can be placed under sleeve flange to obtain proper height. If sleeve stand-out is excessive, check for foreign material under the sleeve flange.

NOTE: Sleeves can be held down with cap screws and large flat washers when checking sleeve flange for proper height.

After making the trial installation without "O" rings, remove sleeves and install the "O" rings in the following order: bottom, center and top. Lubricate the "O" rings with petroleum jelly or equivalent and install sleeves. Be sure the flange of the sleeve is firmly

Fig. 104–Top compression ring should be 0.002-0.015 below ring land for proper fit on D358 and 0.019 below to 0.013 above ring land on DT407 engines.

seated in the crankcase counterbore.

Piston is fitted with two compression rings and one oil control ring.

The top compression ring on D360, D414 and DT414 engines is a full keystone ring and on DT414 engine the second compression ring is a half keystone ring. It is not possible to measure ring side clearance in the groove with a feeler gage. Check fit of top ring on D360 and D414 engines and top two rings on DT414 engine as follows: Install ring in its groove and push ring into groove as far as possible. Measure the distance ring is below ring land. Refer to Fig. 105 for the correct distance. See Fig. 104 for view showing ring fit being checked using IH tools FES 68-3 and dial indicator FES 67.

The second compression is marked TOP be sure that it is facing top of piston.

The oil control ring is a one-piece slotted chrome ring and can be installed either side up.

Additional piston ring information is as follows:

Ring End Gap

    Top compression ring,
        D360 ....0.010-0.020
        D414, DT414 ......0.016-0.026
    Second compression ring
        D360, D414, DT414 0.020-0.030
    Oil control ring
        D360, D414, DT414 0.010-0.020

Ring Side Clearance

    Top compression ring (ring
        drop)
        D360, D414 ....0.000-0.026
        DT414 ....0.000-0.028
    Second compression ring
        D360, D414 ....0.003-0.005
        DT414 (ring drop) 0.0045-0.0215
    Oil control ring
        D360 .... 0.0015-0.0040
        D414, DT414 ......0.002-0.004

PISTON PINS

All Diesel Models

146. The full floating type piston pins are retained in the piston bosses by snap rings. Piston pins are available in 0.005 oversize for model 1026. Specifications are as follows:

Piston pin diameter,
    D358 .... 1.4172-1.4173
    DT407 .... 1.4998-1.5000
    D360, D414, DT414 . 1.6248-1.6250
Piston pin diametral clearance in piston,
    D358 .... 0.0001T-0.0000
    DT407 ....0.0001L-0.0005L
    D360, D414, DT414 . 0.0005-0.0010
Piston pin diametral clearance in rod bushing,
    D358 ....0.0005L-0.0010L
    DT407 ....0.0006L-0.0010L
    D360, D414, DT414 . 0.0006-0.0010

Piston pin bushings are furnished semi-finished and must be reamed or honed for correct pin fit after they are pressed into connecting rods.

CONNECTING RODS AND BEARINGS

All Diesel Models

147. Connecting rod bearings are of the slip-in, precision type, renewable from below after removing oil pan and connecting rod cap. When installing new bearing inserts, make certain the projections on same engage slots in connecting rod and cap and that cylinder identifying numbers on rod and cap are in register and face TOWARDS camshaft side of engine on D358 and DT407. On D360, D414 and DT414 the identifying numbers on rod and cap face AWAY from camshaft. Connecting rod bearings are available in standard size and undersize of 0.010, 0.020 and 0.030 for 358 engines and in standard size and undersize of 0.002, 0.010, 0.020 and 0.030 for DT407 engines. Rod bearings are available for D360, D414 and DT414 engines in standard size and undersize of 0.010 and 0.030. Check the crankshaft, crankpins and connecting rod bearings against the values which follow:

Crankpin diameter,
    D358 .... 2.5185-2.5193
    DT407 .... 2.9980-2.9990
    D360, D414, DT414 . 2.9977-2.9990
Rod bearing diametral clearance,
    D358 .... 0.0023-0.0048
    DT407 .... 0.0018-0.0051
    D360, D414, DT414 . 0.0018-0.0051
Rod side clearance,
    D358 ....0.006-0.010
    DT407 ....0.004-0.015
    D360, D414, DT414 ....0.009-0.015
Rod bolt (or nut) torque,
    D358 .... 45 ft.-lbs.
    DT407 ....105 ft.-lbs.
    D360, D414, DT414 .....130 ft.-lbs.

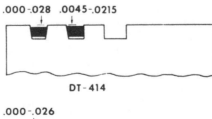

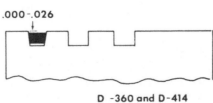

Fig. 105–Chart showing correct distance below ring land.

CRANKSHAFT AND MAIN BEARINGS

Model 826 Diesel

148. Crankshaft is supported in seven main bearings. Main bearings are of the non-adjustable, slip-in, precision type, renewable from below after removing the oil pan and main bearing caps. Crankshaft end play is controlled by the flanged rear main bearing inserts. Removal of crankshaft requires R&R of engine. The crankshaft is counter-balanced by twelve weights bolted opposite to crankshaft throws. The weights are numbered consecutively from 1 to 12 which correspond to numbers stamped on crankshaft. The mounting holes in balance weights are offset. The wide edge goes toward connecting rod bearing. Balance weights are not serviced separately.

Check crankshaft and main bearings against the values which follow:
Crankpin diameter 2.5185-2.5193
Main journal diameter . 3.1484-3.1492
Crankshaft end play 0.006-0.009
Main bearing diametral
    clearance 0.0029-0.0055
Main bearing bolt torque,
    Marked 10K or 10-9 ..... 80 ft.-lbs.
    Marked 12K or 12-9 ..... 97 ft.-lbs.
    Marked 10.9 .... 115 ft.-lbs.
    Marked 12.9 .... 140 ft.-lbs.
Balance weight bolt torque 57 ft.-lbs.

Main bearings are available in standard size and undersizes of 0.010, 0.020, and 0.030. Main bearing caps should be installed with numbered side toward camshaft side of engine.

Model 1026 Diesel

149. Crankshaft is supported in seven main bearings and crankshaft end thrust is taken by the number seven (rear) main bearing. Main bearings are of the non-adjustable, slip-in, precision type, renewable from below after removing oil pan and main bearing caps. Removal of crankshaft requires R&R of engine. Check crankshaft and main bearings against the values which follow:
Crankpin diameter 2.9980-2.9990
Main journal diameter . 3.3742-3.3755
Crankshaft end play 0.007-0.0185
Main bearing diametral
    clearance 0.0018-0.0051
Main bearing bolt torque .. 115 ft.-lbs.

Main bearings are available in standard size and undersizes of 0.002, 0.010, 0.020 and 0.030. Main bearing caps are numbered for correct location and numbered side of caps should be toward camshaft side of engine.

Models 766, 966 and 1066 Diesel

150. Crankshaft is supported in seven main bearings and crankshaft end thrust is taken by the rear main

bearing. Main bearings are of the non-adjustable, slip-in, precision type, renewable from below after removing oil pan and main bearing caps. Removal of crankshaft requires R & R of engine. Check crankshaft and main bearings against the values which follow:

Crankpin diameter 2.9977-2.9990
Main journal diameter . 3.3742-3.3755
Crankshaft end play 0.006-0.012
Main bearing diametral
   clearance 0.0018-0.0051
Main bearing bolt torque ..115 ft.-lbs.

Main bearings are available in standard size and undersize of 0.010 and 0.030. Main bearing caps are numbered for correct location and numbered side of caps should be toward camshaft side of engine.

CRANKSHAFT SEALS
Models 826 and 1026 Diesel

151. **FRONT.** To renew the crankshaft front oil seal, first remove hood, drain cooling system and disconnect radiator hoses. On International models, disconnect forward end of power steering cylinder from center steering arm. On Farmall models, identify and disconnect power steering lines. On all models, identify and disconnect hydraulic oil cooler lines. Plug or cap openings to prevent dirt or other foreign material from entering hydraulic system. Support tractor under clutch housing or hydrostatic drive housing. On Farmall models equipped with wide front axle and all International models, disconnect stay rod ball. Attach hoist to front support, unbolt front support from side rails and roll complete front assembly forward away from tractor. Remove belt, or belts, from crankshaft pulley.

On model 826, remove three cap screws, flat washer and pressure ring from crankshaft. Tap crankshaft pulley with a plastic hammer to loosen pulley, then slide pulley off the wedge rings. Remove wedge rings and oil seal wear ring from crankshaft. Remove and renew oil seal in conventional manner. Renew "O" ring on crankshaft. Inspect wear ring and timing pin for wear or other damage and renew as necessary. Use a non-hardening sealer on timing pin. When installing wear ring, timing pin must engage slot in crankshaft gear. Install one pressure ring on crankshaft with thick end against wear ring. Place wedge rings in pulley bore so the slots in rings are 90 degrees apart. Slide pulley on crankshaft, aligning slot in pulley with timing pin in wear ring. Install pressure ring, flat pressure washer and three cap screws. Tighten the cap screws evenly to a torque of 57 ft.-lbs.

On model 1026, remove crankshaft pulley retaining nut, attach a suitable puller and remove crankshaft pulley. NOTE: Attach puller to tapped holes in pulley. Remove and renew oil seal in conventional manner. Check condition of the crankshaft pulley sealing surface and renew the wear ring if the surface is not perfectly smooth. When reinstalling crankshaft pulley, tighten pulley retaining nut to a torque of 205 ft.-lbs.

On all models reassemble tractor by reversing disassembly procedure.

152. **REAR.** To renew the crankshaft rear oil seal, engine must be detached from clutch housing or hydrostatic drive housing as follows: Disconnect battery cables and remove hood and side panels. Disconnect electrical wiring and tachometer drive cable from engine and lay rearward on fuel tank. Disconnect oil pressure switch wire and pull it rearward. Identify oil cooler hoses and disconnect at rear.

NOTE: It is essential that oil cooler hoses be correctly reinstalled as the oil flowing to oil cooler is maintained at 105 psi by a pressure regulating valve in the multi-valve on hydrostatic drive and by the multiple control valve on gear drive.

Disconnect the two power steering cylinder lines from control (pilot) valve. Shut off fuel and disconnect fuel supply lines at front end. Disconnect controls from injection pump. Disconnect coolant temperature bulb from cylinder head. On model 1026 remove fuel tank left front support and disconnect tank right front support from tank. Install split stand to side rails and support clutch housing or hydrostatic drive housing with a rolling floor jack.

Remove bottom side rail to clutch housing or hydrostatic drive housing cap screw from both side rails and install guide studs. Complete removal of clutch housing or hydrostatic drive housing retaining cap screws and separate tractor.

NOTE: Insert cap screws back in side rails before pulling side rails completely off guide studs.

Unbolt and remove clutch assembly (gear transmission model) or flex plate (hydrostatic drive model), then unbolt and remove flywheel.

On model 826, unbolt and remove seal retainer. Check the depth that old seal is installed in retainer, then remove seal. New oil seal may be installed in retainer in any of three locations; 1/16-inch above flush with retainer (new engine original position), flush with retainer or 1/16-inch below flush with retainer. Location of seal in

retainer will depend on condition of the sealing surface on crankshaft. Use new gasket and install retainer and seal assembly. Tighten seal retainer cap screws evenly to a torque of 14 ft.-lbs.

On model 1026, unbolt and remove oil seal retainer. Remove old seal and press new seal in retainer until front of seal is flush with front (engine side) of seal bore in retainer. To remove oil seal wear sleeve from crankshaft, use a chisel to cut **part way** through the sleeve. This will expand sleeve diameter enough to allow it to be removed. See Fig. 106. When installing wear sleeve or oil seal and retainer assembly, International Harvester recommends the use of wear sleeve and seal installation tool set (IH tool No. FES 68-2). Press the wear sleeve on crankshaft flange. NOTE: One side of wear sleeve is chamfered on the I.D. to aid in starting it on the crankshaft and opposite end is chamfered on the O.D. to aid in sliding oil seal over the sleeve. Install seal and retainer assembly with new gasket, coat cap screws with non-hardening sealer, center seal on crankshaft and tighten cap screws to a torque of 19 ft.-lbs.

On all models install the flywheel, and tighten flywheel retaining cap screws as follows:

826 diesel ....85 ft.-lbs.
1026 diesel ....95 ft.-lbs.

On hydrostatic drive models, bolt flex plate to flywheel. Coat splines of input shaft with Molycote or equivalent and roll tractor together. To aid in engaging shaft splines with flex plate, turn flywheel with a screwdriver through access hole in right side of housing.

On standard (gear) transmission models, place clutch pressure plate and driven disc on shafts in clutch housing. Roll tractor together. Clutch can be bolted to flywheel after tractor is rejoined by working through opening at

Fig. 106–View showing method of expanding rear oil seal wear sleeve for removal from DT407 engines.

bottom of clutch housing.

The balance of reassembly is the reverse of disassembly procedure on all models.

Models 766, 966 and 1066 Diesel

153. **FRONT.** To renew the crankshaft front oil seal, first remove left front hood and left side radiator support panel. Unbolt and move fan shroud to rear. Remove fan belt, fan and shroud. On hydrostatic models remove left side radiator mounting bolts.

Remove the three cap screws and retaining plate from front of crankshaft pulley. Using a puller (IH tool FES 10-17) remove pulley. Unbolt and remove oil pump from front cover and renew seal. Remove the wear sleeve from pulley only it it is worn, scratched or nicked. NOTE: One side of wear sleeve is chamfered on the I.D. to aid in starting it on the crankshaft and opposite end is chamfered on the O.D. to aid in sliding oil seal over the sleeve.

When reinstalling crankshaft pulley, heat in boiling water for a period not to exceed one hour. Then install on crankshaft and secure with retaining plate and torque cap screws to 125 ft.-lbs.

The balance of reassembly is the reverse of disassembly procedure.

154. **REAR.** To renew the crankshaft rear oil seal, engine must be detached from clutch housing or hydrostatic drive housing as follows: Disconnect and remove batteries. Remove left and right front hood, front hood channel and heat baffles. Disconnect electrical wiring and tachometer drive cable from engine and lay rearward on fuel tank. Disconnect oil cooler lines at front of engine, power steering lines and engine oil pressure line. Shut off fuel and disconnect fuel supply lines and filters. Disconnect control from injection pump. Remove coolant temperature bulb from cylinder head. Install split stand to side rails and support clutch housing or hydrostatic

Fig. 107–Splitting and removing rear oil seal on diesel models 766, 966 and 1066.

drive housing with a rolling floor jack.

Remove bottom side rail to clutch housing or hydrostatic drive housing cap screw from both side rails and install guide studs. Complete removal of clutch housing or hydrostatic drive housing retaining cap screws and separate tractor.

NOTE: Insert cap screws back in side rails before pulling side rails completely off guide studs.

Unbolt and remove clutch assembly (gear transmission models) or flex plate (hydrostatic drive models), then unbolt and remove flywheel.

Using a chisel, split the oil seal and remove (1—Fig. 107). NOTE: Do not damage the seal bore in support plate. To remove oil seal wear sleeve from crankshaft, use a chisel to split the sleeve as shown in (1—Fig. 108) being careful not to damage the crankshaft flange.

When installing wear sleeve and/or oil seal, International Harvester recommends the use of wear sleeve and oil seal installation tool set (IH tool No. FES 149-3). Press the wear sleeve on crankshaft flange. NOTE: One side of wear sleeve is chamfered on the I.D. to aid in starting it on crankshaft and opposite end is chamfered on the O.D. to aid in installing oil seal. Install new oil seal. NOTE: When using IH tool to install sleeve and seal, pull them in until the installing tool bottoms against the centering plate.

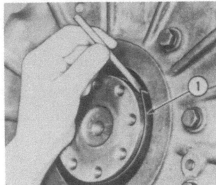

Fig. 108–Use a chisel and split wear sleeve as shown.

Fig. 109–Exploded view of oil pump assembly used on D358 engines.

1. Roll pin dowels
2. Rear cover
3. Pumping drive gear
4. Pumping idler gear
5. "O" rings
6. Pump body
7. Snap ring
8. Spacer
9. Front cover
10. Drive gear and shaft
11. Woodruff keys
12. Seal ring
13. Relief valve body
14. Relief valve piston
15. Spring
16. Plug

Install the flywheel, and tighten flywheel retaining cap screws to a torque of 123 ft.-lbs.

On hydrostatic drive models, bolt flex plate to flywheel. Coat splines of input shaft with Molycote or equivalent and roll tractor together. To aid in engaging shaft splines with flex plate, turn flywheel with a screwdriver through access hole in right side of housing.

On standard (gear) transmission models, place clutch pressure plate and driven disc on shaft in clutch housing. Roll tractor together. Clutch can be bolted to flywheel after tractor is rejoined by working through opening at bottom of clutch housing.

The balance of reassembly is the reverse of disassembly procedure on all models.

FLYWHEEL
All Diesel Models

155. To remove the flywheel, first split tractor as outlined in paragraph 152 or 154. Then unbolt and remove flex drive plate or clutch assembly. Remove the cap screws and lift flywheel from crankshaft. When reinstalling, tighten flywheel retaining cap screws to the following torque values:

D358 . . . . 85 ft.-lbs.
DT407 . . . . 95 ft.-lbs.
D360, D414, DT414 . . . . 123 ft.-lbs.

To install a new flywheel ring gear, heat same to approximately 500 degrees F.

OIL PUMP
Model 826 Diesel

156. The externally mounted gear type oil pump is located on right front side of engine and is driven by the camshaft gear. To remove the oil pump, disconnect the oil cooler lines, then remove two cap screws securing oil filter base to crankcase, then unbolt pump assembly. Remove pump, filter and connecting pipes as an assembly. Slide pump off the pipes.

Remove plug (16—Fig. 109) and

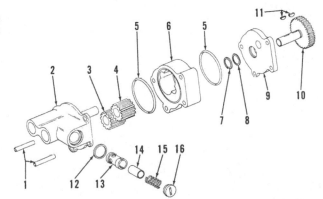

withdraw pressure relief valve components (12 thru 15). Remove rear cover (2); it may be necessary to tap rear cover with plastic hammer to free it from dowel pins (1). Lift out idler gear (4), then withdraw pumping drive gear (3). Remove two Woodruff keys (11), snap ring (7) and spacer (8) from drive gear shaft, then withdraw drive gear and shaft (10) from front cover (9). Separate front cover from pump body (6).

Inspect all parts for scoring, excessive wear or other damage. Pump covers (2 and 9) and pump body (6) are not serviced separately. Check pump against the following specifications:

Drive shaft end play
  (cover installed) 0.000-0.002
Drive shaft running
  clearance 0.001-0.0032
Idler gear to shaft
  clearance 0.001-0.0032
Pumping gears end
  clearance 0.002-0.0038
Pumping gears to body
  radial clearance 0.007-0.012
Pressure regulating spring,
  Spring free length . . . . 2.52 in.
  Spring test and
    length . . . . . 18-20 lbs. at 1.858 in.
Pressure regulating valve,
  Piston diameter 0.825-0.827
  Clearance in bore 0.003-0.007
Oil pressure at 2300 rpm . . . . 35-50 psi
Renew all "O" rings and gaskets when reassembling and reinstalling oil pump and filter assembly.

Model 1026 Diesel

157. The oil pump is an externally mounted unit having the ports so located as to prevent drainback; thus keeping the pump filled and primed. The pump is driven via an idler gear which in turn is driven from the engine camshaft.

To remove the oil pump, first clean area around pump and engine oil cooler. Disconnect tachometer cable from tachometer drive at front end of pump, then unbolt oil pump inlet elbow from cylinder block. Unbolt and remove oil pump with pump to oil cooler tube. Remove outlet tube and inlet elbow from pump. Loosen retaining set screw and withdraw tachometer drive unit. Refer to Fig. 110 and remove retaining screw (10), washer (8), bearing and spacer. Bump shaft (5) forward as necessary, then remove idler gear (6) and forward bearing. Remove cap screws and separate cover (14) from body (4). Withdraw safety relief valve assembly (11) and "O" ring (12). Use a wood dowel as a wedge between pumping gears (16 and 21) and remove Esna nut (2) from forward end of pump drive shaft. Remove dowel wedge and gear (16). Place pump in a press with

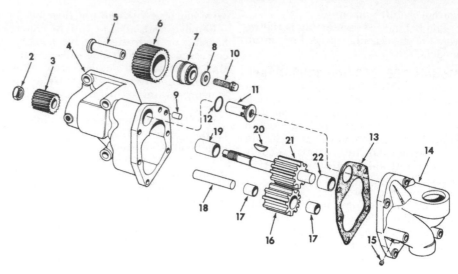

Fig. 110–Exploded view of oil pump assembly used on DT407 engines.

2. Esna nut		18. Shaft
3. Drive gear	8. Washer	19. Bushing
4. Pump body	9. Dowel	20. Woodruff key
5. Idler gear shaft	10. Retaining screw	21. Pumping gear and
6. Idler gear	11. Safety relief valve	shaft
7. Bearing assembly	12. "O" ring	22. Bushing
13. Gasket		
14. Pump cover		
15. Plug		
16. Pumping idler gear		
17. Bushings		

forward end of drive shaft upward. Push drive shaft and gear (21) upward, then wedge between bottom of gear and housing as shown in Fig. 111. Press on drive shaft until shaft has moved a distance equal to the wedge thickness (or until end of key nears bushing), then release press and add a thicker wedge. Continue this operation until key is clear of drive gear, then remove key and complete removal of shaft from gear.

NOTE: It is important that the above removal procedure be followed so that key will not enter the drive shaft bushing (19—Fig. 110) and cause damage to bushing and/or bushing bore. The fit between drive gear and drive shaft may loosen after key is removed so use caution not to let parts fall from body.

Idler gear shaft (5) and shaft (18) can be removed from pump body, if necessary.

Refer to the following values for service information:

Pumping gears backlash . . 0.004-0.014

Pumping gears to bore
  radial clearance 0.004-0.0055
Pumping gears end play . 0.002-0.0057
Drive shaft to bushing
  clearance 0.0025-0.004
Pumping idler gear to
  shaft clearance 0.0025-0.0035
Pump drive gear to idler
  gear backlash 0.002-0.009
Idler gear to camshaft
  backlash 0.002-0.012

Inspect all gears for wear and damaged or chipped teeth and renew as necessary. Inspect all other parts for wear, scoring or other damage. Safety relief valve (11) is available as a unit only; however, it can be disassembled, cleaned and the spring tested. Spring free length is 2.109 inches and should test 27.3 pounds when compressed to a length of 1.338 inches. Safety valve piston must slide freely in valve housing. Safety relief pressure is 90 psi.

NOTE: Pump safety relief valve and the oil pressure regulator valve located under the engine oil cooler rear header are sim-

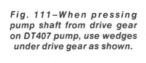

Fig. 111–When pressing pump shaft from drive gear on DT407 pump, use wedges under drive gear as shown.

Wedge (hex nut)

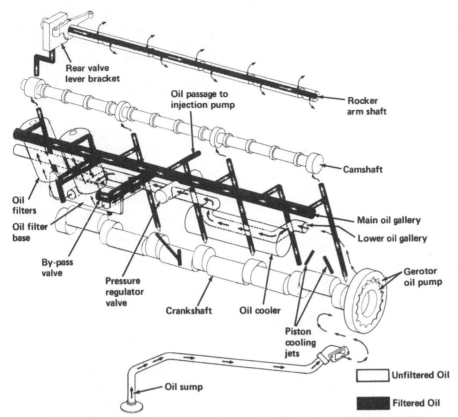

Fig. 112–View of engine lubrication system used on diesel models 766, 966 and 1066.

ilar in appearance but MUST NOT be interchanged. The oil pump safety relief valve can be identified by having an orifice in the piston and by having a heavier spring than the oil pressure regulator valve.

Reassemble pump by reversing the disassembly procedure. Tighten Esna nut (2) on drive shaft to a torque of 95 ft.-lbs. and idler gear bearing retaining screw (10) to a torque of 45 ft.-lbs. Idler gear bearing assembly (7), consisting of two bearing cups, two bearing cones and a spacer is available only as a matched assembly. Use "Plastigage" to check end play of pumping gears. Gasket (13) thickness is equal to 0.007 inch when installed. Use new "O" rings and gaskets and fill (prime) pump with new oil during installation.

Models 766, 966 and 1066 Diesel

158. The oil pump is a GEROTOR type mounted on front cover and driven directly by the crankshaft at engine speed. Refer to Fig. 112 of the lube system. With the exception of the injection pump and turbocharger oil supply,

there is no external piping to direct oil from one component to the next. Two types of pumps are used, one for turbocharged engines and one for naturally aspirated models. The difference being the width of the pump. The wide pump is used on turbocharged engines.

To remove the oil pump, first remove left front hood and left side radiator support panel. Unbolt and move fan shroud to rear. Remove fan belt, fan and shroud. On hydrostatic models remove left side radiator mounting bolts.

Remove the three cap screws and retaining plate from front of crankshaft pulley. Using a puller (IH tool FES 10-17) remove pulley. Unbolt and remove oil pump from front cover.

Remove spacer plate (10—Fig. 113), "O" ring (9) and inner and outer rotor (7 and 8).

Inspect all parts for scoring, excessive wear or other damage. Inner and outer rotor (7 and 8) and wear sleeve and seal (3 and 4) are serviced as matched sets.

Using a feeler gage and refering to Fig. 114, check the radial clearance between housing (1) and outer rotor (2). Specified clearance is 0.0055 to 0.0085 inch. Refer to Fig. 115 and check side clearance using "Plastigage." Specified clearance for inner rotor (2) is 0.0018 to 0.0042 and 0.0014 to 0.0038 inch for outer rotor (1). Oil pressure at rated rpm should be 45-65 psi.

Renew the "O" rings and seal when reassembly and reinstalling oil pump.

Fig. 114–Checking radial clearance between housing and outer rotor. Refer to text.

1. Housing
2. Outer rotor

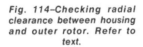

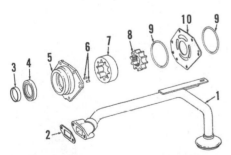

Fig. 113–Exploded view of oil pump assembly used on D360, D414 and DT414 engines.

1. Inlet tube
2. Gasket
3. Wear sleeve
4. Oil seal
5. Housing
6. Dowels
7. Rotor (outer)
8. Rotor (inner)
9. "O" rings
10. Plate

Fig. 115–Using plasitgage to check side clearance. Refer to text.

1. Outer rotor
2. Inner rotor

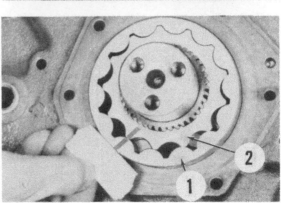

NOTE: Be sure to install special cap screws with the nylon pellet insert.

OIL PRESSURE REGULATOR

Model 826 Diesel

159. On D358 engines, the oil pressure regulator valve is located in the externally mounted oil pump. See Fig. 109. Refer to paragraph 156 for regulator valve and spring specifications. Filter by-pass valve is located in the spin-on type oil filter. Oil pressure at 2300 rpm should be 35-50 psi.

Model 1026 Diesel

160. The oil pressure regulator valve is located under the engine oil cooler rear header. To remove the pressure regulator, drain cooling system, remove drain plug and drain engine oil cooler. Loosen the rear header to oil cooler cap screws and pull them out as far as possible. Disconnect oil filter tube flange from rear header, then unbolt and remove rear header. Pressure regulator can now be withdrawn from its bore in crankcase.

Oil pressure regulator is available as a unit only; however, valve can be disassembled, cleaned and inspected. See Fig. 116. Valve piston must slide freely in valve body. Check regulator valve and spring against the following specifications:

Pressure regulator valve,
Piston diameter 0.621-0.622
Piston clearance in bore 0.002-0.004
Spring free length . . . . 2.056 in.

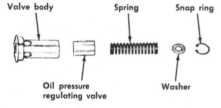

Fig. 116–Exploded view of oil pressure regulator valve assembly used on DT407 engines.

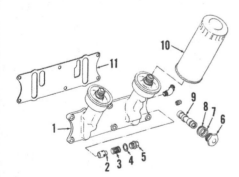

Fig. 117–Exploded view of oil filter assembly used on D360, D414 and DT414 engines.

1. Filter base
2. Valve (bypass)
3. Spring
4. Gasket
5. Plug
6. Cap
7. "O" ring
8. Gasket
9. Valve (pressure regulating)
10. Filter
11. Gasket

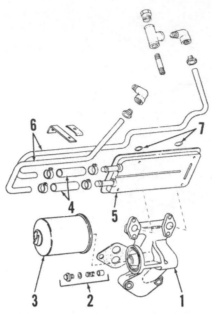

Fig. 118–Exploded view of oil cooler and oil filter used on the D358 engine.

1. Oil filter and oil cooler base
2. Relief valve (oil cooler)
3. Oil filter
4. Hose connection
5. Oil cooler assembly
6. Water pipes to oil cooler
7. "O" rings

Spring test and
  length . . . . . 13.85 lbs. at 1.342 in.
Oil pressure at 2400 rpm . . . . 38-55 psi

NOTE: Oil pressure regulator valve and the oil pump safety relief valve located in oil pump are similar in appearance but MUST NOT be interchanged. Pressure regulator valve can be identified by having a lighter spring than the safety relief valve and by not having an orifice in the valve piston as does the safety relief valve.

Models 766, 966 and 1066 Diesel

161. The oil pressure regulator valve is located at rear of engine oil cooler and behind crankcase breather in main oil gallery. The bypass valve is located in base of oil filters, Fig. 117. Removal of valves is obvious after an examination of assembly.

Oil pressure regulator valve (9) is available as a unit only. Check regulator valve, bypass valve and springs against the following specifications:

Pressure regulator valve,
Piston diameter 0.621-0.622
Piston clearance in bore 0.002-0.004
Spring free length . . . . 2 11/64 in.
Spring test and
  length . . 14.5-16 lbs. at 1 15/64 in.
By-pass valve spring,
Spring free length 2 5/64
Spring test and
  length . . . . . . 6.12 lbs. at 59/64 in.
Oil pressure at 2600 rpm . . 45-65 psi

OIL JET TUBES

Models 1026 and 1066 Diesel

162. DT407 and DT414 engines are equipped with twelve oil jet tubes located in main bearing bosses, which spray oil on the sleeves for added lubrication and cooling of pistons.

When overhauling engine, make certain the jet tubes are open and clean. Oil jet tubes need not be removed unless they are damaged.

OIL COOLER

Model 826 Diesel

163. The engine oil cooler is mounted on oil filter base (1—Fig. 118) which is located on right side of engine. Removal and disassembly of the unit is obvious after an examination of the unit. Inspect oil cooler relief valve assembly (2) for freeness in its bore.

Normal cleaning of the oil cooler consists of blowing out water tubes with compressed air and flushing oil passages with a suitable cleaning solvent. Renew "O" rings and gasket when reassembling.

Model 1026 Diesel

164. The engine oil cooler (heat exchanger) is horizontally mounted between front and rear header as shown in Fig. 119. Service is limited to cleaning and testing although in some cases, soldering of shell seams and hubs (flanges) is permissible.

To remove the oil cooler, first drain cooling system, then remove drain plug from oil cooler and drain cooler. Remove cap screws from front header to oil cooler and the cap screw which re-

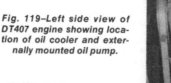

Fig. 119–Left side view of DT407 engine showing location of oil cooler and externally mounted oil pump.

FH. Front header
OC. Oil cooler
OP. Oil pump
RH. Rear header
TD. Tachometer drive

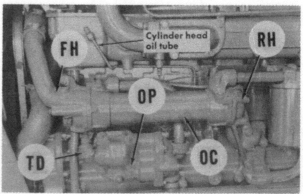

tains front header to cylinder block. Disconnect oil pump and oil filter tube elbows. Loosen rear header to oil cooler cap screws, pull cap screws out as far as possible and remove oil cooler.

Normal cleaning of oil cooler consists of blowing out water tubes with compressed air or the use of a rotary brush or rod of proper diameter. Lubricating oil section of oil cooler can be back flushed with a suitable cleaning solvent or flushing oil. If necessary, oil cooler water tubes can be cleaned by plugging the oil ports and immersing the oil cooler in "Oakite" or similar cleaning solutions.

To test the oil cooler for internal leakage, fabricate two adapter plates with gaskets and block off oil inlet and outlet. Install an air coupling in oil drain hole and attach a regulated air supply. Immerse oil cooler in water of at least 120 degrees F. and allow oil cooler temperature to equalize with water temperature. Apply not more than 80 psi of air pressure. Any leaks present will be readily apparent. Internal leakage will require renewal of oil cooler.

Using new gaskets, reinstall oil cooler by reversing the removal procedure.

Models 766, 966 and 1066 Diesel

165. The engine oil cooler is horizontally mounted to the right side of crankcase.

To remove the oil cooler, first drain cooling system, then remove drain plug (4—Fig. 120) from oil cooler and drain cooler. Unbolt cooler from crankcase and slide off of water tubes.

Normal cleaning of oil cooler consists of blowing out water tubes with compressed air or the use of a rotary brush or rod of proper diameter.

Using new gaskets and "O" rings, reinstall oil cooler by reversing the removal procedure.

OIL PAN

All Diesel Models

166. Removal of oil pan is conventional and on tricycle model tractors can be accomplished with no other disassembly. On Farmall models equipped with adjustable wide tread front axle, the stay rod must be disconnected from stay rod support and stay rod support from side rails before removal of oil pan can be accomplished.

On International and "All Wheel Drive" models, disconnect stay rod (or bracket) from clutch housing or hydrostatic drive housing and front axle from axle support, then raise front of tractor to provide clearance for oil pan removal.

Fig. 120–Right side view of D360, D414 and DT414 engines showing location of oil cooler.

1. Oil cooler
2. Crankcase breather
4. Oil cooler drain

GASOLINE FUEL SYSTEM

CARBURETOR
Models 766 and 826 So Equipped

167. Models 766 and 826 gasoline fuel tractors are equipped with International Harvester built 1⅜ inch updraft carburetors. Disassembly and overhaul is obvious after an examination of the unit and reference to Fig. 121.

Parts data is as follows:

MODEL 826	IH Part No.
Carburetor	533622R91
Gasket package	387454R92
Overhaul package	387459R92
Main jet	533623R1
Idle jet	49798D70
Idle needle	362528R1
Discharge nozzle	384024R1
Venturi	47407D35
Float	47398DX
Float needle and seat	47396DAX40

MODEL 766	IH Part No.
Carburetor	539575R91
Gasket package	387454R93
Overhaul package	387459R92
Main jet	539576R1
Idle jet	49798D70
Idle needle	362528R1
Discharge nozzle	383504R1
Venturi	47407D31
Float	47398DX
Float needle and seat	47396DAX40

168. **LOW IDLE ADJUSTMENT.** Before attempting to adjust the carburetor, first start engine and let it run until thoroughly warmed. Then, adjust idle speed stop screw (2—Fig. 121) to

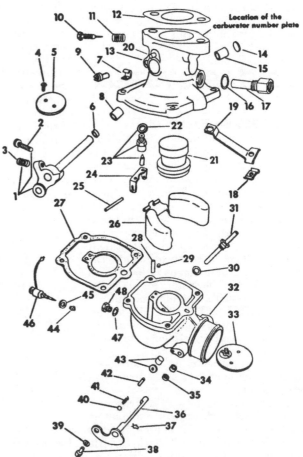

Fig. 121–Exploded view of carburetor used on gasoline engine.

1. Throttle shaft
2. Stop screw
3. Spring
5. Throttle plate
6. Dust seal
7. Clip
8. Bushing
9. Plug
10. Idle adjusting screw
11. Spring
12. Gasket
13. Body
14. Expansion plug
15. Bushing
16. Gasket
17. Screen assembly
18. Clamp
19. Bracket
20. Stop pin
21. Venturi
23. Float needle and seat
24. Pivot support
25. Float pivot pin
26. Float
27. Bowl gasket
28. Idle metering jet
29. Air bleed
31. Metering nozzle
32. Bowl assembly
33. Choke plate
34. Dust seal
35. Dust seal
36. Choke shaft
37. Swivel retainer
38. Swivel
39. Washer
40. Ball
41. Friction spring
42. Stop pin
43. Drip hole filler
44. Fuel adjustment screw seat
46. Fuel shut-off solenoid

obtain a low idle engine speed of 425 rpm. Turn idle mixture screw (10) in or out as required to obtain smoothest idle. Readjust idle speed stop screw, if necessary, to obtain correct low idle speed.

If carburetor has been disassembled, initial setting of idle mixture needle is one turn open.

169. MAIN FUEL ADJUSTMENT. To prevent engine "run-on" or "dieseling" after ignition is switched to "OFF" position, the carburetors are equipped with a solenoid fuel shutoff valve which stops fuel flow through the main jet. The main fuel adjusting screw is located in outer end of solenoid unit (46—Fig. 121). To adjust the main fuel screw, proceed as follows: With ignition switch in "OFF" position, turn adjusting screw in (clockwise) until it just contacts the solenoid plunger. Then, turn adjusting screw out 4½ turns.

170. HIGH IDLE SPEED ADJUSTMENT. The carburetor throttle lever has a stop incorporated; however, engine high idle rpm and governed rpm are controlled by the engine governor. Refer to the governor section for adjustment information.

171. FLOAT SETTING. To check the float setting, remove throttle body from bowl assembly. Invert throttle body and measure distance between top of free ends of float and gasket surface on throttle body. Distance should be 1 5/16 inches. Bend float lever, if necessary, to obtain this setting.

LP-GAS SYSTEM

Model 826 So Equipped

172. An Ensign model CBX 1½ inch updraft carburetor is used on all LP-Gas equipped models which are also equipped with Ensign model RDH regulator-vaporizer. The carburetor is equipped with a diaphragm economizer. Three adjustments; main (load) adjustment, starting adjustment and throttle stop (idle speed) adjustment are located on the carburetor while the idle mixture adjustment is located on the regulator. The system also incorporates a renewable cartridge fuel filter.

173. ADJUSTMENTS. Before attempting to start engine, check and be sure that initial adjustments are as follows:

Starting adjustment ...... 1 turn open
Main adjustment ...... 5⅝ turns open
Idle mixture
 adjustment .... 1⅝ turns open

With initial adjustment made, start engine and run until engine reaches operating temperature. Place throttle control lever in low idle position and

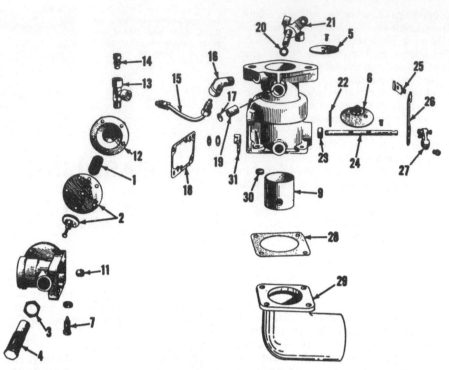

Fig. 122–Exploded view of Ensign model CBX carburetor used on LP-Gas equipped tractors.

1. Economizer spring	7. Starting adjusting screw	16. Elbow	24. Choke shaft
2. Economizer diaphragm	9. Venturi	17. Expansion plug	25. Clamp
3. Lock nut	11. Economizer bleed	18. Gasket	26. Bracket
4. Load adjusting screw	12. Economizer cover	19. Bushing	27. Choke lever
5. Throttle plate	13. Tee	20. Seal	28. Gasket
6. Choke plate	14. Connector	21. Throttle shaft	29. Intake elbow
	15. Balance tube	22. Pin	30. Expansion plug
		23. Dust seal	31. Valve lever

adjust idle speed screw on carburetor to obtain an engine low idle speed of 425 rpm. Turn idle mixture needle on regulator either way as required to obtain the highest and smoothest engine operation. Readjust the carburetor idle speed screw, if necessary, to maintain engine low idle speed of 425 rpm.

With engine low idle adjusted, place throttle control in high idle position. Turn the main (load) adjustment screw (4—Fig. 122) inward until engine begins to falter; then back-out the screw until full power is restored and engine operates smoothly.

NOTE: In some cases, it may be necessary to vary the main fuel adjustment slightly after load is placed on engine.

The initial starting adjustment should provide satisfactory starting performance; however, it may be varied if cold starting is not satisfactory.

174. CARBURETOR OVERHAUL. The carburetor is serviced similar to conventional gasoline type; that is, it can be completely disassembled, cleaned and worn parts renewed. Refer to Fig. 122 for an exploded view of car-

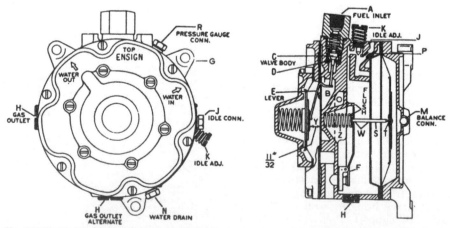

Fig. 123–Sectional views of Ensign model RDH regulator-vaporizer used on LP-Gas equipped tractors.

buretor. Pay particular attention to the economizer diaphragm assembly and make certain that vacuum connections to the economizer chamber do not leak.

175. REGULATOR TROUBLE SHOOTING. To test the regulator on the engine, install a 0-30 psi test gage at the pipe plug opening (R—Fig. 123). Slowly open the vapor service valve on tank. The pressure reading on test gage should raise and hold steady within the range of 8-9 psi. If the pressure is a few pounds over or under this range but remains steady, it is an indication that the high pressure valve lever requires adjustment. If test gage pressure continues to raise beyond 15 psi, the high pressure valve is leaking. High pressure valve (C—Fig. 123) can be serviced without removing or disassembling the regulator unit. Severe leakage at high pressure valve (pressure above 15 psi) will force low pressure valve off its seat, upsetting fuel economy, causing loss of control of idling adjustment and preventing proper starting due to an overly rich starting mixture.

Leakage through the high pressure diaphragm will have an identical effect on operation as a leaking high pressure valve; however, no increase in test gage pressure beyond the normal range (8-9 psi) may be indicated.

If test gage pressure holds steady within the range of 8-9 psi and there is no control of idle adjustment, improper starting is experienced and after standing awhile, frost appears on regulator, the low pressure valve (F) is leaking. A severe leak at low pressure valve can normally be heard.

Leakage through the low pressure diaphragm effects only the sensitive response of low pressure valve to the fuel demand of the engine. Response drops off rapidly with the increase in size of hole in diaphragm.

"Freeze up" of the regulator-vaporizer during engine operation is caused by lack of circulation of hot water through the regulator vaporizer unit. Check for restrictions.

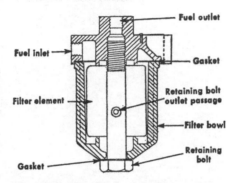

Fig. 125—Sectional view of LP-Gas fuel filter. Some filters may have a center stud and nut instead of the retaining bolt shown.

176. MODEL RDH REGULATOR OVERHAUL. Disassembly of the RDH regulator is obvious after an examination of the unit and reference to Figs. 123 and 124. Wash parts in clean suitable solvent and blow out all passageways with compressed air. When reassembling, renew all gaskets and any parts that show excessive wear or other damage. Adjust pressure valve levers to correct dimensions as follows: High pressure lever (E—Fig. 123) should be set at 11/32-inch from top of lever to face of plate. See dimension (Y). Bend lever, if necessary, to obtain this dimension.

When installing low pressure valve (27—Fig. 124) and spring (25), center the low pressure lever with push pin hole in center of partition plate (30). A rib is provided in recess of body (19) for the purpose of setting low pressure lever height. Top of lever, when valve is seated, should be flush with top of this rib. Bend lever, if necessary, to obtain this setting.

177. LP-GAS FILTER. The filter is designed for a working pressure of 375 psi and filter element should be renewed when filter becomes clogged enough to restrict flow of fuel.

A clogged filter element causes a pressure drop within the filter and generally results in freezing (frost) at the filter as well as a noticeable drop in engine power.

Renewal of filter element is obvious after an examination of the filter and reference to Fig. 125. Use care when handling element to prevent crushing the sides. Make certain gasket surfaces are clean and use all new gaskets when assembling.

DIESEL FUEL SYSTEM

The diesel fuel system of model 826 is equipped with a Robert Bosch injection pump. Model 1026 is equipped with a Roosa-Master injection pump. Models 766, 966 and 1066 are equipped with an American Bosch injection pump. All engines are of the direct injection type.

When servicing any unit of the diesel system, the maintenance of absolute cleanliness is of utmost importance. Of equal importance is the avoidance of nicks or burrs on any of the working parts.

Probably the most important precaution that service personnel can impart to owners of diesel powered tractors, is to urge them to use an approved fuel that is absolutely clean and free from foreign material. Extra precaution should be taken to make certain that no water enters the fuel storage tank.

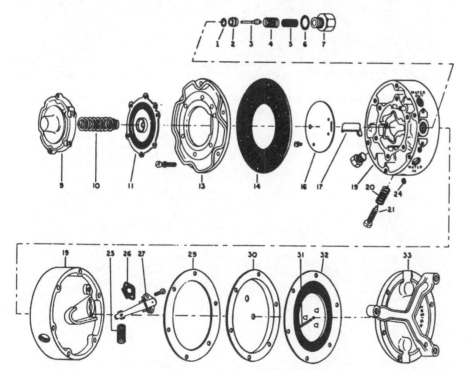

Fig. 124—Exploded view of model RDH regulator-vaporizer.

1. "O" ring	9. Cover	17. Valve lever	27. Low pressure valve
2. Valve seat	10. Spring	19. Body	29. Gasket
3. High pressure valve	11. High pressure diaphragm	20. Spring	30. Plate
4. Spring	13. Cover	21. Idle screw	31. Push pin
5. Valve spring	14. Gasket	24. Bleed screw	32. Low pressure diaphragm
6. Gasket	16. Plate	25. Valve spring	33. Support plate
7. Valve retainer		26. Gasket	

FILTERS AND BLEEDING
All Diesel Models

178. All diesel tractors are equipped with two fuel filters. Model 826 is equipped with a Robert Bosch injection pump and model 1026 is equipped with a Roosa-Master injection pump. Models 766, 966 and 1066 are equipped with American Bosch injection pump. The primary filter is the rear of the two filters. All models are equipped with spin-on cartridge type fuel filters. Filters should be serviced when engine shows signs of losing power.

179. To bleed the fuel system on model 826 D358 engine equipped with Robert Bosch injection pump, refer to Fig. 126 and proceed as follows: Open vent valves on top of primary and final filters (one valve on each). When outflowing fuel is free of air bubbles, close vent valves. Next, open vent screw on injection pump until fuel is flowing free of air bubbles, then retighten vent screw. Fuel system is now free of air.

180. To bleed the fuel system on model 1026 equipped with the Roosa-Master injection pump, refer to Fig. 127 and proceed as follows: Loosen vent valve on top of primary filter and allow fuel to run until a solid stream with no air bubbles appears; then, close vent valve. Repeat this operation on the final fuel filter. Loosen the fuel supply line connection on the injection pump inlet elbow and when bubble free fuel appears, tighten connection. Fuel system is now free of air.

181. To bleed the fuel system on models 766, 966 and 1066 equipped with the American Bosch injection pump, refer to Fig. 128 and proceed as follows: Loosen vent valve on top of primary filter and allow fuel to run until a solid stream with no air bubbles appears; then, close vent valve. Open vent valve on final filter. Then, using the hand primary pump, pump fuel into final filter until fuel coming out of final filter vent is free of air bubbles. Close the vent on final filter and lock the hand pump plunger in place. Fuel system is now free of air.

INJECTION PUMP

All Diesel Models

The Robert Bosch, Roosa-Master and American Bosch injection pumps are all of the rotary distributor type. Because of the special equipment needed, and skill required of servicing personnel, service of injection pumps is generally beyond the scope of that which should be attempted in the average shop. Therefore, this section will include only timing of pump to engine, removal and installation and the linkage adjustments which control the engine speeds.

If additional service is required, the pump should be turned over to an International Harvester facility which is equipped for diesel service, or to some other authorized diesel service station. Inexperienced personnel should NEVER attempt to service diesel injection pumps.

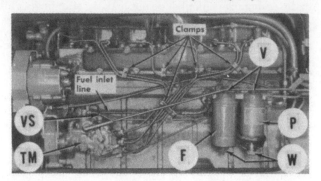

Fig. 126–Left side view of model 826 diesel D358 engine equipped with Robert Bosch injection pump, showing location of diesel fuel system components.

F. Final fuel filter
P. Primary fuel filter
V. Vents
W. Water drain valve
TM. Timing hole cover
VS. Vent screw

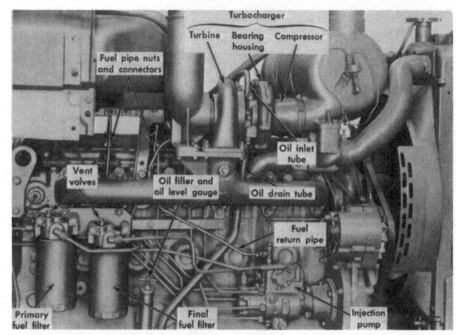

Fig. 127–Right side view of model 1026 diesel DT407 engine, showing location of diesel fuel system components.

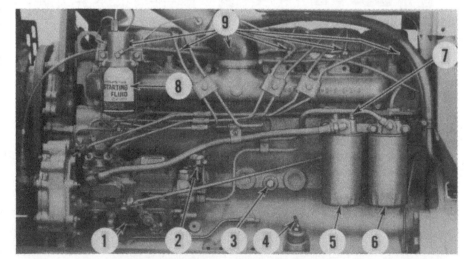

Fig. 128–Left side view of models 766, 966 and 1066 diesel engine equipped with American Bosch injection pump, showing location of diesel fuel system components.

1. Injection pump
2. Hand primer
3. Drain plug (water)
4. Oil filler and level gauge
5. Fuel filter (final)
6. Fuel filter (primary)
7. Vent screw
8. Starting ether
9. Injection lines

REMOVE AND REINSTALL PUMP

Model 826
(Robert Bosch)

182. To remove the Robert Bosch injection pump from the D358 engine, first thoroughly clean injection pump, fuel lines and side of engine. Shut off fuel at tank and remove timing hole cover screws from pump. Rotate cover 90 degrees and remove by prying evenly with two screwdrivers. Remove timing hole cover on right side of clutch housing. Using a bar on flywheel, turn engine in normal direction of rotation until timing line (TL—Fig. 129) on face cam is aligned with timing pointer (TP) in pump.

NOTE: The face cam has two timing lines. Near one of the lines, a letter "L" is etched on face cam. DO NOT use this timing line.

Disconnect throttle control rod and shut-off cable from pump. Disconnect fuel inlet and return lines from pump, then remove high pressure injection lines. Immediately cap or plug all openings. Unbolt and remove the rectangular cover from timing gear cover. Remove nut and washer from pump drive shaft, then remove three cap screws securing hub to drive gear. Remove nuts from pump mounting studs. Install a puller (IH tool No. FES 111-2 or equivalent) to tapped holes in drive gear and force drive shaft from hub. Remove injection pump.

CAUTION: Do not turn crankshaft while pump is removed.

When reinstalling pump, make certain that timing pointer (TP) and timing line (TL) are aligned, then reinstall pump by reversing removal procedure. Align scribe mark on pump mounting flange with punch mark on engine front plate. Tighten pump drive

shaft nut to a torque of 47 ft.-lbs. and the three hub to drive gear cap screws to a torque of 17 ft.-lbs.

Bleed fuel system as in paragraph 179, then check and adjust injection pump timing as in paragraph 186.

Model 1026
(Roosa-Master)

183. To remove the Roosa-Master injection pump from the DT407 engines, first clean injection pump, fuel lines and side of engine. Remove hood and carefully remove turbocharger assembly as outlined in paragraph 201. Unbolt and remove exhaust manifold. Shut off fuel and remove timing hole cover from injection pump. Remove timing hole cover on right side of hydrostatic drive housing. Using a pry bar on flywheel teeth, turn crankshaft in normal direction of rotation until timing marks (TM—Fig. 130) are aligned in pump timing window. Disconnect control rod and the supply and return lines from pump and the lines from injectors. Cap or plug all openings immediately. Wire the throttle lever on pump in high idle (full fuel) position. This will allow governor spring tension to hold governor weights in position. Unbolt injection pump from pump drive adapter, then withdraw pump with injector lines from pump shaft.

If necessary, the injection pump drive shaft can be removed as outlined in paragraph 184. When reinstalling injection pump, align the timing slot on pump drive shaft with the timing pin on the pump rotor. Carefully slide pump into position. Align timing marks (TM—Fig. 130) and tighten the pump mounting nuts. The balance of reassembly is the reverse of disassembly procedure. Bleed fuel system as in paragraph 180 and check and adjust injection pump timing as in paragraph 187.

184. To remove the injection pump drive shaft assembly, first remove injection pump as in paragraph 183. Unbolt and remove pump drive gear cover from front of timing gear cover. Remove the three cap screws securing pump adapter to engine front plate. Remove nut and washer from front end of pump drive shaft and using a suitable puller, force shaft rearward from gear. Withdraw pump shaft and adapter assembly.

Any further disassembly will be obvious after an examination of the unit and reference to Fig. 131.

CAUTION: Do not rotate crankshaft while pump shaft and adapter assembly is removed.

When reinstalling tighten gear retaining nut on drive shaft to a torque of 55 ft.-lbs. Refer to paragraph 183 for pump installation procedure.

Models 766, 966 and 1066
(American Bosch)

185. To remove the American Bosch injection pump from the D360, D414 and DT414 engines, first clean injection pump, fuel lines and side of engine. Remove timing hole cover on right side of clutch housing or hydrostatic drive housing. Using a pry bar on flywheel teeth, turn crankshaft in normal direction of rotation until timing marks on flywheel are aligned at 18° BTDC which is static timing. Disconnect control rod, fuel lines and oil supply line from injection pump. Cap or plug all openings immediately. Remove fuel filter to pump lines, left radiator support panel and pump access cover from front cover. Unbolt and remove pump and adapter plate as an assembly.

NOTE: If same pump is to be reinstalled, DO NOT rotate the pump drive shaft after pump has been removed.

When reinstalling pump with drive gear removed, secure with two cap

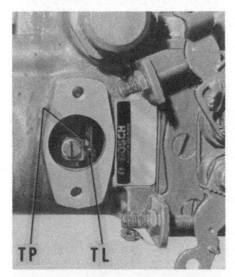

Fig. 129–View of timing pointer (TP) and timing line (TL) on Robert Bosch injection pump.

Fig. 130–Timing marks (TM) on governor drive plate and cam ring on Roosa-Master injection pump.

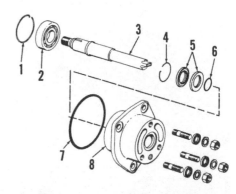

Fig. 131–Exploded view of Roosa-Master injection pump drive shaft assembly used on DT407 engines.

1. Bearing retaining ring
2. Bearing
3. Drive shaft
4. Seal retaining ring
5. Shaft seal
6. "O" ring
7. "O" ring
8. Adapter

screws (3—Fig. 132) with timing mark on flywheel aligned with pointer at 18° BTDC (it does not matter whether it is on No. 1 or No. 6 cylinder.) Turn pump so that line mark (1—Fig. 132 or 2—Fig. 133) on pump hub are aligned with pointer. Install drive gear, meshing it with idler gear. NOTE: Disregard any timing marks on idler gear. Hold pump shaft with socket and tighten cap screws in gear to a torque of 30 ft.-lbs.

Bleed fuel system as in paragraph 181 if engine does not start after fifth or sixth revolution, the pump is out of time. Reposition engine crankshaft to 18° BTDC on flywheel. Then, remove pump drive gear. Rotate pump drive shaft one revolution and reinstall gear. Pump is now in time.

Model 826
(Robert Bosch)

186. **STATIC TIMING.** To check or adjust static timing, first shut off fuel at tank and remove pump timing hole screws. Rotate cover 90 degrees and remove cover by prying evenly with two screwdrivers. Remove timing hole cover from right side of clutch housing or hydrostatic drive housing. Using a pry bar on flywheel teeth, turn crankshaft in normal direction of rotation until number one piston is coming up on compression stroke. Continue turning crankshaft until the 16 degrees BTDC mark on flywheel is aligned with timing pointer on clutch housing. At this time, timing line (TL—Fig. 129) on face cam should be aligned with timing pointer (TP) on roller retainer ring. If not, first make certain that scribe line on pump mounting flange is aligned with punch mark on engine front plate. Then, unbolt and remove rectangular cover from front of timing gear cover. Loosen the three cap screws securing pump drive gear to pump shaft hub. Rotate hub as required to align timing line on face cam with the timing pointer.

Tighten hub retaining cap screws to a torque of 17 ft.-lbs. and install all removed covers.

Model 1026
(Roosa-Master)

187. **STATIC TIMING.** To check or adjust static timing, first shut off fuel at tank and remove timing hole cover from side of injection pump. Remove timing hole cover from right side of hydrostatic drive housing. Using a pry bar on flywheel teeth, turn crankshaft in normal direction of rotation until number one piston is coming up on compression stroke. Continue turning crankshaft until the 6 degrees BTDC mark on flywheel is aligned with timing pointer on hydrostatic drive housing.

At this time, timing marks on injection pump cam ring and governor drive plate should be aligned as shown in Fig. 130. If timing marks are not aligned as shown, loosen the pump mounting stud nuts and rotate pump as required to align the marks. Tighten the mounting nuts and reinstall the removed covers.

Models 766, 966 and 1066
(American Bosch)

188. **STATIC TIMING.** To check or adjust static timing, remove access cover Fig. 132 on early models or plug Fig. 133 on late models. Remove timing hole cover from right side of clutch housing or hydrostatic drive housing. Using a pry bar on flywheel teeth, turn crankshaft in normal direction of rotation until the 18 degrees BTDC mark on flywheel is aligned with timing pointers on clutch housing or hydrostatic drive housing.

At this time, timing mark on injection pump hub and pointer, Fig. 132 or 133, should be aligned as shown. If timing marks are not aligned as shown, remove pump drive gear and

line pump hub and pointer. Reinstall drive gear and torque to 30 ft.-lbs. Install access cover.

Model 826
(Robert Bosch)

189. **LOW & HIGH IDLE.** To adjust low idle speed, start engine and bring to operating temperature. Place throttle control lever in low idle position, loosen jam nut and turn low idle speed stop screw (L—Fig. 134) as required to obtain an engine low idle speed of 600 rpm. Tighten jam nut.

Move throttle control lever to high idle position, loosen jam nut and turn high idle stop screw (H) as required to obtain an engine high idle speed of 2640 rpm. Tighten jam nut.

With engine high idle speed properly adjusted, the rated load engine speed should be 2400 rpm. To adjust rated load rpm, use a dynamometer and load the engine to maintain rated rpm with throttle control lever in high idle position. Adjust the maximum fuel stop screw to obtain approximately 92 (pto) horsepower for gear drive or 84 (pto) horsepower for hydrostatic drive at rated load (2400) rpm. CAUTION: Do not overfuel the engine or attempt to increase horsepower above the rated load.

190. **SHUT-OFF PLUNGER.** The shut-off plunger (S—Fig. 134) provides an excess fuel starting position for the shut-off control lever on injection pump. To adjust the plunger, disconnect shut-off cable, loosen jam nut and back plunger unit out several turns. Operate engine at approximately 900 rpm. Move shut-off lever rearward until engine speed increases. Hold lever in this position, turn plunger unit in until end of plunger just contacts the lever, then tighten jam nut. Move lever fully rearward to depress the spring loaded plunger. Engine must shut off at this position. Connect shut-off cable to control lever.

Fig. 132–View showing the timing pointer and timing mark on pump hub.

1. Line mark on hub
2. Timing pointer
3. Cap screws

Fig. 133–View showing timing pointer on late model tractors. Refer to text.

1. Timing pointer
2. Line mark on hub

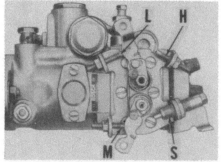

Fig. 134–View of Robert Bosch injection pump showing adjustment points.

H. High idle stop screw	screw
L. Low idle screw	S. Shut-off plunger
M. Maximum fuel stop	adjusting screw

Model 1026
(Roosa-Master)

191. **LOW & HIGH IDLE.** To adjust the engine low and high idle speeds, start engine and bring to operating temperature. Disconnect throttle control rod from pump control lever. Hold pump control lever in high idle position (rearward against stop) and check engine rpm. Engine high idle speed should be 2615 rpm. Adjust high idle stop screw on rear of pump control lever as necessary to obtain correct high idle rpm.

Release the pump control lever and check engine low idle rpm. Low idle speed should be 650 rpm. Adjust engine low idle rpm with the screw on top of pump cover.

With engine operating at low idle rpm, place throttle control lever in low idle position (against dowel pin stop). Adjust length of throttle control rod to hold the pump control lever in low idle position. With control rod connected, place throttle control lever in high idle position. Check to see that pump control lever stop screw is held firmly on its stop by pressure from the control lever spring.

Models 766, 966 and 1066
(American Bosch)

192. **LOW & HIGH IDLE.** To adjust the engine low and high idle speeds, start engine and bring to operating temperature. Disconnect throttle control rod from pump control lever. Hold pump rod lever in high position (rearward against stop) and check engine rpm. Engine high idle should be as follows:

766 (below chassis serial 9000) .. 2600
766 (above chassis serial 9000) .. 2800
966 (all hydrostatic drive) 2600
966 (gear drive below chassis
    serial 17000) 2600
966 (gear drive above chassis
    serial 17000) 2800
1066 (all hydrostatic drive) 2600
1066 (gear drive below chassis
    serial 23000) 2600
1066 (gear drive above chassis
    serial 23000) 2800

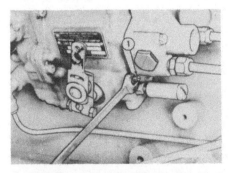

Fig. 135–View showing the high idle stop screw adjustment on American Bosch injection pump.

Adjust high idle stop screw (1—Fig. 135) on rear of pump as necessary to obtain correct high idle rpm.

Release the pump control lever and check engine low idle rpm. Low idle speed should be 700 rpm for all models. Adjust engine low idle rpm with the screw (1—Fig. 136) on rear of pump.

With engine operating at low idle rpm, place the throttle control lever in low idle position (in detent approximately ¾-inch from stop). Adjust length of throttle control rod to hold the pump control lever in low idle position.

INJECTION NOZZLES

Warning: Fuel leaves the injection nozzles with sufficient force to penetrate the skin. When testing, keep your person clear of the nozzle spray.

All Diesel Models

193. **TESTING AND LOCATING FAULTY NOZZLE.** If engine does not run properly and a faulty injection nozzle is suspected, or if one cylinder is misfiring, locate the faulty nozzle as follows: Loosen the high pressure line fitting on each nozzle holder in turn, thereby allowing fuel to escape at the union rather than enter the cylinder. As in checking spark plugs in a spark ignition engine, the faulty nozzle is the one that when its line is loosened least affects the running of the engine.

Remove the suspected nozzle as outlined in paragraph 194, 195 or 196, place nozzle in a test stand and check the nozzle against the following specifications:

Model 826 (D 358)

Opening pressure, new ...... 3000 psi
Opening pressure, used ...... 2900 psi

Nozzle showing visible wetting on the tip after 10 seconds at 2700 psi is permissible. Maximum leakage through return port is 10 drops in 1 minute at 2700 psi.

Model 1026 (DT407)

Opening pressure,
    new .... 3100-3200 psi
Opening pressure,
    used .... 2950-3050 psi

Nozzle showing visible wetting on the tip after 5 seconds at 2500 psi is permissible. Maximum leakage through return port is 10 drops in 1 minute at 1500 psi.

Models 766, 966 and 1066 (D360, D414, DT414)

Opening pressure, new . 3600-3750 psi
Opening pressure, used ...... 2900 psi

Nozzles showing visible wetting on the tip after 5 seconds with pressure held at 100 psi below opening pressure is permissible.

Model 826 (D 358)

194. **R&R NOZZLES.** To remove any injection nozzle, first remove dirt from nozzle injection line, return line and cylinder head. Disconnect leakage return line and high pressure injection line from nozzle and immediately cap or plug all openings. Remove the injector stud nuts and carefully withdraw the injector assembly from cylinder head.

NOTE: It is recommended that cooling system be drained before removing injectors. It is possible that injector nozzle sleeve may come out with injector and allow coolant to enter engine.

When reinstalling, tighten injector stud nuts evenly and to a torque of 8 ft.-lbs.

Model 1026 (DT407)

195. **R&R NOZZLES.** To remove the injector nozzles first remove dirt from injection lines, return line and injector nozzles. Remove turbocharger as outlined in paragraph 201. Then, unbolt and remove the exhaust manifold. Disconnect and remove the leakoff manifold (return line) assembly, then disconnect the high pressure line from injector. Cap or plug all openings immediately. Remove the injector holddown cap screws and lift injector assembly from cylinder head.

NOTE: It is recommended that cooling system be drained before removing injectors. It is possible that injector nozzle sleeve may come out with injector and allow coolant to enter engine.

When reinstalling, tighten nozzle hold-down cap screws evenly (in 2 ft.-lb. increments) to a torque of 10 ft.-lbs.

Models 766, 966 and 1066 (D-360, D414, DT414)

196. **R&R NOZZLES.** To remove any injection nozzle, first remove left front hood. Then remove dirt from nozzle injection line, return line and cylinder head. Disconnect leakage return line and high pressure injection line from nozzle and immediately cap or plug all openings. Remove the injector cap screw and carefully withdraw the injector assembly from cylinder head.

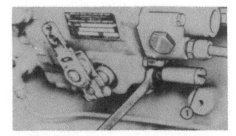

Fig. 136–View showing low idle adjusting screw on American Bosch injection pump.

NOTE: It is recommended that cooling system be drained before removing injectors. It is possible that injector nozzle sleeve may come out with injector and allow coolant to enter engine.

When reinstalling, tighten injector cap screw to a torque of 20 ft.-lbs.

Model 826 (D358)

197. OVERHAUL. To disassemble the injector, place assembly in a vise with nozzle tip pointing upward. Remove nozzle holder nut (13—Fig. 137), then carefully remove nozzle tip (12) and valve (11). Invert the body assembly in vise and remove adjusting lock cap (3). Remove adjusting screw (4), spring seat (5), spring (6), and spindle (7), then remove fuel inlet connector (10).

Thoroughly clean all parts in a suitable solvent. Clean inside the orifice end of nozzle tip with a wooden cleaning stick. The 0.011 diameter orifice spray holes may be cleaned by inserting a cleaning wire of proper size. Cleaning wire should be slightly smaller than spray holes. Clean the fuel return passage in body (8) with a 5/64-inch drill.

When reassembling, make certain all parts are perfectly clean and install parts while wet with clean diesel fuel. To check cleanliness and fit of valve (11) in nozzle tip (12), use a twisting

motion and pull valve about ⅓ of its length out of nozzle tip. When released, valve should slide back to its seat by its own weight.

NOTE: Valve and nozzle tip are mated parts and under no circumstance should valves and nozzle tips be interchanged.

Tighten nozzle holder nut (13) to a torque of 50 ft.-lbs. Connect the assembled injector nozzle to a test pump and flush the valve. Adjust opening pressure by turning adjusting screw (4) in

to increase or out to decrease opening pressure. Opening pressure should be adjusted to 3000 psi. Valve should not show leakage at orifice spray holes for 10 seconds at 2700 psi.

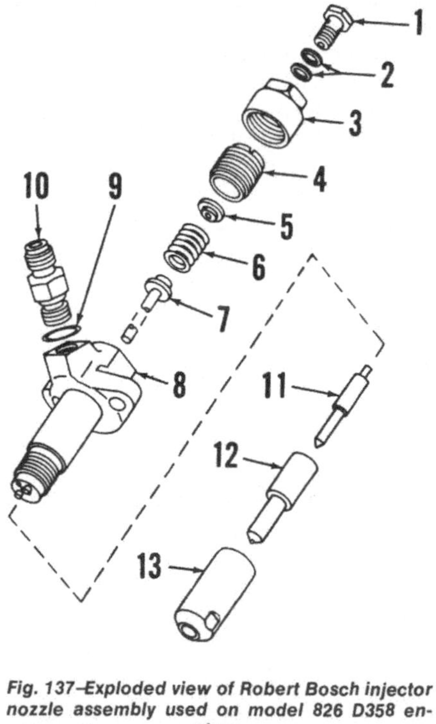

Model 1026 (DT407)

198. OVERHAUL. Place injector assembly in a vise or holding fixture with the nozzle tip pointing upward. Remove holder nut (8—Fig. 138), then carefully remove nozzle (7) and valve (6). Remove body (1) from vise, invert the body and remove intermediate plate (5), spring seat (4), spring (3) and adjusting shim (2). Thoroughly clean all parts in a suitable solvent. Inspect mating surfaces of nozzle, intermediate plate and holder nut for cracks, scratches or other defects. Light scratches can be removed by lapping; however, the sealing surfaces must remain perfectly flat. Clean inside the orifice end of nozzle with a wooden cleaning stick. Clean the "sac" end of nozzle with a 0.050 inch drill. Fuel return passages can be cleaned with a 5/64-inch drill. Orifice spray holes can be cleaned by inserting a cleaning wire of proper size. Cleaning wire should be slightly smaller than spray holes. Orifice spray hole diameter is stamped on nozzle just above orifice tip. Carbon can be removed from exterior surfaces with a soft brass brush and clean diesel fuel.

When reassembling, be sure all parts are perfectly clean and install parts while wet with clean diesel fuel. To check cleanliness and fit of valve (6) in nozzle (7), use a twisting motion and pull valve about ⅓ of its length out of nozzle. When released, valve should slide back to its seat by its own weight.

NOTE: Valve and nozzle (6 and 7) are mated parts and under no circumstance should valves and nozzles be interchanged. Lapping is not recommended on the valve and nozzle seat and faulty units must be renewed.

Reassemble injector by reversing the disassembly procedure. Tighten nozzle holder nut to a torque of 65 ft.-lbs. Adjust opening pressure to 3100-3200 psi. Injector should not show leakage at orifice spray holes for 5 seconds at 2500 psi.

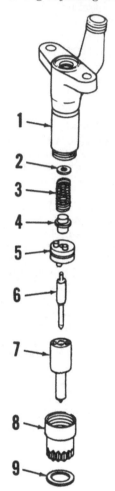

Fig. 138–Exploded view of Robert Bosch injector nozzle assembly used on model 1026 DT407 engines.

1. Body
2. Adjusting shim
3. Spring
4. Spring seat
5. Intermediate plate
6. Valve
7. Nozzle
8. Holder nut
9. Gasket

Fig. 137–Exploded view of Robert Bosch injector nozzle assembly used on model 826 D358 engines.

1. Hollow screw
2. Seal rings
3. Adjusting lock cap
4. Adjusting screw
5. Spring seat
6. Spring
7. Spindle
8. Body
9. Seal ring
10. Fuel inlet connector
11. Valve
12. Nozzle tip
13. Nozzle holder nut

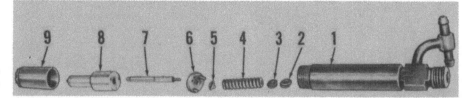

Fig. 139–Exploded view of American Bosch injector nozzle assembly used on diesel models 766, 966 and 1066 tractors.

1. Body
2. Adjusting shims
3. Spacer (2 used .060)
4. Spring
5. Spring seat
6. Valve stop spacer
7. Valve
8. Nozzle
9. Nozzle cap

Models 766, 966 and 1066
(D360, D414, DT414)

199. **OVERHAUL.** Clamp the injector assembly upright in a vise using the injector retainer from the engine. Remove nozzle cap (9—Fig. 139), then carefully remove nozzle (8) and valve (7). Remove body (1) from vise, invert the body and remove valve stop spacer (6), spring seat (5), spring (4), spacer shim (3), adjusting shims (2) and the other spacer shim (3). Thoroughly clean all parts in suitable solvent. Inspect mating surfaces of nozzles, valve stop spacer and nozzle cap for cracks, scratches or other defects. Light scratches can be removed by lapping; however, the sealing surfaces must remain perfectly flat. Orifice spray holes can be cleaned by inserting a cleaning wire of proper size. Cleaning wire should be slightly smaller than spray holes. NOTE: Spray holes on all models are 0.012 except models 966 gear drive below chassis serial 13531 and hydrostatic drive below chassis serial 15356 which have a 0.011 spray holes.

Carbon can be removed from exterior surfaces with a soft brass brush and clean diesel fuel.

When reassembling, be sure all parts are perfectly clean and install parts while wet with clean diesel fuel. To check cleanliness and fit of valve (7) in nozzle (8), use a twisting motion and pull valve about ⅓ of its length out of nozzle. When released, valve should slide back to its seat by its own weight.

NOTE: Valve and nozzle (7 and 8) are mated parts and under no circumstance should valve and nozzles be interchanged. Lapping is not recommended on the valve and nozzle seat and faulty units must be renewed.

Be sure to install one spacer shim (3) on each side of adjusting shims (2). The spacer shims are 0.060 thick.

Reassemble injector by reversing the disassembly procedure. Tighten nozzle cap to a torque of 30 to 35 ft.-lbs. Adjust opening pressure to 3600-3750 psi. Injector should not show leakage at orifice spray holes for 5 seconds at 100 psi below opening pressure.

ETHER STARTING AID
All Diesel Models So Equipped

200. On all diesel engines, it is necessary that ether be used as a starting aid at temperatures below freezing.

To test the ether spray pattern, disconnect ether line at spray nozzle and spray nozzle from manifold air inlet. Reconnect nozzle to ether line. Press ether injection button on dash and observe spray pattern. A good spray pattern is cone-shaped. Dribbling or no spray indicates a blocked spray nozzle or lack of ether pressure. Clean spray nozzle or install new can of ether as needed.

To change the ether fluid container, turn knurled adjusting screw clockwise until container can be removed. Install new container in the bail and tighten adjusting screw (counterclockwise) while guiding container head into position. Rotate container to make certain it is seated properly in injector body, then tighten adjusting screw to hold container firmly in position. CAUTION: Ether must be in twelve ounce containers meeting ICC29 specifications.

NOTE: In warm temperatures, ether container can be removed and a protective plug installed in injector body. DO NOT operate tractor engine without either the ether container or protective plug in position.

DIESEL TURBOCHARGER

Models 1026 and 1066 diesels may be equipped with either an exhaust driven Schwitzer model 3LD turbocharger or an Airesearch model TO-4 turbocharger. The turbocharger consists of the following three main sections. The turbine, bearing housing and compressor.

Engine oil taken directly from the clean oil side of the engine oil filters, is circulated through the bearing housing. This oil lubricates the sleeve type bearings and also acts as a heat barrier between the hot turbine and the compressor. The oil seals used at each end of the shaft are of the piston ring type. When servicing the turbocharger, extreme care must be taken to avoid damaging any of the parts.

NOTE: When the engine has been idle for one month or more, after installation of a new or rebuilt turbocharger or after installing new oil filter elements, the turbocharger must be primed as outlined in paragraph 201 or 202.

REMOVE AND REINSTALL
Model 1026 (DT407)

201. To remove the turbocharger, first remove the hood skirts and hood. Unbolt and remove the exhaust elbow, then remove the turbocharger oil inlet and oil drain tubes. Plug or cap all openings. Loosen clamps and disconnect the cross-over tube and the flexible air intake elbow. Unbolt the turbocharger from exhaust manifold, lift the unit from engine and place it in a horizontal position on a bench.

To reinstall the turbocharger, reverse the removal procedure and prime the turbocharger as follows: With the speed control lever in shut-off position, crank engine with starting motor for approximately 30 seconds. Repeat this operation until the engine oil pressure Tellite goes out. Start engine and operate at 1000 rpm for at least 2 minutes before going to a higher speed.

Model 1066 (DT414)

202. To remove the turbocharger, first remove muffler, left and right front hood and air inlet and outlet pipes. Disconnect the turbocharger oil inlet and oil drain tubes. Plug or cap all openings. Unbolt the turbocharger from exhaust manifold, lift the unit from engine and place it in a horizontal position on a bench.

To reinstall the turbocharger, reverse the removal procedure and prime the turbocharger as follows: With the speed control lever in shut-off position, crank engine with starting motor approximately 30 seconds. Repeat this operation until the engine oil pressure Tellite goes out. Start engine and operate at 1000 rpm for at least 2 minutes before going to higher speed.

203. **OVERHAUL (SCHWITZER)** Remove turbocharger as outlined in paragraph 201 or 202. Before disassembling, place a row of light punch marks across compressor cover, bearing housing and turbine housing to aid in reassembly. Clamp turbocharger mounting flange (exhaust inlet) in a vise and remove cap screws (14—Fig. 140), lock washers and clamp plates (13). Remove compressor cover (3). Remove nut from clamp ring (16), expand clamp ring and remove bearing housing assembly from turbine housing (18).

CAUTION: Never allow the weight of the bearing housing assembly to rest on either the turbine or compressor wheel vanes. Lay the bearing housing assembly on a bench so that turbine shaft is horizontal.

Remove locknut (2—Fig. 140) and slip compressor wheel (1) from end of shaft. Withdraw turbine wheel and shaft (17) from bearing housing. Place bearing housing on bench with compressor side up. Remove snap ring (7), then using two screwdrivers, lift flinger plate insert (6) from bearing housing. Push spacer sleeve (4) from the insert. Remove oil deflector (11), thrust ring (10), thrust plate (9) and bearing (12). Remove "O" ring (8) from flinger plate insert (6) and remove seal rings (5) from spacer sleeve and turbine shaft.

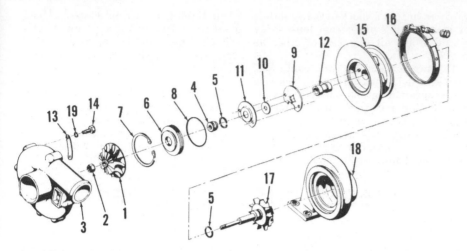

Fig. 140–Exploded view of Schwitzer model 3LD turbocharger assembly used on some diesel models.

1. Compressor wheel	6. Flinger plate insert	11. Oil deflector	16. Clamp ring
2. Lock nut	7. Snap ring	12. Bearing	17. Turbine wheel and
3. Compressor cover	8. "O" ring	13. Clamp plate	shaft
4. Spacer sleeve	9. Thrust plate	14. Cap screw	18. Turbine housing
5. Seal rings	10. Thrust ring	15. Bearing housing	19. Lock washer

Soak all parts in Bendix metal cleaner or equivalent and use a soft brush, plastic blade or compressed air to remove carbon deposits. CAUTION: Do not use wire brush, steel scraper or caustic solution for cleaning, as this will damage turbocharger parts.

Inspect turbine wheel and compressor wheel for broken or distorted vanes. DO NOT attempt to straighten bent vanes. Check bearing bore in bearing housing, floating bearing (12) and turbine shaft for excessive wear or scoring. Inspect flinger plate insert, spacer sleeve, oil deflector, thrust ring and thrust plate for excessive wear or other damage.

Renew all damaged parts and use new "O" ring (8) and seal rings (5) when reassembling. The seal ring used on turbine shaft is copper plated and is larger in diameter than the seal ring used on spacer sleeve. Refer to Figs. 140 and 141 as a guide when reassembling.

Install seal ring on turbine shaft, lubricate seal ring and install turbine wheel and shaft in bearing housing. Lubricate I.D. and O.D. of bearing (12), install bearing over end of turbine shaft and into bearing housing. Lubricate both sides of thrust plate (9) and install plate (bronze side out) on the aligning dowels. Install thrust ring (10) and oil deflector (11), making certain holes in deflector are positioned over dowel pins. Install new seal ring on spacer sleeve (4), lubricate seal ring and press spacer sleeve into flinger plate insert (6). Position new "O" ring (8) on insert, lubricate "O" ring and install insert and spacer sleeve assembly in bearing housing, then secure with snap ring (7). Place compressor

wheel on turbine shaft, coat threads and back side of nut (2) with "Never-Seez" compound or equivalent, then install and tighten nut to a torque of 13 ft.-lbs. Assemble bearing housing to turbine housing and align punch marks. Install clamp ring, apply "Never-Seez" on threads and install nut and torque to 10 ft.-lbs. Apply a light coat of "Never-Seez" around machined flange of compressor cover (3). Install compressor cover, align punch marks, and secure cover with cap screws, washers and clamp plates.

Tighten cap screws evenly to a torque of 5 ft.-lbs.

Check rotating unit for free rotation within the housings. Cover all openings until the turbocharger is reinstalled.

Use a new gasket and install and prime turbocharger as outlined in paragraphs 201 or 202.

204. OVERHAUL (AIRE-SEARCH) Remove turbocharger as outlined in paragraph 201 or 202. Mark across compressor housing, center housing and turbine housing to aid alignment when assembling.

CAUTION: Do not rest weight of any parts on impeller or turbine blades. Weight of only the turbocharger unit is enough to damage the blades.

Remove clamp (1—Fig. 142), compressor housing (2) and diffuser (3). Remove cap screws (11) and lock plates (12) and clamp plates (13); then, remove turbine housing (8). Hold turbine shaft from turning using socket wrench at the center of turbine wheel (14) and remove locknut (22). NOTE: Use a "T" handle to remove locknut in order to prevent bending turbine shaft. Lift compressor impeller (21) off, then remove center housing from turbine shaft while holding shroud (7) onto center housing. Remove back plate retaining cap screws (10) and locks (9). Then, remove back plate (5), thrust bearing (15), thrust collar (16) and spring (4). Carefully remove bearing retainers (19) from ends and withdraw

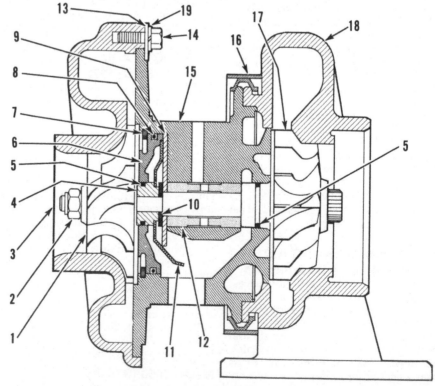

Fig. 141–Cross-sectional view of Schwitzer turbocharger used on some diesel models. Refer to Fig. 140 for legend.

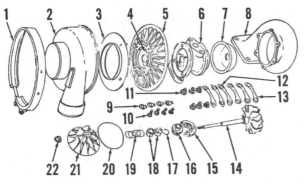

1.	Clamp
2.	Housing (compressor)
3.	Diffuser
4.	Spring
5.	Backplate
6.	Center housing
7.	Turbine shroud
8.	Turbine housing
9.	Lock plates
10.	Backplate cap screws
11.	Turbine housing cap screws
12.	Lock plates
13.	Clamp plates
14.	Turbine wheel and shaft
15.	Thrust bearing
16.	Thrust collar
17.	Seal ring (2 used)
18.	Bearing
19.	Bearing retainers
20.	Seal ring
21.	Compressor impeller
22.	Locknut

Fig. 142–Exploded view of Airesearch model TO-4 turbocharger assembly used on some diesel models.

bearings (18) from center housing. CAUTION: Be careful not to damage bearing or surface of center housing when removing retainers. The center two retainers do not have to be removed unless damaged or unseated. Always renew bearing retainers if removed from groove in housing.

Soak all parts in Bendix metal cleaner or equivalent and use a soft brush, plastic blade or compressed air to remove carbon deposits. CAUTION: Do not use wire brush, steel scraper or caustic solution for cleaning, as this will damage turbocharger parts.

Inspect turbine wheel and compressor wheel for broken or distorted vanes. DO NOT attempt to straighten bent vanes. Be sure all wheel blades are clean, as any deposits left on blades will affect balance. Inspect bearing bores in center housing (6—Fig. 142) for scored surfaces, out of round or excessive wear. Make certain bore in center housing is not grooved in area where seal (17) rides. Oil passage in thrust collar (16) must be clean and thrust faces must not be warped or scored. Ring groove shoulders must not have step wear. Check shaft end play and radial clearance when assembling.

If the bearing inner retainers (19) were removed, install new retainer using IH tool FES 57-3. Oil bearings

(18) and install outer retainers using IH tool FES 57-3. Position the shroud (7) on turbine shaft (14) and install seal ring (17) in groove. Apply a light, even coat of engine oil to shaft journals, compress seal ring (17) and install center housing (6). Install new seal ring (17) in groove of thrust collar (16), then install thrust bearing so that smooth side of bearing (15) is next to the large diameter of collar. Install thrust bearing and collar assembly over shaft, making certain that pins in center housing engage holes in thrust bearing. Install new rubber seal ring (20), make certain that spring (4), if removed, is positioned in back plate (5), then install back-plate making certain that seal ring is not damaged. Install lock plates (9) cap screws (10), tightening screws to 40-60 inch-pounds torque. Install compressor impeller (21) and make certain that impeller is completely seated against thrust collar (16). Install lock nut (22) and tighten to 18-20 inch-pounds torque, then use a

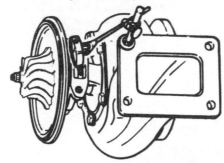

"T" handle to turn lock nut an additional 90°. CAUTION: If "T" handle is not used, shaft may be bent when tightening nut (22). Install turbine housing (8) with clamp plates (13) next to housing, tighten cap screws (11) to 100-130 inch-pounds torque, then bend lock plates (12) up around screw heads.

Check shaft end play and radial play at this point of assembly. If shaft end play (Fig. 143) exceeds 0.004, thrust collar (16—Fig. 142) and/or thrust bearing (15) is worn excessively. End play of less than 0.001 indicates incomplete cleaning (carbon not all removed) or dirty assembly and unit should be disassembled and cleaned. Refer to (Fig. 144) and check turbine shaft radial play. If it exceeds 0.007, unit should be disassembled and bearings, shaft and/or center housing should be renewed.

Make certain that legs on diffuser (3 —Fig. 142) are aligned with spot faces on backplate (5) and install diffuser. Install compressor housing (2) and tighten nut of clamp (1) to 40-80 inch-pounds torque.

Check rotating unit for free rotation within the housing. Cover all openings until the turbocharger is reinstalled.

Use a new gasket and install and prime turbocharger as outlined in paragraphs 201 or 202.

TESTING
Model 1026 (DT407)

205. Before testing the turbocharger, make certain the air filter system is clean and that the fuel injection pump is properly adjusted and delivering the correct amount of fuel to engine. A clogged air filter will cause intake manifold pressure to be low. Excessive fuel delivery will result in high exhaust manifold pressure and low fuel delivery will cause low exhaust manifold pressure.

Connect tractor to a dynamometer or other loading device. Remove ¼-inch plug from intake manifold elbow and install a 60 psi test gage as shown in Fig. 145. Start engine and set speed control lever at high idle position. Load engine to rated load (2400) rpm. Test gage should show intake manifold pressure of 8.5-12.5 psi. Continue to

Fig. 144–Checking radial clearance with dial indicator through oil outlet hole and touching shaft.

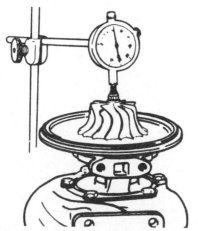

Fig. 143–Checking end play of shaft after unit has been cleaned. Refer to text.

Fig. 145–Install test gage as shown to check intake manifold pressure on turbocharged diesel engine. (IH tool numbers are shown.)

load engine until overload (1800) rpm is reached. Intake manifold pressure should now be 5.8-8.5 psi. Stop engine, remove test gage and install plug.

To check exhaust manifold pressure, remove either ¼-inch plug from center of exhaust manifold. Connect a 60 psi test gage, using a length of tubing to protect gage from high temperature. See Fig. 146. Start engine and set speed control lever at high idle position. Load engine to rated load (2400) rpm. Test gage should show exhaust manifold pressure of 10-13 psi. Continue to load engine to overload (1800) rpm. Exhaust manifold pressure should now be 4.5-7.5 psi. Stop engine, remove test gage and install plug.

Model 1066 Hydrostatic (DT414)

206. Before testing the turbocharger, make certain the air filter system is clean and that the fuel injection pump is properly adjusted and delivering the correct amount of fuel to engine. A clogged air filter will cause intake manifold pressure to be low. Excessive fuel delivery will result in high exhaust manifold pressure and low fuel delivery will cause low exhaust manifold pressure.

Connect tractor to a dynamometer or other loading device. Remove ¼-inch plug from intake manifold and install a 60 psi test gage as shown in Fig. 145. Start engine and set speed control lever at high idle position. Load engine to rated load (2400) rpm. Test gage should show intake manifold pressure of 10-13 psi. Continue to load engine until overload (1800) rpm is reached. Intake manifold pressure should now be 6-9 psi. Stop engine, remove test gage and install plug.

To check exhaust manifold pressure,

remove either ⅛-inch plug from center of exhaust manifold. Connect a 60 psi test gage, using a length of tubing to protect gage from high temperature. See Fig. 146. Start engine and set speed control lever at high idle position. Load engine to rated load (2400) rpm. Test gage should show exhaust manifold pressure of 7-10 psi. Continue to load engine to overload (1800) rpm. Exhaust manifold pressure should now be 3-6 psi. Stop engine, remove test gage and install plug.

Model 1066 Gear Drive (DT414)

207. Before testing the turbocharger, make certain the air filter system is clean and that the fuel injection pump is properly adjusted and delivering the correct amount of fuel to engine. A clogged air filter will cause intake manifold pressure to be low. Excessive fuel delivery will result in high exhaust manifold pressure and low fuel delivery will cause low exhaust manifold pressure.

Models below chassis serial 23000. Connect tractor to a dynamometer or other loading device. Remove ¼-inch plug from intake manifold and install a 60 psi test gage as shown in Fig. 145. Start engine and set speed control lever at high idle position. Load engine to rated load (2400) rpm. Test gage should show intake manifold pressure of 11-14 psi. Continue to load engine until overload (1800) rpm is reached. Intake manifold pressure should now be 8-11 psi. Stop engine, remove test gage and install plug. To check exhaust manifold pressure, remove either ⅛-inch plug from center of exhaust manifold. Connect a 60 psi test gage, using a length of tubing to protect gage from high temperature. See Fig. 146. Start en-

gine and set speed control lever at high idle position. Load engine to rated load (2400) rpm. Test gage should show exhaust manifold pressure of 11-14 psi. Continue to load engine to overload (1800) rpm. Exhaust manifold pressure should now be 6-9 psi. Stop engine, remove test gage and install plug.

Models above chassis serial 23000. Connect tractor to a dynamometer or other loading device. Remove ¼-inch plug from intake manifold and install a 60 psi test gage as shown in Fig. 145. Start engine and set speed control lever at high idle position. Load engine to rated load (2600) rpm. Test gage should show intake manifold pressure of 11-14 psi. Continue to load engine until overload (1700) rpm is reached. Intake manifold pressure should now be 5-9 psi. Stop engine, remove test gage and install plug. To check exhaust manifold pressure, remove either ⅛-inch plug from center of exhaust manifold. Connect a 60 psi test gage, using a length of tubing to protect gage from high temperature. See Fig. 146. Start engine and set speed control lever at high idle position. Load engine to rated load (2600) rpm. Test gage should show exhaust manifold pressure of 10-13 psi. Continue to load engine to overload (1700) rpm. Exhaust manifold pressure should now be 3-7 psi. Stop engine, remove test gage and install plug.

AIR FILTER SYSTEM

All Models

208. All models are equipped with a dry type air cleaner with a safety filter element (2—Fig. 147 or 8—Fig. 148) which should be renewed at least once each year. DO NOT attempt to clean the safety element.

Large filter element (4—Fig. 147 or 148) can be cleaned by directing com-

Fig. 146–Test gage installed to check exhaust manifold pressure on turbocharged diesel engines. Length of tubing is used to protect gage from heat.

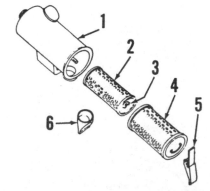

Fig. 147–Exploded view of typical air cleaner assembly used on models 766 and 826 tractors.

1. Housing	4. Main element
2. Safety element	5. Retainer
3. Wing nut	6. Dust unloader

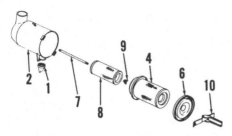

Fig. 148–Exploded view of typical air cleaner assembly used on models 966, 1026 and 1066 tractors.

1. Dust unloader	7. Stud
2. Housing	8. Safety element
4. Main element	9. Wing nut
6. End cover	10. Retainer

pressed air up and down the pleats on the inside of the element. Air pressure must not exceed 100 psi. An element cleaning tool (IH tool No. 407073RI) for use with compressed air, is available from International Harvester. Renew filter element after 10 cleanings or once a year, whichever comes first.

NON-DIESEL GOVERNOR

The governor used on non-diesel engines, is a centrifugal flyweight type and is driven by the crankshaft gear via an idler gear. Before attempting any governor adjustments, check all linkage for binding or lost motion and correct any undesirable conditions.

ADJUSTMENTS
All Non-Diesel Models

209. **SYNCHRONIZING GOVERNOR AND CARBURETOR.** If removal of carburetor, manifold, governor or governor linkage has been performed, or if difficulty is encountered in adjusting engine speeds, adjust the governor to carburetor control rod as follows: Loosen alternator, remove belt, then move alternator away from engine. Pull throttle lever down to put tension on governor spring, then disconnect control rod from governor (if necessary). Hold both carburetor and governor in the wide open (high idle) position and adjust rod clevis until pin slides freely into clevis and rockshaft lever holes, then remove pin and lengthen rod by unscrewing clevis one full turn. Reinstall clevis pin and tighten jam nut. Install and adjust fan belt.

210. **HIGH IDLE SPEED.** Refer to Fig. 149 and back out bumper screw about ½ to ¾-inch. Start engine and run until it reaches operating tempera-

ture. Move throttle lever to the high idle position and check the engine high idle speed which should be approximately 2650 rpm. If engine high idle speed is not as stated, loosen jam nut and turn the high idle stop screw as required.

With engine high idle speed adjusted, move the throttle control lever to the low idle position, then quickly advance it to the high idle position and adjust the bumper screw just enough to eliminate engine surge. Repeat this operation as required until engine will advance from low idle to high idle speed without surging. Do not turn bumper screw in further than necessary as engine low idle speed could be affected.

211. **LOW IDLE SPEED.** With engine running and at operating temperature, place throttle control lever in the low idle position and check the engine low idle speed which should be 425 rpm. If engine low idle speed is not as stated, adjust throttle stop screw on carburetor as required.

If specified engine low idle speed cannot be obtained, recheck the governor to carburetor control adjustment as outlined in paragraph 209. It may also be necessary to vary the length of governor control rod by disconnecting

and adjusting the rod ball joint located on rear end of governor control rod.

OVERHAUL
All Non-Diesel Models

212. To remove the governor unit, remove alternator, disconnect governor control rod and governor to carburetor rod from governor, then unbolt governor from engine front plate.

Disassembly of the removed governor unit will be self-evident after an examination of the unit and reference to Fig. 150.

The governor weights should have an operating clearance of 0.001-0.005 on the pins. Bearings (18 and 25) should be free of any roughness. The rockshaft lever (9) and governor shaft (27) should operate freely in bearings (11 and 12). Make certain that seal (10) is in good condition. The lubricating holes in governor and engine front end plate must be open and clean.

A governor overhaul service package is available and consists of items 8 through 13 and 15 through 25. A governor gear and weight assembly is also available for service.

After assembly and installation are completed, check and adjust the engine speed as outlined in paragraphs 209, 210 and 211.

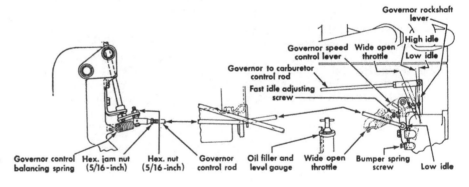

Fig. 149–Schematic view showing non-diesel engine governor linkage and location of bumper screw and high idle adjusting (stop) screw.

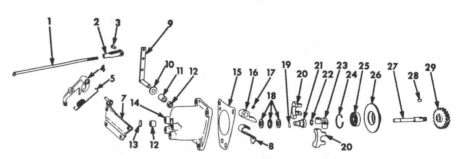

Fig. 150–Exploded view of typical governor used on non-diesel engines.

1. Governor to carburetor rod	9. Rockshaft and lever	16. Fork	23. Weight carrier
2. Clevis	10. Seal	17. Set screw	24. Snap ring
3. Clevis pin	11. Bearing	18. Thrust bearing	25. Bearing
4. Governor lever	12. Bearing	19. Stop ring	26. Bearing carrier
5. Governor spring	13. Expansion plug	20. Weights	27. Governor shaft
7. Lever bracket	14. Housing	21. Thrust sleeve	28. Woodruff key
8. Bumper spring	15. Gasket	22. Snap ring	29. Drive gear

COOLING SYSTEM

RADIATOR

All Models

213. To remove radiator, first drain cooling system, then remove hood skirts, hood and side panels. Remove the air cleaner inlet hose. Disconnect upper and lower radiator hoses from radiator and move them out of the way. Disconnect fan shroud from radiator and remove the radiator drain cock. Disconnect the oil cooler lines bracket. Remove the center cap screw from the four radiator mounts, then lift radiator straight up out of radiator support.

FAN

All Models

214. On models 766 and 826 non-diesel engine the fan is mounted on a shaft and bearing assembly which is attached by a bracket to the thermostat housing. Two belts are used; one to drive the water pump and alternator and one to drive the fan. To remove the fan and fan shaft assembly, first remove hood skirts and hood. Loosen belt adjuster and remove fan belt. Remove cap screws which retain air cleaner bracket and fan belt adjuster to thermostat housing. Pull fan and fan shaft assembly from side after pushing air cleaner bracket upward out of the way.

Refer to Fig. 151 unbolt and remove fan (23), then press shaft and bearing assembly (19) from pulley (21). Remove

retainer (20) and press shaft and bearing from adjuster bracket (18). Shaft and bearing are serviced as an assembly.

When reinstalling, adjust fan belt until a pressure of 25 pounds applied midway between pulleys will deflect belt ⅞-inch.

215. On models 826 and 1026 diesel tractors, the fan is attached to the water pump and one belt drives the fan, water pump and alternator.

To remove the fan, first remove radiator as outlined in paragraph 213. Remove retaining cap screws and lift off the fan.

When reinstalling, adjust fan belt until a pressure of 25 pounds applied midway between water pump and crankshaft pulleys will deflect belt ⅞-inch.

216. On models 766, 966 and 1066 diesel tractors, the fan is attached to the water pump and one belt drives the fan, water pump and alternator.

To remove the fan, first remove left front hood, left radiator support panel and fan belt. Unbolt and slide back the fan shroud. Remove retaining cap screws and remove fan.

When reinstalling, adjust fan belt until a pressure of 25 pounds applied midway between water pump and crankshaft pulleys will deflect belt 13/16-inch.

WATER PUMP
Models 766 and 826 Non-Diesel

217. **R&R AND OVERHAUL.** To remove the water pump assembly, drain cooling system, then remove fan

assembly as outlined in paragraph 214. Loosen alternator mounting bolts and remove drive belt and pulley. Disconnect by-pass hose and water pump inlet hose. Unbolt and remove water pump assembly.

To disassemble the water pump, refer to Fig. 151 and remove plate (2) and gasket (3). Remove retainer (10) from pump body, support pump body and press shaft and bearing assembly (9) from impeller (4) and pump body. Press shaft assembly from hub (11). Remove seal (6).

The shaft and bearing are available only as a pre-lubricated assembly. Water pump overhaul package (IH part No. 374544R94) is available.

Reassemble by reversing the disassembly procedure, keeping in mind the following points: When installing seal (6), press only on the outer diameter. Install hub with smaller diameter facing out. Press impeller on shaft until a clearance of 0.020-0.030 exists between gasket surface of body and face of impeller.

NOTE: If pump is disassembled for leakage and shaft assembly and impeller are in good condition, renew seal (6) in body and face ring (5) in impeller and reassemble pump.

Reinstall pump by reversing the removal procedure. Adjust water pump and alternator belt tension until a pressure of 25 pounds applied midway between water pump and crankshaft pulleys, will deflect belt ⅞-inch. Install fan and adjust fan belt as in paragraph 214.

Model 826 Diesel

218. **R&R AND OVERHAUL.** To remove the water pump, first remove radiator as outlined in paragraph 213. Loosen alternator mounting bolts and remove the drive belt. Unbolt and remove fan (1—Fig. 152), spacer (2) and pulley (3). Remove cap screws securing pump body (7) to water pump carrier, then remove pump.

Disassemble water pump as follows: Remove plastic screw (13) and using a ½x2 inch NC cap screw for a jack screw in rear of impeller, force impeller (12) off rear end of shaft. Using two screwdrivers pry seal assembly (10) out of pump body. Support hub (4) and press out shaft. Press shaft and bearing assembly (5) out front of body. Make certain that body is supported as close to bearing as possible.

When reassembling, press shaft and bearing assembly into body using a piece of pipe so pressure is applied only to outer race of bearing. Bearing race should be flush with front end of body. Install new "O" ring (9) and seal (10).

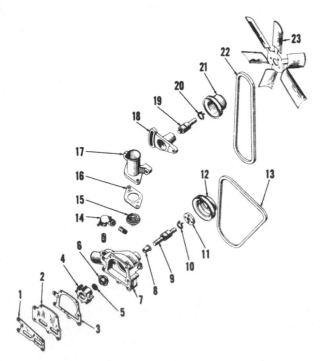

Fig. 151–Exploded view of fan shaft and water pump used on models 766 and 826 non-diesel engines.

1. Gasket
2. Plate
3. Gasket
4. Impeller
5. Impeller face ring
6. Seal assembly
7. Body
8. Slinger
9. Pump shaft & bearing
10. Retainer
11. Hub
12. Pulley
13. Alternator-generator and pump belt
14. By-pass hose
15. Thermostat
16. Gasket
17. Water outlet
18. Fan belt adjuster
19. Fan bearing & shaft
20. Retainer
21. Fan pulley
22. Fan belt
23. Fan

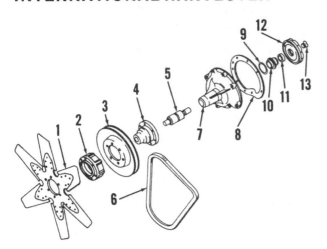

Fig. 152-Exploded view of water pump assembly used on model 826 diesel engine.

1. Fan
2. Fan spacer
3. Pulley
4. Hub
5. Pump shaft & bearing
6. Alternator & water pump belt
7. Body
8. Gasket
9. "O" ring
10. Seal assembly
11. Impeller face ring
12. Impeller
13. Plastic screw

Press only on outer diameter of seal. Support the shaft assembly and press hub on shaft until hub is flush with end of shaft. Install face ring (11) in impeller (12), then press impeller on shaft until there is a clearance of 0.012-0.020 between body and front of impeller (opposite fins). Install plastic screw.

Using a new gasket (8), reinstall pump by reversing the removal procedure. Install pulley, spacer and fan and adjust the belt as outlined in paragraph 215.

Model 1026 Diesel

219. **R&R AND OVERHAUL.** The water pump is comprised of two sections. The rear section (22—Fig. 153) need not be removed for water pump overhaul. Therefore, removal and overhaul procedures will be given only for the front (bearing housing) section of the pump.

To remove the water pump, first drain and remove radiator as outlined in paragraph 213. Loosen alternator mounting bolts and remove drive belt. Unbolt and remove fan (2—Fig. 153) and pulley (5). Unbolt bearing housing (15) from pump rear section (22) and remove the complete front section. Remove Esna nut (4) and using a suitable puller, remove hub (6) and Woodruff key (11). Use a punch on O.D. of seal (8) and collapse seal to remove. Remove bearing snap ring (9). Place pump in a press and force shaft and bearings from impeller (20) and housing (15). Rear bearing (13) may or may not come out with shaft. If bearing remains in housing, it can be pressed out after pump seal (17) is removed. Oil seal (14) will also be removed with the bearing. Front bearing (10) and spacer (7) can also be pressed from shaft (12).

Use all new gaskets and seals when reassembling. Lubricate seals, ceramic washer and "O" ring prior to assembly. Reassemble by reversing the disassembly procedure. Tighten hub retaining Esna nut to a torque of 130 ft.-

lbs. Press impeller on shaft until rear face of impeller hub is 2.030-2.070 inches from mounting flange. Remove plug (16) and fill housing (15) with No. 2 multi-purpose lithium grease, then install plug.

Adjust the drive belt as outlined in paragraph 215.

Models 766, 966 and 1066 Diesel

220. **R&R AND OVERHAUL.** To remove the water pump, first remove fan as outlined in paragraph 216. Remove cap screws securing pump body (4 —Fig. 154) to front cover, then remove pump.

Disassemble water pump as follows: Using a puller remove the fan hub (3). Then, remove snap ring (5) and press the shaft and bearing assembly (6) out of impeller (9) and housing (4). Make certain that body is supported as close to bearing as possible. Tap seal (7) out of housing (4) then, pry rubber bushing and ceramic seal (8) out of impeller (9).

When reassembling, press shaft and bearing assembly into body using a piece of pipe so pressure is applied only to outer race of bearing. Be sure bearing is bottomed in housing. Install snap ring. Press seal (7—Fig. 155) in housing then, assemble the ceramic seal (6) with identification mark to-

ward the rubber bushing (5). Moisten rubber bushing with water, and install rubber bushing and ceramic seal assembly in impeller. Be sure the seal is in the bottom of impeller bore. Support fan hub and press shaft through hub to dimension shown in Fig. 155. Support the hub end of shaft and press impeller on shaft to a clearance as shown in Fig. 155.

Using a new gasket (11—Fig. 154), reinstall pump by reversing the removal procedure. Install pulley, fan and adjust the belt as outlined in paragraph 216.

ELECTRICAL SYSTEM

ALTERNATOR AND REGULATOR
All Models

221. Delco-Remy "DELCOTRON" alternators are used on all models. Some models may be equipped with alternator No. 1100791, 1100805 or 1100883 and a double contact external regulator No. 1119513 or 1119516.

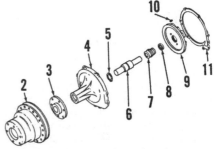

Fig. 154-Exploded view of water pump assembly used on models 766, 966 and 1066 diesel engines.

2. Pulley
3. Hub
4. Housing
5. Snap ring
6. Bearing & shaft assembly
7. Seal
8. Rubber bushing & ceramic seal
9. Impeller
10. Dowel (2 used)
11. Gasket

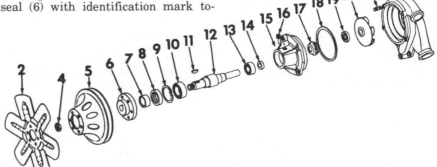

Fig. 153-Exploded view of water pump assembly used on model 1026 diesel engines.

2. Fan
4. Esna nut
5. Pulley
6. Hub
7. Spacer
8. Oil seal
9. Snap ring
10. Bearing
11. Woodruff key
12. Shaft
13. Bearing
14. Oil seal
15. Bearing housing
16. Plug
17. Pump seal
18. "O" ring
19. Ceramic washer
20. Impeller
21. Stud
22. Rear section

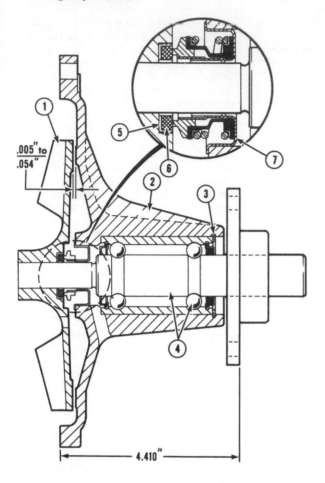

.005" to
.054"

4.410"

Fig. 155—Sectional view of water pump and components. Also dimension for installing fan hub and impeller.

1. Impeller
2. Housing
3. Snap ring
4. Shaft & bearing assembly
5. Rubber bushing
6. Ceramic seal
7. Bearing housing seal

Some models may be equipped with an alternator No. 1100578, 1100587 or 1100593 and a solid state regulator No. 1116387 which is mounted internal and has no provision for adjustment.

CAUTION: Because certain components of the alternator can be damaged by procedures that will not affect a D.C. generator, the following precautions MUST be observed:

   a. When installing batteries or connecting a booster battery, the negative post of battery must be grounded.

   b. Never short across any terminal of the alternator or regulator.

   c. Do not attempt to polarize the alternator.

   d. Disconnect battery cables before removing or installing any electrical unit.

   e. Do not operate alternator on an open circuit and be sure all leads are properly connected before starting engine.

Specification data for the alternators and regulator is as follows:

Alternator 1100805
Field current @ 80° F.,
   Amperes 2.2-2.6
   Volts 12.0
Cold output @ specified voltage,
   Specified volts 14.0
   Amperes at rpm .... 21.0 @ 2200
   Amperes at rpm .... 30.0 @ 5000
Rated output hot,
   Amperes 32.0

Alternator 1100791
Field current @ 80° F.,
   Amperes 2.2-2.6
   Volts 12.0
Cold output @ specified voltage
   Specified volts 14.0
   Amperes at rpm .... 32 @ 2000
   Amperes at rpm .... 50 @ 5000
Rated output hot,
   Amperes 55.0

Alternator 1100883
Field current @ 80° F.,
   Amperes 2.2-2.6
   Volts 12.0
Cold output @ specified voltage
   Specified volts 14.0
   Amperes at rpm .... 33.0 @ 2000
   Amperes at rpm .... 58.0 @ 5000
Rated output hot,
   Amperes 61.0

Alternator 1100578 and 1100593
Field current @ 80° F.,
   Amperes 4.0-4.5
   Volts 12.0
Cold output @ specified voltage,
   Amperes at rpm .... 32.0 @ 5000
Rated output hot,
   Amperes 37.0

Alternator 1100587
Field current @ 80° F.,
   Amperes 4.0-4.5
   Volts 12.0
Cold output @ specified voltage,
   Amperes at rpm .... 67.0 @ 5000

Rated output hot,
   Amperes 72.0

Regulator 1119516 and 1119513
Ground polarity .... Negative
Field relay,
   Air gap 0.015
   Point opening 0.030
   Closing voltage range
      1119516 .... 1.5-3.2
   Closing voltage range
      1119513 .... 3.8-7.2
Voltage regulator
   Air gap (lower points closed) 0.067*
   Upper point opening (lower
      points closed) 0.014
   Voltage setting,
      65° F. 13.9-15.0
      85° F. 13.8-14.8
      105° F. 13.7-14.6
      125° F. 13.5-14.4
      145° F. 13.4-14.2
      165° F. 13.2-14.0
      185° F. 13.1-13.9
*When bench tested, set air gap at 0.067 as a starting point, then adjust air gap to obtain specified difference between voltage settings of upper and lower contacts. Operation on lower contacts must be 0.05-0.4 volt lower than on upper contacts. Voltage setting may be increased up to 0.3 volt to correct chronic battery undercharging or decreased up to 0.3 volt to correct battery overcharging. Temperature (ambient) is measured ¼-inch away from regulator cover and adjustment should be made only when regulator is at normal operating temperature.

222. **ALTERNATOR 1100791, 1100805 OR 1100883 TESTING AND OVERHAUL.** The only tests which can be made without removal and disassembly of alternator are the field current draw and output tests. Refer to paragraph 221 for specifications.

To disassemble the alternator, first scribe match marks (M—Fig. 156) on two frame halves (6 and 16), then remove the four through-bolts. Pry frame apart with a screwdriver between stator frame (11) and drive end frame (6). Stator assembly (11) must remain with slip ring end frame (16) when unit is separated.

NOTE: When frames are separated, brushes will contact rotor shaft at bearing area. Brushes MUST be cleaned of lubricant if they are to be re-used.

Clamp the iron rotor (12) in a protected vise only tight enough to permit loosening of pulley nut (1). Rotor and end frame can be separated after pulley and fan are removed. Check the bearing surfaces of rotor shaft for visible wear or scoring. Examine slip ring surfaces for scoring or wear and windings for overheating or other damage. Check rotor for grounded, shorted or

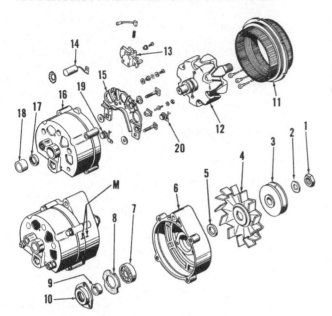

Fig. 156–Exploded view of "DELCOTRON" alternator. Note match marks (M) on end frames.

1. Pulley nut
2. Washer
3. Drive pulley
4. Fan
5. Spacer
6. Drive end frame
7. Ball bearing
8. Gasket
9. Spacer
10. Bearing retainer
11. Stator
12. Rotor
13. Brush holder
14. Capacitor
15. Heat sink
16. Slip ring end frame
17. Felt seal and retainer
18. Needle bearing
19. Negative diode (3 used)
20. Positive diode (3 used)

open circuits using an ohmmeter as follows:

Refer to Fig. 157 and touch the ohmmeter probes to points (1-2) and (1-3); a reading near zero will indicate a ground. Touch ohmmeter probes to the slip rings (2-3); reading should be 4.6-5.5 ohms. A higher reading will indicate an open circuit and a lower reading will indicate a short. If windings are satisfactory, mount rotor in a lathe and check runout at slip rings with a dial indicator. Runout should not exceed 0.002. Slip ring surfaces can be trued if runout is excessive or if surfaces are scored. Finish with 400 grit or finer polishing cloth until scratches or machine marks are removed.

Disconnect the three stator leads and separate stator assembly (11—Fig. 156) from slip ring end frame assembly. Check stator windings for grounded or open circuits as follows: Connect ohmmeter leads successively between each pair of leads. A high reading would indicate an open circuit. NOTE: The three stator leads have a common connection in the center of the windings. Connect ohmmeter leads between each stator lead and stator frame. A very low reading would indicate a grounded circuit. A short circuit within the stator windings cannot be readily determined by test because of the low resistance of the windings.

Three negative diodes (19) are located in the slip ring end frame (16) and three positive diodes (20) in the heat sink (15). Diodes should test at or near infinity in one direction when tested with an ohmmeter, and at or near zero when meter leads are reversed. Renew any diode with approximately equal meter readings in both directions. Diodes must be removed and installed using an arbor press or vise and suitable tool which contacts only the outer edge of the diode. Do not attempt to drive a faulty diode out of end frame or heat sink as shock may cause damage to the other good diodes. If all diodes are being renewed; make certain the positive diodes (marked with red printing) are installed in the heat sink and negative diodes (marked with black printing) are installed in the end frame.

Brushes are available only in an assembly which includes brush holder (13). Brush springs are available for service and should be renewed if heat damage or corrosion is evident. If brushes are re-used, make sure all grease is removed from surface of brushes before unit is reassembled. When reassembling, install brush springs and brushes in holder, push brushes up against spring pressure and insert a short piece of straight wire through hole (W—Fig. 158) and through end frame (16—Fig. 156) to outside. Withdraw the wire after alternator is assembled.

Capacitor (14) connects to the heat sink and is grounded to the end frame.

Capacitor protects the diodes from voltage surges.

Remove and inspect ball bearing (7). If bearing is in satisfactory condition, fill bearing ¼-full with Delco-Remy lubricant No. 1948791 and reinstall. Inspect needle bearing (18) in slip ring end frame. This bearing should be renewed if its lubricant supply is exhausted; no attempt should be made to relubricate and re-use the bearing. Press old bearing out towards inside and press new bearing in from outside until bearing is flush with outside of end frame. Saturate felt with SAE 20 oil and install seal and retainer assembly.

Reassemble alternator by reversing the disassembly procedure. Tighten pulley nut to a torque of 45 ft.-lbs.

NOTE: A battery powered test light can be used instead of ohmmeter for all electrical checks except shorts in rotor windings. However, when checking diodes, test light must not be of more than 12.0 volts.

223. **ALTERNATOR 1100578, 1100587 or 1100593 TESTING AND OVERHAUL.** The only test which can be made without removal and disassembly of alternator is the regulator. If there is a problem with the battery not being charged, and the battery and cable connector have been checked and are good, check the regulator as follows: Operate engine at moderate speed and turn all accessories on and check the ammeter. If ammeter reading is within 10 amperes of rated output as stamped on alternator frame (or refer to paragraph 221 for specifications) alternator is not defective. If ampere output is not within 10 amperes of rated output, ground the field winding by inserting a screwdriver into test hole Fig. 159. If output is then within 10 amperes of rated output, replace the regulator.

CAUTION: When inserting screwdriver in test hole the tab is within ¾-inch of casting surface. Do not force screwdriver deeper than one inch into end frame.

If output is still not within 10 amperes of rated output, the alternator will have to be disassembled. Check

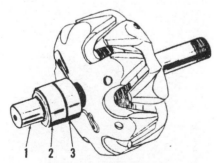

Fig. 157–Removed rotor assembly showing test points when checking for grounds, shorts and opens.

Fig. 158–Exploded view of brush holder assembly. Insert wire in hole (W) to hold brushes up. Refer to text.

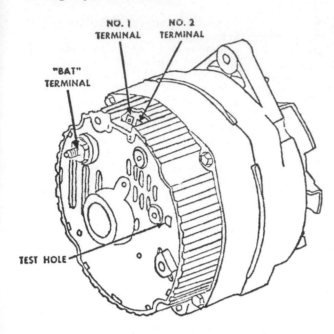

NO. 1 TERMINAL NO. 2 TERMINAL

"BAT" TERMINAL

TEST HOLE

Fig. 159–View showing the terminal and test hole on alternator Nos. 1100578, 1100587 or 1100593. Refer to text.

the field winding, diode trio, rectifier bridge and stator as follows:

To disassemble the alternator, first scribe match marks (M—Fig. 160) on the two frame halves (4 and 16), then remove the four through-bolts. Pry frame apart with a screwdriver between stator frame (12) and drive end frame (4).

Stator assembly (12) must remain with slip ring end frame (16) when unit is separated.

NOTE: When frames are separated, brushes will contact rotor shaft at bearing area. Brushes MUST be cleaned of lubricant with a soft dry cloth if they are to be reused.

Clamp the iron rotor (13) in a protected vise, only tight enough to permit loosening of pulley nut (1). Rotor and end frame can be separated after pulley and fan are removed. Check bearing surface of rotor shaft for visible wear or scoring. Examine slip ring surface for scoring or wear and rotor winding for overheating or other damage. Check rotor for grounded, shorted or open circuits using an ohmmeter as follows:

Refer to Fig. 157 and touch the

ohmmeter probes to points (1-2) and (1-3); a reading near zero will indicate a ground. Touch ohmmeter probes to the slip rings (2-3); reading should be 5.3-5.9 ohms. A higher reading will indicate an open circuit and a lower reading will indicate a short. If windings are satisfactory, mount rotor in a lathe and check runout at slip rings using a dial indicator. Runout should not exceed 0.002. Slip ring surfaces can be trued if runout is excessive or if surfaces are scored. Finish with 400 grit or finer polishing cloth until scratches or machine marks are removed.

Before removing stator, brushes or diode trio, refer to Fig. 161 and check for grounds between points A to C and B to C with an ohmmeter, using the lowest range scale. Then, reverse the lead connections. If both A to C readings or both B to C readings are the same, the brushes may be grounded because of defective insulating washer and sleeve at the two screws. If the screw assembly is not damaged or grounded, the regulator is defective.

To test the diode trio, first remove the stator. Then, remove the diode trio,

noting the insulator positions. Using an ohmmeter, refer to Fig. 162 and check between points A and D. Then, reverse the ohmmeter lead connections. If diode trio is good, it will give one high and one low reading. If both readings are the same, the diode trio is defective. Repeat this test at points B and D and at C and D.

The rectifier bridge (Fig. 163) has a grounded heat sink (A) and an insulated heat sink (E) that is connected to the output terminal. Connect ohmmeter to the grounded heat sink (A) and to the flat metal strip (B). Then, reverse the ohmmeter lead connections. If both readings are the same, the rectifier bridge is defective. Repeat this test between points A and C, A and D, B and E, C and E, and D and E.

Test the stator (12—Fig. 160) windings for grounded or open circuits as follows: Connect ohmmeter leads successively between each pair of leads. A high reading would indicate an open circuit.

NOTE: The three stator leads have a common connection in the center of the windings. Connect ohmmeter leads between each stator lead and stator frame. A very low reading would indicate a grounded circuit. A short circuit within the stator windings cannot be readily determined by test because of the low resistance of the windings.

Brushes and springs are available only as an assembly which included

REGULATOR GROUND SCREW

Fig. 161–Test points for brush holder. Refer to text.

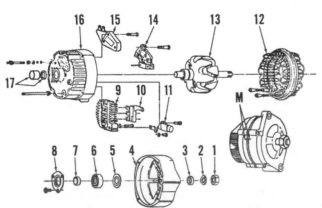

16 15 14 13 12

17

8 7 6 5 4 3 2 1

9 10 11

M

Fig. 160–Exploded view of "DELCOTRON" alternator No. 1100578, 1100587 or 1100593 used with internal mounted solid state regulator. Note match marks (M) on end frames.

1. Pulley nut
2. Washer
3. Spacer (outside drive end)
4. Drive end frame
5. Grease slinger
6. Ball bearing
7. Spacer (inside drive end)
8. Bearing retainer
9. Bridge rectifier
10. Diode trio
11. Capacitor
12. Stator
13. Rotor
14. Brush holder
15. Solid state regulator
16. Slip ring end frame
17. Bearing & seal assembly

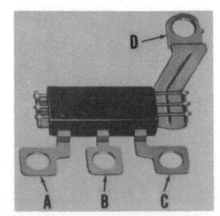

D

A B C

Fig. 162–Diode trio test points. Refer to text.

Fig. 163–Bridge rectifier test points. Refer to text.

brush holder (14—Fig. 160). If brushes are reused, make sure all grease is removed from surface of brushes before unit is reassembled. When reassembling, install regulator and then brush holder, springs and brushes. Push brushes up against spring pressure and insert a short piece of straight wire through the hole and through end frame to outside. Be sure that the two screws at Points A and B (Fig. 161) have insulating washers and sleeves. NOTE: A ground at these points will cause no output or controlled output. Withdraw the wire only after alternator is assembled.

Capacitor (11—Fig. 160) connects to the rectifier bridge and is grounded to the end frame. Capacitor protects the diodes from voltage surges.

Remove and inspect ball bearing (6—Fig. 160). If bearing is in satisfactory condition, fill bearing ¼-full with Delco-Remy lubricant No. 1948791 and reinstall. Inspect needle bearing (17) in slip ring end frame. This bearing should be renewed if its lubricant supply is exhausted; no attempt should be made to relubricate and reuse the bearing. Press old bearing out towards inside and press new bearing in from outside until bearing is flush with outside of end frame. Saturate felt seal with SAE 20 oil and install seal.

Reassemble alternator by reversing the disassembly procedure. Tighten pulley nut to a torque of 50 ft.-lbs.

STARTING MOTORS
All Models

224. Delco-Remy starting motors are used on all models and specification data for these units is as follows:

Starter 1108691
Volts 12.0
Brush spring tension, oz. 35
No-load test,
  Volts . . . . . . : 9.0
  Amperes (min.) 50.0*
  Amperes (max.) 80.0*
  RPM (min.) 5500
  RPM (max.) 9000
  *Includes solenoid

Starter 1113197
Volts 12.0
Brush spring tension, oz. 80
No-load test,
  Volts 9.0
  Amperes (min.) 50.0*
  Amperes (max.) 70.0*
  RPM (min.) 3500
  RPM (max.) 5500
  *Includes solenoid

Starter 1113684
Volts 12.0
Brush spring tension, oz. 80
No-load test,
  Volts 9.0
  Amperes (min.) 130.0*
  Amperes (max.) 160.0*
  RPM (min.) 5000
  RPM (max.) 7000
  *Includes solenoid

Starter 1113685
Volts 12.0
Brush spring tension, oz. 80
No load test,
  Volts 9.0
  Amperes (min.) 75.0*
  Amperes (max.) 105.0*
  RPM (min.) 5000
  RPM (max.) 7000
  *Includes solenoid

Starter 1108334
Volts 12.0
Brush spring tension, oz. 35
No-load test,
  Volts 9.0
  Amperes (min.) 55.0*
  Amperes (max.) 80.0*
  RPM (min.) 3500
  RPM (max.) 6000
  *Includes solenoid

Starter 1113647
Volts 12.0
Brush spring tension, oz. 80
No-load test,
  Volts 11.6
  Amperes (min.) 85.0*
  Amperes (max.) 125.0*
  RPM (min.) 5900
  RPM (max.) 8100
  *Includes solenoid

STARTER SOLENOID
All Models

225. All starting motors are equipped with Delco-Remy solenoid switches. Specification data for these units is as follows:

Solenoid 1114356
Rated voltage 12.0
Current consumption,
  Pull-in winding,
    Volts 5.0
    Amperes 13.0-15.5
  Hold-in winding,
    Volts 10.0
    Amperes 14.5-16.5

Solenoids 1115510 & 1115518
Rated voltage 12.0
Current consumption,
  Pull-in winding,
    Volts 5.0
    Amperes 26.0-29.0
  Hold-in winding
    Volts 10.0
    Amperes 18.0-20.0

STANDARD IGNITION
All Non-Diesel Models

226. Non-diesel engines with standard ignition systems are equipped with IH distributors and the firing order is 1-5-3-6-2-4. Overhaul procedure for the distributor is obvious after an examination of the unit and reference to Fig. 164. Breaker contact gap is 0.020. Breaker arm spring tension should be 21-25 oz.

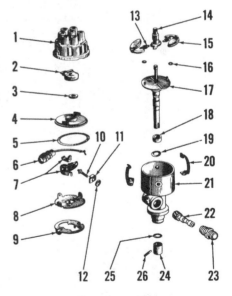

Fig. 164–Exploded view of IH distributor used on non-diesel engines.

1. Distributor cap
2. Rotor
3. Felt seal
4. Cover
5. Gasket
6. Condenser
7. Breaker contact set
8. Breaker plate
9. Weight guard
10. Primary terminal
11. Insulator
12. Insulating washer
13. Weight spring
14. Breaker cams
15. Weight
16. Washer
17. Shaft assembly
18. Oil seal
19. Thrust washer
20. Cap retainer
21. Distributor housing
22. Tachometer gear
23. Housing
24. Collar
25. Thrust washer
26. Pin

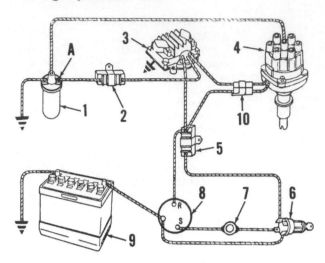

Fig. 165—Typical wiring circuit of magnetic pulse ignition system.

1. Ignition coil
2. Resistor (0.43 ohm)
3. Pulse amplifier
4. Distributor
5. Resistor (0.68 ohm)
6. Ignition switch
7. Starter switch
8. Starter solenoid switch
9. Battery
10. Connector body

standard distributor; however, the internal construction is different. The timer core (4—Fig. 166) and magnetic pickup assembly (5) are used instead of the conventional breaker plate, breaker contact set and condenser assembly. Timer core (4) has the same number of equally spaced projections as engine cylinders. The magnetic

227. DISTRIBUTOR INSTALLATION AND TIMING. With the oil pump properly installed as outlined in paragraph 94 make certain the timing pointer is aligned with the TDC mark on crankshaft pulley or flywheel.

Install the distributor so that rotor arm is in the number one firing position and adjust breaker contact gap to 0.020. Loosen distributor clamp bolts, turn distributor counter-clockwise until breaker contacts are closed; then rotate distributor clockwise until contacts are just beginning to open. Tighten clamp bolts. Attach a timing light and with engine operating at correct high idle, no-load speed, adjust distributor to the following crankshaft

pulley or flywheel degree marks:
Gasoline (C291) . . . . 18° BTDC
Gasoline (C301) . . . . 22° BTDC
LP-Gas (C301) . . . . 24° BTDC

MAGNETIC PULSE IGNITION
Model 826 Non-Diesel

228. Model 826 non-diesel engines may be equipped with the "Delcotronic" transistor controlled magnetic pulse ignition system. This system consists of a special pulse distributor, pulse amplifier, resistors and special ignition coil. Refer to Fig. 165. Ignition switch, starter, solenoid switch and battery are conventional.

The external appearance of the magnetic pulse distributor resembles a

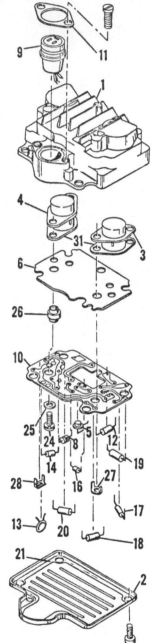

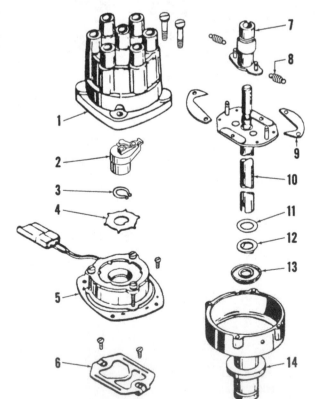

Fig. 166—Exploded view of Delco-Remy distributor used on non-diesel engines equipped with magnetic pulse ignition system.

1. Distributor cap
2. Rotor
3. Snap ring
4. Timer core
5. Magnetic pickup assembly
6. Weight hold-down
7. Advance cam
8. Weight spring
9. Weight
10. Shaft assembly
11. Washer
12. Thrust washer
13. Oil seal
14. Housing

Fig. 167—Exploded view of typical magnetic pulse amplifier.

1. Housing
2. Cover
3. Drive transistor (TR2)
4. Output transistor (TR1)
5. Trigger transistor (TR3)
6. Heat sink
8. Zener diode
9. Wiring connector
10. Printed circuit panelboard
11. Connector clamp
12. Capacitor
13. Capacitor
14. Resistor (10 ohm)
16. Resistor (680 ohm)
17. Resistor (1800 ohm)
18. Resistor (15000 ohm)
19. Resistor (15 ohm)
20. Resistor (150 ohm)
21. Gasket
24. Screw
25. Washer
26. Bushing insulators
27. Clip (round)
28. Clip (rectangular)
31. Insulators

pickup assembly (5) consists of a ceramic permanent magnet, pickup coil and pole piece. The flat metal pole piece has the same number of equally spaced internal projections as the timer core. The timer core which is secured to advance cam (7) is made to rotate around distributor shaft (10) by advance weights (9). This provides a conventional centrifugal advance.

The pulse amplifier (3—Fig. 165) consists primarily of transistors, resistors, capacitors and a diode mounted on a printed circuit panelboard (10—Fig. 167).

229. DISTRIBUTOR INSTALLATION. With oil pump properly installed as outlined in paragraph 94, make certain timing pointer is aligned with the TDC mark on crankshaft pulley or flywheel. Install distributor so that rotor arm is in number one firing position and projections on pole piece and timer core are aligned. Secure distributor with clamps and bolts.

Attach a timing light and with engine operating at correct high idle, no-load speed, adjust distributor to the following crankshaft pulley or flywheel degree marks:

Gasoline .... 22° BTDC
LP-Gas .... 24° BTDC

230. TROUBLE SHOOTING. Ignition problems encountered will be evidenced by one of the following conditions.

A. Engine miss or surge

B. Engine will not run at all.

CAUTION: When trouble shooting the system, use extreme care to avoid accidental shorts or grounds which may cause instant damage to the amplifier. Never disconnect the high voltage lead between coil and distributor and never disconnect more than one spark plug lead unless ignition switch is in "OFF" position. To make compression tests, disconnect wiring harness plug at the pulse amplifier, then remove spark plug leads.

231. If engine misses or surges, and fuel system and governor are satisfactory, check distributor as follows: Make certain that the two distributor leads (solid white and white with green stripe) are connected to connector body as shown in Fig. 168. Disconnect the connector body halves and connect an ohmmeter (step 1) as shown in Fig. 168. Any reading above or below a range of 550-750 ohms indicates a defective pickup coil. Remove one ohmmeter lead from connector body and connect to ground (step 2—Fig. 168). Any reading less than infinite indicates a defective pickup coil. Renew magnetic pickup coil assembly (5—Fig. 166) if necessary.

A poorly grounded pulse amplifier can also cause an engine to miss or surge. To check, temporarily connect a jumper lead from amplifier housing to a good ground. If engine performance improves, amplifier is poorly grounded. Correct as necessary.

232. If engine will not run at all, remove one spark plug lead and hold end of lead about ¼-inch from engine block. Crank engine and check for spark between spark plug lead and block. If sparking occurs, the trouble most likely is not ignition. If sparking does not occur, check ignition system as follows:

With distributor connector and amplifier connector attached, connect a 12-volt test light to connector body as shown (step 1—Fig. 169). Turn ignition switch to "ON" position. If test light bulb does not light, there is an open between ignition switch (6—Fig. 165) and connector body (10), including leads, distributor pickup coil and resistor (5). Check distributor pickup coil as outlined in paragraph 231. If the bulb burns at full brilliance, resistor (5) is not properly connected to ignition switch. If bulb burns at about half brilliance, resistor (5) is properly connected and circuit is satisfactory. Disconnect ignition wire between switch (6) and resistor (5). Turn ignition switch to "ON" position and press starter switch button (7) to crank engine. If bulb does not light, check for open between solenoid switch (8) and resistor (5). If bulb lights, reconnect ignition wire and proceed with tests. Connect the 12-volt test light between connector body (step 2—Fig. 169) and coil input terminal (point A—Fig. 165). Turn ignition switch to "ON" position, press starter switch button and crank engine. If the bulb flickers, the primary circuit is operating normally. Check (in usual manner for standard ignition system) the secondary system, including spark plugs, wiring, ignition coil tower and secondary winding and the distributor cap for evidence of arc-over or leakage to ground. If bulb does not light, check coil ground wire and coil. If test light burns at full brilliance, check for an open between point "A" and connector body. This includes leads, connections and resistor (2). If wiring, connections and resistor are satisfactory, check for poor amplifier ground by connecting a jumper lead between amplifier housing and a good ground. If bulb now burns at half brilliance, amplifier is poorly grounded. Correct as required. If bulb remains at full brilliance, renew or repair pulse amplifier as outlined in paragraph 233.

233. AMPLIFIER TEST AND REPAIR. To check the pulse amplifier for defective components, remove the unit from engine and proceed as follows: Refer to Fig. 167 and remove retaining screws, cover (2) and gasket (21). To aid in reassembly, note location of the three lead connections to the panelboard. See Fig. 170. Remove three panelboard mounting screws and lift the assembly from amplifier housing.

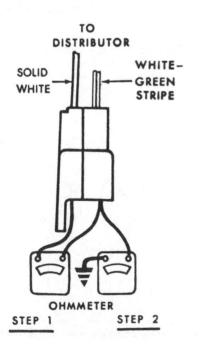

Fig. 168–View showing ohmmeter connections to distributor connector body when checking for defective pickup coil.

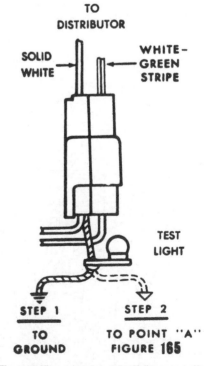

Fig. 169–View showing test light connections when checking for open circuits or defective pulse amplifier.

CAUTION. Drive transistor TR2 (3 —Fig. 167) and output transistor TR1 (4) are not interchangeable and must not be installed in reverse position. Before removing transistors, identify and mark each transistor and their respective installation on the heat sink and panelboard assembly.

Remove the screws securing TR1 and TR2 transistors to panelboard, then separate transistors and heat sink from panelboard. Note the thin insulators (31) between transistors and heat sink and the bushing insulators (26) separating the heat sink from panelboard. Visually inspect the panelboard for defects.

NOTE: To check the panelboard assembly, it is first necessary to unsolder capacitors (C2 and C3—Fig. 170) at location shown in Fig. 171. A 25-watt soldering gun is recommended and 60% tin—40% lead solder should be used when resoldering. DO NOT use acid core solder. Avoid excessive heat which may damage the panelboard. Chip away any epoxy involved and apply new epoxy (Delco-Remy part No. 1966807).

An ohmmeter having a 1½-volt cell is recommended for checking the amplifier components. The low range scale should be used in all tests except where specified otherwise. In all of the following checks, connect ohmmeter leads as shown in Fig. 171, then reverse the leads to obtain two readings. If, during the following tests, the ohmmeter readings indicate a defective component, renew the defective part, then continue checking the balance of the components.

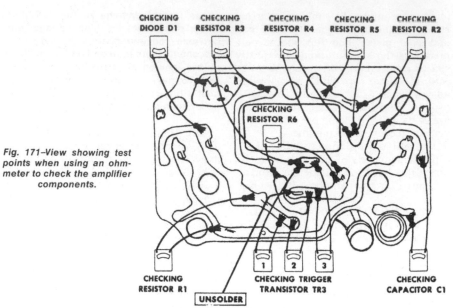

Fig. 171–View showing test points when using an ohmmeter to check the amplifier components.

Trigger Transistor TR3. If both readings in steps 1, 2 or 3 are zero or if both readings in steps 2 or 3 are infinite, renew the transistor.

Diode D1. If both readings are zero or if both readings are infinite, renew the diode.

Capacitor C1. If both readings are zero, renew the capacitor.

Capacitors C2 & C3. Connect ohmmeter across each capacitor. If both readings on either capacitor are zero, renew the capacitor.

Resistor R1. If both readings are infinite, renew the resistor.

Resistor R2. Use an ohmmeter scale on which the 1800 ohm value is within the middle third of the scale. If both readings are infinite, renew the resistor.

Resistor R3. Use an ohmmeter scale on which the 680 ohm value is within the middle third of the scale. If both readings are infinite, renew the resistor.

Resistor R4. Use an ohmmeter scale on which the 15000 ohm value is within the middle third of the scale. If either reading is infinite, renew the resistor.

Resistor R5. Use the lowest range ohmmeter scale. If either reading is infinite, renew the resistor.

Resistor R6. Use an ohmmeter scale on which the 150 ohm value is within the middle third of the scale. If both readings are infinite, renew the resistor.

Transistors TR1 and TR2. Check each transistor as shown in Fig. 172. If both readings in steps 1, 2 or 3 are zero or if both readings in steps 2 or 3 are infinite, renew the transistor.

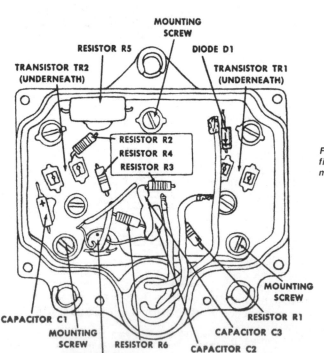

Fig. 170–Typical pulse amplifier with bottom cover removed showing location of components.

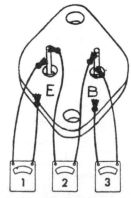

Fig. 172–Test procedure (steps 1, 2 & 3) when checking TR1 and TR2 transistors with an ohmmeter. Transistors must be installed with emitter pin (E) and base pin (B) in their original positions on panelboard.

Reassemble by reversing the disassembly procedure. When installing transistors TR1 and TR2, coat transistor side of heat sink (6—Fig. 167) and both sides of flat insulators (31) with silicone grease. The silicone grease, which is available commercially, conducts heat and thereby provides better cooling.

Delco-Remy part numbers for the pulse amplifier and components parts are as follows:

Pulse amplifier assembly 1115005
Output transistor TR1 1960632
Drive transistor TR2 1960584
Trigger transistor TR3 1960643
Zener diode D1 1960642
Printed circuit panelboard ...1963865
Capacitor C1 1960483
Capacitors C2 and C3 1962104
Resistor R1 (10 ohm) 1960640
Resistor R2 (1800 ohm) 1960639
Resistor R3 (680 ohm) 1960638
Resistor R4 (15000 ohm) 1960641
Resistor R5 (15 ohm) 1963873
Resistor R6 (150 ohm) 1965254

CLUTCH

All Gear Drive Models

234. Models 766, 826 and 966 gear drive tractors are fitted with a 12 inch dry disc clutch. Model 1066 is fitted with a 14 inch dry disc clutch. Spring loaded clutches are standard equipment, however, an over-center clutch is optionally available on International models.

Clutch wear is compensated for by adjusting clutch linkage and when engine clutch is adjusted, it will require that the transmission brake, torque amplifier dump valve (if so equipped) and starter safety switch also be adjusted.

ADJUSTMENT

235. **ENGINE CLUTCH (SPRING LOADED)** To adjust the spring loaded clutch, refer to Fig. 173 and disconnect transmission brake rod from brake lever (M) by removing cotter pin (A)

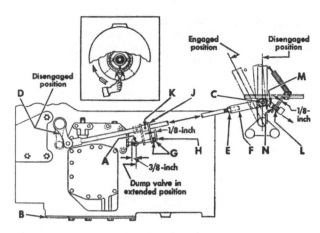

Fig. 174–View of overcenter clutch linkage and points of adjustment. Refer to text for procedure.

A. Pin
B. Bottom cover
C. Clevis pin
D. Clutch lever
E. Jam nut
F. Clevis
G. Jam nuts
H. Pull rod
J. Actuating lever
K. Set screw
L. Lock nuts
M. Boot
N. Safety switch

and pin (B). Loosen jam nuts (F) and rotate clutch rod turnbuckle (G) as required to obtain 11/16-inch pedal free travel for model 826 and ⅝ inch for all other models. Measurement is made between platform and arrow on clutch pedal. Tighten jam nuts (F).

Clutch linkage should be adjusted when pedal free travel has decreased to ⅜-inch.

236. **ENGINE CLUTCH (OVER-CENTER).** To adjust the over-center clutch, be sure the ignition or injection system is in "OFF" position. Place clutch lever in the disengaged position and remove the bottom cover (B—Fig. 174) from clutch housing. Turn engine until adjusting ring lock is on bottom side, then remove lock and turn (tighten) adjusting ring until 30-40 pounds of effort applied to hand lever is required to engage clutch. When properly adjusted, clutch will go over-center with a distinct snap. Install adjusting ring lock and the clutch housing bottom cover. Refer to paragraph 237 for linkage adjustment.

NOTE: When engine is idling, the effort required to engage clutch will be approximately 10 pounds less than when engine is stopped.

237. **LINKAGE ADJUSTMENT.** Whenever the engine clutch has been adjusted, the control linkage should be checked and adjusted, if necessary.

With engine clutch adjusted, the

clutch rod length should be such that clevis pin (C) will freely enter holes of clevis (F) and operating (hand) lever when engine clutch and hand lever are both in the fully disengaged position.

If adjustment of control rod is required, proceed as follows: Engage clutch and loosen jam nut (E). Remove clevis pin (C) and move hand lever to the extreme rear position. This will free clevis so it can be turned and the rod length adjusted as previously stated.

238. **TRANSMISSION BRAKE.** On tractors with spring loaded clutch, the transmission brake should be adjusted each time engine clutch is adjusted. To adjust the transmission brake, disconnect control rod from brake lever (M—Fig. 173), depress the clutch pedal until stop lug (J) contacts platform, move lever (M) rearward as far as possible, then adjust clevis (N) until clevis pin will freely enter clevis and brake lever. Now remove pin and unscrew clevis ½-turn. Reinstall clevis on brake lever and tighten jam nut.

NOTE: If gear clash is experienced, unscrew clevis one turn additional.

239. **DUMP VALVE.** The torque amplifier dump valve should be checked and adjusted each time engine clutch is adjusted. Refer to paragraph 240 for information on tractors with spring loaded clutches and to paragraph 241 for tractors with over-center clutches.

240. To position the torque amplifier dump valve on tractors with spring loaded clutches, depress clutch pedal until stop lug strikes platform, loosen jam nut (V—Fig. 173) and turn operating screw (P) until valve operating lever (R) positions the valve spool pin (T) 9/16-inch in the extended position as shown.

241. On tractors with over-center clutches, position the torque amplifier dump valve as follows: With clutch lever in disengaged position, loosen set screw (K—Fig. 174). Be sure jam nuts

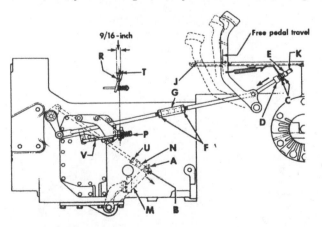

Fig. 173–View of spring loaded clutch linkage and points of adjustment. Refer to text for procedure.

A. Cotter pin
B. Clevis pin
C. Lock nuts
D. Safety switch
E. Boot
F. Jam nuts
G. Turnbuckle
J. Pedal stop lug
K. Operating lever
M. Brake lever
N. Clevis
P. Operating screw
R. Operating lever
T. Valve spool pin
U. Jam nut
V. Jam nut

(G) are tight and that about ⅛ inch of the pull rod extends beyond the rear jam nut. Pull the dump valve spool out ⅜-inch; then, while holding dump valve in this position, slide lever (J) rearward on clutch rod until it is against the jam nuts (G). Tighten set screw (K) to maintain the adjustment.

242. **STARTER SAFETY SWITCH.** The starter safety switch prevents the tractor engine from being started except when clutch is in the disengaged position. To adjust the starter safety switch, refer to Figs. 173 and 174, and proceed as follows: Loosen the two lock nuts which position switch in the bracket and move the switch, in the direction indicated by the arrow, until upper nut contacts plunger boot. On tractors with spring loaded clutches depress clutch pedal until stop lug on pedal strikes platform. On tractors with over-center clutches, pull lever to extreme rear position. Now position switch so that switch plunger is depressed about ⅛-inch and tighten lock nuts to maintain this position.

REMOVE AND REINSTALL
All Gear Drive Models

243. To remove the engine clutch, it is first necessary to separate (split) engine from clutch housing as outlined in paragraph 244 or 245. With engine split from clutch housing, removal of clutch from flywheel is obvious. Refer to paragraph 246 for overhaul data.

TRACTOR SPLIT
Model 826 Gear Drive

244. To split tractor for clutch service, first drain cooling system and remove front hood, rear hood and the steering support housing (cover). Remove batteries, then disconnect all wiring from engine, unclip harness and lay wiring harness rearward on fuel tank. Disconnect tachometer cable at the engine and lay rearward. Disconnect oil pressure switch wire and pull it rearward. Identify the power steering oil cooler hoses and disconnect them at rear.

NOTE: It is essential that the power steering oil cooler hoses be correctly reinstalled as the oil flowing to the oil cooler is maintained at 105 psi by a pressure regulating valve in the multiple control valve and is used as it returns to lubricate the differential before draining back to the main reservoir.

Disconnect the two power steering cylinder lines from control (pilot) valve. Shut off fuel and disconnect fuel supply line at front end. Disconnect the coolant temperature bulb from cylinder head. Disconnect controls from car-

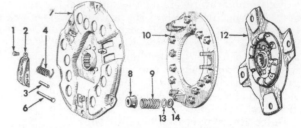

1. Lever adjusting screw
2. Lever
3. Pivot pin
4. Lever spring
5. Lever
6. Lever pin
7. Back plate
8. Spring cup
9. Clutch spring
10. Pressure plate
12. Driven disc
13. Insulating washer
14. Washer retainer

Fig. 175–Exploded view of spring loaded clutch; all models are similar. Refer to text.

buretor, governor or injection pump. Install split stand to side rails and place a rolling floor jack under rear section of tractor. Remove bottom side rail to clutch housing cap screw from both side rails and install guide studs. Complete removal of clutch housing retaining cap screws and separate tractor.

Note: Insert cap screws back in side rails before pulling side rails completely off guide studs.

Rejoin tractor sections by reversing the splitting procedure; however, in order to avoid any difficulty which might arise in trying to align splines of clutch assembly and the transmission input shafts during mating of sections, most mechanics prefer to remove clutch from flywheel and place it over the transmission input shafts. Clutch can be installed on flywheel after tractor sections are joined by working through the opening at bottom of clutch housing.

Models 766, 966 and 1066 Gear Drive

245. To split tractor for clutch service, first disconnect and remove bat-

teries. Remove left and right front hood, front hood channel and heat baffles. Disconnect electrical wiring and tachometer drive cable from engine and lay rearward on fuel tank. Disconnect oil cooler lines at front of engine, power steering lines and engine oil pressure line. Shut off and disconnect fuel supply at front end. Disconnect the coolant temperature bulb from cylinder head. Disconnect controls from carburetor, governor or injection pump. Install split stands to side rails and place a rolling floor jack under rear section of tractor. Remove bottom side rail to clutch housing cap screw from both side rails and install guide studs. Complete removal of clutch housing retaining cap screws and separate tractor.

NOTE: Insert cap screws back in side rails before pulling side rails completely off guide studs.

Rejoin tractor sections by reversing the splitting procedure; however in order to avoid any difficulty which might arise in trying to align splines of clutch assembly and the transmission input shafts during mating of sections, most mechanics prefer to remove clutch from flywheel and place it over

Fig. 176–Exploded view showing components of over-center clutch.

2. Snap ring
3. Snap ring
4. Release bearing
5. Grease fitting
6. Release bearing carrier
7. Connecting link
8. Pin (long)
9. Pin (short)
10. Retainer ring
11. Adjusting ring
12. Washer
13. Washer
14. Return spring (inner & outer)
15. Adjusting lock
16. Back plate
17. Lock screw
18. Release lever
19. Roll pin
20. Pressure plate
21. Spring stud
22. Driven disc
23. Grease fitting
24. Release shaft (RH)
25. Woodruff key
26. Release fork
27. Release shaft (LH)

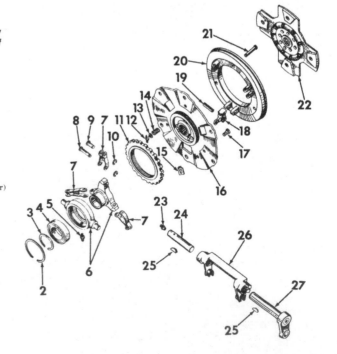

transmission input shafts. Clutch can be installed on flywheel after tractor sections are joined by working through the opening at bottom of clutch housing.

OVERHAUL
All Gear Drive Models

246. The disassembly and adjusting procedure for either clutch will be obvious after an examination of the unit and reference to Figs. 175 and 176. Clutch (driven) discs are available as a unit only.

Specification data is as follows:

Models 826 and 966
(Spring Loaded)
Size, inches 12
Springs:
    Number used 12
    Free length, inches 2½
    Lbs. test @ height
      inches . . . . 140 @ 1 21/32
    Number of coils 9
    Color . . . . 2 bronze stripes
    Lever height, inches 2.301
    Back plate to pressure plate,
      inches. 1.020

Model 766 (Spring Loaded)
Size, inches 12
Springs:
    Number used 12
    Free length, inches 2 19/32
    Lbs. test @ height
      inches . . . . 110 @ 1 21/32
    Number of coils 9½
    Color . . . . 2 purple strips
    Lever height, inches 2.301
    Back plate to pressure plate,
      inches 1.020

Model 1066 (Spring Loaded)
Size, inches 14
Springs:
    Number used 12
    Free length, inches 2 5/16
    Lbs. test @ height
      inches . . . . 170 @ 1 21/32
    Number of coils 8¼
    Color . . . . 2 yellow stripes
    Lever height, inches 2.301
    Back plate to pressure plate,
      inches 1.469

CLUTCH SHAFT
All Gear Drive Models

247. The clutch shaft on tractors not equipped with torque amplifier is the transmission input shaft and will be covered in the transmission section.

The clutch shaft on tractors equipped with torque amplifier is part of the torque amplifier assembly and will be covered in the torque amplifier section.

TORQUE AMPLIFIER AND SPEED TRANSMISSION

The torque amplifier and the speed (forward) transmission are both located in the clutch housing along with the hydraulic pump which supplies the power steering, brakes and torque amplifier. Any service on the torque amplifier requires that the entire speed transmission be disassembled before the torque amplifier can be removed. Therefore, this section will concern both units.

Power from the engine is applied directly to the torque amplifier and during operation, the torque amplifier is locked either in direct drive or torque amplifier (underdrive) position by hydraulic multiple disc clutches. There is no neutral position in either the torque amplifier or speed transmission as the neutral position is provided for in the range (rear) transmission.

During operation in the torque amplifier (underdrive) position, the clutch shaft is locked to the torque amplifier constant mesh output gear by a one-way clutch and an approximate 1/5 speed reduction occurs, resulting in about a 28 percent increase in torque.

LINKAGE ADJUSTMENT
All Gear Drive Models

248. To adjust the torque amplifier linkage, first remove button plug from control lever, then remove the control lever and the steering support cover. Note the locating punch marks on control lever and shaft. Place control lever back on its shaft (without cap screw) and pull control lever rearward until stop lug on pivot shaft contacts stop pin (J—Fig. 177). At this time, distance (C) should measure ¾-inch between center of clevis hole and forward edge of clutch mounting flange as shown. If distance is not as stated, disconnect clevis (F) and adjust as necessary. If valve spool prevents adjustment, disconnect valve lever (horizontal) operating rod, then adjust length of bellcrank (vertical) rod until bellcrank is positioned to the ¾-inch measurement.

With bellcrank (D) position determined, reinstall valve lever operating rod and measure distance (W) which should be 1-9/64 inches. If measurement is not as stated, adjust clevis (B) as required.

With the two previous adjustments made, push control lever forward. At this time, the snap ring located on valve spool should be contacting the multiple control valve body and the stop screw in pivot shaft bracket should be 0.002-0.010 from arm of pivot lever. Adjust stop screw as necessary.

Remove control lever and install steering support cover. Reinstall control lever with aligning punch marks in register and install retaining cap screw and button plug.

To adjust the speed transmission linkage, remove TA control lever and steering support cover. Adjust ball joint linkage at upper end of shift rod to provide accurate positioning of speed transmission lever to the numbers 1-2-3-4 on the quadrant. Shift lever should have clearance at either end of quadrant when transmission is in either first or fourth gear. Reinstall support cover and TA control lever.

REMOVE AND REINSTALL
All Gear Drive Models

249. Removal of the torque amplifier or speed transmission requires the removal of the complete clutch housing from the tractor.

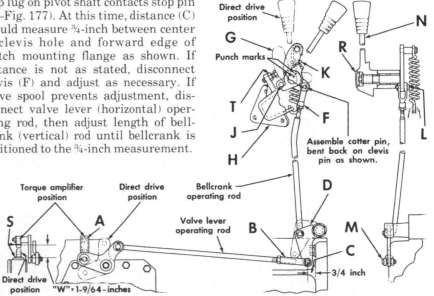

Fig. 177—Schematic view of torque amplifier operating linkage. Refer to text for adjustment procedure.

250. To remove clutch housing from tractor, proceed as follows: Remove front hood. Remove torque amplifier control lever and note match marks on lever and shaft. Remove assist handle, steering support cover and center hood. Remove steering support cover skirts. Remove battery or batteries and tray. Disconnect wiring from engine and starter. Remove fuel supply line, and on diesel models, disconnect the fuel return line. Disconnect temperature bulb from engine. On non-diesel models, remove governor rod and disconnect choke control. On diesel models, remove injection pump control rod.

Disconnect the torque amplifier vertical control rod from bellcrank. Disconnect tachometer cable from engine. Disconnect wires from engine oil pressure switch and transmission oil pressure switch. On model 826 disconnect power steering hand pump lines from control valve. Identify and disconnect power steering oil cooler hoses at rear end. On models 766, 966 and 1066 disconnect power steering lines on right rear side of engine. Identify and disconnect oil cooler hoses at front of engine and lay rearward.

NOTE: It is essential that the power steering oil cooler lines be correctly identified for correct installation as the oil flowing to the cooler is controlled at 105 psi by a regulator valve in the multiple control valve. This pressurized oil returning from the oil cooler is used to lubricate the differential gears and bearings.

The pressurized return oil from the power steering valve is used to operate the brakes and torque amplifier clutches as well as providing lubrication for the torque amplifier clutches and transmission.

Disconnect tank support from top of clutch housing. Remove platform, drive roll pin from speed transmission shifter cam coupling and disconnect clevis from range transmission shifter arm. Remove control link from park lock. Disconnect the four flexible hydraulic lines at their aft ends, if so equipped. Attach a hoist to steering support and secure tank with a support chain from front of tank to hoist.

NOTE: On International model tractors remove the left and right mud shields from platform, fenders and steering support.

Unbolt steering support from top cover of speed transmission. Carefully lift steering support and tank assembly, disengage coupling from shifter shaft and if necessary, disengage any wiring or tubing which may be interfering.

With tank and steering support removed, on model 826 disconnect the power steering lines from control valve. On all models remove the brake control valve line. Remove starter. Remove the torque amplifier control rod, then remove snap ring and the bellcrank and TA valve link. Disconnect clutch rod at both ends, unbolt dump valve pivot bracket and remove rod and bracket assembly. Support front and rear sections of tractor. Unbolt engine and side rails from clutch housing but leave the side rail cap screws in the side rails so they protrude through the engine rear end plate to provide support. Move sections apart, then attach hoist to clutch housing and separate clutch housing from rear frame.

NOTE: Work the IPTO driven gear off the IPTO shaft working through the opening in the bottom of clutch housing as you split the tractor.

OVERHAUL
All Gear Drive Models

251. With the clutch housing removed as outlined in paragraph 250, disassemble the speed transmission and torque amplifier as follows: Disconnect and remove the transmission brake operating rod, then remove snap ring, operating lever and Woodruff key from right end of clutch cross shaft. Loosen the two cap screws in the clutch release fork, bump cross shaft toward left until the two Woodruff keys are exposed and remove Woodruff keys. Complete removal of cross shaft, release fork and throwout bearing assembly.

Unbolt and remove the multiple control valve and pump assembly. Also remove the power steering control valve on model 826. Complete removal of top cover retaining cap screws and lift off the top cover and shifter cam assembly. See Fig. 178. Lift off shifter arms (SA) and be careful not to lose the two cam rollers (CR). Unstake and remove the four Phillips head screws (PS) which retain the shifter rails and forks, then lift out the shifter rails and forks (SF). Unbolt the transmission main shaft bearing cage and as shaft and bearing cage is pulled rearward, pull gears from top of clutch housing. See Fig. 179. Remove the clutch housing bottom cover and brake assembly as shown in Fig. 180. Straighten lock washer on rear end of countershaft and remove the nut. Remove the pto driven gear bearing retainer and the pto driven gear and bearing assembly. Use a thread protector on rear end of countershaft, drive shaft forward and remove gears and spacers from bottom of clutch housing as shaft is moved forward. See Fig. 181.

Fig. 179–Mainshaft, mainshaft bearing assembly and gears shown removed from speed transmission.

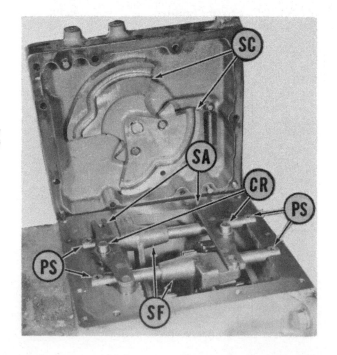

Fig. 178–View of speed transmission (gear drive) with cover removed.

CR. Cam rollers
PS. Phillips screws
SA. Shifter arms
SC. Shifter cam
SF. Shifter forks

Fig. 180–Speed transmission brake assembly is incorporated in clutch housing bottom cover. Brake pad operates against direct drive constant mesh gear.

Pull the three fluid supply tubes from housing as shown in Fig. 182.

At this time, the torque amplifier unit can be removed from clutch housing. However, in order to preclude any damage to sealing rings, or other parts, it is recommended that the torque amplifier unit be held together as follows: Install a cap screw and washer in rear end of clutch shaft to hold the direct drive gear assembly in place. Use a "U" bolt around clutch shaft to hold pto drive gear and carrier assembly in place. Install a lifting eye in front end of clutch shaft. Place clutch housing upright with bell end of housing upward. Attach a hoist to the previously installed lifting eye. Unbolt front carrier and the direct drive gear bearing cage (rear) from housing webs and carefully lift torque amplifier assembly from clutch housing.

252. With the torque amplifier assembly removed from clutch housing, disassemble unit as follows: Remove the previously affixed holding fixtures and pull the pto drive gear and carrier assembly and the direct drive gear assembly from the unit. Unbolt and remove the pto drive shaft bearing cage from carrier assembly and remove bearing cage and pto drive shaft as

Fig. 182–Three fluid supply tubes to torque amplifier can be pulled straight out after removal of multiple control valve. Note "O" ring on each end of tube.

Fig. 183–PTO drive shaft, bearing and bearing cage removed from TA and PTO carrier.

Fig. 184–The one-way clutch is positioned in inner bore of torque amplifier drive gear.

15. Ball bearing
29. TA drive gear
33. One-way clutch

shown in Fig. 183. Pull torque amplifier drive gear and bearing assembly from TA carrier. Remove lubrication baffle and baffle springs from carrier. Remove seal ring, then remove the one-way clutch from drive gear and be sure to note how the unit is installed in the drive gear. See Figs. 184 and 189. It is possible to install the clutch with the wrong side forward and should this happen the torque amplifier would not operate. The quill shaft and output gear can be removed from bearing cage after removing the large internal snap ring. Drive gear and bearing removal from quill gear is obvious. Remove the large internal snap ring from front (lock-up) clutch and remove the clutch plates as shown in Fig. 185. Straighten tabs of lock washers at rear of rear clutch, remove nuts and remove direct

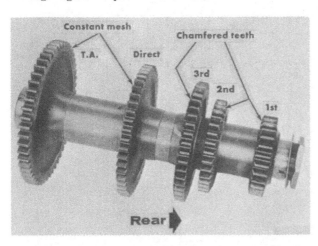

Fig. 181–View of speed transmission countershaft, gears and spacers.

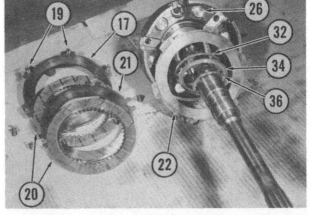

Fig. 185–Front (TA lock-up) clutch disassembled. Note the Teflon seal ring and bronze cup washer.

17. Backing plate	21. Drive plate	32. Teflon seal
19. Springs (6 used)	22. Backing plate	34. Cup washer
20. Driven disc	26. Piston (6 used)	36. Thrust washer

drive (rear) clutch. See Fig. 186. Remove the three cap screws and remove the front (lock-up) clutch carrier. Piston carrier is generally a press fit on clutch shaft and can be removed from clutch shaft after removing rear snap ring and thrust washer. Remove piston from piston carrier. See Fig. 187.

At this time, all parts of the torque amplifier assembly and speed transmission can be inspected and parts renewed as necessary. Procedure for removal of bearings is obvious. Refer to Figs. 188 through 191 for installation dimensions and information. Use new "O" rings and lubricate all torque amplifier parts during assembly.

Reassemble components as follows: Install the six pistons in the front (TA) clutch with the widest edge toward clutch discs (front). Install the large piston in the piston carrier with smooth surface toward inside. Align the oil holes of piston carrier with oil holes of clutch shaft and press carrier on clutch shaft. Install thrust washers on each side of clutch carrier with grooved sides away from carrier and install snap rings. Use new gasket, mount front clutch carrier on piston carrier, tighten the three cap screws to 19-21 ft.-lbs. torque and bend down lock washers. Place the nine aligning pins in rear side of piston carrier and

place the double springs on pins. Start with back plate next to piston, then install a driven plate (internal spline) and alternate with driving plate (external spline).

With clutch plates positioned, place the rear clutch backing plate over discs and install the three long bolts but do not tighten nuts until clutch splines have been aligned with the output gear. With clutch discs aligned and the backing plate positioned, tighten nuts to 53-60 ft.-lbs. torque and bend down lock washers. See Fig. 192 for cross sectional view of assembled clutches. Reassemble the output gear assembly, then install it in the rear clutch and secure it in position with the cap screw and washer used during removal.

Be sure snap ring and baffle is in one-way clutch bore in torque amplifier drive gear, then install the one-way clutch with drag clips toward front as shown in Fig. 189. Check one-way clutch operation before proceeding further. Slide assembly over hub of piston carrier. Turning gear clockwise should lock the one-way clutch. Counterclockwise rotation should cause clutch to over-run.

Remove gear and one-way clutch and using heavy grease, position the bronze cup washer over end of one-way clutch, then install the Teflon seal ring. If the

large ball bearing is installed on drive gear, place clutch discs over drive gear, install gear and position plate (discs) in

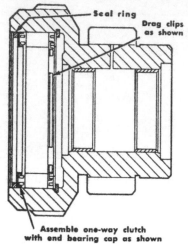

Fig. 189–Assemble one-way clutch assembly in torque amplifier drive gear as shown. A snap ring and baffle are installed in bore ahead of one-way clutch. Note installation of bearing cap (bronze cup washer) and Teflon seal ring.

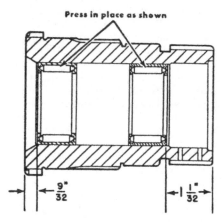

Fig. 190–Needle bearings are installed in direct drive output quill shaft to dimensions shown.

Fig. 186–Rear (direct drive) clutch disassembled. Note inner and outer springs.

37. Piston carrier
43. Driven plate
44. Drive plate
45. Guide plate
46. Inner spring
47. Outer spring
48. Backing plate

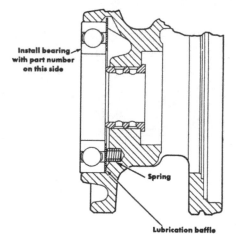

Fig. 188–Install lubrications baffle, baffle springs and torque amplifier drive gear bearing in TA carrier as shown.

Fig. 187–View showing piston removed from piston carrier. Note seal rings on inside and outside diameters of piston. Also note the removed front (lock-up) clutch piston which is installed with widest edge toward front.

25. Clutch carrier
26. Lock-up piston
37. Piston carrier
39. Piston
40. Outer seal
41. Inner seal

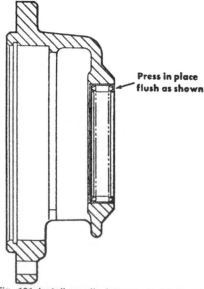

Fig. 191–Install needle bearing in direct drive output bearing cage as shown.

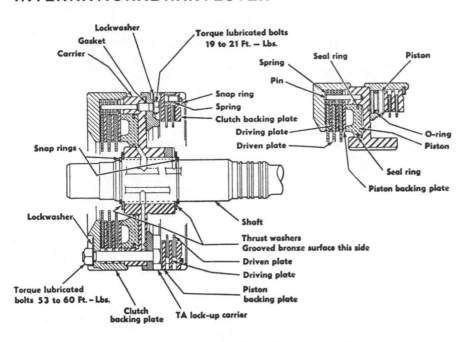

Fig. 192–Cross-sectional view of TA clutches after assembly.

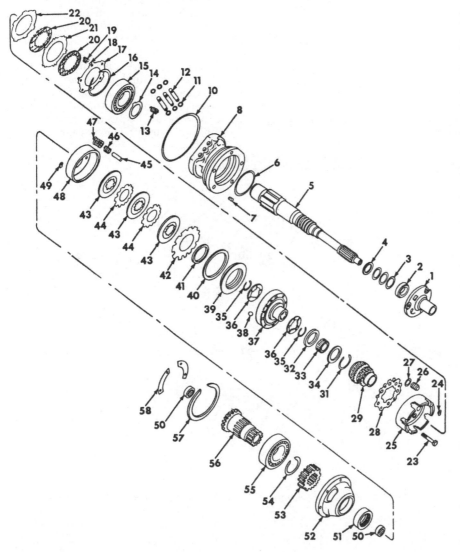

Fig. 193–Exploded view of torque amplifier assembly showing component parts and their relative positions.

clutch carrier. If the bearing is removed from drive gear, install the drive gear and feed clutch discs over drive gear and into clutch carrier. Clutch discs are positioned as follows: Backing plate with no slots in external lugs next to pistons in carrier, driven disc (internal spline), driving disc (with slots in lugs) and driven disc. Place springs through lugs of center driving disc and rest them on lugs of backing plate. Position front backing plate with pins through springs and install the large snap ring.

Renew sealing rings on clutch shaft, install baffle springs and lubrication baffle in TA carrier, with bent tab away from bearing, then install drive carrier over clutch shaft. If removed, install the pto drive shaft and bearing cage, then install the "U" bolt clamp previously used to hold parts in position. Install lifting eye in front end of clutch shaft.

Attach torque amplifier unit to a hoist and lower assembly into clutch housing making sure the cut out portion in direct drive gear bearing cage is on bottom side. Install cap screws in both end bearing cages and tighten securely. Install the three oil inlet tubes. Start speed transmission lower shaft in front of clutch housing and install gears and spacers in the sequence shown in Fig. 181. Reading from left to right, chamfered teeth of number three gear face front, number four face rear and number five face front. Tighten nut on rear of shaft to 300 ft.-lbs. torque. NOTE: On some early model tractors the bearing lock nut is thinner than the nut used on late model tractors. So it may not be possible to torque the thin nut to 300 ft.-lbs. as the socket will slip off the nut. Shaft should rotate freely with no visible end play. If above conditions are not met, recheck installation.

If mainshaft bearing assembly was disassembled, reassemble as follows:

1. Bearing cage	31. Snap ring
2. Oil seal	32. Teflon seal
3. Seal ring	33. One-way clutch
4. Bearing retainer	34. Baffle
5. Clutch shaft	35. Snap ring
6. "O" ring	36. Thrust washer
7. Set screw	37. Piston carrier
8. Carrier	38. Steel ball
10. "O" ring	39. Piston
11. "O" ring	40. Seal, outer
12. Oil inlet tube	41. Seal, inner
13. Spring (2 used)	42. Backing plate
14. Lubrication baffle	43. Driven disc
15. Ball bearing	44. Drive disc
16. Snap ring	45. Guide pin
17. Backing plate	46. Clutch spring, inner
18. Dowel pin	47. Clutch spring, outer
19. Clutch spring	48. Backing plate
20. Driven disc	49. Lock washer
21. Drive plate	50. Needle bearing
22. Piston backing plate	51. Needle bearing
23. Bolt	52. Bearing cage
24. Lock washer	53. Drive gear
25. Clutch carrier	54. Snap ring
26. Piston (TA lock-up)	55. Ball bearing
27. "O" ring	56. Quill shaft
28. Gasket	57. Snap ring
29. Drive gear	58. Lock plate

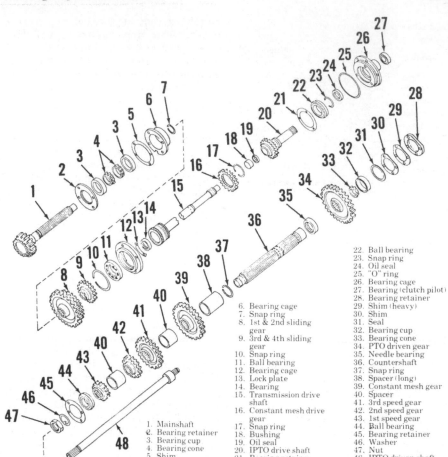

Press forward bearing cup into bearing cage with largest diameter rearward. Place bearing cones in the cage and install rear bearing cup. Place end plate, original shims and bearing assembly over mainshaft. Press both bearings on shaft and install the snap ring. Start shaft in rear of clutch housing and place gears on shaft as shaft is moved forward. Install bearing retainer cap screws and rotate shaft while tightening cap screws. Shaft should turn freely with no visible end play. Vary shims as necessary.

Disassembly and reassembly of transmission top cover and shifter cams is obvious after an examination of the unit.

Balance of reassembly is the reverse of disassembly. Shifter forks are interchangeable. Be sure to stake the shift rail retaining screws at the side of shaft (not front and rear) after installation.

Reinstall the clutch housing assembly by reversing the removal procedure. Note: When reinstalling the rear section on the clutch housing, it will be necessary to work the IPTO driven gear onto the IPTO shaft as the tractor is recoupled. Adjust the clutch, TA dump valve and transmission brake linkage as outlined in paragraphs 235 thru 242 and the torque amplifier linkage and speed transmission linkage as outlined in paragraph 248.

6. Bearing cage
7. Snap ring
8. 1st & 2nd sliding gear
9. 3rd & 4th sliding gear
10. Snap ring
11. Ball bearing
12. Bearing cage
13. Lock plate
14. Bearing
15. Transmission drive shaft
16. Constant mesh drive gear
17. Snap ring
18. Bushing
19. Oil seal
20. IPTO drive shaft
21. Bearing retainer
22. Ball bearing
23. Snap ring
24. Oil seal
25. "O" ring
26. Bearing cage
27. Bearing (clutch pilot)
28. Bearing retainer
29. Shim (heavy)
30. Shim
31. Seal
32. Bearing cup
33. Bearing cone
34. PTO driven gear
35. Needle bearing
36. Countershaft
37. Snap ring
38. Spacer (long)
39. Constant mesh gear
40. Spacer
41. 3rd speed gear
42. 2nd speed gear
43. 1st speed gear
44. Ball bearing
45. Bearing retainer
46. Washer
47. Nut
48. IPTO driven shaft

1. Mainshaft
2. Bearing retainer
3. Bearing cup
4. Bearing cone
5. Shim

Fig. 194—Exploded view of speed transmission showing components used. The one shown is not equipped with torque amplifier.

RANGE TRANSMISSION

The range transmission is located in the front portion of the tractor rear frame. This transmission provides four positions: Hi (direct drive), Lo (underdrive), neutral and reverse in gear drive. The Lo-range position provides a 3½ to 1 speed reduction and the reverse speed is about 25 percent faster than the same forward speed in the Lo-range. The transmission parking lock is also located in the range transmission.

To remove the range transmission gears and shafts, the tractor rear main frame must be separated from the clutch housing and the differential assembly removed. See Fig. 195 or 196 for an exploded view of the range transmission shafts and gears. Refer to paragraph 253 for information on overhaul of the range transmission and to paragraph 258 for linkage adjustment procedures.

Note: The service procedure for gear drive and hydrostatic drive tractors are basically the same. Except the hydrostatic does not use a reverse in the range transmission.

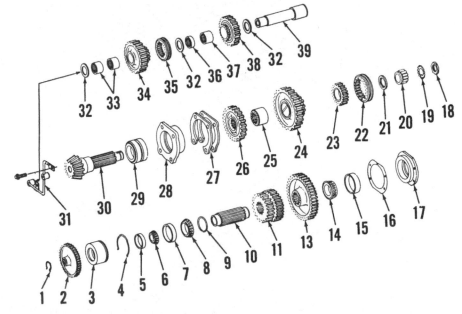

Fig. 195—Exploded view of range transmission shafts and gears used on gear drive models.

1. Retainer
2. Hydraulic pump drive gear
3. Bearing cage
4. Retainer
5. Bearing cup
6. Bearing cone
7. Bearing cup
8. Bearing cone
9. Snap ring
10. Countershaft
11. Lo-drive gear
13. Constant mesh gear
14. Bearing cone
15. Bearing cup
16. Shim
17. Bearing cage
18. Nut
19. Lock washer
20. Roller bearing
21. Spacer
22. Hi-Lo shift collar
23. Shift collar carrier
24. Hi-Lo driven gear
25. Gear carrier
26. Reverse driven gear
27. Shims (.004 & .007)
28. Bearing cage
29. Bearing assembly
30. Bevel pinion shaft
31. Lubrication tube
32. Thrust washers
33. Needle bearings
34. Reverse drive gear
35. Shift collar
36. Needle bearing
37. Needle bearing
38. Reverse idler gear
39. Reverse idler shaft

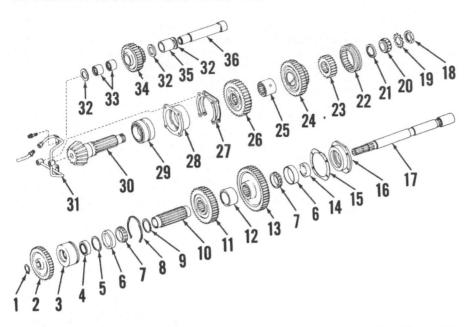

Fig. 196–Exploded view of range transmission shafts and gears used on hydrostatic drive models. Items 32 thru 36 are used on all Wheel Drive. Item 36 is a lube tube if not equipped with All Wheel Drive.

1. Retaining ring	10. Countershaft	19. Lock washer	29. Bearing assembly
2. Hydraulic pump drive gear	11. Lo-drive gear	20. Roller bearing	30. Bevel pinion shaft
3. Bearing cage	12. Spacer	21. Spacer	31. Lubrication tube
4. Bearing	13. Constant mesh gear	22. Hi-Lo shift collar	32. Thrust washers
5. Retaining ring	14. IPTO oil seal	23. Shift collar carrier	33. Needle bearings
6. Bearing cup	15. Shim	24. Hi-Lo driven gear	34. Reverse drive gear (all Wheel Drive)
7. Bearing cone	16. Bearing cage	25. Gear carrier	35. Spacer
8. Retaining ring	17. Shaft (PTO driven)	26. Reverse driven gear	36. Reverse drive shaft or lube tube
9. Snap ring	18. Nut	27. Shims (.004 & .007)	
		28. Bearing cage	

R&R AND OVERHAUL
All Models

253. If proper split stands are available, the rear main frame can be separated from the clutch housing before any disassembly is done. However, if range transmission is to be disassembled, it is usually as satisfactory to

remove the pto unit, hydraulic lift assembly and the differential before main frame is separated from clutch housing. Thus the main frame can be easily supported for the preliminary disassembly and after the above mentioned components are removed, the main frame can be removed and placed on a bench for service.

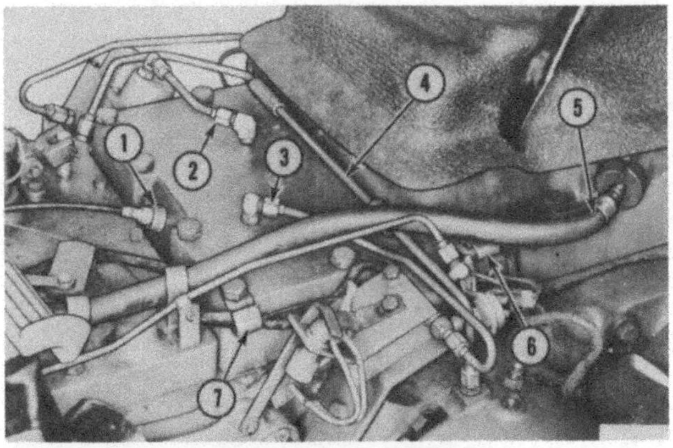

254. To disassemble the range transmission, proceed as follows: Drain both sections of the rear frame. Remove platform and seat. Support tractor under clutch housing and place a rolling floor jack under rear frame. Remove rear wheels and both fenders. Remove both brake housings and brake assemblies. Remove all brake lines.

NOTE: If tractor is equipped with hydrostatic drive, disconnect the speedometer drive cable (1—Fig. 197) and oil cooler hose (5). Remove the foot-n-inch valve dump tube (3) and the hydraulic brake tubes. Attach hoist to final drive assemblies and remove final drives from rear frame. See Fig. 198. Disconnect actuating link from the pto control valve, attach hoist to pto unit, then unbolt and remove the pto unit and extension shaft.

NOTE: As pto unit is withdrawn, be sure to tip front of unit upward as shown in Fig. 199, otherwise damage to the oil inlet tube and screen will result. Also, be sure to leave the two short cap screws in place so unit will not separate during removal.

Remove the hitch upper link and disconnect lift links from rockshaft lift arms. Disconnect the rear break-away coupling bracket from rear frame. Unbolt hydraulic lift housing from rear frame, attach hoist and lift unit from rear frame. Attach hoist to differential, remove the differential bearing retainers and lift differential from rear frame. Note that bearing on the bevel gear side of differential is larger than the opposite side so be sure to keep bearing retainers in the proper relationship.

NOTE: At this time, most mechanics prefer to remove drawbar, or hitch

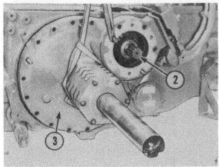

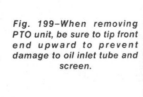

Fig. 197–View showing the lines that have to be removed, when removing top cover of rear frame on hydrostatic drive tractors.

1. Speedometer cable
2. Hydraulic brake tube
3. Foot-N-Inch dump valve tube
4. L.H. brake tube
5. Oil cooler supply hose

Fig. 198–Right final drive assembly being removed from rear frame. Use guide pins when reinstalling.

Fig. 199–When removing PTO unit, be sure to tip front end upward to prevent damage to oil inlet tube and screen.

lower links, to facilitate handling of the rear frame.

Disconnect spring and remove the park lock operating rod (turnbuckle assembly). Disconnect clevis from range transmission shifter arm or arms. Attach hoist to rear frame, unbolt from clutch housing and move rearward until pto drive shaft clears the clutch housing.

NOTE: On gear drive tractors work the pto driven gear off the pto shaft, working through the opening in the bottom of the clutch housing as the rear section is moved rearward.

255. The range transmission can now be disassembled as follows: Lift park lock shaft about ½-inch and support in this position. Remove top cover cap screws, lift top cover slightly from rear frame and work reverse fork shaft (10 —Fig. 200) out of its bore as top cover is removed. Remove nut from front end of mainshaft and using a split collar and puller, remove and discard the mainshaft front bearing. Remove the lubrication tube at rear of mainshaft, then

complete removal of the mainshaft bearing retaining cap screws, pull the mainshaft rearward and lift gears from top of housing. Remove cover from left front of rear main frame, pull the reverse shift fork shaft from fork, then turn fork slightly (top inward) and lift from groove of shift collar. Screw a slide hammer in the threaded hole in front of reverse idler shaft, bump shaft forward and remove thrust washers, gears and shift collar from side of main frame. Remove the small retainer at rear of the hydraulic pump drive gear, then attach slide hammer to front of pto drive shaft and bump shaft forward and out of transmission countershaft. Remove the large retainer at front of countershaft rear bearing cage and remove the bearing cage and bearings. Remove countershaft front bearing retainer and bump shaft rearward. Remove gears from top of main frame.

256. Clean and inspect all parts and renew any which show excessive wear or damage. Refer to Figs. 202 and 203 when renewing needle bearings in the

reverse idler and reverse drive gears. The mainshaft rear bearing assembly is available only as a package of mated parts which will provide the correct operating clearance. Package contains both bearing cups, both bearing cones and the center spacer. If the mainshaft rear bearing assembly is renewed, proceed as follows: Press both bearing cups into bearing retainer, with smallest diameters toward center, until they bottom. Press rear bearing on mainshaft, with largest diameter toward gear, until it bottoms. Place bearing cage over mainshaft with flange toward gear (rear), then place spacer

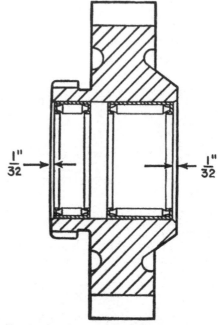

Fig. 202–Bearings in reverse idler gear are installed as shown on gear drive tractors.

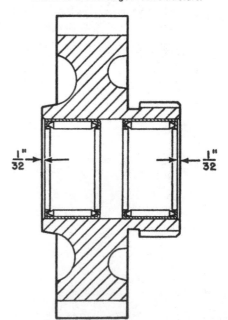

Fig. 203–Bearings in reverse drive gear are installed as shown, used on gear drive or hydrostatic with All Wheel Drive.

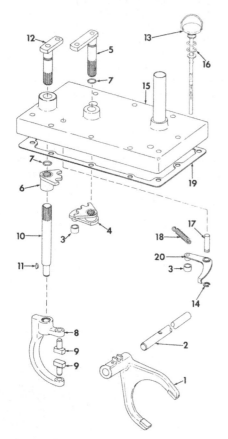

Fig. 200–Exploded view of range transmission top cover and shifting mechanism used on all gear drive tractors.

1. Hi-Lo shift fork	11. Woodruff key
2. Shifter shaft	12. Reverse pivot shaft
3. Detent roller	13. Oil level gage
4. Hi-Lo pivot arm	14. Snap ring
5. Hi-Lo pivot shaft	15. Cover
6. Reverse detent sector	16. "O" ring
7. "O" ring	17. Pivot pin
8. Reverse shift fork	18. Spring
9. Fork pads	19. Gasket
10. Reverse fork shaft	20. Control arm

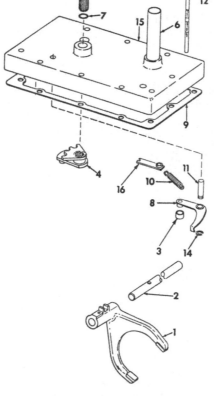

Fig. 201–Exploded view of range transmission top cover and shifting mechanism used on all hydrostatic drive tractors.

1. Hi-Lo shift fork	
2. Shifter shaft	9. Gasket
3. Detent roller	10. Spring
4. Hi-Lo pivot arm	11. Pivot pin
5. Hi-Lo pivot shaft	12. "O" ring
6. Guide oil level	13. Oil level gage
7. "O" ring	14. Snap ring
8. Control arm	15. Cover

over shaft. Press front bearing on shaft with largest diameter toward front and as bearing cone enters bearing cage, rotate the cage to insure alignment of parts.

Any disassembly or service required on transmission top cover assembly will be obvious after an examination of the unit and reference to Figs. 200, 201 and 207.

257. Reassembly sequence of the range transmission is countershaft, reverse idler (gear drive) shaft, mainshaft and pto drive shaft. To install countershaft, install snap ring (9—Fig. 195 or 196) on rear of shaft, then press on bearing cone (7—Fig. 196 or 8—Fig. 195) with largest diameter forward.

NOTE: Heat the rear bearing cone in Hy-Tran oil to 275° F. before installing on countershaft. Do not install bearing cold.

Note that rear bearing cone is narrower than front bearing cone on gear drive tractors. On hydrostatic drive the bearings are the same. Start shaft into housing at rear and install Lo drive gear (11—Fig. 195 or 196) with smallest gear rearward (on gear drive) and constant mesh gear (13) with hub rearward on the shaft. Obtain a bolt 8 inches long and two large washers and with bolt inserted through countershaft, pull front bearing cone on front of countershaft. Large diameter of bearing cone is toward rear. Press both bearing cups (5 and 7—Fig. 195) on gear drive into bearing cage (3) with large diameter toward front. Note on hydrostatic drive you have a ball bearing (4—Fig. 196), retainer ring (5) and bearing cup (6) which is installed the same as gear drive. Place the small bearing cone (6—Fig. 195 or 7—Fig. 196) in cage, then install cage and secure it with the horseshoe retainer. On

Fig. 205–Whenever tractor is split between rear main frame and clutch housing, renew oil suction tube "O" ring and retainer. Hydrostatic drives are similar.

O. "O" ring
R. Retainer
S. Spacer
25. Bearing
27. Nut
28. Reverse idler shaft
40. PTO shaft

gear drive tractors install bearing cup (15—Fig. 195) in bearing cage (17) with large diameter rearward. On hydrostatic drive install oil seal (14—Fig. 196) with lip rearward and bearing cup (6) in bearing cage (16).

Place original shims (16—Fig. 195 or 15—Fig. 196) on bearing cage, or if new parts were installed, use one shim as a starting point. Shims are 0.009 thick. Install bearing cage and tighten cap screws finger tight; then, rotate shaft and tighten cap screws evenly until shaft binds or bearing cage bottoms. If shaft binds, add shims as required, or if shaft has excessive end play, remove shims. Shaft should turn freely with no end play. Tighten cap screws to 35 ft.-lbs. torque. See Fig. 204 for an assembly view.

On some late production model 1066 gear drive, the bearing cage (17—Fig. 195 or 16—Fig. 196) has two gage slots for direct measurement and is adjusted as follows:

Install bearing cage without shims and torque two opposite bolts to 150 in.-lbs. while rotating the countershaft. Then loosen the two bolts and retorque to 75 in.-lbs. while rotating countershaft. Measure through the two gage slots the distance from housing to outer surface of bearing cage and average the readings. Then remove the bearing cage and measure the thickness of bearing cage next to the slots and average the readings. The desired shim pack is the difference of the two measurements plus 0.016 inch, ± 0.001 inch. With shim pack installed torque the bolts to 35 ft.-lbs. and measure the countershaft end play with dial indicator the end play should be 0.001 to 0.006 inch, if not add or remove shims to obtain the correct end play.

To install reverse idler shaft and gears, on gear drive, start shaft (39—Fig. 195) in its bore in front of main frame. As shaft is moved rearward install thrust washer (32), idler gear (38) with shift collar teeth rearward, shift collar (35) and center thrust washer (32), reverse drive gear (34) with shift collar forward and the rear thrust washer (32). Align dowel of shaft with notch in rear frame and bump shaft into position.

NOTE: As shaft is bumped rearward BE SURE rear thrust washer does not catch the rear step of shaft or thrust washer will be damaged. When properly installed, front of shaft must be flush with front sur-

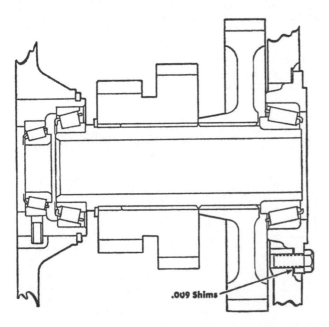

Fig. 204–Cross-sectional view showing installation of range transmission countershaft.

.009 Shims

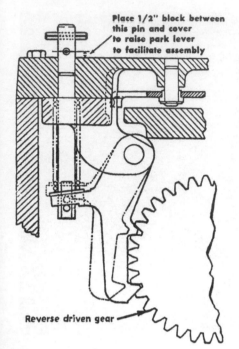

Place 1/2" block between this pin and cover to raise park lever to facilitate assembly

Reverse driven gear

Fig. 206–When installing range transmission top cover, use a ½-inch spacer positioned as shown to hold park lock in disengaged position.

FEELER GAGE

FES 68-13   FES 68-13

Fig. 208–Use IH tool No. 68-13 and feeler gage when checking pinion shaft setting.

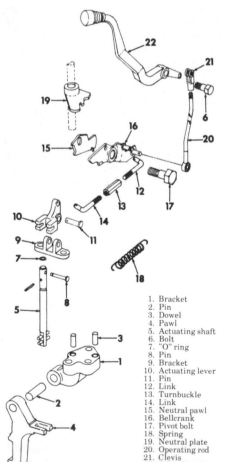

1. Bracket
2. Pin
3. Dowel
4. Pawl
5. Actuating shaft
6. Bolt
7. "O" ring
8. Pin
9. Bracket
10. Actuating lever
11. Pin
12. Link
13. Turnbuckle
14. Link
15. Neutral pawl
16. Bellcrank
17. Pivot bolt
18. Spring
19. Neutral plate
20. Operating rod
21. Clevis
22. Operating lever

Fig. 207–Exploded view of typical park lock and linkage. Park lock pawl (4) engages teeth on reverse driven gear in range transmission.

face of rear frame. Clutch housing rear surface retains shaft in position.

To install the mainshaft, install rear bearing assembly as outlined in paragraph 256. Place original shims (27—Fig. 195 or 196) on bearing cage (28) and start mainshaft in rear frame. Place reverse driven gear (26) on shaft. Place carrier (25) on shaft. NOTE: Be sure to align oil holes in mainshaft with holes in carrier (25). Then install Hi-Lo driven gear (24) on carrier with shift collar teeth forward. Place shift collar (22) and collar carrier (23) on shaft. Install spacer (21) and a new bearing (20). Install lock washer (19) and nut (18), then tighten nut to a torque of 100 ft.-lbs. Do not install lubrication tube at this time. Install and tighten mainshaft rear bearing retaining cap screws to a torque of 85 ft.-lbs. Check rotation of shaft. Shaft should turn freely and if binding exists, recheck rear bearing assembly.

Install pinion shaft (mainshaft) locating tool (IH tool No. FES 68-13) as shown in Figs. 208 and 209. A cone setting number is etched on rear end of pinion shaft. They are marked with etched setting numbers 60 thru 90. Use this setting number and refer to chart in Fig. 210 to determine correct feeler gap (A—Fig. 209). Add or remove shims at pinion shaft rear bearing to obtain the determined gap (A) plus or

minus 0.001. For example: If a number 73 is etched on pinion shaft, feeler gap (A) would be 0.045 plus or minus 0.001. Shims are available in thicknesses of 0.004 and 0.007.

After pinion shaft (mainshaft) has been adjusted, remove locating tool and install the lubrication tube (31—Fig. 195 or 196).

Insert the pto drive shaft through the countershaft and into its rear bearing cone or ball bearing.

NOTE: Although pto shaft can be bumped through the rear bearing, it is recommended that a cap screw, washer and a pipe spacer of proper diameter be used to pull the shaft through bearing cone.

Install hydraulic pump drive gear (2) and the small horseshoe retainer (1) on rear of pto shaft.

Install pads in reverse shifter fork (if so equipped), position shifter fork in groove of reverse shift collar, then install the fork shaft and Woodruff key.

Reinstall differential assembly in rear main frame and check carrier bearing preload as outlined in paragraph 283 and backlash as outlined in paragraph 286.

Place spacer on forward end of pto shaft, renew hydraulic pump suction tube "O" ring and retainer (Fig. 205) and join rear main frame to clutch housing.

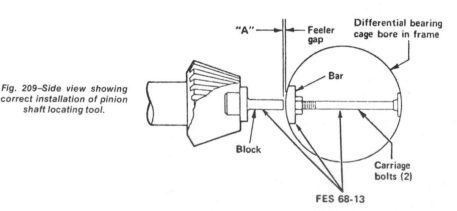

"A"   Feeler gap   Differential bearing cage bore in frame

Bar

Block

Carriage bolts (2)

FES 68-13

Fig. 209–Side view showing correct installation of pinion shaft locating tool.

Complete reassembly of tractor by reversing the disassembly procedure. Installation of the range transmission top cover will be simplified if a ½-inch spacer is positioned between lower pin of park lock shaft and top cover as shown in Fig. 206.

LINKAGE ADJUSTMENT

258. To adjust the range transmission and park lock linkage, first move range transmission lever to neutral position. On gear drive tractors remove the TA control lever. On hydrostatic

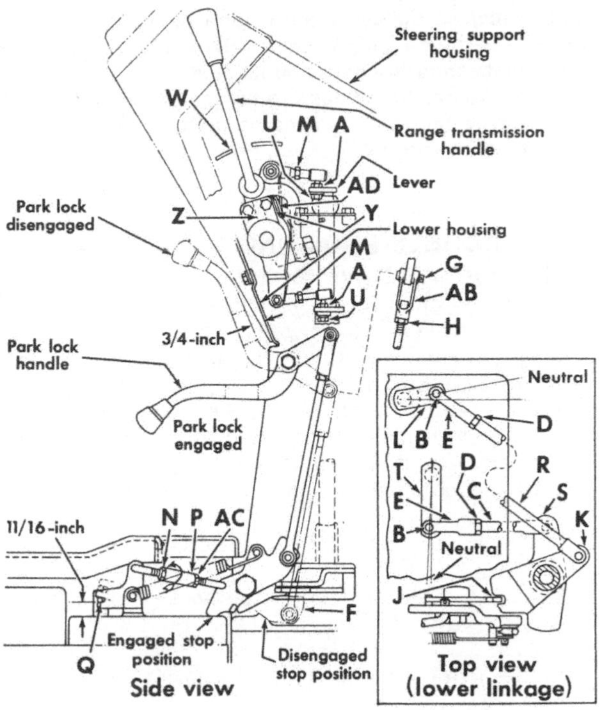

Fig. 211–View showing range transmission and park lock linkage adjustment. Refer to text for procedure.

Setting Numbers	Gap "A"
60	.006
61	.009
62	.012
63	.015
64	.018
65	.021
66	.024
67	.027
68	.030
69	.033
70	.036
71	.039
72	.042
73	.045
74	.048
75	.051
76	.054
77	.057
78	.060
79	.063
80	.066
81	.069
82	.072
83	.075
84	.078
85	.081
86	.084
87	.087
88	.090
89	.093
90	.096

Fig. 210–Chart used in determining feeler gap "A". Setting numbers (60 thru 90) are etched on pinion shaft.

drive tractor remove the range control lever. And on all models remove steering support top and rear cover. Refer to Fig. 211 and disconnect the Hi-Lo ball joint linkages at stud (A) and on gear drive disconnect the reverse ball joint. Remove transmission lower linkage pins (B), then remove linkage (R) on gear drive only. With park lock operating lever in disengaged (up) position, remove pin (G) and loosen jam nut (H). Hold park lock lever approximately ¾-inch from support housing as shown and with bellcrank (F) in disengaged stop positon (against support base), adjust clevis (AB) until pin (G) can be installed. Secure pin with cotter pin and tighten nut (H).

Move park lock lever to engaged position so that pawl (J) passes through neutral notches in plates (K and S). Plate (K) is used on gear drive tractor only. With lever (L) in neutral position (gear drive only) and pawl (J) centered in neutral notch in plate (K), adjust clevis (E) on reverse linkage (R) until pin (B) can easily be installed through clevis and lever (L). Secure pin and tighten jam nut (D).

Then on all models, with lever (T) in neutral position and pawl (J) centered in neutral notch in plate (S) adjust clevis (E) on Hi-Lo linkage (C) until pin (B) can be installed through clevis and lever (T). Secure pin and tighten jam nut (D).

On gear drive tractors, with range transmission control lever centered in neutral gate of shift pattern cover (W), align slots in Hi-Lo shift hub (AD) and reverse shift hub (Y) with shift pin (Z) on shift lever. Loosen jam nuts (M) and adjust both ball joint linkages so that stud (A) aligns with holes in levers on vertical shafts. Secure with nuts (U) and tighten jam nuts (M).

With park lock operating lever in engaged position, disconnect the actuating spring. Back off jam nuts (N and AC). Make certain that bellcrank (F) is contacting transmission cover (engaged stop position), then while rocking rear wheel slightly, rotate turnbuckle (P) in direction of arrow until finger tight. Tighten jam nuts (N and AC) and connect the actuating spring.

When park lock is properly adjusted and is in engaged position a dimension of approximately 11/16-inch should exist between range transmission cover and pin (Q) in actuating shaft. Reinstall steering support covers and TA or Range control lever.

HYDROSTATIC DRIVE

Models 826, 966 and 1066 tractors are available with the optional hydrostatic drive. All Model 1026 tractors are equipped with the hydrostatic drive. The hydrostatic drive consists of a variable volume reversible swashplate axial piston **pump, a variable displacement axial piston motor, a center section which houses the shuttle valve, relief valves and check valves, two servo cylinders, external control valve and control linkage. The pump input shaft is driven by a flex**

plate, which is bolted to the engine fly-wheel. The motor output shaft drives a two-speed range transmission located in the rear frame. The Speed-Ratio (hand control) type (Fig. 212) is used on all hydrostatic drive models.

TROUBLE SHOOTING
All Hydrostatic Drive Models

259. Some of the troubles and their possible causes which may occur during operation of the hydrostatic drive are as follows:

1. Tractor will not always move forward or will not pull the load after decelerating or going in reverse.
   a. Check ball leaking in drive control valve.
   b. Check ball leaking in deceleration valve.
2. Tractor fails to move after the Foot-N-Inch valve is depressed and released.
   a. Check for sticking high pressure relief valve.
3. Tractor will not move forward (reverse ok) until the S-R lever is moved several inches and then lurches forward.
   a. Check snap ring on shuttle valve in center section.
   b. Shuttle valve binding or sticking in center section.
4. Tractor continues to go into reverse momentarily after moving the S-R lever into forward.
   a. The servo pressure is low and should be adjusted.
5. Tractor is very slow to accelerate, has low torque and no distinct acceleration above the blue line.
   a. Plugged fixed orifice in motor servo.
   b. Leaking teflon seal at the variable orifice of motor servo.
   c. Ball plug out of drilled passage in motor servo.
6. Tractor maximum speed is about 4 MPH in low range and 10 MPH in high range.
   a. Variable orifice plugged in motor servo.
   b. Piston seal rings leaking in motor servo.
7. Tractor travels in forward anytime the S-R lever is moved from center of the neutral slot.
   a. Fixed orifice is plugged in pump servo.
   b. Excessive leak at pump servo variable orifice or at "O" ring under orifice adjusting block.
8. Tractor will propel itself but lacks pulling ability.
   a. Misadjusted or faulty Foot-N-Inch valve.
   b. Faulty drive control valve or linkage.

ADJUSTMENTS
All Hydrostatic Drive Models

260. **NEUTRAL OR SERVO ADJUSTMENT.** There is a small neutral zone on the servo cam, during which no movement of the servo cylinders occurs. Correct adjustment of the movable pump variable orifice in this neutral zone is necessary to assure proper transmission control. Prior to making adjustments, operate tractor long enough to obtain a transmission oil temperature of at least 50 degrees F.

Using a jack under right rear axle housing, raise tractor until right rear wheel is free to rotate. With range transmission in gear and Foot-N-Inch pedal up, operate engine at 1700 rpm and check Speed-Ratio lever response. If the S-R lever is approximately centered in neutral slot and wheel does not

move or if wheel begins to move when lever is moved the width of the lever in forward or reverse slots, neutral adjustment is O.K. If not, proceed as follows: With engine operating at 1700 rpm, move S-R lever to a position on quadrant where rear wheel does not move. Leave lever in this POSITION and stop engine. Pump swashplate is now in vertical (zero output) position. Move S-R lever into neutral slot. Adjust length of the servo control rod so that pin (A—Fig. 213) is even with index mark on side of the clutch housing. This will synchronize the mechanical stop of the servo cam and S-R lever on quadrant. Drain 5 gallons of oil from transmission and remove access cover from left side of hydrostatic drive housing. Mark the adjustable, variable orifice block to cylinder with a pencil (Fig. 214). Loosen the lock bolts. Then, by turning the ad-

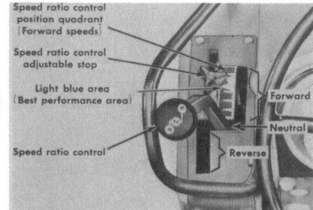

Fig. 212–View showing Speed-Ratio control and quadrant used on hydrostatic drive models.

Fig. 213–View showing index mark for adjusting servo control rod. Refer to text.

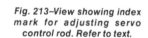

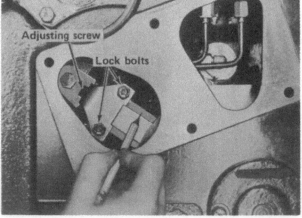

Fig. 214–View showing the marking of the orifice block and the pump servo variable orifice adjusting screw. Refer to text.

justing screw, a 1/16-inch movement of the orifice block will reposition the S-R lever approximately ¼-inch. If hydraulic neutral was on reverse side of S-R quadrant, orifice must be moved to the rear. Tighten orifice block lock bolts. Reinstall access cover and transmission oil.

Operate engine at 1700 rpm, cycle the S-R lever and check axle rotations. The wheel should just start to move when lever is moved the width of the lever in forward and reverse. If it is not correct repeat the adjustment.

261. **FOOT-N-INCH VALVE.** The Foot-N-Inch valve is a pedal operated needle type valve, located below left side of platform. To adjust valve linkage, refer to Fig. 215 and proceed as follows: Push pedal down until lug (D) is against the platform. Loosen lock nut on valve stem (A), then turn valve stem (A) into clevis to give maximum extension of the valve without pulling the pedal away from the stop. Then back valve stem out of clevis one turn. Tighten lock nut. NOTE: The pedal will now hit its stop before the valve hits the internal stop. Return pedal to up position and there should be a minimum of 1/16-inch clearance at (B). If not back out valve stem to obtain the clearance.

To adjust the safety starting switch, first depress pedal until stop lug (D) strikes platform. With pedal in this position loosen lock nuts (C) and adjust switch until plunger is depressed ⅛-inch. Tighten lock nuts (C).

PRESSURE CHECKS
All Hydrostatic Drive Models

262. **DRIVE PRESSURE.** To check the forward and reverse drive pressure, first support rear of tractor so both rear wheels are free to rotate. Operate tractor until transmission oil is warmed to a temperature of 135 degrees F. Install a 10,000 psi test gage (IH tool number FES 96-2 or equivalent) at the drive control valve as shown in Fig. 216.

Operate engine at high idle rpm, place range transmission in high range and Foot-N-Inch valve pedal up. Slowly move S-R lever forward while applying both brakes. NOTE: When applying brakes to build up drive pressure, allow wheels to rotate so that hydrostatic motor is turning. The forward drive relief valve pressure of 6000 ± 50 psi should be reached before the S-R lever is in the blue zone.

Check reverse drive pressure by moving the S-R lever rearward. Reverse drive relief valve pressure should also be 6000 ± 50 psi.

Place range transmission in low range. Move S-R lever to the blue line

and release brakes enough to recover rated load engine rpm. The pressure should read approximately 2600 psi on models 826 and 966 and 3300 psi on models 1026 and 1066. Check reverse by moving the S-R lever completely rearward and with engine operating at rated load, pressure should read 2600 psi on models 826 and 966 and 3300 psi on models 1026 and 1066.

NOTE: If engine will not stall in high range with S-R lever all the way forward and holding the wheels with the brakes, unit is not developing enough pressure. If tractor will stall

before reaching relief valve pressure with S-R lever below blue line, tractor engine horsepower should be checked.

If drive pressure is low, checks can be made to determine if the problem is external or internal by eliminating the Foot-N-Inch and drive control valves from the circuit. To protect the system when eliminating the two valves, two isolating valves (Dump valves, IH number 398 715R91) and a 10,000 psi gage will have to be installed as shown in Fig. 217.

CAUTION: When the Foot-N-Inch valve is eliminated from the circuit, depressing the pedal will not stop the tractor. The rear

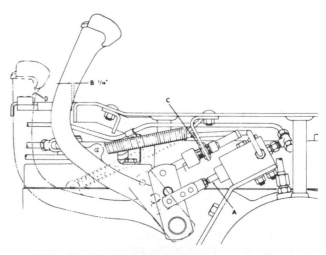

Fig. 215–View showing adjustment of Foot-N-Inch valve linkage and safety starting switch on hydrostatic drive tractors.

Fig. 216–When checking drive pressure, connect 10,000 psi gage on control valve as shown.

Fig. 217–View showing the isolating valves and gage for checking drive pressure with Foot-N-Inch valve and control valve disconnected. (F) is forward drive and (R) is reverse drive.

of tractor should be supported so both rear wheels are free to rotate.

Check the drive pressure as before with gage in one side and then the other, being sure to cap the opposite tee to the one that is being checked.

If pressures are low, check the internal high pressure lines and the high pressure relief valves. If pressures at this point are ok, the problem will be in the drive control valve or Foot-N-Inch valve.

Normally if drive pressure is low in only forward or reverse, the problem is in the drive control valve and it will have to be overhauled as outlined in paragraph 279. If drive pressure is low in both forward and reverse, the problem is in the Foot-N-Inch valve and/or drive control valve.

To test the Foot-N-Inch valve, remove the isolating valves and connect the high pressure lines. Then, install gage in drive control valve as shown in Fig. 216. Disconnect the dump line from Foot-N-Inch valve and make pressure check as before. If fluid comes out of valve when attempting to build drive pressure (with pedal up), the Foot-N-Inch valve will have to be overhauled as outlined in paragraph 280. If there is no leak at Foot-N-Inch valve the problem is in the drive control valve.

263. SERVO PRESSURE. To check servo pressure, connect a 600 psi test gage and snubber (IH-FE 94-6) as shown in Fig. 218. With engine operating at 1700 rpm and higher, servo pressure should be approximately 400 psi.

To adjust servo pressure, turn adjusting screw in to raise and out to lower (Fig. 218). Allow a few seconds to stabilize the system after making an adjustment before proceeding. NOTE: Do not adjust servo pressure with S-R lever in neutral.

264. DECELERATION PRESSURE. To check deceleration pressure, install a 10,000 psi test gage in the port on drive control valve (Fig. 216). Support rear of tractor so right rear wheel is free to rotate. Remove clevis pin and disconnect S-R control lever rod from cam actuating rod. Place range transmission in low range. Move S-R lever into forward which will allow fluid in the reverse circuit to go past the reverse check ball and be exposed to deceleration valve. Then pull up on control cam link which will put pump in reverse stroke. This will simulate deceleration, as pressure is built up in reverse side when decelerating. The gage reading should be from 1500 to 1800 psi.

265. CHARGE PUMP PRESSURE AND FLOW. Adequate charge pump pressure and flow are necessary

for an efficient operation of the hydrostatic drive unit.

To check charge pressure and flow, support rear of tractor so both rear wheels are free to rotate. Operate tractor until transmission oil is warmed to 135 degrees F. Then, with tractor stopped install a 600 psi test gage at the tee on the rear of multiple control valve (Fig. 219), this will give charge pressure. NOTE: DO NOT ATTEMPT TO MEASURE CHARGE FLOW AT THIS TEE. With engine operating at 1700 rpm, the pressure should be 230 to 280 psi for model 826, 275 to 425 psi for model 1026 and 440 to 500 psi for models 966 and 1066 in

neutral. With S-R lever in forward or reverse on models 826 and 1026 the pressure should be 130 to 170 psi, and 170 to 215 psi for models 966 and 1066. The hydrostatic motor must be turning when checking charge pressure in forward or reverse.

These pressures are not adjustable; if charge pressure is low, proceed as follows before making charge flow check:

Determine if pressure is low in all S-R lever positions. If pressure is low in neutral only, there could be a leak at the maximum regulating valve in the Multi-Valve. If pressure checks low in forward and reverse only, there could be a leak at the minimum regulating

Fig. 218–To check and adjust servo pressure, connect 600 psi test gage to check port as shown.

Fig. 219–View showing 600 psi gage connected for checking charge pressure.

600 PSI gauge

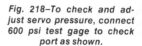

Fig. 220–View showing Flo-Rater connected for checking the charge flow.

1. Oil cooler hoses
2. Adapter and union (FES 501-1 and FES 2-35)
3. Return line (oil cooler)
4. Multi-valve "out" pipe (to oil cooler)
5. 10,000 psi gage (FES 96-2)
6. Auxiliary valve control
7. Flo-Rater (FES 51-D)

valve in the Multi-Valve. If pressure is low in all positions, it could be a lack of output from charge pump, or damaged "O" rings on check valve manifold charge tube.

If none of the above is responsible, cause for low charge pressure is probably due to excessive leakage in center section area, such as scored valves or bearing plates. Refer to paragraph 266 and make charge flow check.

266. When making charge flow test, an auxiliary valve must be placed on demand (6—Fig. 220) to divert the hitch pump oil away from oil cooler by-pass valve so ONLY charge flow is measured.

Disconnect oil cooler hoses (1) from pipes (3 and 4) on left side of tractor. Using a Flo-Rater or equivalent, connect the "out" pipe from multi-valve (4) to the inlet of Flo-Rater. Connect the outlet of Flo-Rater to return pipe (3). Install a 10,000 psi gage (5) in the port on drive control valve. NOTE: Restrictor valve on Flo-Rater must be all the way open, so back pressure in hoses and Flo-Rater will not exceed 60 psi. If back pressure is too high, the oil cooler by-pass valve will open and a true reading on the flow cannot be obtained.

Refer to Fig. 221 for the typical charge flow.

Flow under load (4—Fig. 221) should be within one or two gallons of the flow in item 3 (zero forward, brakes applied). If the charge flow is in this range, it would indicate that there is nothing wrong with the tractor.

A drop of more than two gallons would indicate a leak past the valve and bearing plates.

HYDROSTATIC DRIVE UNIT

Some components of the hydrostatic drive, other than external valves, can be removed and serviced without removing the hydrostatic drive unit from the tractor. The control cam, variable orifices, servo cylinders and check valve manifold can be removed after the multiple control valve, bottom plate and left hand side plate are removed. If check valve manifold is to be removed, the tractor must be in reverse before stopping engine. This will extend the pump servo cylinder to give additional clearance. The forward

and reverse drive high pressure relief valves can be removed after removing left side cover and the rectangular cover on the right side. The shuttle valve can be removed after disconnecting the high pressure relief lines and plug on left side of drive housing. However, if failure of these components is due to contaminated oil (dirt, water or metal particles), it is recommended that the complete hydrostatic drive system be removed and disassembled, and all components cleaned and repaired as outlined in the following paragraphs.

All Hydrostatic Drive Models

267. **REMOVE AND DISASSEMBLE.** To remove the hydrostatic drive assembly, first drain fluid from range transmission, differential and hydrostatic drive housing. Then, split (detach) engine from hydrostatic drive housing as follows: Disconnect battery cables, remove batteries and battery supports. Unbolt and remove hood and side panels. Disconnect electrical wiring and tachometer drive cable from engine and lay rearward on fuel tank. Identify oil cooler hoses and disconnect them at rear on models 826 and 1026. On models 966 and 1066 identify oil cooler hoses and disconnect them at front of engine.

NOTE: It is essential that oil cooler hoses be correctly reinstalled as the oil flowing to oil cooler is maintained at 105 psi by a pressure regulating valve in the multi-valve.

Models 826 and 1026 disconnect the two power steering cylinder lines from control (pilot) valve. On models 966 and 1066 disconnect power steering lines on right rear side of engine. Shut off fuel and disconnect fuel supply lines at front end. Disconnect controls from carburetor, governor or injection pump. Disconnect coolant temperature bulb from cylinder head. On models 1026 and 1066 remove fuel tank left front support and disconnect tank right front support from tank. Install split stand to side rails and support hydrostatic drive housing with a rolling floor jack.

Remove bottom side rails to hydrostatic drive housing cap screw from both side rails and install guide studs. Complete removal of hydrostatic drive

housing retaining cap screws and separate tractor.

NOTE: Insert cap screws back in side rails before pulling side rails completely off guide studs.

Remove front center cover on platform and rear steering support cover. Disconnect wires to the starter solenoid. Disconnect drive control valve lines, drive control valve drain line, S-R control rod clevis, drive control valve to Foot-N-Inch valve line and the hi-lo shift rod. Disconnect power steering lines from control (pilot) valve. Disconnect wiring cable from neutral start switch and rear wiring harness and bring forward. Disconnect speedometer cable and move forward. Disconnect park lock actuating spring, operating rod and bellcrank pivot bolt. Attach a hoist to fuel tank and steering support, unbolt and lift assembly from tractor.

Attach a hoist to the hydrostatic drive housing and support rear frame with blocks or jacks. Unbolt housing from rear frame and remove the assembly.

To disassemble the removed assembly, first remove the charge filter and filter base assembly. Disconnect power steering and brake lines from multiple control valve and remove the multiple control valve. Remove the multi-valve assembly (1—Fig. 222) and housing top cover (2). Remove the lube block and line and side cover on left hand side of hydrostatic drive housing. Remove snap ring at rear of driven pto gear. Remove pto front cover (63—Fig. 223) and shims (61). The pto driven shaft is drilled and tapped and can be pulled using a slide hammer. Remove front bearing cover (3—Fig. 224), using jack screws (1). Remove pump servo anchor bolt (1—Fig. 225). Install a horseshoe type puller (IH tool number FES 123-3 or equivalent) between pto drive gear and pump swashplate as shown in Fig. 226. Tighten puller cap screws just enough to release the spring pressure on pump trunnion bearings. Support pump assembly, and remove trunnions

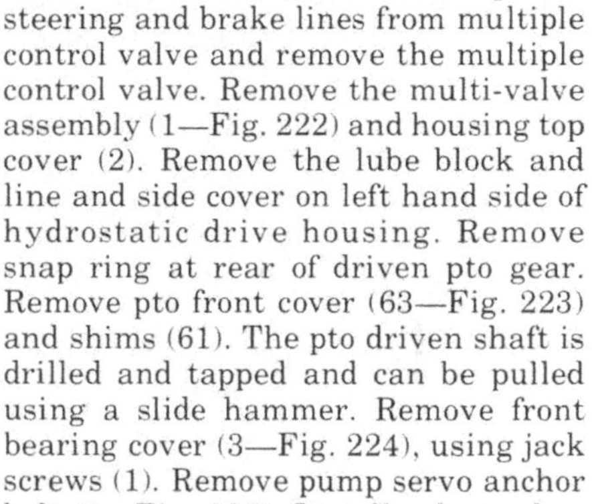

Fig. 222–View of the Multi-Valve and top cover, on the hydrostatic drive housing.

1. Multi-valve
2. Top cover

Typical Charge Flow	
1200 RPM Engine Speed and Auxiliary Valve on Demand	
1. S-R lever in zero-neutral	7.3 gpm
2. S-R lever in zero-forward	9.0 gpm
3. S-R lever in zero-forward-brakes applied	7.5 gpm
4. S-R lever in forward-brakes applied – 4000 psi drive pressure (1200 rpm engine speed must be held)	7.0 gpm

Fig. 221–Chart of the typical charge flow. Refer to text.

(8—Fig. 223). NOTE: Keep shims with the right hand trunnion for use in reassembly. Then, being extremely careful not to drop the brass bearing plate from end of pump, remove pump and swashplate assembly and lay it on a clean area for later disassembly.

To remove the center section, first remove high pressure lines (4 and 5—Fig. 227) and servo tube (7). NOTE: Disconnect servo tube at the fitting end first, being careful not to bend the tube in the snap ring area when removing from housing. Remove check valve manifold (6) and charge tube (1). Collapse center section connector tube (2) by prying off the retaining clip (56—Fig. 223), and then push tube upward. NOTE: Some models may be equipped with a one piece tube and it will be necessary to remove the tube from the top side of hydrostatic drive housing. Then, unbolt the center section using an aligning dowel. NOTE: When removing the center section, use extreme care not to drop the motor brass bearing plate or the steel valve plate. Lay center section on a clean, protected surface to prevent the highly machined surface from being damaged. If valve and bearing plates stay on the motor block when center section is removed,

Fig. 224–Use jack screws as shown to remove front bearing cover of hydrostatic drive.

Fig. 225–View showing the pump anchor bolt (1) and the pump servo (2).

Fig. 226–Use horseshoe type puller between pto drive gear and pump swashplate to release spring pressure on pump trunnion bearings.

Fig. 227–View showing the center section and components.

1. Charge tube	5. Forward high
2. Connector tube	pressure tube
3. Center section	6. Check valve manifold
4. Reverse high	7. Servo supply tube
pressure tube	

Fig. 228–View of motor cylinder block and bearing retainer cap screw and washer.

1. Cap screw	3. Motor cylinder block
2. Bearing retaining washer	4. Output shaft front bearing

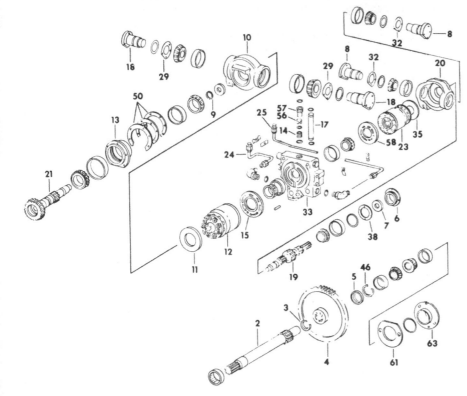

Fig. 223–Exploded view of the hydrostatic drive unit.

2. Pto driven shaft	12. Hydrostatic motor	21. Output shaft	38. Shims
3. Retainer ring	13. Output shaft housing	23. Hydrostatic pump	46. Retainer ring
4. Pto driven gear	14. Tube connector sleeve	24. Reverse pressure tube	50. Output housing shims
5. Spacer	15. Motor valve plate	25. Forward pressure tube	56. Tube retainer clip
6. Input shaft retainer	17. Charge tube	29. Shims	57. Center section connector
7. Oil seal	18. Motor swashplate trunnion	32. Shims	58. Pump valve plate
8. Pump swashplate trunnion	19. Input shaft	33. Center section	61. Shims
9. Retainer ring	20. Pump swashplate	35. Thrust plate	63. PTO front bearing cover
10. Motor swashplate			
11. Thrust plate			

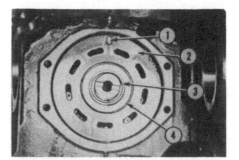

Fig. 229–View of motor cylinder block with cap screw, washer and bearing removed.

1. Dowel	
2. Motor cylinder block	
3. Spring guide	4. Brass bearing plate centering ring

do not attempt to pry them loose with screw drivers or similar tools as they will be scratched. Use a hardwood dowel to separate plates. Lay the assembly in a clean area for later disassembly.

With center section and bearing plates removed, remove the cap screw (1—Fig. 228), retaining washer (2) and bearing cone (4) from front of output shaft. Remove brass bearing plate centering ring (4—Fig. 229) and spring

guide (3). Now the motor assembly may be removed. Be sure to protect the machined surface. Lay the motor assembly in a clean area for later disassembly.

To remove the servo cylinders, rotate pump servo (9—Fig. 230) and remove the orifice (5). Remove the cam pivot (2 —Fig. 231), and the servo control cam and actuating rod assembly (1). Remove plug (3) so the motor servo anchor bolt can be removed. Remove retaining pins (Fig. 232), then withdraw the servo anchor shaft. NOTE: Thread a 5/16 × 4 inch N.C. cap screw into end of anchor shaft to aid in removal. Hold servos inward while removing shaft to prevent binding.

Remove the Allen head screws in the output shaft bearing cage and withdraw shaft assembly, keeping any shims with the assembly. Remove

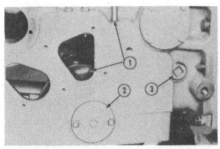

Fig. 230–View of the servo cylinders and components.

1. Cam actuating rod
2. Cam plate
3. Anchor shaft
4. Motor orifice
5. Pump orifice
6. Motor servo
7. Sealing rings
8. Anchor shaft bushing
9. Pump servo
10. Wavy spring washer
11. Charge tube fitting
12. Connector
13. Fitting nut
14. Servo charge tube
15. Retainer ring
16. "O" ring
17. Pump shoulder bolt
18. Motor shoulder bolt

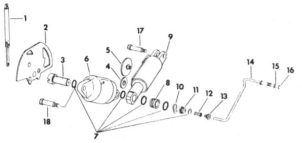

Fig. 231–View showing plate removed from left hand side of hydrostatic drive housing.

1. Servo control cam and actuating rod
2. Cam pivot
3. Plug

Fig. 233–Exploded view of the multi-valve.

1. Multi-valve body
2. By-pass valve (oil cooler)
3. Spring
4. "O" ring
5. Plug
6. Bushing (poppet valve)
7. Poppet valve
8. Spring (min.)
9. Spring (max.)
10. Guide
11. "O" ring
12. Plug

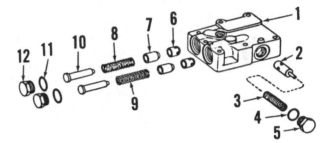

motor swashplate trunnions (18—Fig. 223), and remove swashplate (10). NOTE: Keep shims with the right hand trunnion for use in reassembly.

268. **OVERHAUL COMPONENTS.** Disassemble the components, wash parts in clean diesel fuel and lubricate with IH "Hy-Tran" fluid during reassembly as outlined in the following paragraphs. Wrap each cleaned and repaired component in clean paper until time of installation.

269. **MULTI-VALVE.** The multi-valve contains the minimum charge pressure regulator valve and the maximum regulator valve, along with the oil cooler by-pass valve. To disassemble the unit, refer to Fig. 233 and remove plugs (12) and "O" rings (11). Withdraw guide pins (10), maximum regulator spring (9), minimum regulator spring (8), poppet valves (7) and valve bushing (6). Remove plug (5) with "O" ring (4), spring (3) and by-pass valve (2).

Clean and inspect all parts and renew any showing excessive wear or damage. All parts are serviced separately. Lubricate with clean "Hy-Tran" fluid and reassemble by reversing disassembly procedure.

270. **CHARGE AND SERVO PUMPS.** There are two pumps mounted on inside of multiple control

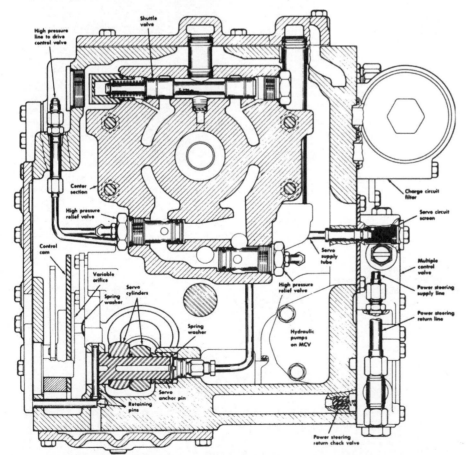

Fig. 232–Cross sectional view of the center section and related parts (viewed from rear).

valve. One is a 12 gpm pump which is referred to as pump No. 1. It delivers fluid to the power steering and to the charge circuit. The other pump is 4.5 gpm pump which is mounted "piggy back" on the No. 1 pump. This pump is referred to as pump No. 2 and it delivers fluid to the servo cylinders. The excess oil of both pumps is used for cooling and lubrication. To overhaul pump No. 1, refer to paragraph 46 and 48. Overhaul procedure on pump No. 2 is as follows:

To disassemble the **Cessna** 4.5 gpm pump, refer to Fig. 234 and remove cover retaining cap screws and separate cover (1) and body (9). Remove "O" rings (2, 3 and 4), diaphragm seal (5), back-up gasket (6), diaphragm (7) and the gears and shafts (8). There is an "O" ring and gasket package available. Pump body and cover or gears and shafts are not serviced separately.

Refer to the following table for dimensional information.

O.D. of shafts at bushings 0.6875
I.D. of bushing in body and
    cover 0.689
Gear width 0.312
I.D. of gear pockets 1.716

To disassemble the **Thompson** 4.5 gpm pump, refer to Fig. 235 and remove cover retaining cap screws and separate cover (1) and body (9). Remove "O" rings (2 and 3) from pump cover. Remove gears and shafts (4), bearings (5), "O" rings (6), "O" ring retainers (7) and pressure plate spring (8) from the body.

Pump cover, body, bushings, gears and shafts are not available separately.

Refer to the following table for dimensional information.
O.D. of shafts at bushings 0.625
I.D. of bearings and cover
    bushings 0.6277
Gear width 0.4395
I.D. of gear pockets 1.449

When reassembling either pump, use all new "O" rings and gaskets. Lubricate gears and shafts prior to installation, reinstall bearings in their original positions. Tighten body to cover screws evenly and check rotation of pump. Pump will have a slight drag

but should turn evenly with no tight spots.

271. CENTER SECTION AND CHECK VALVE MANIFOLD. The center section (Fig. 236) contains the shuttle valve and the forward and reverse drive high pressure relief valves. Forward and reverse check valves are located in the manifold. To disassemble the center section, remove plug (1) and "O" rings (2). Withdraw shuttle valve spool assembly. Remove snap ring (3), washers (4 and 6) and centering spring (5) from spool (7). Unscrew and remove forward and reverse high pressure relief valves (11). Remove check valves (17) from manifold (20), install a small screw into end of check valve to assist in removal.

Clean and inspect all parts. Check spool (7) and spool bore in body (10) for excessive wear or scoring. NOTE: The valve spool and body are stamped with letter A, B, C, D, E or F indicating the bore size. Order new spool with letter that matches letter on body. Centering spring (5) should have a free length of 1 59/64 inches and should test 15 pounds when compressed to a length of 1 19/64 inches. Check valves (17) and relief valves (11) are available only as assemblies.

When reassembling, use all new "O" rings and back-up rings and lubricate with clean "Hy-Tran" fluid. Tighten relief valves (11) to a torque of 20 ft.-lbs. and manifold (20) to a torque of 47 ft.-lbs. The balance of assembly is the reverse of disassembly. IMPORTANT:

Be careful not to scratch or damage the machined surfaces of center section.

272. HYDROSTATIC PUMP AND MOTOR. The pump and motor are of similar construction, and the motor is the larger of the two units. After input shaft and swashplate are removed from the pump, disassembly procedure is the same for both pump and motor.

To remove input shaft from pump, clamp input shaft in a vise and attach a puller as shown in Fig. 237. Place cardboard (3) between puller and finished surface of pump, then remove rear bearing. Remove assembly from vise, invert the assembly and carefully remove input shaft, horseshoe tool, swashplate and thrust plate. Front bearing on input shaft can be renewed in conventional manner if necessary.

To disassemble pump or motor, grasp outer diameter of slipper retainer (1—Fig. 238) and remove retainer with nine pistons (13). Remove ball guide (2) and withdraw nine slipper retainer

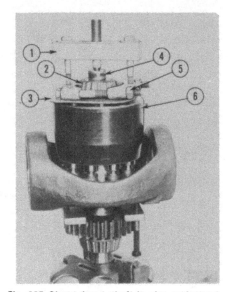

Fig. 237–Clamp input shaft in vise and attach puller as shown to remove input (pump) shaft rear bearing.

1. Puller (OTC 515-A)	5. Split puller (OTC 950)
2. Rear bearing	6. Pump assembly
3. Cardboard	
4. Step plate	

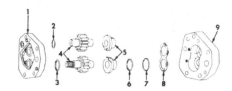

Fig. 235–Exploded view of Thompson 4.5 gpm pump used in hydrostatic drive tractors.

1. Cover	5. Bearings
2. "O" ring	6. "O" ring
3. "O" ring	7. "O" ring retainer
4. Gears and shafts assembly	8. Pressure spring
	9. Body

Fig. 236–Exploded view of hydrostatic drive center section and check valve manifold.

1. Shuttle valve plug
2. "O" ring
3. Snap ring
4. Washer
5. Centering spring
6. Washer
7. Shuttle valve
8. Orifice plug
9. Orifice screen
10. Center section (body)
11. High pressure relief valves
12. "O" rings
13. Back-up washers
14. "O" rings
15. Sealing ring (2 used)
16. Spacer (2 used)
17. Check valve (2 used)
18. "O" ring (2 used)
19. Back-up washer (2 used)
20. Check valve manifold

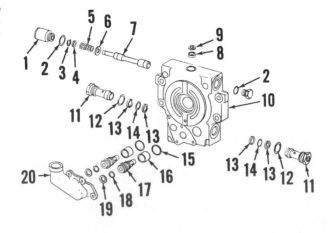

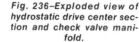

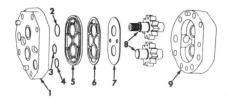

Fig. 234–Exploded view of Cessna 4.5 gpm pump used in hydrostatic drive tractors.

1. Cover	6. Back-up gasket
2. "O" ring	7. Pressure diaphragm
3. "O" ring	8. Gears and shafts assembly
4. "O" ring	9. Body
5. Diaphragm seal	

springs (12). Place cylinder block in a press on wood blocks. Using a step plate on spring retainer (9), apply pressure to compress spring (5). Remove retaining ring (6), release press and remove spring retainer (9), spring (5) and spring seat (4).

Clean all parts and inspect for excessive wear or other damage. Check pistons and bores in clyinder block for linear scratches and excessive wear. Check cylinder block face for shiny streaks, indicating high pressure leakage between cylinder block and brass bearing plate. Inspect piston slippers for scratches, imbedded material or other damage. Light scratches in slippers can be removed by lapping. All nine slippers must be within 0.002 thickness of each other in pump or in motor. Check springs against the following specifications:

Pump cylinder block spring (826, 966, 1026 and 1066),

    Number of coils 3½
    Test length ....1 inch
    Test load (min.) .... 160 lbs.
    Test load (max.) .... 195 lbs.

Motor cylinder block spring (826 and 966),

    Number of coils 3½
    Test length ....1 3/16 inches
    Test load (min.) .... 246 lbs.
    Test load (max.) .... 300 lbs.

Motor cylinder block spring (1026 and 1066),

    Number of coils 3½
    Test length .... 1 21/64 inches
    Test load (min.) ....272.5 lbs.
    Test load (max.) .... 332 lbs.

Lubricate all parts with "Hy-Tran" fluid and reassemble by reversing the disassembly procedure. Spring seat (4) must be installed with beveled side away from cylinder block spring (5) and spring retainer (9) installed with retaining ring groove away from spring. When installing ball guide (2), align the wide tooth of internal splines

to the two ground off teeth of internal splines on cylinder block.

NOTE: Before installing input shaft

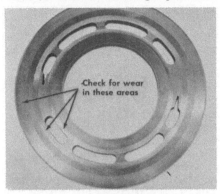

Fig. 239–Inspect steel valve plates for wear or scoring in the areas shown.

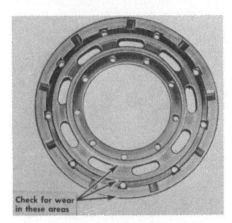

Fig. 240–Inspect brass bearing plates for wear or scoring in the areas shown.

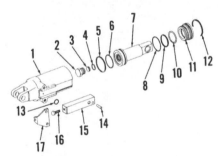

Fig. 241–Exploded view of pump servo cylinder assembly.

1. Cylinder body	10. "O" ring
2. Fixed orifice plug	11. Piston guide
3. Orifice screen	12. Retaining ring
4. "O" ring	13. "O" ring
5. Teflon seal ring	14. Dowel pin
6. "O" ring	15. Adjusting block
7. Piston	16. Adjusting bolt
8. "O" ring	17. Block support
9. Teflon seal ring	bracket

Fig. 242–Exploded view of motor servo cylinder assembly.

1. Cylinder body
2. Fixed orifice plug
3. Orifice screen
4. "O" ring
5. Teflon seal ring
6. "O" ring
7. Piston
8. "O" ring
9. Teflon seal ring
10. "O" ring
11. Piston guide
12. Retaining ring

and swashplate on pump, adjust pump swashplate trunnion bearings as follows: Install pump swashplate, trunnion bearings and trunnions in position in hydrostatic drive housing. Do not install "O" rings or shims at this time. Tighten cap screws in left hand trunnion to 35 ft.-lbs. Torque two cap screws opposite to one another in right trunnion to 25 in.-lbs. while rocking the swashplate. Then, without rocking swashplate, tighten cap screw to 50 in.-lbs. Using a feeler gage, measure gap between right trunnion flange and housing beside each of the two cap screws. Select a shim pack equal in thickness to the average of the two measurements plus or minus 0.002. Remove both trunnions, bearings and swashplate. Keep each bearing with its trunnion and keep the selected shim pack with the right trunnion and set them aside for later installation.

Install input shaft and swashplate with thrust plate on pump by reversing the removal procedure. Be sure to position the horseshoe tool between pto drive gear and swashplate with open side of tool opposite from servo cylinder attaching ears. Install spring guide (10) and bearing plate centering ring (8) before pressing input shaft rear bearing into position against shoulder on shaft.

273. VALVE PLATES AND BEARING PLATES. Inspect steel valve plates and brass bearing plates for excessive wear or scoring in the areas shown in Fig. 239 and 240. Although bearing plates and valve plates are available separately, a valve plate and bearing plate should be renewed as a set for either the pump or the motor. This will assure proper sealing between plates which is necessary for efficient operation.

274. SERVO CYLINDERS. The servo cylinders (Fig. 241 and Fig. 242) are of similar construction except for the variable orifice adjusting block used on the pump servo. Unbolt and remove block support bracket (17—Fig. 241) and adjusting block (15) from pump servo cylinder. Then, on either servo cylinder, clamp cylinder body (1 —Fig. 241 or Fig. 242) in a vise. Using

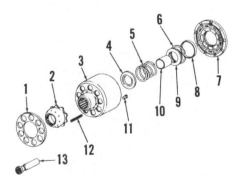

Fig. 238–Exploded view of hydrostatic pump or motor.

1. Slipper retainer	8. Centering ring
2. Ball guide	9. Spring retainer
3. Cylinder block	10. Spring guide
4. Spring seat	11. Aligning dowel
5. Spring	12. Slipper retainer
6. Retaining ring	spring (9 used)
7. Brass bearing plate	13. Piston (9 used)

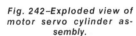

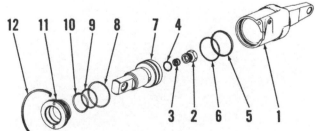

a spanner wrench, rotate piston guide (11) and remove retaining ring (12). Remove piston guide and piston assembly from cylinder body. Slide piston guide from piston. Remove external "O" ring (8), internal Teflon seal ring (9) and "O" ring (10) from piston guide. Remove seal ring (5) and "O" ring (6) from piston. Clamp flat sides of piston in a brass jawed vise and remove fixed orifice plug (2) and orifice screen (3).

Clean and inspect all parts. Check servo cylinder body and piston for scoring and renew as necessary. Renew all "O" rings and Teflon seal rings and reassemble by reversing removal procedure. NOTE: Soak Teflon seal ring (5) in water heated to 200 degrees F. for a few minutes before installing on piston.

275. CONTROL CAM AND VARIABLE ORIFICES. Inspect the control cam plate for excessive wear, scoring or other damage. The cam plate must be straight and smooth as it controls the flow through the variable orifices. The face of variable orifices must also be free of scratches or burrs. A spring washer is used on the pump servo variable orifice and must be installed as shown in Fig. 243. Install new Teflon seal rings on the orifices.

276. OUTPUT SHAFT AND BEARINGS. To remove the bearings from the output shaft, remove the snap ring and press shaft out of front bearing and bearing cage. Using a punch through access holes in rear of shaft, drive rear bearing far enough to get a puller behind it. Clean and inspect bearings and shaft and renew parts as necessary.

NOTE: This is an adjustable bearing made by selection of proper thickness snap ring to provide 0.0025 preload to 0.0025 end play. But by actually measuring each snap ring at assembly, the setup range can be reduced to 0.001 preload to 0.001 end play.

When reassembling press rear bearing cone into position on shaft. Place bearing cage, with bearing cups, on the rear bearing. Place the assembly in a press and using a piece of pipe over the shaft, carefully press front bearing cone into place. Using a cord and spring scale on bearing cage as shown

in Fig. 244, rotate bearing cage while slowly applying pressure on the bearing until a rolling torque of 4 to 20 pounds is indicated on scale. This will give a preload rolling torque of 10 to 60 inch-pounds. Remove assembly from press and install the thickest snap ring which will go into the groove. Snap rings are available in thicknesses of 0.156 to 0.186 in increments of 0.003. Place assembly in the press with support bars under edge of bearing cage. Apply pressure to front end of shaft to seat front bearing against snap ring. Remove assembly from press and lay it aside for later installation.

277. REASSEMBLE AND REINSTALL HYDROSTATIC UNIT. To reassemble the hydrostatic drive unit, make certain the housing is thoroughly cleaned and proceed as follows: Install output shaft and rear bearing cage assembly, into drive housing without shims. Install three bolts equally spaced in rear bearing cage and torque to 35 ft.-lbs. as shown in Fig. 245 without center section. Using a depth micrometer (4), measure the distance from machined surface of drive housing

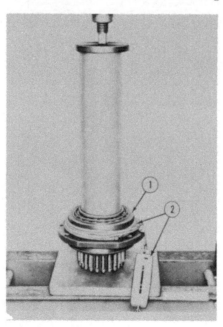

Fig. 244–Use a spring scale to check rolling torque of output shaft rear bearings. Refer to text.

to the bearing cage flange (3). Record this measurement, and then remove the output shaft assembly. Install center section (1—Fig. 245) with bearing cups installed, and torque cap screws to 110 ft.-lbs. Put front bearing (2) on output shaft and install without shims. Install three bolts equally spaced in rear bearing cage, and torque to 25 in.-lbs. while rotating the shaft. Then, without rotating the shaft, tighten cap screws to 50 in.-lbs. Measure from drive housing to bearing cage flange (3) with a depth micrometer (4). Subtract this reading from the one obtained without the center section. The correct shims is the difference between the two readings plus 0.009 inch plus or minus 0.002. Using a dial indicator, check end play in shaft assembly which should be 0.003 to 0.007 inch. Unbolt and remove output shaft assembly and lay it aside with the selected shim pack. Unbolt and remove the center section.

Install motor swashplate, trunnion bearings and trunnions in position in hydrostatic drive housing. Install new "O" ring on left hand trunnion. Do not install "O" ring or shims on right trunnion. Tighten cap screws in left hand trunnion to a torque of 35 ft.-lbs. Install two cap screws in right trunnion opposite to one another, and tighten them to a torque of 25 in.-lbs. while rocking the swashplate. Then without rocking swashplate, tighten cap screws to 60 in.-lbs. Using a feeler gage, measure gap between right trunnion flange and housing beside each of the two cap screws. Select a shim pack equal in thickness to the average of the two measurements plus or minus 0.002. Remove right trunnion and reinstall with shim pack and new "O" ring. Tighten cap screws to a torque of 35 ft.-lbs.

Reinstall output shaft with the selected shim pack and torque the five

Teflon seal (396 109 R1)

Spring (403 989 R1) washer

Fig. 243–View showing correct installation of spring washer on pump servo variable orifice. Spring washer is not used on motor servo variable orifice.

Fig. 245–Use a depth micrometer to measure for right shim pack when adjusting output shaft bearings. Refer to text.

cap screws to 35 ft.-lbs. NOTE: It is recommended that the range transmission and differential housing (rear frame) be flushed with clean solvent. To clean the rear frame, remove upper link bracket and spray flush the final drive. Also refer to SYSTEM CLEAN-UP in paragraph 281.

Recouple hydrostatic drive housing to rear frame, using new gasket. Torque bolts to 110 ft.-lbs. Then, reassemble as follows: Install motor servo cylinder through the opening in the bottom of hydrostatic housing, connect servo cylinder to the swashplate using the special shoulder bolt. Tighten motor servo bolt to 75 ft.-lbs. Install pump servo in hydrostatic housing. Then, using new seals and "O" rings install the anchor shaft, using extreme care not to damage the seal rings. Install the retaining pins, servo control cam and actuating rod assembly (1—Fig. 246) and cam pivot (2). Then, rotate pump servo cylinder and install the orifice and spring. Lubricate motor swashplate and thrust plate with "Hy-Tran" fluid and install thrust plate with chamfered edge to the rear. Align indexing tooth on motor assembly with missing tooth on output shaft and install motor. NOTE: When the motor is properly installed, it should be recessed approximately ½-inch below the machine surface of the housing. Install spring guide (3—Fig. 247) and bearing plate centering ring (4). Lubricate and install brass bearing plate, making certain the flats on dowel pin are aligned to allow bearing plate to fully contact the motor cylinder block.

Lubricate steel valve plate and install on center section (motor side). Being careful not to drop the valve plate, install center section. While rotating the output shaft, tighten center section cap screws to a torque of 110 ft.-lbs. Install bearing retaining washer and cap screw making certain the tang on washer goes into the mating slot in the end of output shaft. Tighten cap screw to a torque of 75 ft.-lbs. using a chain wrench around output gear to

hold shaft. Install connector tube from top side of housing into the center section.

Install check valve manifold and charge tube, torque manifold cap screws to 47 ft.-lbs. Install servo tubes and high pressure tubes.

Lubricate steel valve plate and install on center section (pump side) making certain the plate is fitted down over flats on the dowels. Lubricate pump brass bearing plate and install on pump cylinder block. Be sure slot in bearing plate is aligned with flats on dowel pin. Then, install pump assembly using care not to allow bearing plate to drop from pump during installation. Install the swashplate trunnion bearings. Tighten the two cap screws in horseshoe tool until trunnions can be slipped easily into the bearings. Using new "O" rings on both trunnions and the shim pack selected in paragraph 272 on right hand trunnion, install trunnions and torque cap screws to 35 ft.-lbs. Remove the horseshoe tool. Using the special shoulder bolt, connect pump servo to swashplate and torque to 47 ft.-lbs.

Adjust input shaft bearing assembly as follows: Install front cover (2—Fig. 248) with new seal ring and tighten cap

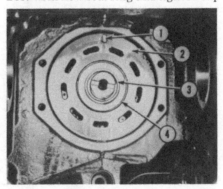

Fig. 247–View of motor cylinder block installed in housing.

1. Dowel
2. Motor cylinder block
3. Spring guide

4. Brass bearing plate centering ring

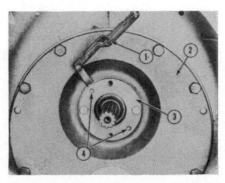

Fig. 248–Using a feeler gage to measure gap when adjusting input shaft bearings. Refer to text.

1. Feeler gage
2. Front cover

3. Bearing retainer (input)

screws. Without the "O" ring and using the old seal install bearing retainer (3) using two bolts as shown in Fig. 248. Torque bolts to 25 inch-pounds while rotating input shaft. Then, without rotating the shaft, torque bolts to 50 inch-pounds. Measure distance between front cover and bearing retainer with a feeler gage (1) as shown at (4). Take the average of the two readings and add 0.013 ± 0.001 to determine correct shim pack. Remove bearing retainer and install a new oil seal and "O" ring. NOTE: Lip of oil seal should face inward or to the rear of housing. Reinstall bearing retainer and tighten cap screws. With pump swashplate at zero angle and without pto driven gear installed the input shaft rolling torque should be 10 to 18 ft.-lbs.

Install pto driven gear, driven shaft and retaining ring (2, 3 and 4—Fig. 223), with rear bearing cup and bearing assembly. Install bearing cup in bearing cover (2—Fig. 249) and adjust bearings as follows: Install the bearing cover with two cap screws, and torque to 25 inch-pounds while rotating pto shaft. Then without rotating shaft tighten to 50 inch-pounds. Measure the gap at two places as shown in (3—Fig. 249). The correct shim pack is the average of the two measurements plus 0.010 plus or minus 0.002. Remove cover and install "O" ring and shim pack. Reinstall cover and tighten cap screws.

Install top cover and multi-valve on hydrostatic housing. Install multiple control valve and connect the lines. Install charge filter base and filter, the left hand side cover, bottom cover and lube block and tube. NOTE: Be sure to install the recirculating screen and plug in the bottom of hydrostatic housing.

Lubricate flex drive plate splines and input shaft splines with Molycote or equivalent. Use aligning dowels and assemble rear section of tractor to the engine. Make certain that input shaft splines are engaged in flex drive plate splines by having engine and bell

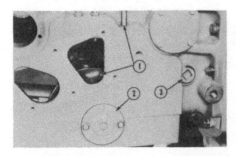

Fig. 246–View showing location of the control cam and cam pivot on left side of hydrostatic drive housing.

1. Servo control cam and actuating rod

2. Cam pivot
3. Plug

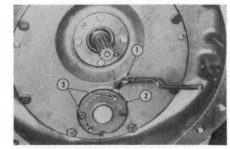

Fig. 249–Using a feeler gage to measure gap when adjusting pto shaft bearings. Refer to text.

1. Feeler gage
2. Pto bearing cover

housing within ½-inch of each other before installing cap screws.

NOTE: To aid in engaging splines, turn flywheel by engaging the ring gear with a screwdriver through access hole in right side of bell housing. Install and tighten cap screws.

Install fuel tank and steering support assembly by reversing the removal procedure. Connect hydraulic lines and electrical wiring and complete the balance of reassembly.

Refer to SYSTEM CLEAN-UP in paragraph 281 and ADJUSTMENTS starting with paragraph 260.

278. **OVERHAUL EXTERIOR CONTROL VALVES.** If trouble shooting checks (paragraph 259) indicate a faulty drive control valve or Foot-N-Inch valve, remove the valve, clean and repair or renew the faulty valve as outlined in the following paragraphs.

279. DRIVE CONTROL VALVE. Remove rear cowling and the S-R lever assembly. Then, disconnect high pressure line and dump line, then unbolt and remove valve.

To disassemble the valve, refer to Fig. 250 and remove plug (1) with "O" ring (2). Withdraw valve pin (6), spring (5) and anti-coast valve (7). Unscrew plug (8) and remove the deceleration and check valve assembly (items 8 thru 19). This assembly is referred to as capsule "B" and normally is withdrawn as an assembly. However, if the capsule comes apart during removal, lay parts out in order. Spring (10 and 15) and balls (11 and 16) must not be interchanged.

Unscrew plug (20) and remove combination check valve assembly (items 20 thru 32) which is referred to as capsule "A". If capsule "A" comes apart during removal, lay parts out in order as they are removed. Springs (22 and 28) and balls (23 and 29) must not be interchanged. Remove actuating lift pin (33) from valve body.

Remove end plug (36) with "O" ring (37), then withdraw valve spool (42) with centering spring assembly. Remove snap ring (38), spring (40) and washers (39 and 41). Remove oil seal (34) from valve body.

Clean and inspect all parts for excessive wear or other damage. Spool (42), anti-coast valve (7) and valve pin (6) are selected fits and are not serviced separately. The valve pin is a loose, but matched fit in the anti-coast valve. Spool (42) is a close fit in valve bore but must be free to move without binding. Capsule "A" (items 20 thru 32) and capsule "B" (items 8 thru 19) are available as assemblies.

Lubricate parts with "Hy-Tran" fluid, use new oil seal, "O" rings and

back-up rings and reassemble by reversing the disassembly procedure. NOTE: Make certain the four check balls are installed in proper positions. On production valves, three balls (11, 23 and 29) are ¼-inch in diameter and ball (16) is 7/32-inch in diameter. On service capsules, balls (11 and 23) are 5/16-inch, ball (16) is 7/32-inch and ball (29) is ¼-inch in diameter.

Install valve by reversing the re-

moval procedure. Adjust valve linkage as outlined in paragraph 260.

280. FOOT-N-INCH VALVE. The Foot-N-Inch valve is located below left side of platform. Removal of valve is obvious after examination of the unit.

To disassemble valve, refer to Fig. 251 and remove guide (12) with "O"

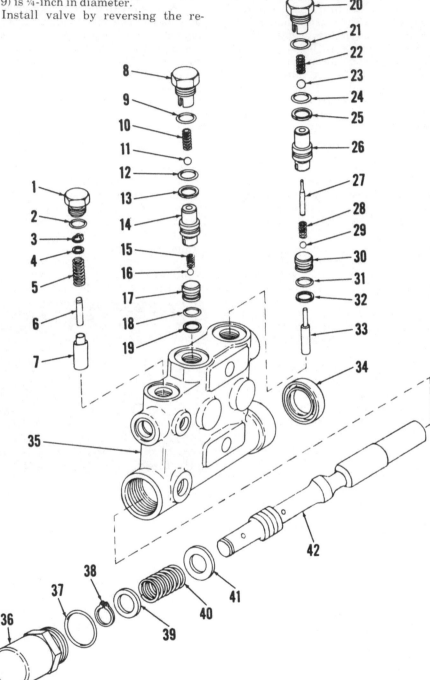

Fig. 250—Exploded view of drive control valve.

1. Plug	22. Spring	32. Back-up ring
2. "O" ring	23. Steel ball	33. Actuating lift pin
3. Snap ring	13. Back-up ring	24. "O" ring	34. Oil seal
4. Washer	14. Deceleration and	25. Back-up ring	35. Valve body
5. Spring	check valve body	26. Combination check	36. Plug
6. Valve pin	15. Spring	valve body	37. "O" ring
7. Anti-coast valve	16. Steel ball	27. Follower lift pin	38. Snap ring
8. Plug	17. Seat	28. Spring	39. Washer (outer)
9. "O" ring	18. "O" ring	29. Steel ball	40. Centering spring
10. Spring	19. Back-up ring	30. Seat	41. Washer (inner)
11. Steel ball	20. Plug	31. "O" ring	42. Valve spool
12. "O" ring	21. "O" ring		

ring (9), gasket (14) and oil seal (13). Withdraw piston (6), plate (7) retaining ring (8), "O" ring (10), shims (11), spring (5) and poppet valve (4). Remove valve seat (2) and "O" ring (3).

Clean and inspect all parts. Check valve seat and poppet for excessive wear or imbedded material. The seat and fitting is a special dimension from the shoulder to the seat and must not be altered. When reassembling use new gasket, "O" rings and oil seal. Install the correct spring and shim pack. Do not attempt to increase relief pressure above 6000 psi.

Reinstall valve and adjust linkage as outlined in paragraph 261.

281. **SYSTEM CLEAN-UP.** To clean-up the system after overhauling hydrostatic drive unit, range transmission or differential, remove oil cooler and lines and back flush with solvent and compressed air. Replace oil cooler and lines reversing the lines at the multi-valve to back flush the cooler during clean-up procedure. Install a new main filter, with dummy by-pass valve in the regular hydraulic filter so all the fluid must pass through the filter. A regular by-pass valve can be reworked by soldering a plate over the by-pass hole. Fill transmission with "Hy-Tran" fluid. Support rear of tractor so both rear wheels are free to rotate. Operate tractor for two hours in both forward and reverse. Operate power steering, hitch and auxiliary valves.

CAUTION: Operate S-R lever up to the blue line but not above.

After two hours, replace the charge filter and main filter, then install the original by-pass valve. Reconnect oil cooler lines in their normal positions.

LUBRICATION AND FILTER

All Models

282. The tractor rear frame serves as a common reservoir for all hydraulic and lubrication operations. Only IH "Hy-Tran" fluid should be used and level should be maintained at "FULL" mark on dipstick. The dipstick is located on top right front of rear frame. The fluid should be drained and new "Hy-Tran" fluid installed every 1000 hours of operation or once a year, whichever occurs first. The refill capacity is approximately 28 gallons on models 826 and 1026 and 27 gallons on models 966 and 1066.

When checking the fluid level, operate engine with range transmission and hydrostatic drive in neutral position until engine is at normal operating temperature. Then, with engine operating at 1600 rpm, check the fluid level which should be at "FULL" mark.

The main filter and charge filter element should be renewed at 10 hours, 100 hours, 250 hours and then every 250 hours of operation thereafter.

MAIN DRIVE BEVEL GEARS AND DIFFERENTIAL

The differential is carried on tapered roller bearings. The bearing on the bevel gear side of the differential is larger than that on the opposite side and therefore, it is necessary to keep the bearing cages in the proper relationship.

ADJUSTMENT

All Models

283. **CARRIER BEARING PRELOAD.** The carrier bearings can be adjusted by either of two methods; however, in either case the hydraulic lift assembly should be removed as outlined in paragraph 322, and the final drives removed as outlined in paragraph 291.

NOTE: After adjusting carrier bearings, readjust bull pinion bearings as outlined in paragraph 293.

284. To use the direct measurement method, proceed as follows: Install differential in rear frame with no shims behind bearing cages and tighten left bearing cage cap screws to 73 ft.-lbs. torque. Rotate the differential and tighten the right hand bearing cage cap screws to 100 in.-lbs. torque. Now loosen the right hand bearing cage cap screws, rotate differential and retighten cap screws to 20 in.-lbs., then without rotating differential, tighten bearing cage cap screws to 50 in.-lbs. torque. Use a depth gage through puller bolt holes and measure between the surface of rear main frame and outside of bearing cage and average the readings. Measure the thickness of bearing cage flange at puller bolt holes and average the readings. Subtract second reading from first reading which will give the required thickness of shim pack. Shim pack must be within plus or minus 0.002 of the determined shim pack thickness. Shims are available in thicknesses of 0.003,

0.004, 0.007, 0.012 and 0.030. Shims can be divided between the two bearing cages to provide the proper backlash as outlined in paragraph 286.

285. To use the rolling torque method, use the original shim packs behind bearing cages and rotate the differential as bearing cage cap screws are tightened. Be sure there is some backlash maintained between bevel gear and pinion. Place the range transmission in neutral, then loosen the right bearing cage so there is no preload on bearings (cap screws finger tight). Wrap a cord around differential, attach to a spring scale and note the pounds pull required to keep differential and transmission in motion. Tighten the bearing cage cap screws securely and recheck the rolling torque. Now vary the shims until 2 to 7 pounds more pull is required to keep differential and transmission in motion than when no preload was applied to bearings. With bearing preload determined, refer to paragraph 286 to set backlash between bevel gear and pinion.

286. **BACKLASH ADJUSTMENT.** With the differential carrier bearing preload determined as outlined in paragraph 284 or 285, the backlash between bevel gear and drive pinion should be checked and adjusted as follows: Mount a dial indicator and while holding drive pinion forward, check backlash in at least three places during a revolution of the differential. Correct backlash is 0.005-0.015. If the backlash is not as stated, shift bearing cage shims from one side to the other as required. NOTE: Do not add or remove shims as the previously determined bearing preload will be changed. Shifting 0.010 shim thickness from one side to the other will change backlash approximately 0.0075.

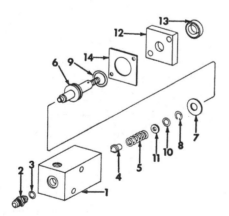

Fig. 251—Exploded view of the Foot-N-Inch valve.

1. Valve body	8. Retaining ring
2. Seat	9. "O" ring
3. "O" ring	10. "O" ring
4. Poppet valve	11. Shims
5. Spring	12. Guide
6. Piston	13. Oil seal
7. Plate	14. Gasket

R&R BEVEL GEARS
All Models

287. The main drive bevel pinion is also the range transmission mainshaft. The procedure for removing, reinstalling and adjusting pinion setting is outlined in the range transmission section (paragraphs 253 thru 257).

To remove the bevel ring gear, follow the procedure outlined in paragraph 288 for R & R of differential. The ring gear is secured by the differential case bolts which should be tightened to a torque of 112 ft.-lbs.

R&R DIFFERENTIAL
All Models

288. All tractors are equipped with a four pinion type differential. To remove differential, remove final drives as outlined in paragraph 291 and the hydraulic lift assembly as outlined in paragraph 322.

With final drives and hydraulic lift assembly removed, attach a hoist to differential assembly, remove both differential bearing cages and carefully lift differential from rear frame.

To disassemble differential, use a puller and remove the carrier bearings and note that bearing on bevel gear side of differential has a larger O.D. and contains more rollers than the opposite side bearing. Remove the differential case bolts and separate differential assembly. Any further disassembly is obvious. Oil seals and bearing cups in bearing cages can also be renewed at this time and procedure for doing so is obvious. For all models except 1066 gear drive oil seals are installed with lip toward inside and within 0.010 of being flush with outside surface of bore. See Fig. 253.

For model 1066 gear drive see Fig. 254 for installing seal. NOTE: Some model 1066 gear drive are equipped with pressure lubrication to the differential and a cup is used in place of the seal Fig. 255.

When reassembling differential, tighten the differential case bolt nuts to 112 ft.lbs. Refer to paragraphs 283 thru 286 for carrier bearing preload and backlash adjustments.

DIFFERENTIAL LOCK

All Models So Equipped

289. The differential lock is located on outside of right hand brake housing, and is hydraulically actuated. When operator holds the control valve in down position, this allows hydraulic oil pressure to actuate the clutch assembly so that the two bull pinion shafts are locked together which causes the differential to act as a solid hub. Fig. 256. When control valve is released, the clutch assembly will disengage.

The differential lock is not intended for continuous use or when slippage is not a problem and therefore the unit is disengaged whenever the control valve is released.

290. **R&R AND OVERHAUL.** To remove the differential lock, remove the cover and piston housing (Fig. 256). Any further disassembly required will be obvious. Clean and inspect all parts. Renew any parts showing excessive wear or other damage.

Reinstall by reversing the removal procedure.

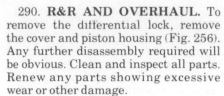
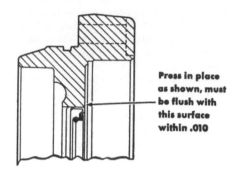

Fig. 253–Install oil seals in differential bearing cages as shown.

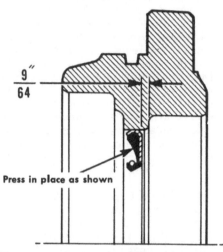

Fig. 254–Install oil seal in differential bearing cages as shown on all model 1066 tractors without pressure lubrication to differential.

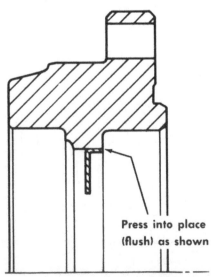

Fig. 255–Install cup in differential bearing cages as shown on all model 1066 tractors with pressure lubrication to differential.

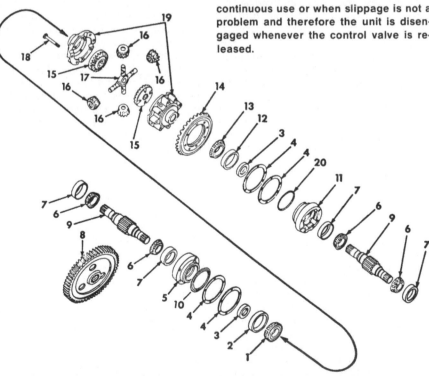

Fig. 252–Exploded view of differential, bull pinions and associated parts.

1. Bearing cone (LH)
2. Bearing cup (LH)
3. Oil seal
4. Bearing cage shims
5. Differential bearing cage (LH)
6. Bearing cone
7. Bearing cup
8. Bull gear
9. Bull pinion shaft
10. Oil baffle (LH)
11. Differential bearing cage (RH)
12. Bearing cup (RH)
13. Bearing cone (RH)
14. Bevel ring gear
15. Bevel gears
16. Pinion gears
17. Spider
18. Case bolts (12 used)
19. Differential case

The differential lock pressure for gear drive tractors is 235 to 280 psi.

The pressure for hydrostatic drive tractors is 240 to 300 psi.

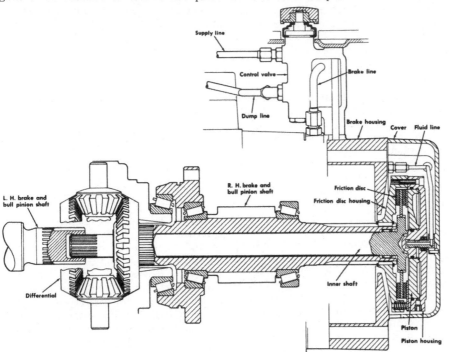

Fig. 256–Sectional view of differential lock.

pedal springs. The brake valve and pedals are removed with axle carrier. Remove outer brake housing and brake disc assembly. Remove bolt circle from final drive housing, attach hoist and remove final drive assembly from main frame as shown in Fig. 257.

Reinstall by reversing the removal procedure.

All Models

293. OVERHAUL. With final drive removed as outlined in paragraph 292, disassemble as follows: Remove brake piston housing and adjusting shims, then remove the bull pinion. Using a suitable puller, remove bearing cone (2 —Fig. 258) from inner end of axle shaft. Remove snap ring (17), on all models except 1066 gear drive. On model 1066 gear drive remove spacer, bull gear and snap ring (3). Unbolt axle outer bearing retainer and withdraw axle shaft assembly from carrier. Any further disassembly required is evident after an examination of the unit.

When reassembling, press rear axle outer bearing on axle using a pipe having a diameter which will contact inner race only. Press bearing on axle shaft until it bottoms against shoulder. Install grease shield in carrier (if equipped), if it was removed, then install axle. If oil seal in outer bearing retainer is renewed, install with spring loaded lip toward inside and inner surface flush with inner surface of bore as

FINAL DRIVE

The final drive assemblies consist of the rear axle, bull gear and bull pinion and can be removed from the tractor as a unit.

All Models Except High Clearance

291. REMOVE AND REINSTALL. With the exception of the brake valve assembly, removal of either final drive assembly is the same. Therefore, removal procedure will be given for the right final drive assembly and will include the necessary brake lines removal procedure. NOTE: On hydrostatic drive, when removing the left axle carrier, disconnect the foot-n-inch valve lines.

292. To remove the right final drive, first remove drain plug and drain housing. Remove fender, then support tractor under main frame and remove the tire and wheel assembly. Disconnect the hydraulic lines and brake

Fig. 257–Right final drive assembly being removed from rear frame.

1. Sling   2. Aligning dowel   3. Rear axle carrier

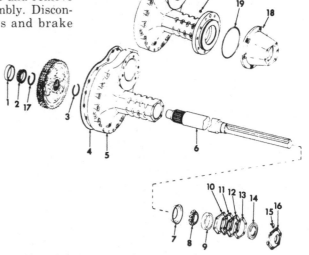

Fig. 258–Exploded view of final drive assembly. Note carrier (21) and extension (18) used when tractor has wide tread rear axles.

1. Bearing cup	8. Bearing cone	15. Grease fitting
2. Bearing cone	9. Bearing cup	16. Cap
3. Snap ring	10. Shim (light)	17. Snap ring
4. Gasket	11. Shim (medium)	18. Extension
5. Carrier	12. Shim (heavy)	19. "O" ring
6. Rear axle shaft	13. Shim (ex. light)	20. Plug
7. Grease shield	14. Oil seal	21. Carrier (wide tread)

shown in Fig. 259. Use shim stock or seal sleeve when installing seal retainer over axle. Install retainer cap screws but do not tighten them at this time. Install snap ring, bull gear, snap ring on all models except 1066 gear drive, which has a spacer and inner bearing on inner end of axle shaft. Place the bull pinion in position on bull gear and install brake piston housing without shims or "O" ring. Use aligning dowels and install the final drive assembly to the main frame and torque axle carrier cap screws to 170 ft. lbs. NOTE: Hold bull pinion in position while installing the assembly so seals and bearings will not be damaged.

Install three cap screws through brake piston housing and tighten them evenly to a torque of 100 in.-lbs. while rotating the bull pinion shaft to align and seat the bearings. Loosen the cap screws, then retighten them evenly to a torque of 50 in.-lbs. while rotating the shaft. Next, without rotating the bull pinion shaft, further tighten the cap screws to a torque of 100 in.-lbs. Using a feeler gage, measure the gap adjacent the cap screws. The correct shim pack is the average measured gap plus 0.015. Shims are available in the following thicknesses: 0.003, 0.004, 0.007, 0.012 and 0.035. Remove brake piston housing, install new "O" ring and oil seal, then reinstall brake piston housing with the shim pack.

To adjust the axle shaft bearings, tighten three evenly spaced cap screws retaining axle outer cap (seal retainer) to axle carrier to a torque of 150 in.-lbs. while rotating the axle. Then loosen the cap screws and retighten to 20 in.-lbs. while rotating the axle. Then, without rotating axle, tighten **evenly** to 50 in.-lbs. Measure distance between outer cap and carrier next to the three tightened cap screws and average the three readings. Make up a shim pack that will be within a plus or minus 0.002 of the average measured distance and install between axle outer cap and

carrier. NOTE: The cap screw in the opposite cap should be loosened so there is no preload on the bearing. After adjusting the axle shaft bearings, tighten all cap screws securely and complete reassembly of tractor.

High Clearance Models

294. On high clearance model tractors, the complete final drive assembly is removed in the same manner as outlined in paragraph 292 although the drop housing (18—Fig. 260) should also be drained.

With unit removed, proceed as follows: Remove pan (19) from housing (18). Remove bearing cap assembly (items 21, 22, 23 and 26), then remove inner bearing (25) and spacers (27 and 35) from end of axle (24). Remove outer bearing cap assembly (items 32, 34, 33 and 31), then bump axle outward and remove driven gear (28) from bottom of housing. Remove bearing cap assembly (items 13, 14, 16 and 17) from housing (18), then unbolt and remove housing (18) from carrier (1). Remove bearing (12), grease retainer (11) and drive gear (10), then remove snap ring (9) and the oil seal retainer (6) and oil seal (5) assembly. Remove bearing and snap ring from inner end of axle and bump

axle out of bull gear. Any further disassembly required will be obvious after an examination of the unit and reference to Fig. 260.

Clean and inspect all parts and renew as necessary. Be sure to use all new "O" rings and gaskets and reassemble by reversing the disassembly procedure. However, delay the final adjustment of shaft bearings until unit is installed on tractor. Refer to paragraph 295 for procedure.

295. To adjust the drive gear axle bearings, remove all the shims (17—Fig. 260) from behind bearing cap (14). Install three cap screws (one every other hole) in bearing cap and tighten evenly to 150 in.-lbs. Bump bearing cap to insure bearings are seated, loosen the three cap screws, then retighten to 50 in.-lbs. Measure distance between bearing cap and drop housing next to the three cap screws, then select and install a shim pack equal to the averaged reading. This will provide the shaft bearings with the recommended zero end play. Shims are available in 0.004, 0.005, 0.007, 0.012 and 0.0299 thicknesses.

Repeat the above operation to adjust bearings for the rear axle shaft (24).

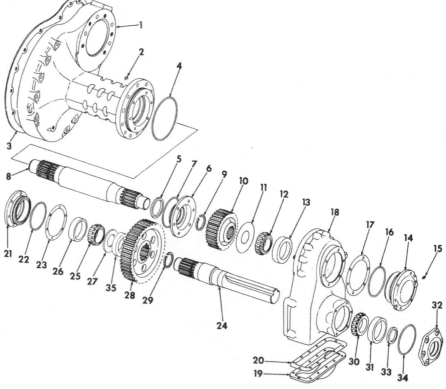

Fig. 260–Exploded view of final drive assembly used on high clearance tractors.

1. Carrier	10. Drive gear	18. Housing	27. Axle spacer
2. Plug	11. Grease retainer	19. Pan	28. Driven gear
3. Gasket	12. Bearing cone	20. Gasket	29. Snap ring
4. "O" ring	13. Bearing cup	21. Bearing cap	30. Outer bearing cone
5. Oil seal	14. Bearing cap	22. "O" ring	31. Outer bearing cup
6. Seal retainer	15. Grease fitting	23. Shim	32. Bearing cap
7. "O" ring	16. "O" ring	24. Rear axle shaft	33. Oil seal
8. Drive gear axle	17. Shim	25. Inner bearing cone	34. "O" ring
9. Snap ring		26. Inner bearing cup	35. Gear spacer

Press flush in place as shown

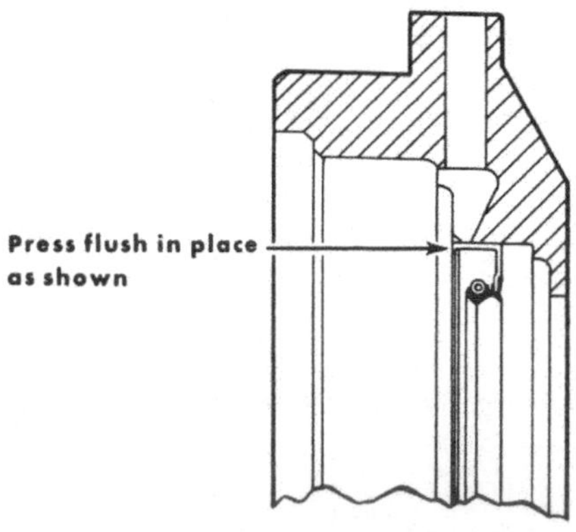

Fig. 259–When installing rear axle cap (retainer) seal, position as shown with lip toward inside.

NOTE: Seals (5) and (33) are installed with lips facing toward inside. Use shim stock or a seal protector when installing retainer (6) or cap (32).

Complete reassembly of tractor and bleed brakes as outlined in paragraph 298.

BRAKES AND CONTROL VALVE

Brakes on all models are actuated hydraulically and are a self-adjusting, double disc type. Brake operation can be accomplished with engine inoperative because of a one-way check valve located on top rear of the multiple control valve. This check valve closes when hydraulic pressure ceases and thus provides a closed circuit which permits operating the brake control valve with the oil trapped within the circuit.

Service (foot) brakes MUST NOT be used for parking or any other stationary job which requires the tractor to be held in position. Even a small amount of fluid seepage would result in brakes loosening

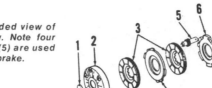

Fig. 263—Exploded view of brake assembly. Note four self-adjusters (5) are used on each brake.

1. Cup plug
2. Housing
3. Disc
4. Intermediate plate
5. Brake adjuster assy.
6. Primary plate
7. Insulator (brake piston)
8. Piston
9. Piston seal ring (outer)
10. Piston seal ring (inner)
11. Dowel
12. Shim
13. Brake piston housing
14. Connector (bleeder valve)
15. Bleed screw
16. "O" ring
17. Oil seal

and severe damage to equipment or injury to personnel could result. USE PARK LOCK when parking tractor.

BRAKE ADJUSTMENT
All Models

296. The only external adjustments that can be made on brakes are the brake pedal maximum travel and the brake control valve adjustment.

297. To make the brake pedal maximum travel adjustment, refer to Fig. 261 and proceed as follows: Disconnect brake springs to simplify operation. Remove cotter pin (G) and pin (H), then move pedal forward until it is stopped by stop screw (F). At this time, rear of pedal arm should be 3½ inches forward of the platform stop (C) as shown. If measurement (T) is not as stated, loosen jam nut and turn stop screw as required. Tighten jam nut and reinstall pin (H) and cotter pin (G). Repeat operation on other pedal. Do not reconnect return springs until brake valve is adjusted as follows:

To adjust brake valve, refer to Fig. 262 and loosen cap screw (A). Use a small rod in hole of valve spool (B) and turn spool out of clevis until about ⅛-inch clearance appears between brake pedal arm and platform stop (C—Fig. 261). Now, turn spool into clevis until pedal arm just contacts the pedal stop, then give valve spool an additional ¼-turn in the same direction. Tighten clamp screw (A). Repeat operation for

other pedal and reconnect brake return springs.

BLEED BRAKES
All Models

298. To bleed brakes, attach a length of hose to bleeder valve (D—Fig. 262) on top of brake housing and place open end in a container. Start engine and run at low idle rpm. Depress brake pedal and while holding in this position open bleeder valve and when a solid flow of oil appears, close valve. Repeat operation on opposite brake. Check brake pedal feel. If brake pedal operation feels spongy rather than having a solid feel, repeat the bleeding operation.

NOTE: With engine not running, braking action should occur in last ⅓ of the pedal travel.

BRAKE ASSEMBLIES
All Models

299. **R&R AND OVERHAUL.** Removal of either brake is accomplished by removing cap screw and spacer located between platform and top of brake housing, disconnecting the brake line, then removing the housing retaining cap screws. Remove brake assembly (1 thru 6—Fig. 263). Carefully withdraw brake piston housing (13). Do not lose shims (12) as these shims control bull pinion bearing adjustment.

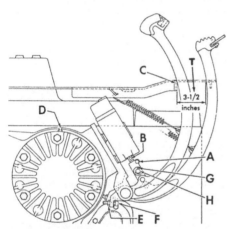

Fig. 261—View showing brake pedals, control valve and adjusting points. Distance (T) is measured with clevis pin (H) removed.

A. Clamp screw
B. Valve spool
C. Pedal return stop
D. Bleed valve
E. Jam nut
F. Pedal stop screw
G. Cotter pin
H. Clevis pin

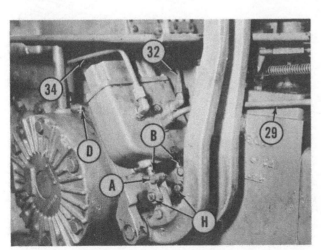

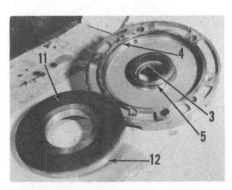

Fig. 262—View showing typical installation of brake valve on tractor.

A. Clamp screw
B. Valve spool
D. Bleed valve
H. Clevis pin
29. Pressure line
32. Left brake line
34. Right brake line

Fig. 264—Brake piston removed from piston housing. Note insulator (11) on piston.

3. Oil seal
4. Outer seal ring
5. Inner seal ring
11. Insulator
12. Piston

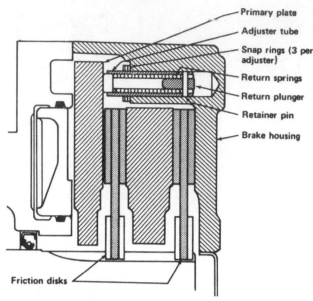

Fig. 265–Cross-sectional view of brake self-adjuster assembly.

Piston can be removed from piston housing by bumping edge of housing on work bench, or by applying air pressure to the line fitting of housing. However, when using air pressure to remove piston, use caution not to "blow" piston out which could result in damage to parts or injury to personnel. Should piston become cocked during

removal, it will be necessary to straighten it to complete removal. See Fig. 264.

With brake assembly disassembled, clean and inspect all parts including cylinder cavity in piston housing. Renew any parts showing excessive wear or other damage. Install new oil seal (17—Fig. 263) with lip toward inside (away from brake), then install new piston seal rings (9 and 10). Install piston (8) with insulator (7) in piston housing (13). Install brake piston housing (13) using new "O" ring (16) and shim pack (12).

Before installing the balance of brake parts, check and reset self-adjuster assemblies (5—Fig. 263) as follows: Refer to Fig. 265 and measure the distance the plunger protrudes from end of adjuster tube. This distance should be 0.025 to provide correct brake disc clearance when brakes are released. Reset the three snap rings so that inner snap is ¾-inch from end of adjuster tube. As brake linings wear, the primary plate will force adjuster

tube through snap rings and deeper into bore in brake housing to maintain brake adjustment.

With self-adjusters (3—Fig. 266) reset and installed in cover, stack the balance of brake parts in brake housing (2—Fig. 263) as follows: Place outer brake disc (3) in housing, install intermediate plate (4), and inner brake disc (3), then install primary plate (6). Make certain that bosses on intermediate plate and primary plate straddle the dowel pins as shown in Fig. 266 and 267. Install assembly on the tractor and tighten cap screws.

When assembly is completed, check external adjustments as outlined in paragraph 296, start engine and bleed brakes as in paragraph 298.

BRAKE CONTROL VALVE
All Models

300. **R&R AND OVERHAUL.** Remove brake control valve as follows: Disconnect all brake lines which in some cases will require removal of platform before dump line can be disconnected from valve body cap. Disconnect brake pedal return springs. Remove clevis pin from outer pedal clevis, then turn clevis one quarter turn to give clearance for removal of inner pedal clevis pin. Unbolt and remove control valve from mounting bracket.

301. With brake control valve removed, refer to Fig. 268 and disassemble valve as follows: Loosen the clevis clamp screws, hold valve spools with a small rod and count turns as each clevis is removed. This will eliminate a considerable amount of adjusting after control valve is reinstalled. Remove cap from valve body, then identify each spool with its bore and remove spool assemblies and springs. Place small rod, or pin punch, through hole in valve spool and remove the retaining cap screw (7) and reaction spring (3). The square cut snap

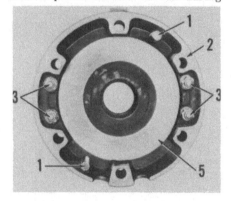

Fig. 266–View showing the self-adjusters and dowel pins in the brake housing.

1. Dowel pins	3. Self-adjusters
2. Brake housing	5. Intermediate plate

Fig. 267–With the intermediate and primary plates installed on the dowel pins as shown, there must be clearance between the plates and housing at the cap screw bosses.

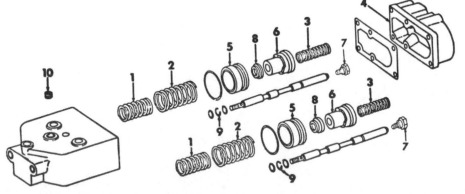

Fig. 268–Exploded view of brake control valve. Gasket and "O" rings which are not numbered are available only in a kit, IH part No. 384 502R92. Valve body and spools are not serviced separately.

1. Return spring	4. Body cap	8. Spring seat
2. Large piston spring	5. Large piston	9. Snap ring
3. Reaction spring	6. Small piston	10. Orifice
	7. Retaining cap screw	

ring (9) can also be removed if necessary. Remove the orifice (10).

Clean and inspect all parts. Pay particular attention to pistons (5 and 6) and their bores. Be sure orifice (10) is clean as well as all other oil passages. Renew springs (1 and 2) if any doubt exists as to their condition. Use all new "O" rings and gaskets and reassemble by reversing disassembly procedure.

NOTE: The retaining cap screw (7) is sealed with "Loctite" when installed and a new screw and sealant is available as a service package under IH part number 530888R92.

After valve is installed, bleed brakes as outlined in paragraph 298 and adjust brake pedals as outlined in paragraph 297.

BELT PULLEY

Belt pulley attachment is available for models 766, 826 and 966. The belt pulley housing is attached to the PTO housing and the unit is driven by the 1000 rpm PTO output shaft.

REMOVE AND REINSTALL
All Models So Equipped

302. To remove the belt pulley assembly, disconnect inlet (top) hose from the orifice connector located at hole in PTO valve linkage support, remove orifice connector, then plug hole with ⅜-inch plug (with "O" ring). Disconnect outlet (bottom) hose from upper left side of PTO housing and plug hole with the ½-inch plug and "O" ring which is stored in a retainer located below the hole. Connect the disconnected ends of the two hoses together and tighten securely to retain oil in belt pulley housing. Remove two diagonally op-

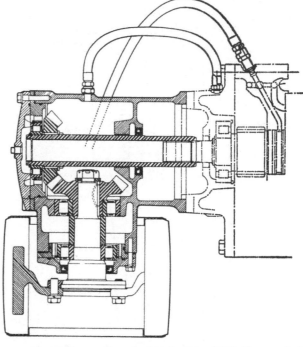

Fig. 270—Cross-sectional view of installed belt pulley attachment. Note lubrication hoses connected to pto.

posite retaining cap screws and install guide studs. Attach hoist to unit, remove the two remaining cap screws and slide unit rearward off the PTO output shaft.

Reinstall by reversing the removal procedure.

OVERHAUL
All Models So Equipped

303. With belt pulley removed, disassemble as follows: Refer to Fig. 269. Remove retainer (1) and pull the shaft, gear and bearing assembly from housing. Bushing (12) and oil seal (13) can now be removed from housing. Remove pulley (26) from pulley shaft (25). Remove cotter key, nut and washer (Fig. 270) from inner end of

pulley shaft (25—Fig. 269), then unbolt bearing retainer (22) and remove shaft and bearings assembly from housing. Any further disassembly required will be obvious.

Clean and inspect all parts. Pay particular attention to the straight roller bearings to see that they have no rough spots or are not excessively worn. Bushing (12) is pre-sized and if carefully installed should require no final sizing. Seals (13) and (23) are installed with lips toward inside of housing. Plug (7) in aft end of input shaft need not be removed unless damaged. Items (3,4,8 and 10) are available only as an assembly.

To reassemble, start with pulley shaft and reverse the disassembly procedure . Use all new "O" rings and seals. Vary shims (6) and (18) as required to provide 0.002-0.010 backlash between the gears. Shims are available in light, medium and heavy thicknesses.

POWER TAKE-OFF

The power take-off used on all models is an independent type. The pto is available as a dual speed unit having both 1000 and 540 rpm output shafts, or as a 1000 rpm unit which is convertible to the dual speed unit by adding the necessary 540 rpm shafts and gears. The pto unit for all models incorporates its own hydraulic pump which furnishes approximately 3 gpm of oil which is used to actuate the pto clutch and provide lubrication for the pto,

Fig. 269—Exploded view of belt pulley attachment.

1. Bearing retainer	10. Input shaft
2. "O" ring	11. Housing
3. Snap ring	12. Bushing
4. Roller bearing	13. Oil seal
5. Bearing cage	14. Driven gear
6. Shim	15. Roller bearing
7. Plug	16. Spacer
8. Drive gear	17. "O" ring
9. Snap ring	

18. Shim	27. Plug storage retainer
19. Bearing cage	28. Flared cap
20. Roller bearing	29. Orifice connector
21. Oil seal	30. "O" ring
22. Bearing retainer	31. Inlet hose
23. Oil seal	32. Outlet hose
24. Woodruff key	33. "O" ring
25. Pulley shaft	34. Plug
26. Pulley	35. "O" ring

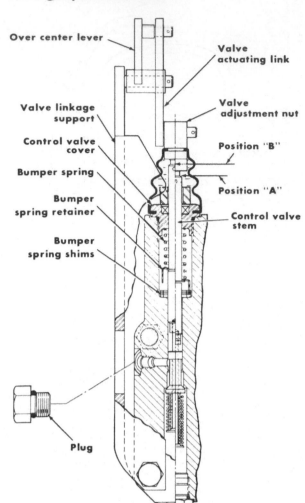

Over center lever

Valve actuating link

Valve linkage support

Control valve cover

Bumper spring

Bumper spring retainer

Bumper spring shims

Valve adjustment nut

Position "B"

Position "A"

Control valve stem

Plug

Fig. 271–Cut-away view of the pto unit control valve. Note positions "A" and "B". Refer to text for adjustment.

and on tractors so equipped, the belt pulley unit. Operation of the pto unit is controlled by a spool type valve located in a bore in left side of pto housing.

OPERATING PRESSURE
All Models So Equipped

304. To check operating pressure, run tractor until hydraulic fluid is approximately 150 degrees F. Refer to Fig. 271 and remove the test port plug which is located at hole in linkage support (left side) and attach either an IH Flow Rater, or a test gage capable of registering at least 300 psi. If necessary, loosen linkage support cap screws and push support downward as far as possible and retighten cap screws. Unscrew valve from adjusting nut and swing nut and actuating link up out of the way. Start engine with pto engaged, operate engine at 2100 rpm (1000 pto rpm). Pull valve stem up until stem contacts bumper spring retainer (position "A"—Fig. 271) and hold in this position. Note: This is partial engagement position. If gage reads 41-46 psi, pressure is satisfactory at this point. If pressure is below 41 psi, remove control valve cover, bumper

spring and bumper spring retainer and install bumper spring shims as required. Each shim will change pressure about 5 psi. Reposition the actuating link and adjusting nut and screw valve stem into nut to approximately the original position. Move control handle and pull the over-center link into up position (position "B"—Fig. 271) and check pressure. Gage should read 235 psi. If gage pressure is not as stated, turn valve stem into adjusting nut to increase or out of nut to decrease pres-

sure. Note: This is the full engagement position.

With these adjustments made, reduce engine speed to low idle rpm (pto engaged) at which time the pressure must not drop more than 40 psi from the 235 psi reading. Disengage pto and check operation of pto brake (anticreep). Brake must stop pto rotation within a maximum of 3 seconds.

If above conditions cannot be met, remove and overhaul pto unit as outlined in paragraph 305.

R&R AND OVERHAUL
All Models

305. To remove pto unit, drain the rear main frame, then disconnect control lever rod and remove linkage support. Attach hoist to unit, complete removal of retaining cap screws, then pull unit rearward, tip rear end downward to prevent damage to suction tube assembly and remove unit from tractor main frame. See Fig. 272.

NOTE: If pto extension shaft remains with pto unit, it will be necessary to remove it from the pto unit input shaft before withdrawing unit from main frame.

Leave the two short cap screws which retain rear cover to housing in place so the pto sections will be held together.

306. With pto unit removed from tractor, use Fig. 273 as a reference and disassemble unit as follows:

NOTE: This procedure is for the dual speed pto unit. Procedure for the single speed unit will remain basically the same except no idler gear and shaft, or 540 rpm driven gear and shaft, are used in the single speed unit.

Remove safety shield. Remove retainer (77—Fig. 273), pump housing and suction tube assembly from carrier (67). See Fig. 274. Remove pump idler gear from housing and the drive gear and Woodruff key from input shaft. If necessary, bearing (76) and suction tube (6) can be removed from pump

Fig. 272–Pto unit being removed. Note rear end being tipped downward to prevent damage to suction tube.

housing. Note position of carrier (67) on housing (8—Fig. 273), then unbolt and remove carrier. Remove the large snap ring (60), retainer (59) and the clutch discs (57 and 58). Place assembly in a press and using a straddle tool, depress clutch spring retainer (62), remove snap ring (63), then lift out retainer and clutch release spring (61). Grasp hub of piston (54) and work piston out of clutch cup (51). Place a bar across clutch cup and again apply

press to relieve pressure on the Truarc snap ring (53). Remove snap ring and pull clutch cup from the 1000 rpm shaft (46). Cover (37) can now be separated from housing (8) by removing the two short cap screws. See Fig. 275. Shafts and gears are now available for service. Brake (anti-creep) springs and pistons can be removed from rear side of housing as shown in Fig. 276. If control valve is to be disassembled, remove adjusting nut (45—Fig. 273), stop (32)

and valve cover (33) and seal (34) assembly. Remove plug (14) from bottom of housing and remove valve assembly. Pull spring (28), sleeve (31), washer (36), any shims (29) which may be present and spacer (30) from top of housing. Any further disassembly required will be obvious.

Fig. 274–Oil pump assembly which is mounted on front end of carrier assembly.

6. Suction tube	72. Drive gear
67. Carrier	73. Idler gear
71. Pump housing	76. Bearing

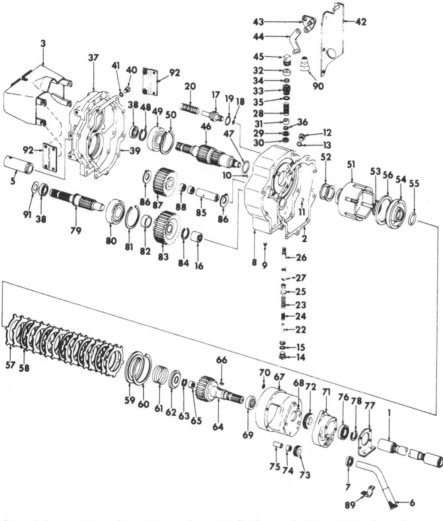

Fig. 273–Exploded view of the dual speed pto unit. Single speed units are basically similar except they do not include idler gear assembly (85 thru 88) or 540 rpm output shaft and gear unit (79 thru 84).

Fig. 275–View after cover is removed showing arrangement of gears and shafts.

46. 1000 rpm shaft	85. Idler shaft
79. 540 rpm shaft	86. Thrust washer
83. 540 rpm gear	87. Idler gear

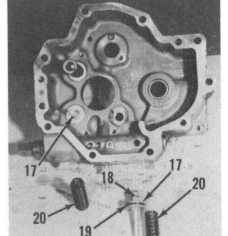

Fig. 276–Rear side of housing showing brake pistons and springs.

17. Piston	19. "O" ring
18. "O" ring	20. Spring

1. Extension shaft	26. Guide stem	48. Snap ring	71. Pump housing
2. Gasket	27. Dowel pin	49. Bearing	72. Drive gear
3. Safety shield	28. Bumper spring	50. Snap ring	73. Idler gear
5. Protection tube	29. Shim	51. Clutch cup	74. Bearing
6. Suction tube	30. Sleeve spacer	52. Seal (Teflon)	75. Shaft
7. Seal	31. Sleeve	53. Snap ring (Truarc)	76. Bearing
8. Housing	32. Stop	54. Piston	77. Bearing retainer
9. Plug	33. Cover	55. "O" ring	78. Snap ring
10. Steel ball	34. Seal	56. Seal	79. Output shaft (540 rpm)
11. Steel ball	35. "O" ring	57. Driven disc	80. Bearing
12. Plug	36. Washer	58. Drive disc	81. Snap ring
13. "O" ring	37. Cover	59. Retainer	82. Spacer
14. Plug	38. Seal	60. Snap ring	83. Driven gear (540 rpm)
15. "O" ring	39. Gasket	61. Release spring	84. Snap ring
16. Bearing	40. Plug	62. Spring retainer	85. Idler shaft
17. Brake piston	41. "O" ring	63. Snap ring	86. Thrust washer
18. "O" ring	42. Linkage support	64. Drive shaft	87. Idler gear
19. "O" ring	43. Over-center lever	65. Bearing	88. Bearing
20. Brake spring	44. Actuating link	66. Woodruff key	89. Support
22. Valve guide	45. Adjusting nut	67. Carrier	90. Dust boot
23. Spring, light	46. Output shaft (1000 rpm)	68. Dowel	91. Seal shield
24. Spring, heavy		69. Bearing	92. Extension
25. Control valve	47. "O" ring	70. "O" ring	

307. With unit disassembled, clean and inspect all parts and renew if necessary. Pay particular attention to the clutch discs which should be free of scoring or warpage. Use all new "O" rings, seals and gaskets during reassembly.

While reassembly is the reverse of disassembly, the following points are to be considered during reassembly.

When installing rear cover (37—Fig. 273) to housing (8), the brake springs (20) will hold the sections apart so it will be necessary to use longer cap screws to pull the sections together. After sections are mated, the original cap screws can be installed. Be sure step on end of idler shaft (85) is toward front of pto unit as shaft mates with edge of carrier (67).

If the Teflon seal rings (52) used on hub of clutch cup are renewed, stretch the seals on the clutch cup. After the seals are installed, clamp a ring compressor around the seals to force them back to their original size.

When installing clutch piston (54) in clutch cup (51) be sure "O" ring is in inner bore of piston and the outer seal

(56) has the lip facing away from piston hub. Use the following method to install piston in clutch cup. Obtain a piece of shim stock 4 inches wide, 18 inches long and not more than 0.002 thick. Smooth all edges of the shim stock to preclude any possibility of injuring piston outer seal, then roll the shim stock and place it in clutch cup and against bottom of cup. Lubricate piston and inside surface of shim stock liberally with oil, then start piston into clutch cup and as piston is moved into position, carefully maintain the shim stock as nearly cylindrical as possible. Remove shim stock after piston is bottomed.

Clutch pack consists of 12 steel driven discs and 5 bronze drive discs. Install discs as follows: Start with two steel discs, then add one bronze disc. Repeat this 2 steel, 1 bronze assembly until the 12 steel and 5 bronze discs are installed. When properly assembled, clutch pack will start and end with two steel discs. Install disc retainer (59) and snap ring (60). Use input shaft (64) to align discs.

When renewing bearings and seals, refer to Figs. 277 thru 282 for dimensional information.

PTO DRIVEN GEAR
All Gear Drive Models

308. The pto driven gear, located at front of clutch housing, receives its drive from a hollow shaft which is splined to the backplate (cover) of the engine clutch. Removal of the pto driven gear requires that the tractor be split between rear frame and clutch housing; at which time, gear and bearing assembly can be removed from bottom of clutch housing.

After tractor has been rejoined, bearing of the pto driven gear should be checked, and if necessary, adjusted. Refer to Fig. 283 for a schematic view showing arrangement of pto shafts. To adjust bearing, use only two cap screws and install bearing retainer without shims or seal. Turn gear and shaft and tighten the cap screws evenly to 10 in.-lbs. torque; then without turning gear and shaft, tighten the cap screws to a torque of 20 in.-lbs. Now take three measurements around bearing retainer and average these readings. See Fig. 284. Correct shim pack is gap (Z) plus 0.018-0.023. Shims are available in thicknesses of 0.007 and 0.028.

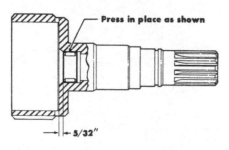

Fig. 277–Needle bearing is pressed in bore of input shaft to dimension shown.

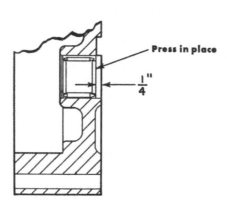

Fig. 279–The 540 rpm shaft front bearing is pressed into housing to the dimension shown.

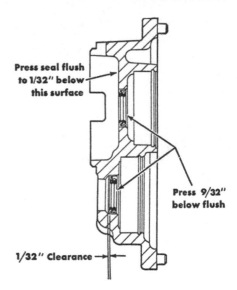

Fig. 281–View showing seal installation in pto cover. Lips are toward front.

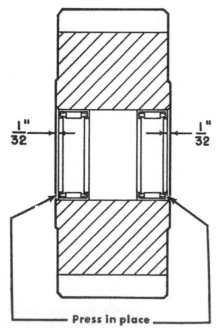

Fig. 278–Bearings are pressed into pto idler gear 1/32-inch below flush.

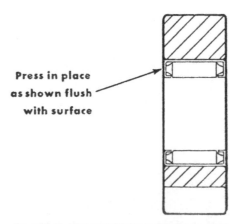

Fig. 280–Bearing installation in idler gear of pto hydraulic pump.

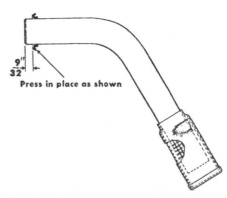

Fig. 282–Seal installation on suction tube and screen assembly.

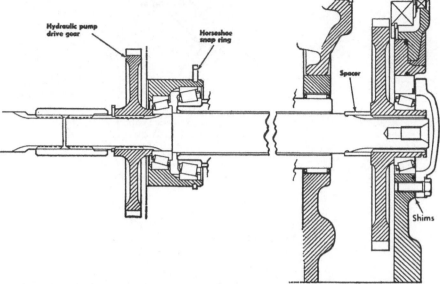

Fig. 283–Schematic view showing pto driven gear and shaft on gear drive models. Note location of adjusting shims.

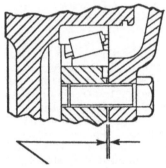

"Z" (Average of 3 readings)

Fig. 284–When adjusting bearing of pto driven gear on gear drive models, measure gap (Z) as shown. Refer to text for correct shim pack.

Remove retainer, install shims and seal ring and tighten retainer cap screws.

All Hydrostatic Drive Models

309. The pto driven gear, located at front of hydrostatic housing, is driven by the input shaft which is splined to the flex drive plate. Removal of the pto driven gear requires that tractor be split between hydrostatic housing and engine. Remove multiple control valve, bottom (front) cover and front pto bearing cover and shims. Remove the snap ring from groove on pto driven shaft at rear of pto driven gear. The driven shaft is drilled and tapped in the

Fig. 285–Using a feeler gage to measure gap when adjusting pto shaft bearings. Refer to text.

1. Feeler gage
2. PTO bearing cover

end, and can be removed using a slide hammer. Driven gear can be removed from bottom of hydrostatic housing.

Reinstall pto driven gear, driven shaft and snap ring, with rear bearing cup and bearing assembly. To adjust bearings use only two cap screws and install bearing cover without shims and "O" ring. Torque cap screws to 25 inch-pounds while rotating pto shaft. Then without rotating shaft tighten to 50 inch-pounds. Measure the gap at two places as shown (3—Fig. 285). The correct shim pack is the average of the two measurements plus 0.010 plus or minus 0.002. Remove cover, install shims and "O" ring and tighten cover cap screws.

Complete reassembly of tractor by reversing the disassembly procedure.

HYDRAULIC LIFT SYSTEM

The hydraulic lift system provides load (draft) and position control in conjunction with either a 2-point or 3-point hitch. Note: 1066 is available with only a 3-point hitch. Load control is taken from the lower links and transferred through a torsion bar and sensing linkage to the main control valve located in the hydraulic lift housing. Torsion bar and sensing linkage are located in the rear main frame. The hydraulic lift housing, which also serves as the cover for the differential portion of the tractor

rear main frame, contains the work cylinder, rockshaft, valving and the necessary linkage. Mounted on the top right side of the lift housing is a seat support which contains the system relief valve and unloading valve along with control quadrant and levers. Also attached to the inside surface of the seat support, if tractor is equipped with an auxiliary system, are the auxiliary valves (either one or two) which control the hydraulic power to trailed or front mounted equipment. The pump which supplies the hydraulic system is attached to a plate which is mounted on left side of tractor rear main frame. The pump is driven by a gear located in the rear compartment of main frame on aft end of the pto driven shaft. All oil used by the hydraulic components of the tractor is drawn through a filter located in the rear main frame directly across from the hydraulic lift system pump.

TROUBLE SHOOTING
All Models

310. The following are symptoms which may occur during the operation of the hydraulic lift system. By using this information in conjunction with the Check and Adjust information and the R&R and Overhaul information, no trouble should be encountered in servicing the hydraulic lift system.

1. Hitch will not lift. Could be caused by:
   a. Unloading valve orifice plugged or piston sticking.
   b. Unloading valve ball not seating or seat loose.
   c. Faulty main relief valve.
   d. Faulty cushion relief valve.
   e. Faulty or disconnected internal linkage.
2. Hitch lifts when auxiliary valve is actuated. Could be caused by:
   a. Unloading valve orifice plugged.
   b. Unloading valve piston sticking.
   c. Unloading valve ball not seating or valve seat loose.
3. Hitch lifts load too slowly. Could be caused by:
   a. Unloading valve seat leaking.
   b. Excessive load.
   c. Faulty main relief valve.
   d. Faulty cushion relief valve.
   e. Scored work cylinder or piston or piston "O" ring faulty.
4. Hitch will not lower. Could be caused by:
   a. Drop piston sticking or "O" ring damaged.
   b. Control valve spool sticking or spring faulty.
   c. Drop check valve piston sticking.
5. Hitch lowers too slowly. Could be caused by:
   a. Action control valve spool sticking.

b. Action control valve linkage maladjusted.

c. Drop check valve "O" ring damaged or pilot ball cage maladjusted.

6. Hitch lowers too fast with position control. Could be caused by:
   a. Action control valve malfunctioning.

7. Hitch will not maintain position. Could be caused by:
   a. Drop check valve in main control valve leaking.
   b. Work cylinder or piston scored or piston "O" ring damaged.
   c. Check valve pilot valve leaking or ball cage maladjusted.
   d. Cushion valve leaking or damaged.

8. Hydraulic system stays on high pressure. Could be caused by:
   a. Linkage maladjusted, broken or disconnected.
   b. Auxiliary valve not in neutral.
   c. Mechanical interference.

9. Hitch over-travels (depth variation). Could be caused by:
   a. Torsion bar linkage sticking and needs lubrication.
   b. Unloading valve orifice partially plugged.

10. Draft (load) sensing too rapid in slow action position. Could be caused by:
    a. Action control linkage improperly adjusted.

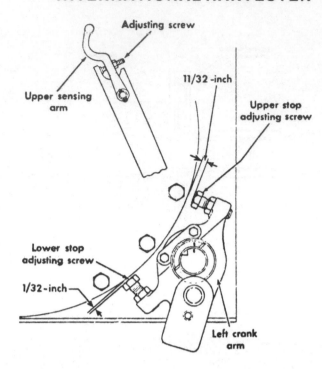

Fig. 286—Torsion bar left crank arm showing stop screw adjustments for 826 and 1026.

TEST AND ADJUST

All Models

Before proceeding with any testing or adjusting, be sure the hydraulic pump is operating satisfactorily, hydraulic fluid level is correct and filter is in good condition. All tests should be conducted with hydraulic fluid at operating temperature which is normally 120-180 degrees F. Cycle system if necessary to insure that system is completely free of air.

311. **RELIEF VALVE.** On tractors equipped with an auxiliary hydraulic system, the relief valve can be tested as follows: Attach an IH Flo-Rater, or similar flow rating equipment, to any convenient outlet from an auxiliary control valve and be sure outlet hose from test unit is securely fastened in the hydraulic system reservoir. Start engine and run at rated speed. Manually hold auxiliary control valve in operating position, close valve of test unit and note the gage reading which should be 1450-1700 psi for model 826, and 1900-2100 psi for models 766, 966, 1026 and 1066. If pressure is not as stated, renew relief valve which is available only as a unit.

At this time, the hydraulic lift system pump delivery and the auxiliary control valve detent (latching) mechanism can also be checked.

At the engine rated rpm the hydraulic lift pump should deliver 12 gpm at 1450-1700 psi for model 826, and 12 gpm at 1900-2100 psi for models 766, 966, 1026 and 1066.

To check the auxiliary control valve detent assembly, run engine at low idle rpm, pull control valve lever into operating position until it latches, then slowly close shut-off valve of the test unit and observe the pressure gage. Valve control lever should unlatch and return to neutral at not less than 1150 psi nor more than 1350 psi for model 826, or not less than 1550 psi nor more than 1750 psi for models 766, 966, 1026 and 1066. If detent assembly does not operate properly, refer to paragraph 337.

NOTE: On tractors which are not equipped with auxiliary control valves, it will be necessary to remove the relief valve and bench test it in order to check its condition. To check the relief valve in this manner will require use of a hydraulic hand pump, gage and a test body such as an IH FES64-7-1 test body. Bear in mind that a relief valve tested in this manner will show a test pressure that will be on the low side of the pressure range due to the low volume of oil being pumped.

312. **QUADRANT LEVERS (EARLY 826-1026).** To test and adjust system, first check quadrant levers (load and position control). These levers should require 4-6 pounds of force, applied at knob, to move levers. If adjustment is necessary, adjust load control lever with nut on outer end of lever shaft. This nut is reached through a hole at bottom of quadrant. Adjust po-

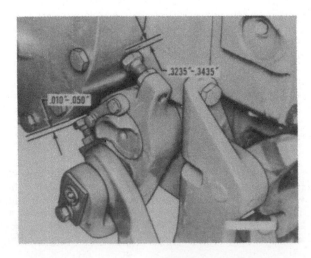

Fig. 287—Torsion bar left crank arm showing stop screw adjustments for 766 and 966.

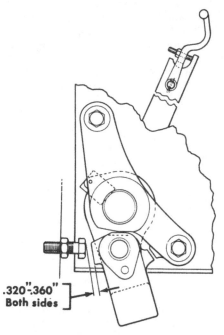

Fig. 288–Torsion bar adjustment for model 1066.

.320"-.360"
Both sides

sition control lever by working through opening in bottom of quadrant support. Loosen lock (ring) nut and turn the inner nut as required. A small punch can be used in the holes, provided in nuts to make the position control lever adjustment.

313. QUADRANT LEVERS (LATE 826-1026 AND ALL OTHER MODELS). To test and adjust system, first check quadrant levers (load and position control). These levers should require 4-8 pounds of force, applied at knob, to move levers. If adjustment is necessary, adjust the levers by tightening or loosening the cap screws in the bearing support, which are located between the levers at the end of quadrant support.

314. CONTROL LINKAGE. To adjust the control linkage, first check and adjust, if necessary, the clearance for the torsion bar crank arm stop bolts. Refer to Figs. 286, 287 and 288, and adjust lower stop screw until the distance between head of stop screw and final drive housing is as shown.

Fig. 289–View showing location of turnbuckle (A) in right hand seat support.

Adjust upper stop screw to obtain a clearance as shown.

Make certain there is no weight on hitch, then proceed as follows: Move position control lever forward to "LOWER" position (at offset in quadrant before action control section). Remove upper link bracket (3-point hitch) or plate (2-point hitch) to allow access to the sensing arm. Turn sensing arm adjusting screw in (clockwise) 3 or 4 turns. Place the load (draft) control lever 2⅞ inches on models 766, 826 and 966 and 2 inches on models 1026 and 1066 from rear of slot in quadrant. Start engine and turn sensing arm adjusting screw out (counter-clockwise) until the hitch starts to raise. NOTE: Hitch will raise all the way. Install upper link bracket or plate.

Move load control lever to full forward position in slot on quadrant. Slowly move position control lever from rear of quadrant forward until hitch reaches fully lowered position. Remove seat support side cover and adjust turnbuckle (A—Fig. 289) until position control lever is at the offset "LOWER" position on the quadrant. CAUTION: With seat support side cover removed, run engine at low idle rpm only. Install cover.

With load control lever fully forward, move position control lever toward "RAISE" position until the following measurements between centerline of hole in outer right torsion bar crank

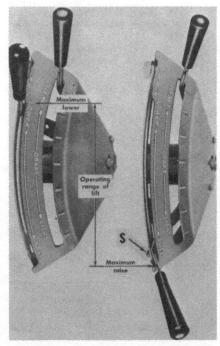

Fig. 290–When maximum raise measurements are correct, secure stop (S) against position control lever.

arm and centerline of hole in rockshaft lift arm are obtained:

2-point hitch . . . . 37⅞ inches
3-point hitch . . . . 39¾ inches
3-point hitch (1066) . . . 39 3/16 inches

When correct measurement is obtained, adjust stop (S—Fig. 290) against control lever and tighten securely.

Move position control lever to bottom of action control section (fully forward). Remove cover from front of lift housing. Adjust nut (A—Fig. 291) until bushing is bottomed against valve body and continue turning nut (A) until position control lever moves ⅛-inch rearward away from front end of slot. Install the cover.

315. DROP CHECK VALVE SEAT ADJUSTMENT. The drop valve plug (30—Fig. 292) is bottomed in the seat and must be left in that position.

Check and adjust the drop valve seat as follows: Attach an implement, or weights, to hitch and start engine. Place the load control lever in the heavy (forward) position. Move the position control lever toward raise until hitch has taken the attached weight, then move lever about 1-inch more in the same direction and mark the position of lever on quadrant. Now slowly move lever toward lower position until hitch just starts to lower. Measure distance the position control lever has moved. This distance should be ⅜-inch. If measurement is not ⅜-inch, remove front cover from lift housing and adjust drop valve seat as follows: Loosen lock nut and turn valve seat in (clockwise) if measurement was more than ⅜-inch; or turn seat out (counter-clockwise) if measurement was less than the ⅜-

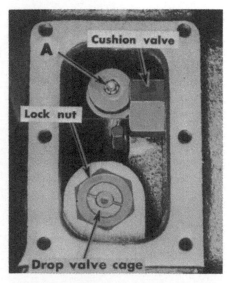

Fig. 291–With cover removed from front of left housing, units shown are available for adjustment or service. (A) is esna nut on forward end of action control valve actuating rod.

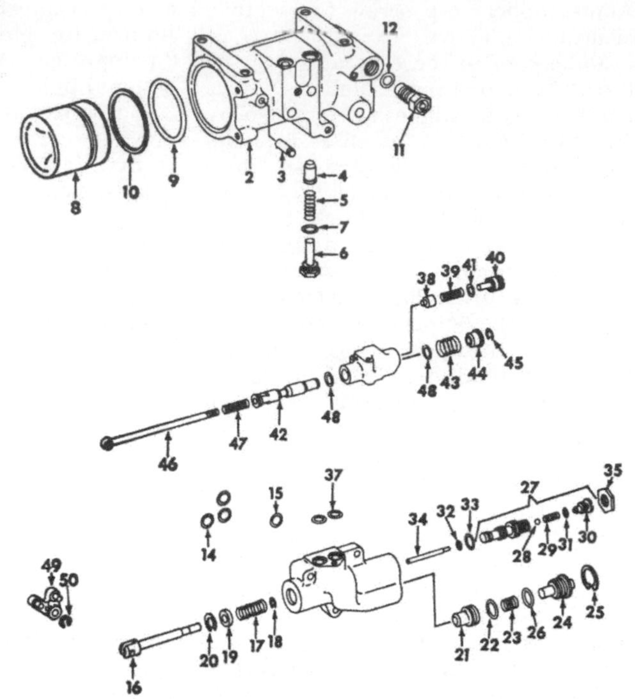

Fig. 292–Exploded view of work cylinder, action control valve and main control valve assemblies.

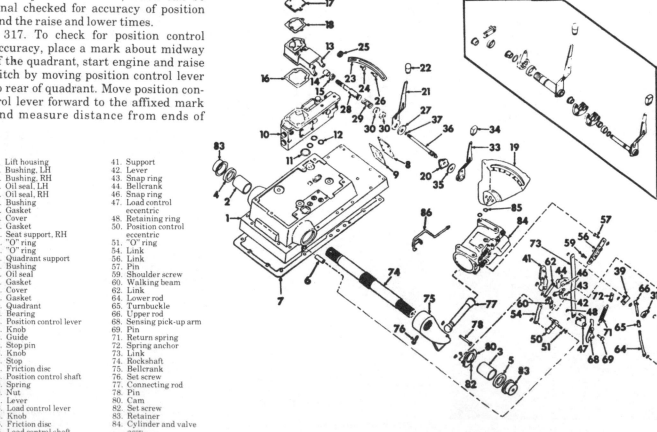

2. Cylinder
3. Lever pin
4. Check valve poppet
5. Spring
6. Plug
7. "O" ring
8. Piston
9. "O" ring
10. Back-up washer
11. Cushion valve
12. "O" ring
14. "O" ring
15. "O" ring
16. Spool actuator
17. Spring
18. Retaining ring
19. Spring retainer
20. Snap ring
21. Drop poppet valve
22. "O" ring
23. Spring
24. Valve plug
25. Snap ring
26. "O" ring
27. Drop valve seat assy.
28. Ball
29. Spring
30. Plug
31. "O" ring
32. "O" ring
33. "O" ring
34. Actuating rod
35. Lock nut
37. "O" ring
38. Piston
39. Spring
40. Plug
41. "O" ring
42. Variable orifice spool

43. Return spring
44. Bushing
45. Snap ring
46. Actuating rod

47. Spring
48. "O" ring
49. Bellcrank
50. Snap ring

inch. Turn seat in increments of ¼-turn and note that the drop valve seat has the two slots on its outer diameter.

316. POSITION CONTROL AND CYCLE TIME. With the above tests and adjustments made, hitch can be final checked for accuracy of position and the raise and lower times.

317. To check for position control accuracy, place a mark about midway of the quadrant, start engine and raise hitch by moving position control lever to rear of quadrant. Move position control lever forward to the affixed mark and measure distance from ends of

lower links to ground. Repeat the operation and again measure the distance from ends of lower links to ground. These two measurements should not

vary more than ⅛-inch. Now push position control lever forward to lower hitch, then move it rearward to the affixed mark. The measurement between ends of lower links and ground should not vary more than 1-inch from the measurements obtained in the first test. If differences are excessive, refer to paragraph 321.

318. To check hitch raise and lower times, be sure hydraulic fluid is at operating temperature and load hitch with an implement or weights. To check raise time, start engine and run at high idle rpm. Be sure hitch is in lowest position, then quickly move position control lever rearward to raise position. Hitch should reach full raise in three seconds or less.

319. To check the minimum lowering (drop) time, move the load control lever to the light load (rear) position and the position control lever forward to offset of quadrant. Now move the load control lever forward to the heavy position. Hitch should completely lower in two seconds or less.

320. To check the maximum lowering (drop) time, repeat the minimum time operation given in paragraph 319 except the position control lever should be in the extreme forward slow action position. Hitch drop time should be six seconds or more.

1. Lift housing
2. Bushing, LH
3. Bushing, RH
4. Oil seal, LH
5. Oil seal, RH
6. Bushing
7. Gasket
8. Cover
9. Gasket
10. Seat support, RH
11. "O" ring
12. "O" ring
13. Quadrant support
14. Bushing
15. Oil seal
16. Gasket
17. Cover
18. Gasket
19. Quadrant
20. Bearing
21. Position control lever
22. Knob
23. Guide
24. Stop pin
25. Knob
26. Stop
27. Friction disc
28. Position control shaft
29. Spring
30. Nut
31. Lever
33. Load control lever
34. Knob
35. Friction disc
36. Load control shaft
37. "O" ring
39. Lever

41. Support
42. Lever
43. Snap ring
44. Bellcrank
46. Snap ring
47. Load control eccentric
48. Retaining ring
50. Position control eccentric
51. "O" ring
54. Link
56. Link
57. Pin
59. Shoulder screw
60. Walking beam
62. Link
64. Lower rod
65. Turnbuckle
66. Upper rod
68. Sensing pick-up arm
69. Pin
71. Return spring
72. Spring anchor
73. Link
74. Rockshaft
75. Bellcrank
76. Set screw
77. Connecting rod
78. Pin
80. Cam
82. Set screw
83. Retainer
84. Cylinder and valve assy.
85. Seal ring
86. Lubrication tube

Fig. 293–Exploded view of hydraulic lift housing showing relative position of component parts. Inset showing controls used on late 826-1026 and all other models.

321. If position control accuracy and the hitch raise and lower times are not satisfactory, remove lift housing from rear main frame and inspect internal linkage. A visual inspection is sufficient to find any linkage defects. Renew any linkage or pins showing excessive wear or damage. Also check the main valve return spring and the main valve spool for binding. Spring can be renewed; however, if valve spool is defective, renew the spool and body assembly.

HYDRAULIC LIFT HOUSING

All Models

322. **R&R AND OVERHAUL.** To remove the lift housing, remove seat and platform, then if tractor is so equipped, disconnect the flexible hoses from the auxiliary valve lines. Disconnect lift links from rockshaft arms and remove rockshaft arms. Disconnect the rear break-away coupling bracket and the pto lever control rod. Remove attaching cap screws and attach a hoist to housing so that unit will balance as it is removed. Carefully lift assembly from main frame and if available, mount unit in an engine stand.

NOTE: At this time, quadrant control levers and sensing pick-up arm (68 —Fig. 293) are susceptible to damage. If no engine stand is available, it is recommended that the right seat support, control lever quadrant assembly and the auxiliary control valves and tubing (if so equipped) be removed as a unit. The lower ends of control lever rods must be disconnected before assembly can be removed from lift housing. Also remove the left seat support.

323. Disconnect and remove sensing (return) spring (71), then remove pickup arm (68). Disconnect link (56) from main control valve spool. The complete valve and work cylinder assembly can now be unbolted and removed from housing.

With assembly removed, the action control valve can be removed from main control valve and main control valve removed from the cylinder and the procedure for doing so is obvious.

When reinstalling cylinder and valve assembly in housing, install the mounting cap screws loosely; then use a small bar to hold cylinder assembly against mounting boss in housing and tighten the mounting cap screws securely. Complete reassembly by reversing the disassembly procedure and adjust system as outlined in paragraphs 311 through 320.

324. **WORK CYLINDER AND PISTON.** To disassemble the cylinder assembly after lift housing is removed as outlined in paragraph 322, first remove the lubrication tube, then remove piston (8—Fig. 292) by bumping open end of cylinder against a wood block. Cushion valve (11) and the check valve assembly (items 4, 5, 6 and 7) can also be removed. If necessary, lever pin (3) can also be renewed.

Inspect cylinder and piston for scoring or wear. Small defects may be removed using crocus cloth. Pay particular attention to the piston "O" ring and back-up ring as well as the check valve poppet. If any doubt exists as to the condition of these items, be sure to renew them during assembly. Cushion valve (11) can be bench tested by using a hydraulic hand pump, gage, test body IH No. FES64-7-1, adapter IH No. FES 64-7-2 and petcock IH No. FES 64-7-4. Valve should test 1600-1800 psi. Cushion valve adjusting screw is heavily staked in position and valve cannot be disassembled. If valve does not meet the above specifications renew the complete valve.

NOTE: When checking valve with hand pump bear in mind that the relief pressure obtained will be on the low side of the pressure range due to the low volume of oil being pumped.

Coat parts with IH "Hy-Tran" fluid before reassembly. Piston "O" ring is installed nearest closed end of piston.

325. **MAIN CONTROL VALVE.** To disassemble the main control valve after lift housing is removed as outlined in paragraph 322, first remove plug (30—Fig. 292), spring (29) and ball (28) from end of check valve ball seat. Loosen lock nut (35), then while counting the number of turns, remove the valve seat. Remove snap ring (20) and pull spool assembly (16) from valve body. Compress return spring, remove snap ring (18) and disassemble spool assembly. Remove snap ring (25) and pull plug (24), spring (23) and poppet valve (21) from valve body.

Clean and inspect all parts. Spool return spring (17) has a free length of 2 21/64 inches and should test 18.4-21.6 pounds when compressed to a length of 1 29/32 inches. Check ball spring (29) has a free length of 59/64-inches and should test 3.5-4.1 pounds when compressed to a length of ¾-inch. Drop poppet valve spring (23) has a free length of 1 11/32-inch and should test 5 pounds when compressed to a length of 63/64-inch. If valve spool or spool bore show signs of scoring or excessive wear, renew complete valve assembly. All other parts are available for service.

Use all new "O" rings, dip all parts in IH "Hy-Tran" fluid and reassemble by reversing the disassembly procedure, keeping the following points in mind. Retainer (19) is placed on actuator (16)

with relieved (chamfered) side toward spring (17). Drop valve seat is turned into valve body the same number of turns as were counted during removal. Tighten the lock nut (35) to 45 ft.-lbs. torque and plug (30) to 10 ft.-lbs. torque.

After assembly, mount valve on cylinder and piston assembly.

326. **ACTION CONTROL VALVE.** To disassemble the action control valve after lift housing is removed as outlined in paragraph 322, count the turns and remove the esna nut from end of actuating rod (46—Fig. 292) and remove rod and spring (47). Remove retainer ring (45), then remove bushing (44), return spring (43) and variable orifice spool (42). The "O" ring (48) is located in I.D. of bushing bore. Remove plug (40), spring (39) and the drop control piston (38). Remove "O" ring from I.D. of orifice spool bore.

Clean and inspect all parts and renew any which show signs of excessive scoring or wear. Spool (42) and piston (38) should be a snug fit yet slide freely in their bores.

Use all new "O" rings, dip all parts in IH "Hy-Tran" fluid and reassemble by reversing the disassembly procedure. Count the turns when installing the esna nut on end of actuating rod so nut will be installed to original position.

After assembly, mount action control valve on the main control valve.

327. **ROCKSHAFT.** Renewal of rockshaft or rockshaft bushings requires removal of the hydraulic lift housing; however, if only the rockshaft seals are to be renewed, it is possible to pry seals

Fig. 294–View of hydraulic lift assembly bottom side after removal.

out of their bores after removing the rockshaft lift arms. Seals are installed with lips toward inside and are driven into bores until bottomed. Dip seals in IH "Hy-Tran" fluid prior to installation.

With the lift housing removed, the rockshaft can be removed as follows: Remove set screws (76 and 82—Fig. 293) and slide rockshaft out right side of housing. Bushings (2 and 3) can now be driven out of housing.

Rockshaft O. D. at bearing surface is larger on right side than on left. Rockshaft bearing surface O. D. is 3.020-3.022 for right side and 2.780-2.782 for left side. Rockshaft bushing I. D. is 3.023-3.028 for right side and 2.783-2.788 for left side. Rockshaft has 0.001-0.008 operating clearance in bushings. Use a piloted driver when installing bushings and install bushings with outer ends flush with bottom of seal counterbore.

When installing rockshaft, start rockshaft in right side of housing and align master splines of rockshaft, actuating hub (80) and bellcrank (75). Tighten hub and bellcrank set screws.

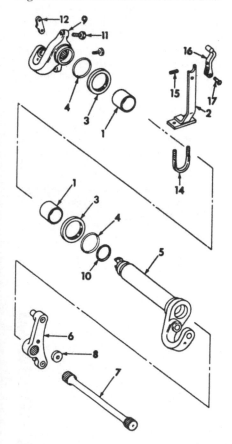

Fig. 295—View of lower load (draft) sensing mechanism used on all models except 966 hydrostatic and all 1066 tractors.

1. Bushing assembly	9. Crank arm, LH
2. Lower sensing arm	10. "O" ring
3. "O" ring retainer	11. Stop screw
4. "O" ring	12. Retainer
5. Crank arm, RH	14. "U" bolt
6. Anchor bracket	15. Adjusting screw
7. Torsion bar	16. Upper sensing arm
8. Retainer	17. Pin

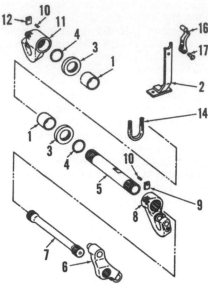

Fig. 296—View of lower load (draft) sensing mechanism used on models 966 hydrostatic and all 1066 tractors.

1. Bushing assembly	9. Retainer
2. Lower sensing arm	10. Pin
3. Retainer	11. Crank arm L.H.
4. "O" ring	12. Block L.H.
5. Shaft	14. "U" bolt
6. Anchor bracket	16. Upper sensing arm
7. Torsion bar	17. Pin
8. Crank arm R.H.	

328. TORSION BAR ASSEMBLY. To remove the torsion bar assembly, it is necessary to drain the rear main frame and remove the pto unit, if so equipped. Remove "U" bolt (14—Fig. 295 or 296) and remove sensing arm assembly. Remove retainer (12) and left hand crank arm (9). Model 1066 tractors will have a roll pin and retainer plate. Remove anchor bracket (6) and be sure not to lose retainer (8). On model 1066, remove roll pin and retainer plate. Right hand crank (5) and torsion bar (7) assembly can now be withdrawn from rear frame. Do not let torsion bar slide out of right hand

crank. Any chips, dents or scratches could set up stress points which might cause a torsion bar failure. Bushing case and bushing (1), "O" ring retainer (3) and "O" ring (4) renewal can now be accomplished and procedure for doing so is obvious.

Reassemble by reversing the disassembly procedure.

329. SEAT SUPPORT. The right hand seat support contains the system relief valve and an unloading valve and in most cases, these components can be removed for service without removing seat support.

To remove the unloading valve, remove plug (18—Fig. 297), unloading valve seat (19) and piston (21). The seat, ball, follower and spring can be removed from carrier for cleaning; however, unit is available only as an assembly. Relief valve (28) can be removed at any time. Relief valve adjusting plug is heavily staked in place and cannot be removed. Faulty relief valves must be renewed and valves can be identified by having a "1600" stamped on head of valve used on model 826 or a "2000" stamped on head of valve used on models 766, 966, 1026 and 1066.

Check the unloading valve piston (21) as follows: Shake piston. Piston must not rattle nor should rod be loose or easy to turn. Place unit in a soft jawed vise and compress. Piston unit must compress at least 0.045. Blow through orifices in piston to insure that they are open and clean. If piston does not meet all of these conditions, renew the complete unit.

Reassemble by reversing the disassembly procedure.

330. QUADRANT AND LEVER ASSEMBLY (EARLY 826-1026). The quadrant and lever assembly can be

Fig. 297—Exploded view of right seat support when tractor is equipped with load control hydraulic lift system. Item 33 is used for simultaneous lift.

4. Plug (without simultaneous lift)	
8. Seat support	
9. "O" ring	
10. Pipe plug	
11. "O" ring	
14. "O" ring	
15. "O" ring	
16. Cover	
17. Gasket	
18. Plug	
19. Unloading valve seat	
20. "O" ring	
21. Piston	
22. Plug	
25. "O" ring	
26. "O" ring	
27. Back-up washer	
28. Relief valve	
29. "O" ring	
33. Plug (with simultaneous lift)	

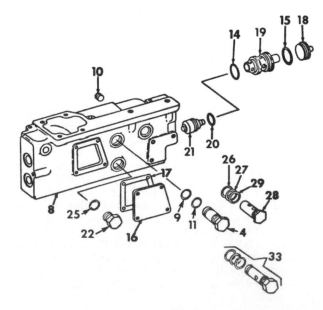

removed and disassembled after removal of tractor seat, and if desired, the right fender.

Remove lever knobs, then unbolt quadrant and slide it off levers. Use two small punches to loosen lock nuts (B—Fig. 298) to relieve spring tension. Remove cover plate (C) and side plate (E). Move lever to place rod (F) in uppermost position, then use a pair of vise grips through side opening to hold rod in this position. Remove retaining rings (G). Remove the two cap screws (D) and Allen head screws located under cover (C). Control lever assemblies can now be pulled out for service. Do not mix friction discs or alter position control linkage adjustment. If bushing or seal in quadrant support

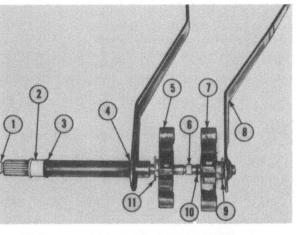

Fig. 300–Exploded view of the control handle assemblies removed from quadrant support. Note the cap screws in the bearing supports (5 and 7) used to adjust the pull on the control levers.

1. Bearing
2. Bearing
3. "O" ring
4. Position control handle
5. Bearing support
6. Bearing
7. Bearing support
8. Draft control handle
9. Bearing
10. "O" ring
11. Bearing

need renewal, remove support (J) from seat support.

When reassembling, note all marked splines and adjust friction disc tension as outlined in paragraph 312.

331. QUADRANT AND LEVER ASSEMBLY (LATE 826-1026 AND ALL OTHER MODELS). The quadrant and lever assembly can be removed as follows: Remove seat support rear cover, control handle knobs, quadrant sector and the cover from quadrant support. Refer to Fig. 299, and drive roll pin (1) out of draft control lever being very careful not to drop pin. Remove retaining rings (2 and 5) and the screws securing quadrant support (6) to seat support. Remove link (3) from draft control lever (4) and then remove lever and spacer. Remove control quadrant assembly from seat support. Remove the position control link, spacer and pull control handle assemblies out of quadrant support. Refer to Fig. 300 and loosen the cap screws in bearing supports (5 and 7) and disassemble the control handles. Inspect shaft, bearings and "O" rings and renew if necessary.

Reassemble by reversing the disassembly procedure. Be sure and note all marked splines and adjust friction tension on levers as outlined in paragraph 313.

HYDRAULIC LIFT PUMP
All Models

332. The pump for the hydraulic lift system is located in the left forward end of the differential portion of tractor rear main frame. Pump is driven from a gear mounted on aft end of the pto driven shaft. See Fig. 301. Pump can be either 12 gpm or 17 gpm capacity on models 826 and 1026 and only a 12 gpm on models 766, 966 and 1066 tractors and the basic difference between the two pumps is the width of gear teeth. The 17 gpm pump is generally installed on International model tractors which are being used for industrial purposes.

333. R&R AND OVERHAUL. To remove the hydraulic pump, drain the tractor rear main frame, then unbolt and remove mounting plate, spacer and pump. Pump and spacer can now be separated from mounting plate by removing the attaching cap screws. See Fig. 302.

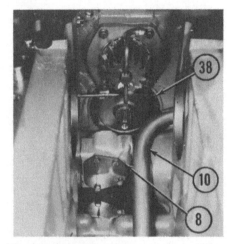

Fig. 301–Hydraulic lift system pump (8) is mounted at left front of differential compartment. Differential has been removed for illustrative purposes.

10. Suction tube   38. Drive gear

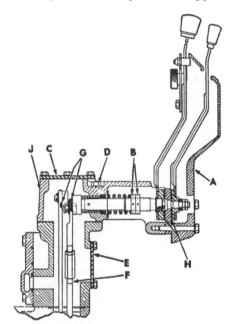

Fig. 298–Cross-sectional view of the quadrant and support assembly used on early 826-1026 tractors.

A. Quadrant
B. Lock nuts
C. Cover
D. Cap screw
E. Cover
F. Lower rod
G. Retainer rings
H. Friction discs
J. Support

Fig. 299–View showing the quadrant support and linkage connecting points used on late 826-1026 and all other models.

1. Roll pin
2. Retaining ring
3. Eccentric (to lever link)
4. Draft control lever
5. Retaining ring
6. Quadrant support

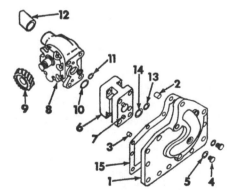

Fig. 302–View of hydraulic lift pump, spacer and mounting flange.

1. Mounting flange
2. Dowel
3. Dowel
4. Plug
5. "O" ring
6. Spacer
7. Dowel
8. Pump
9. Drive gear
10. "O" ring
11. "O" ring
12. Suction tube
13. "O" ring
14. "O" ring
15. Gasket

334. OVERHAUL CESSNA. With pump removed from spacer, proceed as follows: Remove pump drive gear and key, then unbolt and remove covers (2 and 14—Fig. 303). Balance of disassembly will be obvious after an examination of the unit.

Pump specifications are as follows:

12 GPM Pump

O.D. of shafts at bushings
(min.) 0.810
I.D. of bushings in body and
cover (max.) 0.816
Thickness (width) of gears
(min.) 0.572
I.D. of gear pockets (max.) 2.002
Max. allowable shaft to
bushing clearance 0.006

17 GPM Pump

O.D. of shafts at bushings
(min.) 0.810
I.D. of bushings in body and
cover (max.) 0.816
Thickness (width) of gears
(min.) 0.813
I.D. of gear pockets (max.) 2.002
Max. allowable shaft to
bushing clearance 0.006

When reassembling, use new diaphragms, gaskets, back-up washers, diaphragm seal and "O" rings. With open part of diaphragm seal (5) towards cover (2), work same into grooves of cover using a dull tool. Press protector gasket (6) and back-up gasket (7) into the relief of diaphragm seal. Install check ball (3) and spring (4) in cover, then install diaphragm (8) inside the raised lip of the diaphragm seal and be sure bronze face of diaphragm is toward pump gears. Dip gear and shaft assemblies in oil and install them in cover. Position wear plate (15) in pump body with the bronze side toward pump gears and cut-out portion toward inlet (suction) side of pump. Install pump body over gears and shafts and install retaining cap screws. Torque cap screws to 20 ft.-lbs. for the 12 gpm pump or to 25 ft.-lbs. for the 17 gpm pump.

Check pump rotation. Pump will have a slight amount of drag but should turn evenly.

335. OVERHAUL THOMPSON. With pump removed from spacer, proceed as follows: Remove pump drive gear and key (8—Fig. 304), then unbolt and remove pump covers (2 and 15). Bearings (7), pressure plate spring (6), "O" ring retainers (5), "O" rings (4), back-up washers (3) and oil seal (1) can now be removed from cover. Note location of bearings (7) so they can be reinstalled in the same position. Remove "O" rings (11 and 12), wear plate (16) and the pump gears and shafts (9) from pump body. Wear plate (16) is installed

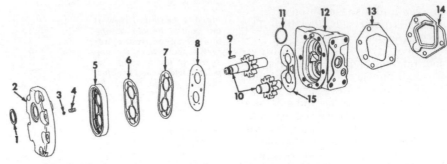

Fig. 303–Exploded view of Cessna pump. Pump may be either 12 or 17 gpm capacity.

1. Oil seal	5. Diaphragm seal	8. Pressure diaphragm	12. Body
2. Cover	6. Protector gasket	9. Key	13. Gasket
3. Ball	7. Back-up gasket	10. Gears and shafts	14. Rear cover
4. Spring		11. "O" ring	15. Wear plate

with reliefs toward pressure side of pump.

Pump gears and shafts, as well as the pump shaft bearings, are available only in sets. Except for suction port "O" rings (12), none of the gaskets or "O" rings are available separately. However, gaskets, seal and "O" rings are available in a seal kit.

Pump specifications are as follows:

12 GPM Pump

O.D. of shafts at bearings (min.) 0.812
I.D. of bearings in body and
cover (max.) 0.816
Thickness (width) of gears
(min.) 0.7765
I.D. of gear pockets (max.) 1.772
Max. allowable shaft to bearing
clearance 0.004

17 GPM Pump

O.D. of shafts at bearings (min.) 0.812
I.D. of bearings in body and
cover (max.) 0.816
Thickness (width) of gears
(min.) 1.072
I.D. of gear pockets (max.) 1.772
Max. allowable shaft to bearing
clearance 0.004

Lubricate all parts during assembly, use all new gaskets and seals and be sure bearings in cover are reinstalled in their original positions, if same bearings are being installed. Tighten cover to body cap screws to a torque of

20 ft.-lbs. for the 12 gpm pump, or 30 ft.-lbs. for the 17 gpm pump.

Check pump rotation. Pump will have a slight amount of drag but should turn evenly.

AUXILIARY CONTROL VALVE
All Models

336. Tractors may be equipped with either single or double control valve auxiliary hydraulic systems. Control valves are mounted on inner side of the right seat support and all valves used are similar. Therefore, only one valve will be discussed.

Hydraulic flow for the auxiliary valves is supplied by the same pump that provides fluid for the implement hitch. All of the hydraulic fluid returning to the reservoir passes through the center ports of the auxiliary valve when valve is in neutral. However, when valve is actuated, the flow through this port is stopped, the fluid directed to one of the power beyond outlets and the system immediately goes on pressure.

337. R&R AND OVERHAUL. To remove the auxiliary control valve, or valves, first remove seat and base as an assembly. Remove the front and rear support covers and the left seat support. Remove unions and separate

Fig. 304–Exploded view of Thompson pump. Pump may be either 12 or 17 gpm capacity.

1. Oil seal
2. Cover
3. Back-up washer
4. "O" ring
5. Retainer
6. Pressure plate spring
7. Bearings
8. Key
9. Gears and shafts
11. "O" ring
12. "O" ring
13. Body
14. Gasket
15. Rear cover
16. Wear plate

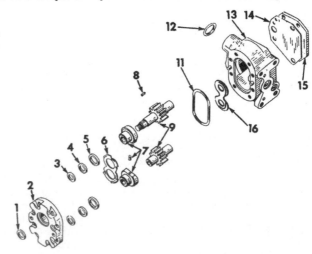

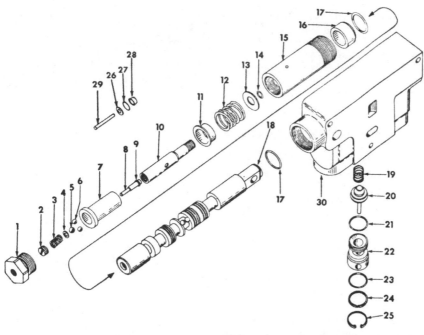

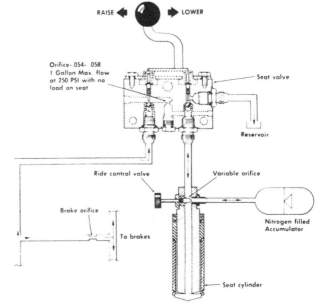

bling these valves, sleeve (15) must be removed before removing actuator (10).

Detent (3, 4, 5 and 6) can be disassembled after removing plug (2). Push unlatch piston (8) out of actuator (10) with a long thin punch.

NOTE: Unlatch piston (29), back-up washer (26), "O" ring (27) and plug (28) are used on models 766, 966, 1026 and 1066 instead of unlatch piston (8).

Inspect all parts for nicks, burrs, scoring and undue wear and renew parts as necessary. Spool (18) and body (30) are not available separately. Check detent spring (3) and centering spring (12) against the following specifications:

Detent spring
   Free length-inches 1 1/16
   Test load lbs. at
     length-in ...... 23.5-28.5 @ 45/64
Centering spring
   Free length-inches 2 5/16
   Test load lbs. at
     length-in. .... 26.5-33.5 @ 1-7/64

Use all new "O" rings and reassemble by reversing the disassembly procedure. Detent unlatching pressure is adjusted by plug (2). Unit must unlatch at not less than 1150 psi nor more than 1350 psi for model 826, or not less than 1550 psi nor more than 1750 psi for models 766, 966, 1026 and 1066.

Fig. 305–Exploded view of auxiliary control valve. Unlatching piston (29), back-up washer (26), "O" ring (27) and plug (28) are used on all models except 826.

1. Cap	9. "O" ring	16. "O" ring retainer	23. "O" ring
2. Adjusting plug	10. Actuator	17. "O" ring	24. Back-up washer
3. Detent spring	11. Spring retainer	18. Spool	25. Snap ring
4. Washer	12. Centering spring	19. Poppet spring	26. Back-up washer
5. Actuating ball	13. Washer	20. Poppet	27. "O" ring
6. Detent ball (3 used)	14. "O" ring	21. "O" ring	28. Plug
7. Sleeve	15. Sleeve	22. Check valve retainer	29. Unlatching piston
8. Unlatching piston			30. Valve body

tubing from top of valve. Disconnect linkage from valve spool. Remove the attaching cap screws, hold tubing up away from valve and remove valve from tractor.

When reinstalling, tighten the mounting cap screws to 20-25 ft.-lbs. torque in 5 ft.-lbs. increments. All

mounting cap screws must be tightened exactly the same.

To disassemble, use Fig. 305 as a guide. Remove end cap (1), then unscrew the actuator (10) and remove the actuator and detent assembly. Remove sleeve (15) and pull balance of parts from body. Check valve retainer (22) and poppet (20) assembly can be removed after removing snap ring (25).

NOTE: Some valves may not include the detent assembly. When disassem-

CHECK VALVE
All Models

338. A double acting check valve is used with auxiliary system rear outlet which checks the flow of fluid in both directions and precludes the possibility of equipment dropping either during transport or while parked.

Removal and disassembly of the unit will be obvious upon examination of the unit and reference to Fig. 306.

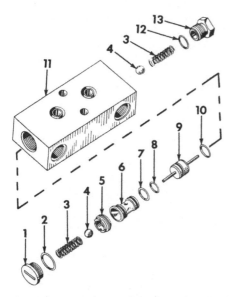

Fig. 306–Exploded view of double acting check valve assembly.

1. Plug	
2. "O" ring	8. "O" ring
3. Spring	9. Piston
4. Ball	10. "O" ring
5. Nut	11. Block
6. Retainer	12. "O" ring
7. "O" ring	13. Plug

Fig. 307–Schematic view of hydraulic circuit, control valve and cylinder of the hydraulically controlled seat attachment.

HYDRAULIC SEAT

All Models So Equipped

339. A hydraulic controlled seat attachment is available. Fluid is supplied from a tee connection in the brake supply line. See Fig. 307. Fluid flows to the seat control valve only when seat is being raised. The speed of raise is controlled by a 0.054-0.058 drilled orifice in the control valve. Return oil from the single action cylinder is dumped back in reservoir. Seat ride control is adjustable and is controlled by rotating the ride control valve knob. This needle valve adjusts the variable orifice opening between the seat cylinder and a nitrogen filled accumulator.

340. **CONTROL VALVE.** To disassemble the control valve, remove pivot pin and control lever (1—Fig. 308). Unbolt and remove plate (14), then withdraw caps (13), springs (12) and pistons (11) with "O" rings (10). Unscrew connectors (2) and remove washers (3), springs (5), retainers (6) and balls (7). Clean and inspect all parts for excessive wear or other damage.

Renew all "O" rings and reassemble by reversing the disassembly procedure.

341. **CYLINDER.** Disassembly of the single action seat cylinder is obvious after an examination of the unit and reference to Fig. 309. Clean and inspect all parts for excessive wear or other damage.

When reassembling, use new "O" rings (3, 6, 8 and 10) and new wiper ring (4).

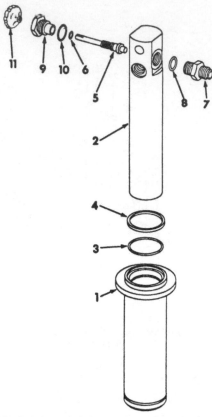

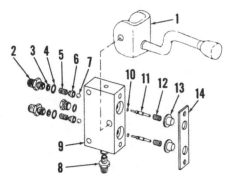

Fig. 308–Exploded view of hydraulic seat control valve.

1. Control lever	8. Connector
2. Connector	9. Body
3. Washer	10. "O" ring
4. "O" ring	11. Piston
5. Spring	12. Spring
6. Retainer	13. Cap
7. Ball	14. Plate

Fig. 309–Exploded view of single action seat cylinder.

1. Cylinder barrel	6. "O" ring
2. Cylinder ram	7. Connector
3. "O" ring	8. "O" ring
4. Wiper ring	9. Retainer
5. Ride control	10. "O" ring
adjusting screw	11. Knob

Wiring Diagrams are located at the back of this manual.

INTERNATIONAL
HARVESTER

Models ■ 786 ■ 886 ■ 986 ■ 1086

Previously contained in I&T Shop Manual No. IH-52

SHOP MANUAL

INTERNATIONAL HARVESTER

SERIES
786 — 886 — 986 — 1086

Engine serial number is stamped on right side of engine crankcase on Models 786 and 886 equipped with D-358 engines and on left side of engine crankcase on Model 886 equipped with D-360 engine and all 986 and 1086 models. Tractor serial number is stamped on name plate attached to right side of rear frame. Cab/ROPS serial and model number is stamped on plate on right side of cab.

INDEX (By Starting Paragraph)

INDEX CONT.

CONDENSED SERVICE DATA

	786, 886 (Late)	886 (Early)	986	1086 (Early)	1086 (Late)
GENERAL					
Engine Make	IH	IH	IH	IH	IH
Engine Model	D-358	D-360	D-436	DT-414	DT-414B
Number of Cylinders	6	6	6	6	6
Bore – Inches	3.875	3.875	4.300	4.300	4.300
Stroke – Inches	5.060	5.085	5.004	4.754	4.754
Displacement – Cubic Inches	358	360	436	414	414
Main Bearing, Number of	7	7	7	7	7
Cylinder Sleeves	WET	WET	WET	WET	WET
Forward Speeds without T.A.	8	8	8	8	8
Forward Speeds with T.A.	16	16	16	16	16
Alternator and Starter Make	Delco-Remy, Motorola or Niehoff				
TUNE-UP					
Compression Pressure (2)	315-340	385-415	385-415	335-365	335-365
Firing Order	1-5-3-6-2-4	1-5-3-6-2-4	1-5-3-6-2-4	1-5-3-6-2-4	1-5-3-6-2-4
Valve Tappet Gap (Hot)					
Intake	0.012	0.012	See Para. 64	0.020	See Para. 64
Exhaust	0.012	0.021	0.025	0.025	0.025
Valve Seat Angle (Degrees)					
Intake	45	30	30	30	30
Exhaust	45	45	45	45	45
Timing Mark Location	Flywheel	Flywheel	Flywheel	Flywheel	Flywheel
Injection Pump Make	Robert Bosch	Amer. Bosch	Amer. Bosch	Amer. Bosch	Amer. Bosch
Injection Pump Timing	See Para. 114	18° BTDC	18° BTDC	18° BTDC	18° BTDC
Battery Terminal Grounded	Negative				
Engine Low Idle Rpm	700	700	700	700	700
Engine High Idle Rpm, No Load	2640	2640	2640	2640	2640
Engine Full Load Rpm	2400	2400	2400	2400	2400

(2) Approximate psi, at sea level, at cranking speed

	786, 886 (Late)	886 (Early)	986	1086 (Early)	1086 (Late)
SIZES – CAPACITIES – CLEARANCES					
Crankshaft Main Journals Diameter, Inches	3.1484-3.1492	3.3742-3.3755	3.3742-3.3755	3.3742-3.3755	3.3742-3.3755
Crankpin Diameter, Inches	2.5185-2.5193	2.9977-2.9990	2.9977-2.9990	2.9977-2.9990	2.9977-2.9990
Camshaft Journal Diameter, Inches					
No. 1 (Front)	2.2823-2.2835	2.2824-2.2835	2.2824-2.2835	2.2824-2.2835	2.2824-2.2835
No. 2	2.2823-2.2835	2.2824-2.2835	2.2824-2.2835	2.2824-2.2835	2.2824-2.2835
No. 3	2.2823-2.2835	2.2824-2.2835	2.2824-2.2835	2.2824-2.2835	2.2824-2.2835
No. 4	2.2823-2.2835	2.2824-2.2835	2.2824-2.2835	2.2824-2.2835	2.2824-2.2835
Piston Pin Diameter, Inches	1.4172-1.4173	1.6248-1.6250	1.6248-1.6250	1.6248-1.6250	1.6248-1.6250
Valve Stem Diameter, Inch					
Intake	0.3919-0.3923	0.3718-0.3725	0.3718-0.3725	0.3718-0.3725	0.3718-0.3725
Exhaust	0.3911-0.3915	0.3718-0.3725	0.3718-0.3725	0.3718-0.3725	0.3718-0.3725
Main Bearing Diametral Clearance, Inch	0.0029-0.0055	0.0018-0.0051	0.0018-0.0051	0.0018-0.0051	0.0018-0.0051
Rod Bearing Diametral Clearance, Inch	0.0023-0.0048	0.0018-0.0051	0.0018-0.0051	0.0018-0.0051	0.0018-0.0051
Piston Skirt Diametral Clearance, Inch	0.0039-0.0047	0.0045-0.0065	0.0045-0.0065	0.0045-0.0065	0.0045-0.0065
Crankshaft End Play, Inch	0.0060-0.0090	0.006-0.012	0.006-0.012	0.006-0.012	0.006-0.012
Camshaft Bearing Diametral Clearance, Inch	0.0009-0.0033	0.002-0.0066	0.002-0.0066	0.002-0.0066	0.002-0.0066
Camshaft End Play, Inch	0.0040-0.0180	0.005-0.013	0.005-0.013	0.005-0.013	0.005-0.013
Cooling System Capacity – Quarts	24	24	26	26	27
Crankcase Oil – Quarts	12	12	14	20	20
Transmission and Differential – Gallons (Approximate) Standard Transmission	23	23	22.5	22.5	22.5
Front Differential Housing (All Wheel Drive Tractor) Quarts	10	10	10	10	10

FRONT SYSTEM TRICYCLE TYPE

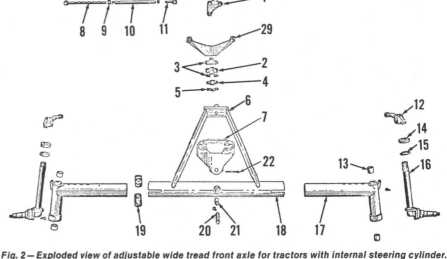

All models except Model 786, are available with dual wheel tricycle type suspension as shown in Fig. 1.

DUAL WHEELS

Tractors So Equipped

1. The pedestal for the dual tricycle wheels is bolted to the steering pivot shaft (12–Fig. 1). The pedestal is available as a pre-riveted assembly (13), or the pedestal and axle are available as

Fig. 2—Exploded view of adjustable wide tread front axle for tractors with internal steering cylinder.

1. Steering arm (center)		17. Axle extension
2. Ball socket	7. Axle support	18. Axle main member
3. Shim	8. Tie rod extension	12. Steering arm	19. Axle clamp
4. Socket cap	9. Clamp	13. Bushing	20. Pivot pin
5. Lock plate	10. Tube	14. Thrust bearing	21. Pivot bushing
6. Stay rod	11. Tie rod end	15. Felt washer	29. Socket support
		16. Steering knuckle	

separate repair parts. Inspect wheel bearings, seals and front axle and renew as necessary. Wheel bearings are adjusted with a slight preload.

FRONT SYSTEM AXLE TYPE

AXLE MAIN MEMBER

Tractors So Equipped

2. For all tractors equipped with an adjustable wide tread front axle, refer to Fig. 2 and 3. The axle main member (18) pivots on pin (20) which is pinned in the axle support (7). The two pivot pin bushings (21) are pressed into axle main member and should be reamed after installation to provide a free fit for the pivot pin.

To remove the axle main member, disconnect tie rods from steering arms, then remove axle clamps (19–Fig. 2) and on Fig. 3 remove the pin and loosen clamp bolts and withdraw axle extension, knuckle and wheel assemblies. Remove cotter pin and pivot pin retaining pin, then disconnect stay rod ball

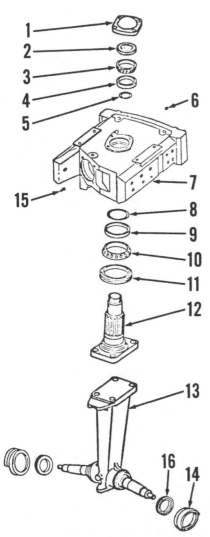

Fig. 1—Exploded view of front support, tricycle type front axle and components available for all models except Model 786.

1. Cover	9. Bearing cup
2. Shaft nut	10. Bearing cone
3. Bearing cone	11. Oil seal
4. Bearing cup	12. Steering pivot shaft
5. "O" ring	13. Pedestal assembly
6. Plug	14. Dust shields
7. Front support	15. Grease fitting
8. "O" ring	16. Seals

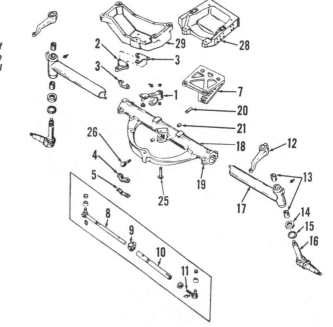

Fig. 3—Exploded view of standard adjustable wide tread front axle with external steering cylinder.

1. Steering arm (center)
2. Ball socket
3. Shim
4. Socket cap
5. Lock plate
7. Axle support
8. Tie rod extension
9. Clamp
10. Tube
11. Tie rod end
12. Steering arm
13. Bushing
14. Thrust bearing
15. Felt washer
16. Steering knuckle
17. Axle extension
18. Axle main member
19. Clamp
20. Pivot pin
21. Pivot bushing
25. Pin
26. Ball
28. Upper bolster
29. Stay rod support

from its support. Save shims (3–Fig. 2 or 3) located between socket (2) and socket cap (4). On models with external cylinder, disconnect and cap the hydraulic lines. Drive pivot pin (20) out of axle support and axle main member and remove axle main member. Reinstall by reversing the removal procedure and if necessary, adjust front wheel toe-in as outlined in paragraph 5.

Hi-Clear Models 986-1086

3. High clearance tractors are equipped with a front axle as shown in Fig. 4.

Removal procedure for this axle is the same as given in paragraph 2 except that the auxiliary stay rods (31–Fig. 4) should be disconnected when removing axle extension and wheel assemblies. Refer to paragraph 5 for toe-in adjustment.

STAY ROD AND BALL

All Models

4. The stay rod ball is available as an individual part, except on heavy duty adjustable wide front axle with internal steering cylinder (Fig. 2). If stay rod ball is renewed, ball must shoulder against stay rod and nut must be torqued to 185 ft.-lbs. Clearance between stay rod ball and socket can be adjusted by adding or subtracting shims which are located between socket and cap. Torque stay rod ball socket cap screws to 85-100 ft.-lbs. and bend lock plate tabs against the flats of the cap screws. Models equipped with adjustable wide tread front axle have a socket support bolted to side rails and any service required on support is obvious.

TIE RODS AND TOE-IN

All Models

5. Removal of tie rods on all models is obvious after an examination of the units. Tie rod ends are non-adjustable and faulty units will require renewal.

Adjust toe-in on all models to ¼-inch, plus or minus 1/16-inch. Adjustment is made by varying the length of tie rods. Both rods should be adjusted an equal amount with not more than one turn difference when adjustment is complete.

STEERING KNUCKLES

All Models

6. Removal of steering knuckles from axle is obvious after an examination of the unit and reference to Fig. 5.

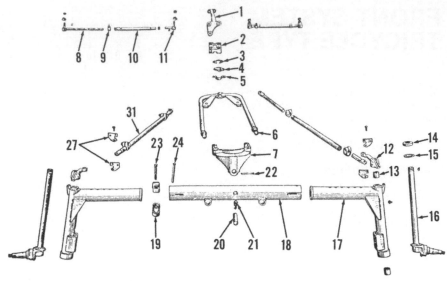

Fig. 4 – Exploded view of adjustable wide front axle used on Models 986 and 1086 Hi-Clear tractors.

1. Steering arm (center)	10. Tube	19. Clamp
2. Ball socket	11. Tie rod end	20. Pivot pin
3. Shim	12. Steering arm	21. Pivot bushing
4. Socket cap	13. Bushing	22. Pin
5. Lock plate	14. Thrust bearing	23. Clamp bolt
6. Stay rod	15. Felt washer	24. Clamp pin
7. Axle support	16. Steering knuckle	27. Bracket
8. Tie rod extension	17. Axle extension	31. Auxiliary stay rod
9. Clamp	18. Axle main member	assembly

When renewing steering knuckle bushings (4–Fig. 5) be sure to align lube hole in bushing with lube hole in axle or axle extension. Install bushing so outer ends are below the counterbore on axle extension. Ream bushings after installation to provide an operating clearance of 0.001-0.006 inch. Torque steering arm retaining nut to 400 ft.-lbs. and bend locking tab against flat of nut.

FRONT SYSTEM ALL-WHEEL DRIVE

All models are available as 4-wheel drive (All-Wheel Drive) units. Models 986

Fig. 5 – Cross sectional view of typical steering knuckle used on all models.

1. Steering arm
2. Nut (400 ft.-lbs. torque)
3. Lock plate
4. Bushings
5. Spindle
6. Bearing
7. Felt washer
8. Oil seal
9. Inner bearing
10. Hub
11. Outer bearing

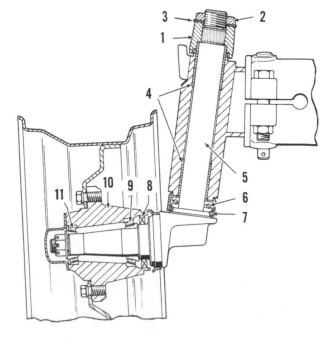

and 1086 with internal steering cylinder use American Coleman units. All models equipped with external steering cylinder use Elwood 4-wheel drive units.

American Coleman front axle assembly consists of a one-piece center housing having flanged ends to which stub axle ends are bolted. Wheel spindles and wheel hubs are carried on tapered roller bearings. The axle center housing incorporates a straddle mounted pinion and a four pinion differential gear unit. Full floating axles extend outward from the differential and attach the Cardan type universal joints located inside wheel hubs.

Elwood front axle assembly consists of a one-piece center housing having ball type pivot ends to which steering knuckles are attached. The steering knuckles and end housings are carried on tapered roller bearings. The axle center housing incorporates a straddle mounted pinion and a four pinion differential gear unit. Drive shafts extend outward from the differential and attach to planetary type end housings. The shafts pivot on universal joints located inside the steering knuckles. The planetary gears are carried on needle bearings.

Power for the front axle assembly is taken from a gear reduction unit (transfer case) which is bolted to left side of tractor rear frame and driven by an idler gear on reverse idler shaft of the range transmission. The reduction unit and front axle are connected by a drive shaft fitted with two conventional universal joints of which the rear has a slip joint that compensates for oscillation of front axle.

The gear reduction unit (transfer case) has a shifting mechanism which permits shifting to neutral, thus disconnecting power to the front axle. Gear sets with various ratios are available for the reduction unit to match front and rear tire size combinations. Therefore, it is essential that correct front and rear tire sizes be used with the correct gear set. If a change in tire sizes or gear sets is contemplated, contact International Harvester Company for information concerning proper combinations.

WHEEL AND PIVOT BEARINGS (AMERICAN COLEMAN)

Models So Equipped

10. **WHEEL BEARINGS.** Wheel bearings should be removed, cleaned and repacked annually. Removal of inner wheel bearing requires removal of hub assembly.

To remove both wheel bearings, refer to Fig. 7 and proceed as follows: Support axle assembly, then remove wheel cover (51) and the wheel and tire assembly. Clip and remove lock wire from axle retaining cap screws, then remove cap screws and tapered bushings (47) and remove axle by pulling straight outward as shown in Fig. 8.

NOTE: Use caution when removing axle shaft not to damage the oil seal (21 — Fig. 7) located in center housing as shown in Fig. 9. Swing power yoke (41 — Fig. 8) aside, straighten tabs of retainer plates (25 — Fig. 10), then unbolt and remove bearing plate (34) and shims (24). Outer wheel bearing, power yoke, compensating ring and hub can now be removed from spindle. See Fig. 11.

To remove inner wheel bearing (33 – Fig. 7) from spindle (22), loosen jam nut and adjusting screw (23), then position a small punch against outer end of adjusting wedge (30 – Fig. 11) and bump wedge inward until clamp ring (31) is free and remove clamp ring. Insert a pin punch through knock-out holes provided in the spindle and bump on in-

Fig. 8 – View showing axle (46) being removed. Item (41) is power yoke. Refer also to Fig. 9.

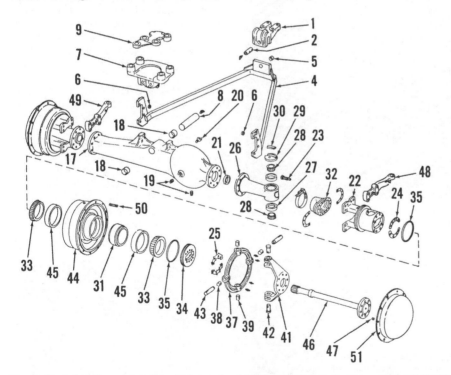

Fig. 7 – Exploded view of American Coleman front drive axle used on tractors equipped with "All Wheel Drive". Axle shown is used on tractors with internal steering cylinder.

1. Stay bar support	20. Vent	29. Bearing cap	41. Power yoke
2. Rear pivot pin	21. Oil seal	30. Adjusting wedge	42. Yoke pin
4. Stay rod	22. Spindle	31. Clamp ring	43. Hub pin
5. Bushing	23. Wedge adjusting	32. Boot	44. Hub
6. Gusset washer	screw	33. Wheel bearing	45. Bearing cup
7. Axle support	24. Shims	34. Bearing plate	46. Axle shaft
8. Pivot pin	25. Lock plates	35. Seal	47. Tapered bushing
9. Center steering arm	26. Axle stub	37. Compensating ring	48. Steering arm L.H.
17. Housing	27. Bearing cup	38. Bushing	49. Steering arm R.H.
18. Bushing	28. Pivot bearing	39. Bushing	50. Wheel stud
19. Plugs			51. Wheel cover

Fig. 9 – When removing axle use caution not to damage the oil seal (21) shown. Differential and carrier have been removed for clarity.

ner side of bearing inner race until bearing is about ½-inch from inner flange of spindle, then attach puller, if necessary, and complete removal of bearing. Be sure to keep bearing straight while removing or damage (scoring) to spindle could result.

NOTE: In some cases, the top pivot bearing cap may come out of bore in spindle when clamp ring and inner wheel bearing are removed and the spindle may drop and rest on the stub axle. If this occurs, proceed as follows when reassembling. Place a wood block under spindle, then with a hydraulic jack under block, raise spindle into its proper position and place upper pivot bearing and cap in its bore. Start inner wheel bearing on spindle and over upper bearing cap. This will hold parts in position. Start clamp ring over

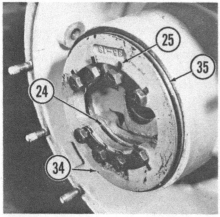

Fig. 10 — View showing bearing adjusting plate and shims. Power yoke and compensating ring are removed for clarity.

24. Shims	34. Bearing plate
25. Lock plates	35. Seal

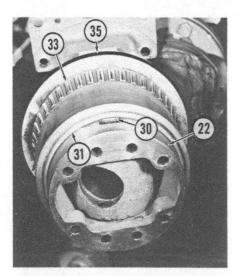

Fig. 11 — Spindle assembly with outer wheel bearing and hub removed. Note outer end of pivot bearing adjusting wedge (30).

22. Spindle	
30. Adjusting wedge	33. Inner bearing
31. Clamp ring	35. Seal

spindle and complete installation of both bearing and clamp ring.

11. Clean and inspect all parts for wear, excessive scoring or other damage and renew as necessary. Repack the wheel bearings with a good grade of multi-purpose lithium grease.

12. Reassemble and adjust wheel bearings and pivot bearings as follows: Install inner wheel bearing on spindle with largest diameter toward inside. Place clamp ring over spindle and against inner wheel bearing and make sure it does not contact the cage of inner wheel bearing. Tighten adjusting wedge until it feels solid, then measure distance from end of wedge to end of slot in pivot bearing cap. This distance (D–Fig. 12) should be at least 1⅜ inches. If measured distance is less than 1⅜ inches, remove spindle and add shims under bottom pivot bearing cone. Shims can be made from shim stock and each 0.012 inch shim will change the measured distance about ⅛-inch.

NOTE: This operation is to insure that a satisfactory adjustment can be made for the pivot bearings.

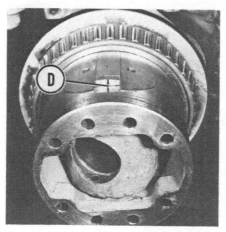

Fig. 12 — A distance (D) of 1⅜ inches between end of adjusting wedge and end of slot in pivot bearing cap should be maintained. Refer to text.

Fig. 13 — Attach spring scale as shown to check preload ov pivot bearings. Refer to text.

With clamp ring installed as outlined above and a new seal on spindle, install the hub, compensating ring, power yoke assembly and outer wheel bearing. With new seal on bearing plate, install original shims, bearing plate and lock plates. Tighten cap screws securely and check rotation of wheel hub. Hub should rotate with a slight drag. Add or subtract shims under bearing plate to obtain correct bearing adjustment. Shims are available in thicknesses of 0.002, 0.010 and 0.030 inch. Lock cap screws with lock plates when adjustment is complete.

Disconnect tie rod from steering arm, then tighten wedge adjusting screw until a slight drag is felt when moving spindle through its full range of travel. Move spindle back and forth several times, then using a spring scale attached to tie rod ‘hole of steering arm, check the pounds pull required to keep spindle in motion. See Fig. 13. Pull should not exceed 12 pounds and must be read while spindle is in motion. Tighten adjusting screw jam nut when adjustment is complete. Reinstall axle using caution not to damage oil seal located in center housing. Install tapered bushings and the axle retaining cap screws and tighten cap screws securely. Lock wire the axle cap screws. Reinstall wheel and tire and wheel cover and mate holes of wheel cover with grease fittings of compensating ring if tractor is equipped with this type wheel cover.

13. **PIVOT BEARINGS.** To remove the pivot bearings, the wheel bearings must be removed as outlined in paragraph 10.

With wheel bearings removed, disconnect tie rod from steering arm, then remove pivot bearing cups from their recesses in the stub axle by driving on a punch inserted in the knockout holes (H–Fig. 14) provided in outer end of stub axle. Use caution during this operation not to let punch slip past cups and

damage bearings. With bearing cups free of their recesses, tilt inner end of spindle upward and remove from stub axle. Do not use force to remove spindle. If spindle does not come off readily, it is probable that bearing cups are not completely free of their recesses. The lower pivot bearing can be removed from spindle by driving on a punch inserted through the knock-out holes provided in spindle. Upper pivot bearing can be removed from upper bearing cap in the same manner. See (H – Fig. 15).

14. Clean and inspect all parts. Check bearings for roughness, damage or undue wear. Renew parts if any doubt exists as to their condition.

15. To reinstall pivot bearings, first lubricate lower bearing, drive it into position in spindle and place lower bearing cup over lower bearing. Place spindle over stub axle and support it with a jack and wood block as shown in Fig. 16. Align bearing cup with its recess in stub axle, raise jack and press lower bearing cup into its recess.

NOTE: If necessary, bearing cup can be seated by driving on top side of stub axle while spindle is supported.

Drive upper bearing cup into its recess on top side of stub axle. Install upper pivot bearing on top cap and install assembly in spindle with shallowest end of adjusting wedge slot toward outside. Be sure adjusting wedge is installed with rounded surface upward.

Wheel bearings can now be installed and pivot bearings adjusted as outlined in paragraph 12.

FRONT DRIVE AXLE (AMERICAN COLEMAN)

Models So Equipped

16. **R&R AXLE ASSEMBLY.** To remove the "All-Wheel Drive" front axle

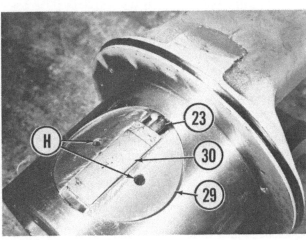

Fig. 15 — View of pivot bearing top cap (29) showing location of knock-out holes (H). Note also adjusting wedge (30) and wedge adjusting screw (23).

Fig. 16 — Support spindle on a wood block during installation of pivot bearings.

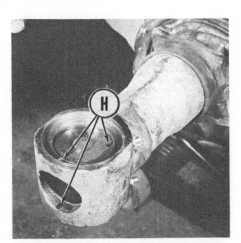

Fig. 14 — Knock-out holes (H) are provided for removal of pivot bearing cups.

and wheels as an assembly, first disconnect tie rods from steering arms. Remove "U" bolts from drive shaft front universal joint and separate universal joint. Disconnect stay rod from bracket under clutch housing and position a rolling floor jack under stay rod to support assembly. Support front of tractor, drive out pivot pin, then raise front of tractor and roll the axle and wheels assembly forward away from tractor.

17. **OVERHAUL.** Overhaul of the axle assembly can be considered as two operations and both operations can be accomplished without removing the axle assembly from the tractor. One operation concerns the stub axle along with the hub and the parts which make up the outer end of the axle. The other operation concerns the differential and carrier which is carried in the axle center housing. With the exception of removing the driving axles, which is involved in work of either operation, overhaul of either portion of the "All-Wheel Drive" axle can be accomplished without disturbing the other section.

18. **AXLE OUTER END.** Outer end of axle can be disassembled as follows: Remove wheel cover (51 – Fig. 7) and the wheel and tire assembly. Clip lock wire and remove axle retaining cap screws and tapered bushings (47) and pull axle from housing. Use caution when removing axle not to damage oil seal (21) located in axle center housing. Also see Fig. 9. Remove the two plugs and pins (43 – Fig. 7) and remove power yoke (41) and compensating ring (37) assembly from hub.

NOTE: Pins (43) are tapped so cap screws can be used to aid in removal as shown in Fig. 17.

Fig. 17 — Use cap screws as shown to pull hub to compensating ring pins.

Remove roll pins and yoke pins (42 – Fig. 7) and separate power yoke and compensating ring. Straighten lock plate tabs and remove bearing adjustment plate (34), shims (24), outer wheel bearing (33) and hub (44).

NOTE: Identify and keep removed shims in their original relationship.

Loosen jam nut and wedge adjusting screw (23), then with a small punch positioned in slot of upper pivot bearing cap (29), bump adjusting wedge (30) inward until clamp ring (31) is free. See Fig. 11. Use punch in the knock-out holes provided in spindle and bump inner wheel bearing about ½-inch toward outer end of spindle, then attach a puller, if necessary, and complete removal of bearing.

NOTE: Be sure to keep bearing straight during removal as damage (scoring) to spindle could result.

Use punch in knockout holes (H – Fig. 14) provided in stub axle and bump pivot bearing cups from recesses in stub axle, then tilt inner end of spindle upward and pull assembly from stub axle.

NOTE: Do not use force to remove spindle from stub axle. If spindle cannot be removed freely it is probably that one or the other pivot bearing cups are not completely free of stub axle.

If necessary, stub axle (26 – Fig. 7) can now be removed from center housing (17).

19. Clean and inspect all parts. Pay particular attention to wheel bearings and pivot bearings in regard to roughness, damage or wear and renew any which are in any way doubtful. Wheel bearing cups can be bumped from hub, if necessary. Compensating ring bushings are available for service and renewal procedure is obvious. Be sure also that wear on driving pins (42 and 43) is not excessive. Check dowel pins in hub and roll pins in yoke pins to see that they are straight and not unduly worn.

Reassemble by reversing the disassembly procedure and adjust pivot bearings and wheel bearings as outlined in paragraph 12.

20. DIFFERENTIAL UNIT. To remove and overhaul the differential unit, first drain differential housing and disconnect left tie rod from left steering arm. Remove "U" bolts from front universal joint and separate universal joint. Remove both front wheel covers, then remove both front axles. Unbolt carrier (1 – Fig. 18) and pull carrier and differential assembly from axle center housing. See Fig. 19.

With unit removed, remove bearing caps (2 – Fig. 18), then remove differential, bearing cups (24) and bearing adjusters (4) from carrier. Differential carrier bearings (25) can be removed from differential at this time, if necessary. Remove cap screws retaining oil seal retainer (17) and bearing cage (10) to carrier and remove pinion and bearing assembly. Identify and save bearing cage shims (13, 14 and 15). Remove nut (23), washer (22), yoke (21) and seal retainer (17) from pinion (7), then press pinion from outer bearing (12) and bearing cage (10). Spacer (16), inner bearing (12) and pilot bearing (6) can now be removed from pinion although pilot bearing (6) should be unstaked from pinion prior to removal. Bearing cups (11) can be driven from bearing cage (10) after pinion shaft is out. Match mark the differential case halves (26 and 27), remove case retaining bolts, then separate case halves and remove spider (34), pinions (32), thrust washers (33), side gears (30) and side gear thrust washers (31). If not already done, bearings (25) can now be removed from differential case halves. If necessary, lubricator (29) can also be removed. If bevel drive gear (8) is to be renewed, remove rivets (9) by drilling and punching.

Clean all parts and inspect for undue wear or scoring, chipped teeth or other damage and renew parts as necessary.

Reassembly is the reverse of disassembly, however, consider the follow-

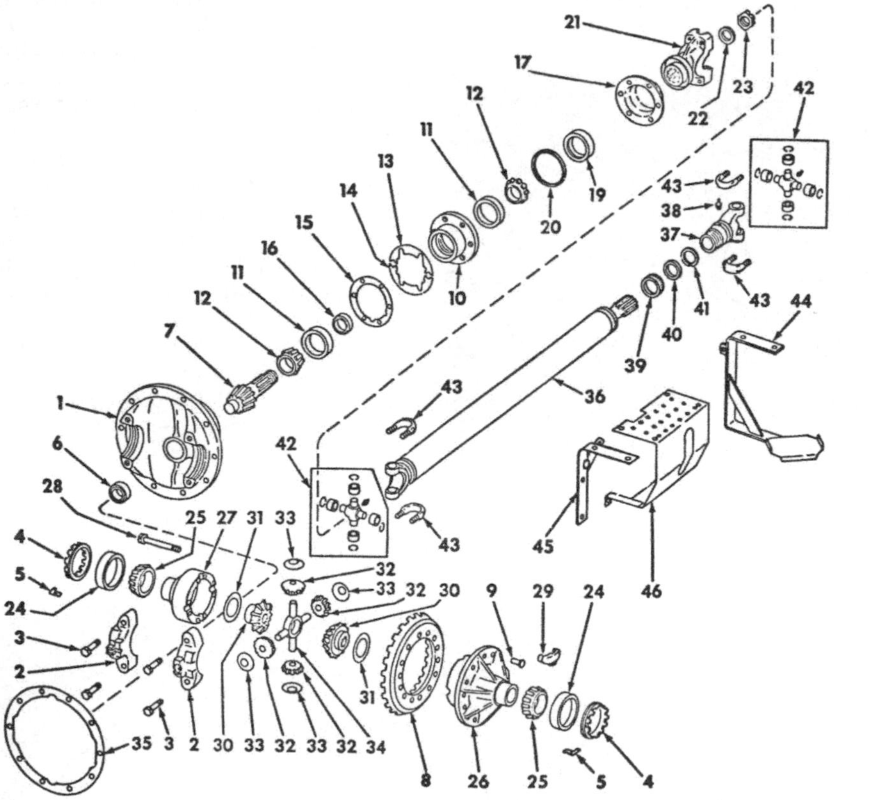

Fig. 18 – Exploded view of American Coleman "All Wheel Drive" front axle differential and carrier assembly showing component parts and their relative positions.

1. Carrier	12. Bearing	24. Bearing cup	36. Stub (drive) shaft
2. Bearing cap	13. Shim (upper)	25. Bearing	37. Sleeve yoke
3. Cap screw	14. Shim (lower)	26. Case half (flanged)	38. Grease fitting
4. Bearing adjuster	15. Shim	27. Case half (plain)	39. Dust cap
5. Adjuster lock	16. Spacer	29. Lubricator	40. Cork washer
6. Pilot bearing	17. Seal retainer	30. Side gear	41. Steel washer
7. Pinion	19. Oil seal	31. Thrust washer	42. "U" joint package
8. Bevel gear	20. Cork washer	32. Pinion gear	43. "U" bolt
9. Rivet	21. Yoke	33. Thrust washer	44. Support
10. Bearing cage	22. Washer	34. Spider	45. Shield support
11. Bearing cup	23. Nut	35. Gasket	46. Shield

Fig. 19 – View showing differential carrier and differential unit removed from center housing.

ing information during assembly. Shims (13 and 14) are available in 0.003 inch thickness. Shims (15) are available in thicknesses of 0.005, 0.010 and 0.030 inch. Spacer (16) is available in widths of 0.506 through 0.526 inch in increments of 0.001 inch. Pinion and bevel ring gear are available only as a matched set. Note also that axle oil seals (21 – Fig. 7) can be renewed when carrier and differential assembly is out.

21. To reassemble differential unit, proceed as follows: Place inner bearing cone (12 – Fig. 18) over pinion shaft (7) with largest diameter toward gear and press bearing on shaft until it bottoms. Invert pinion shaft and press pilot bearing (6) on end of pinion shaft and stake in at least four positions. Install bearing cups (11) in bearing cage (10) with smallest inside diameters toward center. Insert pinion shaft and inner bearing in bearing cage and install spacer (16), outer bearing cone (12), seal assembly retainer (17), yoke (21), washer (22) and nut (23). Attach a holding fixture to yoke and while turning bearing cage (10), tighten nut (23) to a torque of 255 ft.-lbs. With nut tightened, check the rotation of the pinion shaft which should require a rolling torque of 8-15 in.-lbs. If shaft preload is not as stated, change the spacer (16) as required. Spacers are available in thicknesses of 0.506 through 0.526 inch in increments of 0.001 inch. Use original shims (13, 14 and 15) as a starting point and install pinion shaft reassembly in carrier.

Reassemble the differential assembly by reversing the disassembly procedure. Place differential in carrier, install bearing caps and bearing adjusters and tighten bearing adjusters until differential bearings have zero end play. Differential assembly can be moved left or right as required by loosening one bearing adjuster and tightening the opposite. Move the differential toward the pinion

shaft until backlash is approximately 0.006-0.012 inch. Paint ten or twelve teeth of bevel ring gear with red lead or prussian blue and turn pinion in direction of normal rotation. Tooth contact pattern should be located approximately midway of both length and width (depth) of bevel ring gear teeth. If pattern is too far toward toe of bevel gear teeth, turn adjusters as required to move gear away from pinion. If pattern is too far toward heel, turn adjusters (4) as required to move gear toward pinion. If pattern is too far toward root of bevel ring gear teeth, vary shims (13 and 14) to move pinion away from bevel drive gear. If pattern is too near top of bevel ring gear teeth, vary shims (13 and 14) to move pinion toward bevel drive gear. Continue this operation until tooth pattern is centered on the bevel ring gear teeth and backlash is 0.006-0.012 inch between pinion and bevel gear.

When a satisfactory pattern is obtained, slightly preload differential carrier bearings by tightening each bearing adjuster one notch. Lock both bearing adjusters.

Complete reassembly by reversing the disassembly procedure.

DRIVE HOUSING ASSEMBLY (AMERICAN COLEMAN TRANSFER CASE)

Models So Equipped

The drive housing which transmits power to the front axle is mounted on the left front side of the tractor rear frame and is driven by an idler gear on the reverse idler shaft of the range transmission.

22. **REMOVE AND REINSTALL.** To remove drive housing, remove shield, then remove "U" bolts of universal joint located closest to drive housing and separate the universal joint. Use tape or

some other suitable means, to retain universal bearing cups on spider. Disconnect shifter rod from shifter shaft lever, drain rear frame, then unbolt and remove drive housing from tractor rear frame.

Before installing the unit on tractor, first determine the number and thickness of shim gaskets to be used between drive housing and rear frame as follows: Place a punch mark on "U" joint yoke two inches from center line of output shaft as shown in Fig. 20. Using a dial indicator as shown, record the backlash between the small spool gear and output gear. Next, engage the transmission park lock and measure the backlash between park lock, reverse driven gear and reverse drive gear as shown in Fig. 21. Record the backlash. Then, with park lock engaged, install drive housing assembly using one thin shim gasket. With dial indicator tip on the previously installed punch mark on yoke as shown in Fig. 22, check and record the backlash. This backlash reading should be 0.003-0.007 inch greater than the sum of the two previous backlash readings. If the backlash increase is less than 0.003 inch, install one thick gasket or a combination of shim gaskets to increase the backlash. Gaskets are available in two thicknesses, (thin) 0.011-0.019 inch and (thick) 0.016-0.024 inch.

Installation of drive housing unit to tractor rear main frame will be facilitated if two ½-inch guide studs are used.

23. **OVERHAUL.** With unit removed as outlined in paragraph 22, wedge gears and remove the Esna nut which retains the universal joint yoke (28 – Fig. 23) to output shaft and remove yoke from shaft. Loosen jam nut and lock screw (24) in shifter fork and remove shifter shaft (25), spacer (27) and fork (23). Position housing with open side up,

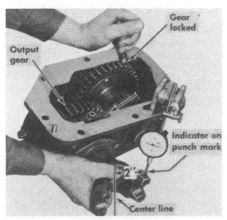

Fig. 20 – Install punch mark on yokes 2 inches from center line of output shaft. Check backlash as shown.

Fig. 21 – With park lock engaged check backlash between park lock, reverse driven gear and reverse drive gear.

Fig. 22 – Install unit using one thin shim gasket and check total backlash. Refer to text.

remove the roll (spring) pin (8) which retains spool gear shaft in housing, then remove spool gear shaft (5), spool gear (9) and thrust washers (12). Needle bearings in spool gear can be renewed at this time. Remove output shaft oil seal cage (20) and gasket. Lift rear snap ring (16) from its groove in output shaft and slide rearward. Slide output shaft gear (17) rearward and catch detent balls (14) and spring (15) as they emerge from bore of output shaft. Complete removal of shaft and lift gear from housing. Ball bearing (18) can be removed from shaft after removing snap ring (19) and needle bearing (3) can be removed from bore in housing. Any further disassembly is obvious.

Clean and inspect all parts for scoring, undue wear, chipped teeth or other damage and renew as necessary. Reassemble by reversing the disassembly procedure and refer to paragraph 22 for installation information.

DRIVE SHAFT
Models So Equipped

The drive shaft between drive housing and the front axle differential is conventional and can be removed and serviced as outlined in paragraph 24.

24. **R&R AND OVERHAUL.** To remove drive shaft from tractor, first remove shield, then remove "U" bolts from front and rear universal joint yokes and lift shaft from tractor. The four exposed bearing cups can now be removed from the universal joint spiders. Unscrew dust cap from sleeve yoke, pull sleeve yoke from drive shaft and remove dust cap, steel washer and cork washer. Remove bearing cup retaining snap rings from yokes, then remove bearing cups by driving spider first one way then the other.

Individual parts available for service

are sleeve yoke, drive shaft, dust cap, steel washer and cork washer. Universal joint bearing cups and spider are available only as a package.

Reassembly is the reverse of disassembly, however be certain that bearing cups are packed with grease (either by hand during assembly or through grease fitting after assembly.) Bearing cups can usually be pressed into yokes simultaneously by using a bench vise. Small spacers can be used to complete installation of cups and allow installation of snap rings. Be sure ends of spider do not catch ends of bearing needles when pressing in the bearing cups.

WHEEL AND PIVOT BEARINGS (ELWOOD)
Models So Equipped

25. **WHEEL BEARINGS.** Wheel bearings should be removed, cleaned and repacked annually. Removal of wheel bearings requires removal of wheel hub and planetary assembly.

To remove wheel bearings, refer to Fig. 24 and proceed as follows: Support tractor front axle assembly and remove front wheels. Using a center punch, index wheel hub (31 – Fig. 24) to planetary housing flange (43). Drain oil from planetary housing then remove retaining nuts from planetary housing flange. Using jackscrews, split planetary assembly from hub. Remove tapered dowels (40), then remove planetary assembly from hub. Remove snap ring (45) from axle shaft end, then remove sun gear (46), spacer (47) and ring gear (48). Attach a lifting eye and hoist to wheel hub, remove locknuts from spindle shaft flange (3), and remove hub (31) from steering knuckle (8).

Loosen cap screw on bearing locknut (25 – Fig. 24) and remove locknut from shaft.

NOTE: Locknut is positioned with recess holes away from bearing. Recess holes are for tightening purposes using a spanner wrench or punch. See Fig. 28.

Support wheel hub across two wood blocks and drive spindle shaft (3 – Fig. 24) out of hub using a wood block. Lift inner bearing (26) out of hub. Remove oil seal (34) and outer bearing (33) from hub. Using a brass drift punch, drive out inner and outer bearing cups. Inspect spindle shaft oil seal (17) and bushing (4) and renew if necessary.

NOTE: Use a blind hole puller to remove brass spindle shaft bushing.

Clean and inspect all parts and renew as necessary. If renewal of spindle shaft bushing (4 – Fig. 24) is necessary, bushing must be driven in until it bottoms in the counterbore.

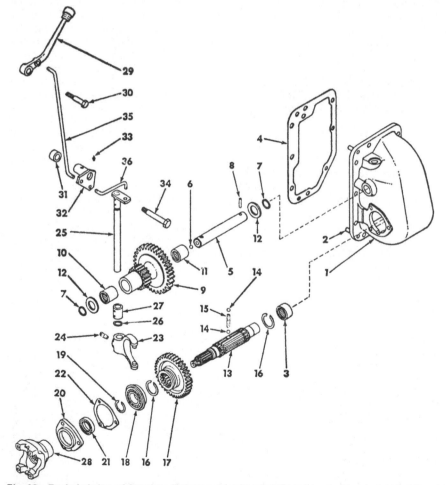

Fig. 23 – Exploded view of American Coleman drive housing (transfer case) used when tractors are equipped with "All Wheel Drive".

1. Housing	10. Needle bearing	19. Snap ring	28. Yoke
2. Dowel	11. Needle bearing	20. Oil seal cage	29. Shift lever
3. Needle bearing	12. Thrust washer	21. Oil seal	30. Lever bolt
4. Gasket	13. Output shaft	22. Gasket	31. Spacer (No. T.A.)
5. Spool gear shaft	14. Poppet ball	23. Shifter fork	32. Bellcrank
6. Steel ball	15. Poppet spring	24. Lock screw	33. Grease fitting
7. "O" ring	16. Snap ring	25. Shifter shaft	34. Bellcrank pivot
8. Spring (roll) pin	17. Output gear	26. "O" ring	35. Actuating rod
9. Spool gear	18. Ball bearing	27. Spacer	36. Shifter rod

26. Reassemble and adjust spindle and wheel bearings as follows: Repack wheel bearings with a good grade of multi-purpose lithium grease. Install outer wheel hub bearing (33–Fig. 24) and oil seal (34). Install spindle shaft (3), oil deflector (1) and spacer (2). See Fig. 29. Install inner wheel hub bearing (26–Fig. 24) and tighten bearing nut (25) to preload bearings to a rolling torque of 8 to 16 pounds, using a length of cord and a scale as shown in Fig. 30.

Mount spindle and hub assembly on axle housing flange and seal with gasket material such as (Dow Corning RTV-732 or equivalent). Install nuts and torque nuts to 65-70 ft.-lbs. Place ring gear,

Fig. 27 – View showing removal of retaining ring that secures ring gear, sun gear and spacer in wheel hub.

1. Ring gear
2. Retaining ring
3. Sun gear

spacer and sun gear in hub housing and install snap ring. Install planetary assembly into hub and use new gasket (28–Fig. 24). Install tapered dowels on studs, then install nuts and tighten to a torque of 65-70 ft.-lbs. Turn hub until arrow is pointing down and fill housing with lubricant until it runs out of the filler hole. Replace filler hole plug.

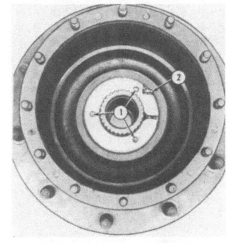

Fig. 28 – View showing bearing locknut with recess holes facing away from bearing. Recess holes (1) are for tightening purposes using a punch and hammer or spanner wrench.

1. Recess holes
2. Cap screw

Fig. 29 – Use wood block and hammer to install spindle shaft (1), spacer (2) and oil deflector (3) into wheel hub. Refer to text.

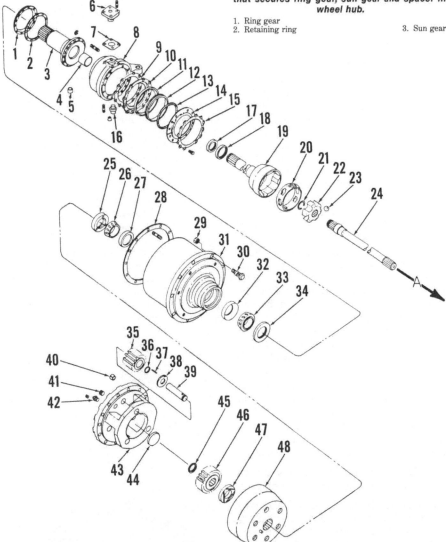

Fig. 24 – Exploded view of Elwood "All-Wheel Drive" axle end housing and steering knuckle. Elwood "All-Wheel Drive" is available on all models equipped with external steering cylinder.

1. Retainer (oil slinger)	13. Tension spring	25. Locknut	37. Needle bearing
2. Retainer (spacer)	14. Oil seal retainer	26. Bearing cone	38. Washer
3. Spindle shaft	15. Oil seal retainer	27. Bearing cup	39. Planet gear pin
4. Bushing	16. King pin	28. Gasket	40. Tapered dowel
5. Steering knuckle	17. Oil seal	29. Nut	41. Plug
bushing	18. Oil seal	30. Stud	42. Relief fitting
6. Top bearing cap	19. Shaft	31. Hub	43. Planetary drive flange
7. Shim	20. Bearing cage	32. Bearing cup	44. Thrust washer
8. Steering knuckle	21. Snap ring	33. Bearing cone	45. Retaining ring
9. Oil seal gasket	22. Race	34. Oil seal	46. Sun gear
10. Oil seal retainer	23. Ball	35. Planet gear	47. Flat spacer
11. Felt seal	24. Shaft	36. Washer	48. Ring gear
12. Oil seal			

Fig. 30 – View showing use of cord and hand scale to measure bearing preload by means of rolling torque. Refer to text.

27. PIVOT BEARINGS. To remove pivot bearings, the planetary wheel hub assembly must be removed as outlined in paragraph 25.

With hub assembly removed, pull axle out of axle housing. Remove cap screws on steering knuckle and separate retainers, seals and gaskets of steering knuckle as shown in Fig. 32.

NOTE: On right side only, disconnect power steering cylinder lines from cylinder and pull pin securing rod end to knuckle. Plug all open hydraulic ports.

Remove steering knuckle axle oil seal (2–Fig. 33). Remove top bearing cap (7–Fig. 34) and shims (6), keeping shims together. Lift out upper bearing (10). Disconnect tie rod (9) from steering arm (16). Remove nuts and tapered dowels from steering arm (1–Fig. 33) and remove steering arm, keeping shims together. Remove steering knuckle (5–Fig. 34) from trunnion.

NOTE: Be sure to catch lower bearing (15 – Fig. 34) to avoid damage to bearing.

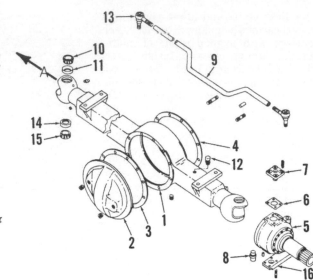

Fig. 34—Exploded view of Elwood "All-Wheel Drive" center axle section.

1. Axle housing
2. Carrier cover
3. Gasket
4. Gasket
5. Steering knuckle
6. Shim
7. Upper bearing cap
8. King pin
9. Tie rod
10. Bearing cone (upper)
11. Bearing cup (upper)
12. Air vent
13. Tie rod end
14. Bearing cup (lower)
15. Bearing cone (lower)
16. Steering arm (lower bearing cap)

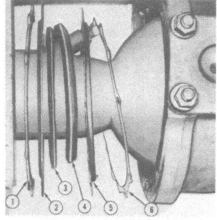

Fig. 32 – Disassembled view of steering knuckle and trunnion showing placement order of retainers, seals and gaskets.

1. Outer split retainer
2. Seal retainer
3. Oil seal
4. Felt seal
5. Felt retainer
6. Gasket

Fig. 33 – View of steering knuckle with spindle shaft and drive axle removed.

1. Steering arm (lower bearing cap)
2. Axle shaft oil seal
3. Upper bearing cap

28. Inspect bearing cones and cups for wear and renew as necessary. Lubricate pivot bearings and install in steering knuckle. Install steering arm over lower bearing and upper bearing cap over bearing using original shim packs for each bearing. Torque upper and lower bearing cap retaining nuts to 60-70 ft.-lbs. Check adjustment of bearing preload by placing torque wrench on upper bearing cap nut and swing steering knuckle from side to side. Proper adjustment should require 15 to 25 ft.-lbs. of torque to pivot steering knuckle. Add or remove upper or lower bearing shims until specified torque is obtained. Shims are available in thicknesses of 0.005 and 0.010 inch. Bearing preload should be checked without ball seals or hub components installed.

Install new axle shaft oil seal until it bottoms in counterbore. Install gasket, flange, steering ball felt, seal with spring, split retainer ring and retaining ring halves on rear of knuckle housing. See Fig. 32. Install lockwashers and mounting bolts and torque bolts to 10 to 15 ft.-lbs. Pack inside of ball with good grade multi-purpose lithium grease and install axle, spindle and hub assembly.

NOTE: When installing axle shaft be careful not to damage axle shaft oil seal.

When assembling tie rod ends to steering arms, tighten slotted nut securely until slot in nut aligns with hole in pin, then install cotter pin.

Remainder of assembly is reverse of disassembly procedure.

FRONT DRIVE AXLE (ELWOOD)

Models So Equipped

29. **R&R AXLE ASSEMBLY.** To remove the "All-Wheel Drive" front axle and wheels as an assembly, first disconnect tie rods from steering arms and steering cylinder, then remove tie rods. Remove "U" bolts from drive shaft front universal joint and separate universal joint. Remove hydraulic lines to steering cylinder and plug all hydraulic openings. Using two rolling floor jacks, place one under differential housing, the other under steering cylinder mounting bracket. Support front of tractor with suitable jackstands. Remove mounting bolts securing axle assembly to upper bolster and lower assembly. Roll assembly out from under tractor.

30. **OVERHAUL.** Overhaul of axle assembly can be accomplished in three separate operations; planetary assembly, drive axle universal joint and

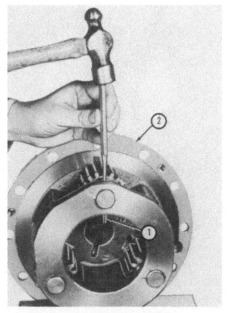

Fig. 36—View showing removal of roll pin from planet gear pin. Refer to text.

1. Roll pin
2. Planetary flange assembly

differential. The planetary assembly and drive axle operations can be accomplished without removal of complete axle assembly from tractor. The differential overhaul will be easier if the center section is removed from tractor.

31. PLANETARY ASSEMBLY. Planetary assembly can be overhauled as follows: Remove planetary housing (end cover) from wheel hub as outlined in paragraph 25. Tap roll pin (1 – Fig. 36) from planet gear pin with punch. Remove planet pin and gear.

NOTE: The planet pin serves as the inner race for needle bearings inside the planet gear, therefore the planet pin must be absolutely smooth. See Fig. 37.

Remove needle bearings and spacer from planet gear.

NOTE: The planet spacer is located between sets of needle bearings inside the planet gear.

Clean and inspect all parts. Renew worn or damaged parts as necessary. Assembly is the reverse of disassembly. Torque planetary housing retaining nuts to 65-70 ft.-lbs.

32. AXLE DRIVE SHAFT. Overhaul of axle drive shaft universal joint is as follows: Remove planetary assembly, wheel hub and shaft spindle as outlined in paragraph 25. Pull axle shaft from center section, being careful not to damage axle shaft oil seal. Place axle in a soft jawed vise. Hold shaft end and strike backside of joint housing with a soft faced hammer as shown in Fig. 38.

Remove lock ring from shaft and discard. Use new lock ring when assembling. See Fig. 39.

Tilt inner race as shown in Fig. 40 until one ball bearing can be removed. Repeat this procedure removing all ball bearings. Rotate universal joint cage 180° to the outer race bell. Roll cage until the two elongated slots in cage align with opposite teeth of outer race bell. See Fig. 41. Lift out ball cage.

Turn inner race in the ball cage at a right angle. Align notched tooth of inner race with the elongated slot in ball cage and roll out inner race. See Fig. 42.

Clean and inspect all parts. Renew any worn or damaged parts. Repack universal joint with good lithium based grease. Reassembly is reverse of disassembly procedure.

33. DIFFERENTIAL UNIT. To remove differential unit, refer to Fig. 43 and proceed as follows: Support tractor front with suitable jack stands, remove front wheels and attach hoist to wheel hub. Remove steering knuckle pivot ball retainers, gaskets and seals. Remove upper and lower pivot bearing caps, then remove bearing cones. Pull wheel hub assembly and axle out of center section being careful not to damage axle shaft seal.

Disconnect drive shaft from differential. Disconnect tie rods from steering cylinder. Use two floor jacks, one under the differential housing, the other under steering cylinder mounting bracket.

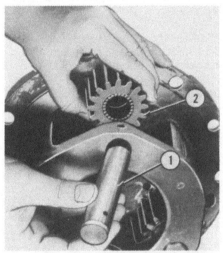

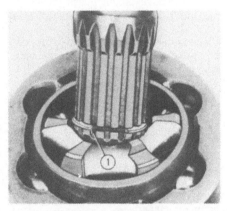

Fig. 39 — View of drive axle universal joint showing retaining ring (1). Renew ring upon reassembly. Refer to text.

Fig. 41 — View showing rotation of ball cage in bell for removal. Refer to text.

1. Ball cage   2. Outer race bell

Fig. 37 — View showing removal of planet gear and planet gear pin. Pin must be absolutely smooth. Note needle bearings inside planet gear.

1. Planet gear pin   2. Planet gear

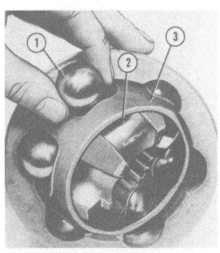

Fig. 38 — Tap back face of universal joint on axle shaft (1) to gain access to retaining ring.

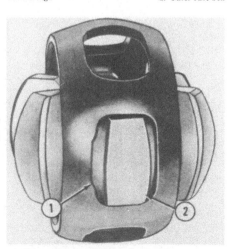

Fig. 40 — Rotate bearing cage in race bell and remove ball as shown. Repeat rotation for removal of all balls. Refer to text.

1. Ball
2. Inner race   3. Bearing cage

Fig. 42 — View of ball cage showing small tooth of inner race in elongated slot of ball cage. Refer to text.

1. Elongated slot   2. Small tooth of inner race

Fig. 43 — View showing removal of wheel hub, steering knuckle and drive axles as one complete assembly. Refer to text.
1. Steering arm (lower bearing cap)
2. Axle end and knuckle assembly
3. Upper bearing cap
4. Retainers and seals

Fig. 45 — Index differential bearing caps using a center punch and hammer. Refer to text.
1. Bearing cap   2. Index mark

Remove mounting bolts securing axle assembly to mounting saddle. Lower axle and roll from under tractor.

With axle center unit removed, remove differential hex nuts and tapered dowels. Using a suitable hoist, lift differential assembly from axle housing. Remove thrust screw and bronze thrust pad (2 and 3 – Fig. 44) from side of housing.

NOTE: Bronze pad is a separate serviceable item.

Using a center punch, index bearing cap and carrier for correct reassembly as shown in Fig. 45. Remove adjusting lockclips (25 – Fig. 44) from top of bearing cap. Remove bearing cap mounting bolts and lockwashers, then lift off bearing caps.

Place a piece of pipe or a heavy metal bar through differential carrier and hoist out assembly as shown in Fig. 46. Place unit on clean work bench and remove bearing adjusting nuts (4 – Fig. 44) and bearing cups (15). Remove differential bearings (16) using a suitable puller. Index differential halves to ring gear to assure correct wear pattern on assembly (3 – Fig. 46). Remove mounting bolts securing differential halves and ring gear, then separate differential halves. Remove pinion gears (20 – Fig. 44), side gears (18), thrust washers (19) and spider unit (21).

Remove slotted hex nut (8 – Fig. 44) and yoke (6). Index pinion bearing flange (5) to housing and remove. Remove oil seal (9). Renew oil seal upon reassembly.

Tap rear side of pinion shaft with brass drift punch and hammer to start pinion shaft out of carrier.

NOTE: Hold pinion shaft in hand while tapping to prevent shaft from falling out. Due to weight and fit of pinion shaft, it could easily fall out unexpectedly and become damaged.

If bearings need renewal, use a suitable puller to remove. When installing front pinion bearing (10 – Fig. 44), drive on inner race only, until collar on inner race seats against pinion head. Press rear pinion bearing (12) on shaft until chamfered side of inner race seats against shoulder on pinion shaft. After rear bearing is seated, install lock ring (13).

Clean and inspect all bearings, gears and shafts for wear or damage. Clean and inspect differential case assembly for cracks, damage or distortion. Inspect splined ends of axle shafts for twisting or cracks. Use new gaskets

Fig. 46 — Hoist differential and ring gear assembly out of carrier using a steel pipe and sling hoist. Refer to text.
1. Steel pipe
2. Sling hoist   3. Index marks

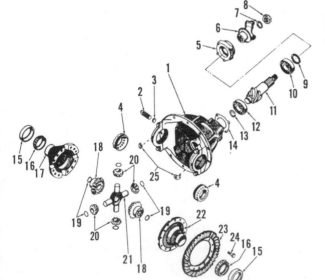

Fig. 44 — Exploded view of Elwood "All-Wheel Drive" differential assembly and carrier housing.
1. Carrier housing
2. Thrust pad screw
3. Thrust pad
4. Adjusting ring nuts
5. Bearing retainer flange
6. Yoke
7. Washer
8. Slotted nut
9. Oil seal
10. Outer pinion bearing
11. Pinion shaft
12. Inner pinion bearing
13. Snap ring
14. Gasket
15. Bearing cup
16. Bearing cone
17. Differential case cover
18. Side gears
19. Thrust washers
20. Pinion gears
21. Spider
22. Differential case cover
23. Ring gear
24. Cap screw
25. Adjusting ring lock clips

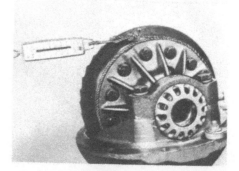

Fig. 47 — Attach spring scale and cord to measure rolling torque of bearing preload. Refer to text.

when reassembling. Renew any of the above mentioned items which show signs of wear or damage.

34. **INSTALLATION AND ADJUST-MENT.** Install pinion shaft assembly into differential carrier. Install pinion bearing flange and pinion shaft oil seal over pinion shaft and torque bolts to 45-50 ft.-lbs. Install yoke assembly, washer and slotted nut. Tighten nut to 160-180 ft.-lbs. torque. Noting position of index marks, assemble differential halves. Tighten ring gear bolts and lock washers evenly until ring gear is flush with case flange. Tighten ring gear bolts to 85-95 ft.-lbs. torque.

Place differential in carrier housing. Index and install bearing caps, tightening cap bolts until lockwashers just flatten out. Tighten adjusting ring nuts until zero end play is obtained on differential bearings. Wrap a cord around O.D. of differential and use a hand spring scale to measure rolling torque. See Fig. 47. Tighten each ring nut, one notch at a time, until an effort of 10 to 16 pounds is required to steadily rotate differential.

Next, install a dial indicator as shown in Fig. 48 and check ring gear radial

runout. If runout exceeds 0.004 inch, remove differential and check for warpage or other causes.

Install dial indicator as shown in Fig. 49 and check ring and pinion backlash. Backlash should be 0.008 to 0.010 inch. Adjust backlash by turning both adjusting ring nuts in same direction, equal amounts until backlash specifications are obtained.

NOTE: When original ring and pinion is installed, the wear patterns must be considered in backlash adjustment. Gears that have long term wear patterns must not have the backlash changed greatly. Backlash should not be reduced so that gears mesh deeper than they did when new.

After rolling torque and backlash are set, tighten bearing cap bolts to 190-220 ft.-lbs. torque, then recheck backlash and rolling torque. Install adjusting ring nut locks.

Inspect bronze thrust pad and renew if necessary. Install thrust pad and bolt, turning in until pad engages back face of ring gear. Back off bolt 1/12-turn, install and tighten locknut to 125-140 ft.-lbs. torque.

NOTE: Be sure that thrust bolt does not turn when tightening locknut.

This provides 0.005 to 0.007 inch clearance between thrust pad and ring gear. Remainder of installation is reverse of disassembly procedure. Tighten nuts securing differential assembly into center axle section to 75-85 ft.-lbs. torque.

DRIVE HOUSING ASSEMBLY TRANSFER CASE (ELWOOD)

Models So Equipped

The drive housing which transmits power to the front axle is mounted on left front side of tractor rear frame and is driven by an idler gear on reverse idler shaft of the range transmission.

35. **REMOVE AND REINSTALL.** To remove drive housing, remove shield, then remove "U" bolts of universal joint located closest to drive housing and separate the universal joint. Use tape or some other suitable means to retain universal bearing cups on spiders. Disconnect shifter rod from shifter shaft lever, drain rear frame, then unbolt and

Fig. 48—Set up dial indicator as shown to measure differential bearing side play and radial runout.

1. Dial indicator   2. Pry bar

Fig. 49—Set up dial indicator as shown to measure ring and pinion backlash. Refer to text.

1. Dial indicator

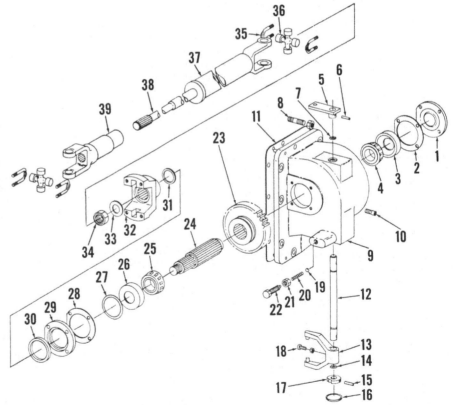

Fig. 50—Exploded view of Elwood "All-Wheel Drive" transfer case and drive shaft.

1. Bearing cap	11. Gasket	21. Nut
2. Gasket	12. Shift shaft	22. Adjusting bolt
3. Bearing cup	13. Shifter fork	23. Gear
4. Bearing cone	14. "O" ring	24. Output shaft
5. Shift lever	15. Pin	25. Bearing cone
6. Pin	16. Expansion plug	26. Bearing cup
7. "O" ring	17. Shift lock	27. Shim
8. Stud	18. Set screw	28. Gasket
9. Transfer case	19. Ball 7/16-in.	29. Seal retainer
10. Allen head cap screw	20. Spring	30. Seal
		31. Spacer
		32. Yoke
		33. Washer
		34. Nut
		35. "U" bolt
		36. Spider
		37. Shaft
		38. Shaft
		39. Slip yoke

remove drive housing from tractor rear frame.

NOTE: Back plate (1 – Fig. 50) will have to be removed to allow Allen head cap screw (10) to be removed to accomplish housing removal.

When reinstalling the unit on tractor, use a new gasket (11). Gaskets are all the same thickness therefore gear backlash settings are pre-set without further measurements or adjustments. Install Allen head cap screw (10), then install rear bearing cap (1). Remainder of installation is the reverse of removal procedure. Refill rear frame to proper level with Hy-Tran fluid.

36. **OVERHAUL.** With unit removed as outlined in paragraph 35, wedge gears and remove Esna nut which retains universal joint yoke (32 – Fig. 50) to output shaft and remove yoke from shaft. Remove seal cap (29), spacer (31) and gasket (28).

NOTE: Keep shims (27) together for reassembly.

Loosen control lever adjusting bolt (22) and press shaft (24) out rear of housing. This will also press bearing cup (3) out of housing. Using a brass drift press bearing cone (4) from shaft. Using a brass drift, drive out bearing cup (26). Remove roll pin (6), then remove control lever (5). Loosen locknut on jam bolt (18) and drive shift shaft (12) through fork (13) and out of housing. Remove roll pin (15) and remove lock (17). Remove and discard "O" rings (7 and 14).

Clean and inspect all parts for scoring, undue wear, chipped teeth or other damage and renew as necessary.

Reassemble by reversing the disassembly procedure. Use new "O" rings, seals and gasket. Set bearings to zero end play, shimming as required to achieve specified bearing preload. Installed shaft (24) should have 5 to 8 in.-lbs. rolling torque.

NOTE: Leave back plate (1) off to allow installation of Allen head cap screw when installing on tractor.

Tighten gear shift tension adjusting bolt (22) so that 35 to 40 ft.-lbs. torque are required to shift lever position.

When installing housing to tractor, use new gasket and brass washers on end plate (1). Install drive shaft with slip joint end on gear box yoke. Add Hy-Tran fluid to rear frame to proper level.

DRIVE SHAFT

Models So Equipped

The drive shaft between drive housing and the front axle differential is conventional and can be removed and serviced as outlined in paragraph 24.

POWER STEERING SYSTEM

NOTE: The maintenance of absolute cleanliness of all parts is extremely important in the operation and servicing of the hydraulic power steering system. Of equal importance is the avoidance of nicks or burrs on any working parts.

OPERATION

All Models

37. Power steering is standard equipment on all tractors and, except for steering cylinders, components for all tractors are similar. Refer to Figs. 51 and 52 for views showing the general layout of component parts.

The pressurized hydraulic fluid used for power steering, power brakes and the driven clutches in the torque amplifier is supplied by a 12 gpm pump mounted on the inner cover plate of the multiple control valve (MCV). The 12 gpm pump output passes through a flow control mechanism or flow divider located in the multiple control valve body. This sends a five gpm priority flow to operate the power steering, brakes and torque amplifier driven clutches while the remaining seven gpm flow is used to cool the fluid and lubricate the differential gears before returning to the main reservoir.

The five gpm priority flow to the power steering has return oil flow controlled by the pressure regulating valve located in the MCV. A secondary priority of one gpm is taken to operate the brakes when applied. The remaining five gpm is used to operate hydraulic pistons which engage or disengage clutches in the torque amplifier. Once either clutch is engaged, there is no flow of fluid therefore continuing the five gpm flow to do other work. A regulating valve in the MCV once again reduces pressure of the five gpm flow and directs it to lubricate the torque amplifier clutches before returning to the main reservoir. On tractors without torque amplifier the five gpm flow goes directly to lubrication.

LUBRICATION AND BLEEDING

All Models

38. The tractor rear frame serves as the main reservoir for all hydraulic and lubricating operations. The filter, shown in Fig. 53, should be renewed at 20, 100 and 200 engine hours and every 200 hours of operation thereafter. The tractor rear frame should be drained and new fluid added every 1000 hours or once a year, whichever occurs first.

Only IH "Hy-Tran" fluid should be used and the level should be maintained at the "FULL" mark shown on dip stick which is located at the back of rear frame on all models.

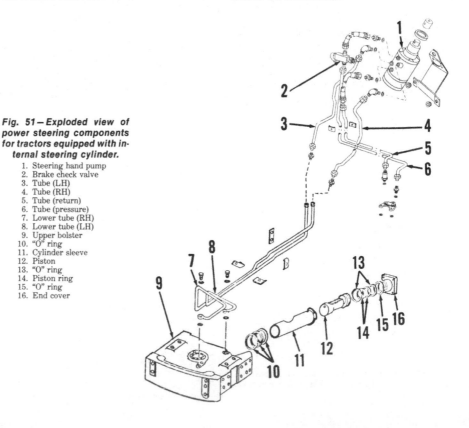

Fig. 51 – Exploded view of power steering components for tractors equipped with internal steering cylinder.

1. Steering hand pump
2. Brake check valve
3. Tube (LH)
4. Tube (RH)
5. Tube (return)
6. Tube (pressure)
7. Lower tube (RH)
8. Lower tube (LH)
9. Upper bolster
10. "O" ring
11. Cylinder sleeve
12. Piston
13. "O" ring
14. Piston ring
15. "O" ring
16. End cover

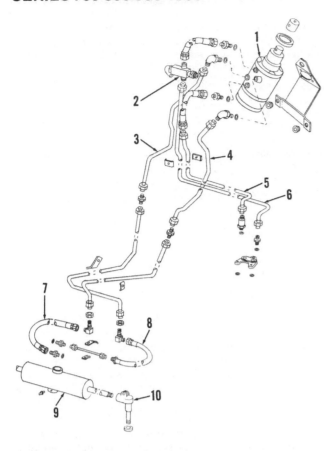

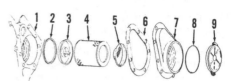

Fig. 53 — Exploded view of hydraulic filter assembly used on all models.

1. Rear frame
2. Retainer seal
3. Retainer
4. Cartridge
5. By-pass valve
6. Gasket
7. Filter frame
8. "O" ring
9. Filter frame cover

Fig. 52 — Exploded view of power steering components for tractors equipped with external steering cylinder.

1. Steering hand pump
2. Brake check valve
3. Tube (LH)
4. Tube (RH)
5. Tube (return)
6. Tube (pressure)
7. Hose (RH)
8. Hose (LH)
9. Steering cylinder
10. Ball joint end

equipped, and connect inlet hose of a 14-51D Flo-Rater or equivalent. See Fig. 56. Connect outlet hose of Flo-Rater to any auxiliary coupler for fluid to return to system. Start engine and operate until hydraulic fluid reaches approximately 100 to 150 degrees F. With engine running at 2400 rpm turn steering wheel until it stops, then continue holding wheel against stop while restricting Flo-Rater to 1200 psi. Multiple control valve pump should meet a minimum requirement of nine gpm. If specification is not met check for internal leakage or faulty pump. Continue holding steering wheel against stop and close down Flo-Rater restrictor valve. System relief pressure should read 1900 to 2350 psi for tractors equipped with external steering cylinder and 1600 psi for tractors with internal steering cylinder. If pressures are not as specified, renew relief valve (8 – Fig. 55) located at the bottom of multiple control valve. Relief valve is available as a unit only.

OPERATIONAL TESTS

All Models

41. The following tests are valid only when the power steering system is completely void of any air. If necessary, bleed system as outlined in paragraph 38 before performing any operational tests.

42. **MANUAL PUMP.** With multiple control valve (MCV) pump inoperative

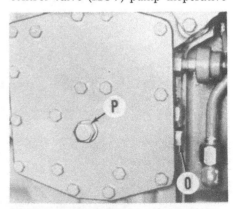

Fig. 54 — View of multiple control valve cover. Remove plug "O" for installation of Flo-Rating equipment when checking operating and system relief pressures. If equipped, use chamber "P" for Flo-Rater installation. Refer to text.

Whenever power steering lines have been disconnected, or fluid drained, start engine and cycle power steering system by turning steering wheel from stop to stop several times to bleed air from the system, then if necessary, check and add fluid to the reservoir.

TROUBLESHOOTING

All Models

39. The following table lists some of the troubles which may occur in the operation of the power steering system. When the following information is used in conjunction with the information in the Power Steering Operational Tests section (paragraphs 41 through 46), no trouble should be encountered in locating system malfunctions.

1. No power steering or steers slowly.
   a. Binding mechanical linkage.
   b. Excessive load on front wheels and/or air pressure low in front tires.
   c. Steering cylinder piston seal faulty or cylinder damaged.
   d. Faulty power steering supply pump.
   e. Faulty commutator in hand pump.
   f. Flow divider valve spool sticking or leaking excessively.
   h. Circulating check ball not seating.
   i. Flow control valve orifice plugged.

2. Will not steer manually.
   a. Binding mechanical linkage.
   b. Excessive load on front wheels and/or air pressure low in front tires.
   c. Pumping element in hand pump faulty.
   d. Faulty seal on steering cylinder or cylinder damaged.
   e. Pressure check valve leaking.
   g. Check valve in clutch housing inlet tube stuck in closed position.

3. Hard steering through complete cycle.
   a. Low pressure from supply pump.
   b. Internal or external leakage.
   c. Faulty steering cylinder.
   d. Binding mechanical linkage.
   e. Excessive load on front wheels and/or air pressure low in front tires.
   f. Cold hydraulic fluid.

4. Momentary hard or lumpy steering
   a. Air in power steering circuit.

OPERATING PRESSURE AND RELIEF VALVE

All Models

40. System operating pressure and relief valve operation can be checked as follows: Remove small orifice chamber plug (O – Fig. 54) which is the bottom plug located on rear side of multiple control valve, or test plug (P – Fig. 54) if

(engine not running), attempt to steer manually in both directions.

NOTE: Manual steering with MCV pump not running will require high steering effort. If manual steering can be accomplished with MCV pump inoperative, it can be assumed that the manual steering pump will operate satisfactorily with the MCV pump operating.

43. STEERING WHEEL SLIP (CIRCUIT TEST). Steering wheel slip is the term used to describe the inability of the steering wheel to hold a given position without further steering movement. Wheel slip is generally due to leakage, either internal or external, or a faulty hand pump or steering cylinder. Some steering wheel slip, with hydraulic fluid at operating temperature, is normal and permissible. A maximum of one revolution per minute is acceptable. By using the steering wheel slip test and a process of elimination, a faulty unit in the power steering system can be located.

However, before making a steering wheel slip test to locate faulty components, it is imperative that the complete power steering system be completely free of air before any testing is attempted.

To check for steering wheel slip (circuit test), proceed as follows: Check reservoir (rear frame) and fill to correct level, if necessary. Bleed power steering system, if necessary. Bring power steering fluid to operating temperature, cycle steering system until all components are approximately the same temperature and be sure this temperature is maintained throughout the tests. Remove

steering wheel cap (monogram), then turn front wheels until they are against stop. Attach a torque wrench to steering wheel nut.

NOTE: Either an inch-pound, or a foot-pound wrench may be used. Advance hand throttle until engine reaches rated rpm, then apply 72 inch-pounds (6 foot-pounds) to torque wrench in the same direction as the front wheels are positioned against the stop. Keep this pressure (torque) applied for a period of one minute and count the revolutions of the steering wheel. Use same procedure and check the steering wheel slip in the opposite direction. One revolution per-minute in either direction is acceptable and system can be considered as operating satisfactorily. If, however, the steering wheel revolutions per minute exceed the maximum, record the total rpm for use in checking the steering cylinder or hand pump.

44. STEERING CYLINDER TEST. If steering wheel slip, as checked in paragraph 43 exceeds the maximum one revolution per minute, proceed as follows: Be sure fluid is at operating temperature, then disconnect and plug the hydraulic lines to steering cylinder and repeat wheel slip test as outlined in paragraph 43. If steering wheel slip is less than that recorded from previous test, overhaul or renew steering cylinder.

45. FLOW DIVIDER. When checking flow divider operation, also check orifice (7 – Fig. 55) to see that it is open and clean. This orifice is the unit that meters the five gpm used to operate the power steering.

To check operation of the flow divider and orifice proceed as follows: Disconnect steering hose (D-Fig. 57) coming from top of brake check valve (B) and install an in-line "T" fitting (C). Connect inlet hose of a 14-51D Flo-Rater or equivalent to the "T" fitting. Connect outlet hose of Flo-Rater to return port of any auxiliary coupler. With engine operating at 2400 rpm and hydraulic fluid at an operating temperature of 100 to 150 degrees F., hold steering wheel against the stop and restrict the Flo-Rater to 1200 psi. The flow from flow divider to steering hand pump should read approximately 4.5 gpm. If fluid flow is below specification, service flow divider as outlined in paragraph 52. If flow is not as specified after flow divider service, check multiple control valve pump. Bleed power steering system after reconnecting lines.

46. HYDRAULIC PUMP. To check hydraulic pump operating pressure, refer to procedure outlined in paragraph 40. To check hydraulic pump free flow on all models, proceed as follows: Remove orifice chamber plug (O – Fig. 54) then remove orifice (7 – Fig. 55) using orifice plug tool 14-556-3.

NOTE: If orifice plug tool is not available, removal of orifice will be simplified if a small wire is attached to a screwdriver and extended about two inches beyond the end of bit. Insert wire through orifice hole and wire will support orifice during removal from multiple control valve.

Disconnect lower oil cooler line from front side of multiple control valve and place a container under opening to catch fluid. Connect inlet hose of a 14-51D Flo-Rater to the flow divider port and connect outlet hose to any auxiliary valve

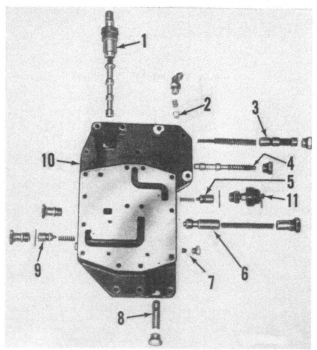

Fig. 55 — Multiple control valve with front cover removed and showing control valves removed from their bores.

1. Drive selector valve
2. Brake check valve
3. Pressure regulator valve
4. Clutch dump valve
5. Lubrication regulator valve
6. Flow divider valve
7. Flow control orifice
8. Safety relief valve
9. Oil cooler by-pass valve
10. Multiple control valve body
11. Pressure switch

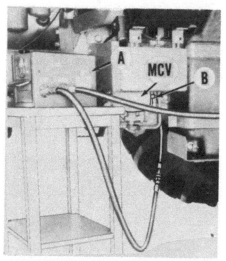

Fig. 56 — View showing the Flo-Rater connected for checking operating and relief pressures and pump flow.

A. 14-51D Flo-Rater
B. Orifice chamber connection
MCV. Multiple Control Valve

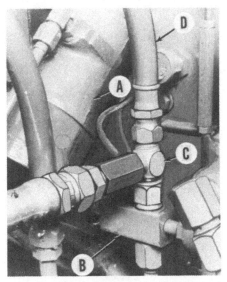

Fig. 57—View of firewall showing brake check valve with "T" connector installed for Flo-Rater hook-up to check steering flow from flow divider.

A. Steering hand pump
B. Brake check valve
C. "T" connector
D. Steering pressure supply line

return coupler. If Flo-Rater is not available use a piece of hose connected to the flow divider port and place other end in a suitable container. With engine operating at 2400 rpm and Flo-Rater restrictor valve fully open, flow should be 12 gpm, with a slight leakage from the open oil cooler port.

If pump free flow is not as specified, remove and service pump as outlined in paragraph 47.

PUMP

All Models

47. **REMOVE AND REINSTALL.** To remove the power steering pump, drain hydraulic fluid from clutch hous-

ing. Disconnect torque amplifier operating rod from the operating lever (T–Fig. 58), brake pressure supply line (B) and the wire from transmission oil pressure switch (P). Disconnect clutch operating rod from clutch pedal and clutch release shaft. Disconnect operating link from the torque amplifier selector valve. Unbolt and remove dump valve pivot bracket (C). Disconnect supercharge tube (S) and oil cooler lines (O). Remove all multiple control valve mounting cap screws except for pump retaining screws (B–Fig. 59) and remove multiple control valve and pump assembly. See Fig. 60.

NOTE: During removal of pump assembly be very careful not to lose or damage the small check valve and spring located in clutch housing as shown in Fig. 61 or (7–Fig. 62). This check valve allows fluid to be drawn into power steering circuit when steering with engine inoperative.

Pump can now be removed from multiple control valve by removing the four cap screws (B–Fig. 59).

Before installing pump, measure the gasket located between multiple control valve and clutch housing with a micrometer, to determine which thickness gasket to use. These gaskets are available in light (0.011-0.019 inch) and heavy (0.016-0.024 inch) thicknesses. In addition to sealing, the gasket also controls backlash between pump drive gear and the driving gear.

When reinstalling pump and valve assembly, use new gasket of same thickness as original and reinstall assembly by reversing removal procedure.

48. Overhaul procedure for the Cessna 12 gpm power steering pump will be covered in the following paragraph.

Fig. 60—Multiple control valve removed from tractor showing the power steering pump.

49. **OVERHAUL (CESSNA 12 GPM).** With pump removed as outlined in paragraph 47, remove stop nut holding pump gear, then remove gear and key. Remove cover (2–Fig. 63). Do not pry cover off. The pump body and cover will damage easily. All parts are now accessible for inspection and/or renewal. Items 5, 6, 7, 8 and 15 are not available separately and must be ordered in a package which includes all the "O" rings and seals.

When reassembling, use new diaphragms, gaskets, back-up washers, diaphragm seal and "O" rings. Press new drive shaft oil seal (1) into front cover (2). Be sure lip of seal faces **inward** (note position of original seal). Stake the cover to prevent seal from loosening. Install diaphragm seal (5), protector gasket (6) and back-up gasket (7) into cover making sure seals and gaskets are fully worked into the grooves of cover. Install wear plate (15) in pump body with bronze side toward pump gears and cut-out portion toward inlet (suction) side of pump. Place check ball and spring (3 and 4) in suction side of cover and place pressure diaphragm (8) on cover with

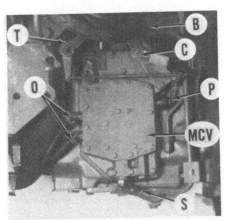

Fig. 58—View of multiple control valve (MCV) for 86 series tractors showing control linkage and hydraulic connections.

B. Brake pressure tube
C. Torque amplifier valve operating pivot
P. Transmission oil pressure switch
S. Supercharge line
T. Dump valve operating lever
O. Oil cooler lines

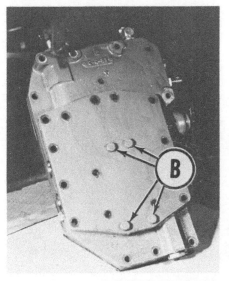

Fig. 59—Multiple control valve showing the four cap screws at (B) retaining hydraulic pump to inner face.

Fig. 61—View showing tractor with multiple control valve removed. Use caution not to lose or damage the small check valve and spring shown at (CV). Pump is driven by the pto driven gear (G).

INTERNATIONAL HARVESTER

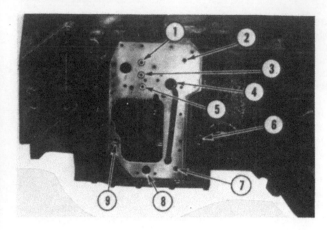

Fig. 62—View of clutch housing with multiple control valve and pump removed showing hydraulic ports for various controls of the MCV.

1. TA lubrication
2. Steering circuit return
3. Direct drive
4. To differential lube
5. Torque amplifier (TA)
6. Hydraulic clutch assist return
7. Sump check valve
8. Pump supply
9. To steering

Fig. 64—Exploded view of telescoping steering column.

1. Cap
2. "O" ring
3. Wheel
4. Nut (35 ft.-lbs.)
5. Collar
6. "O" ring
7. Bushing
8. Washer
9. Shaft
10. Tube
11. Bushing

bronze side toward gears. Install drive gear and driven gear (10) in cover and place pump body over the gears and on-to front cover. Pump drive gear must be in bore nearest suction port.

IMPORTANT: The shaft oil seal must be "worked" over the shoulder of the drive gear shaft to prevent the lip from being pushed outward as the shaft is installed.

Torque cap screws to 45 ft.-lbs. After pump has been completely assembled, pump shaft should turn freely by hand without tight spots. A slight drag is always present because of the shaft seal. Fill pump with oil prior to mounting on multiple control valve. Torque stop nut holding drive gear to 60 ft.-lbs.

Refer to the following table for dimensional information.

Cessna 12 gpm Pump

O.D. of shafts at bushings (min.)	0.810 in.
I.D. of bushing in body and cover (max.)	0.816 in.
Thickness (width) of gears (min.)	0.572 in.
I.D. of gear pockets (max.)	2.002 in.
Max. allowable shaft to bushing clearance	0.006 in.

HAND PUMP

All Models

50. **REMOVE AND INSTALL.** To remove the power steering hand pump, first remove adjusting collar on telescop-

ing steering column (Fig. 64) and pull steering wheel and shaft out of housing. Remove knobs from throttle and torque amplifier (TA) handles. Instrument cowling and dash panel assemblies (Fig. 65) are then removed carefully to avoid damaging necessary wiring to the panel. The housing and nylon bushing are removed by pulling the shaft up. The pump is located on the firewall. Remove left hand hood sheet and clean the area around pump and hydraulic lines thoroughly (Fig. 66). Disconnect necessary hydraulic lines at pump and cap all lines. Remove the three mounting bolts and remove pump. Reassemble by reversing disassembly procedure. To remove steering wheel from steering column, place shaft in brass jawed vice, remove cap (monogram) from wheel, remove nut, attach puller and remove wheel.

51. **OVERHAUL STEERING MOTOR.** To disassemble the removed steering motor and control valve assembly, refer to Fig. 67 and proceed as follows: Install a fitting in one of the four ports in valve body (25), then clamp fitting in a vise so that input shaft (17) is pointing downward. Remove cap screws (39) and remove end cover (38).

NOTE: Lapped surfaces on end cover (38), commutator set (33 and 34), manifold (32), stator-rotor set (31), spacer (29) and valve body (25) must be protected from scratching, burring or any other damage as sealing of these parts depends on their finish and flatness.

Fig. 65—View showing removal of instrument panel and steering column bushing for access to steering hand pump.

C. Column bushing
H. Throttle lever
P. Instrument panel (removed)
T. Torque amplifier lever

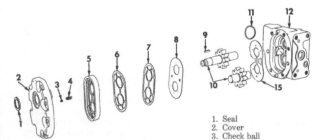

Fig. 63—Exploded view of Cessna 12 gpm power steering pump used on 86 series tractors.

4. Check spring
5. Diaphragm seal
6. Protector gasket
7. Back-up gasket
8. Pressure diaphragm
9. Key
10. Gears and shafts
11. "O" ring
12. Body
15. Wear plate

1. Seal
2. Cover
3. Check ball

Fig. 66—View of firewall from left side showing steering hand pump, hydraulic lines and mounting bolts.

A. Hydraulic lines
B. Mounting bolts
C. Steering hand pump

Remove seal retainer (35) and seal (36), then carefully remove wear washer (37), commutator set (33 and 34) and manifold (32). Grasp spacer (29) and lift off the spacer, drive link (30) and stator-rotor set (31) as an assembly. Separate spacer and drive link from stator-rotor set.

Remove unit from vise, then clamp fitting in vise so that input shaft is pointing upward. Remove water and dirt seal (2) and felt seal (3). Place a light mark on flange of upper cover (9) and valve body (25) for aid in reassembly. Unbolt upper cover from valve body, then grasp input shaft and remove input shaft, upper cover and valve spool assembly. Remove and discard seal ring (10). Slide upper cover assembly from input shaft and remove Teflon spacer (16). Remove shims (12) from cavity in upper cover or from face of thrust washer (14) and note number of shims for aid in reassembly. Remove snap ring (4), stepped washer (5), brass washer (6), Teflon washer (7) and seal (8). Retain stepped washer (5)

and snap ring (4) for reassembly. Do not remove needle bearing (11) as it is not serviced separately.

Remove snap ring (13), thrust washers (14) and thrust bearing (15) from input shaft. Drive out pin (18) and withdraw torsion bar (21) and spacer (20). Place end of valve spool on top of bench and rotate input shaft until drive ring (19) falls free, then rotate input shaft clockwise until actuator ball (23) is disengaged from helical groove in input shaft. Withdraw input shaft and remove actuator ball. Do not remove actuator ball retaining spring (24) unless renewal is required.

Remove plug (28) and recirculating ball (26) from valve body.

Thoroughly clean all parts in a suitable solvent, visually inspect parts and renew any showing excessive wear, scoring or other damage.

If needle bearing (11) is excessively worn or otherwise damaged, renew upper cover assembly (9) as bearing is not serviced separately.

Using a micrometer, measure thickness of the commutator ring (33 – Fig. 67) and commutator (34). If commutator ring is 0.0015 inch or more thicker than commutator, renew the matched set.

Place the stator-rotor set (31) on the lapped surface of end cover (38). Make certain that vanes and vane springs are installed correctly in slots of the rotor.

NOTE: Arched back of springs must contact vanes. (See inset X – Fig. 67). Position lobe of rotor in valley of stator as shown at (V – Fig. 69). Center opposite lobe on crown of stator, then using two feeler gages, measure clearance (C) between rotor lobes and stator. If clearance is more than 0.006 inch, renew stator-rotor assembly. Using a micrometer, measure thickness of stator and rotor. If stator is 0.002 inch or more thicker than rotor, renew the assembly. Stator, rotor vanes and vane springs are available only as an assembly.

Before reassembling, wash all parts in clean solvent and air dry. All parts, unless otherwise indicated, are installed dry. Install recirculating ball (26 – Fig. 67) and plug (28) with new "O" ring (27) in valve body and tighten plug to a torque of 10-14 ft.-lbs. Clamp fitting (installed in valve body port) in a vise so that top end of valve body is facing upward. Install thrust washer (14), thrust bearing (15), second thrust washer (14) and snap ring (13) on input shaft (17). If actuator ball retaining ring (24) was removed install new retaining ring. Place actuator ball (23) in its seat inside valve spool (22). Insert input shaft into valve spool, engaging the helix and actuator ball with a counter-clockwise motion. Use the mid-section of torsion bar (21) as a gage between end of valve spool and thrust washer, then place the assembly in a vertical position with end of input shaft resting on a bench. Insert drive ring (19) into valve spool until drive ring is engaged on input shaft spline. Remove torsion bar gage. Install spacer (20) on torsion bar and insert the

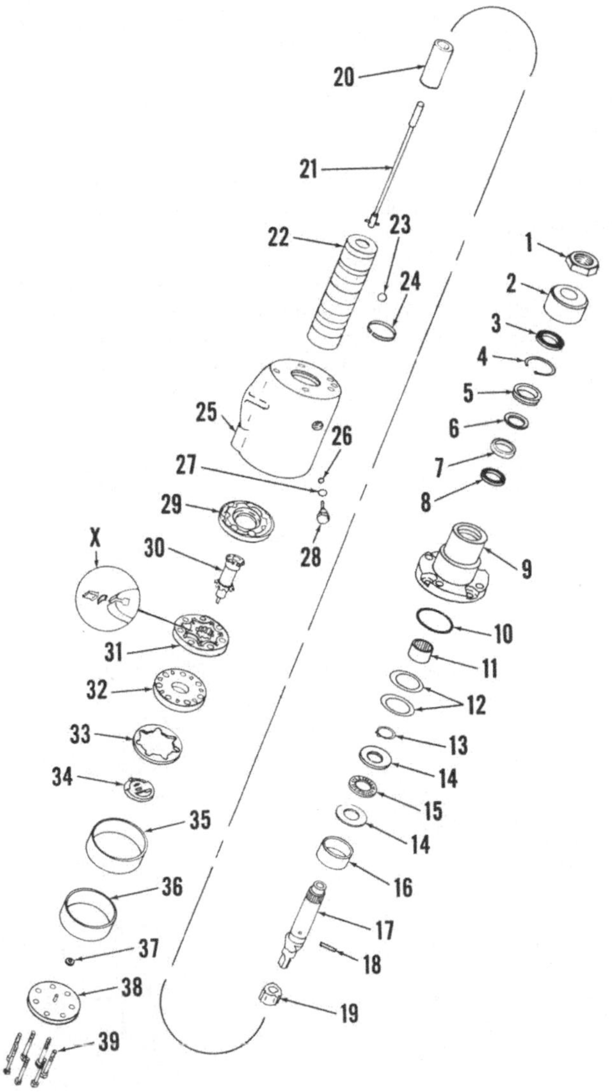

Fig. 67 – Exploded view of Ross steering motor and control valve assembly used on all models. Inset "X" shows vane and vane spring used in slot on each rotor lobe.

1. Nut
2. Water and dirt seal
3. Felt seal
4. Snap ring
5. Stepped washer
6. Brass washer
7. Teflon washer
8. Seal
9. Cover (upper)
10. Seal ring
11. Needle bearing
12. Shims
13. Snap ring
14. Thrust washers
15. Thrust bearing
16. Teflon spacer
17. Input shaft
18. Pin
19. Drive ring
20. Spacer
21. Torsion bar
22. Valve spool
23. Actuator ball
24. Retaining ring
25. Valve body
26. Recirculating ball
27. "O" ring
28. Plug
29. Spacer plate
30. Drive link
31. Stator rotor
32. Manifold
33. Commutator ring
34. Commutator
35. Seal retainer
36. Seal
37. Washer
38. End cover
39. Cap screws

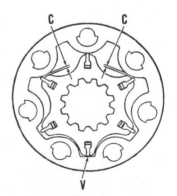

Fig. 69 – With rotor positioned in stator as shown, clearances (C) must not exceed 0.006 inch. Refer to text.

assembly into valve spool. Align cross-holes in torsion bar and input shaft and install pin (18). Pin must be pressed into shaft until end of pin is about 1/32-inch below flush. Place spacer (16) over spool and install spool assembly into valve body. Position original shims (12) on thrust washer (14), lubricate new seal ring (10), place seal ring in upper cover (9) and install upper cover assembly. Align the match marks on cover flange and valve body and install cap screws finger tight. Tighten a worm drive type hose clamp around cover flange and valve body to align the outer diameters, then tighten cap screws to a torque of 18-22 ft.-lbs.

NOTE: If either input shaft (17) or upper cover (9) or both have been renewed, the following procedure for shimming must be used. With upper cover installed (with original shims) as outlined above, invert unit in vise so that input shaft is pointing downward. Grasp input shaft, pull downward and prevent it from rotating. Engage drive link (30) splines in valve spool and rotate drive link until end of spool is flush with end of valve body. Remove drive link and check alignment of drive link slot to torsion bar pin. Install drive link until its slot engages torsion bar pin. Check relationship of spool end to body end. If end of spool is within 0.0025 inch of being flush with end of body, no additional shimming is required. If not within 0.0025 inch of being flush, remove cover and add or remove shims (12) as necessary. Reinstall cover and recheck spool to valve body position.

With drive link installed, place spacer plate (29) on valve body with plain side up. Install stator-rotor set over drive link splines and align cap screw holes. Make certain vanes and vane springs are properly installed. Install manifold (32) with circular slotted side up and align cap screw holes with stator, spacer and valve body. Install commutator ring (33) with slotted side up, then install commutator (34) over drive link end making certain that link end is engaged in the smallest elongated hole in commutator. Install seal (36) and retainer (35). Apply a few drops of hydraulic fluid on commutator. Use a small amount of grease to stick wear washer (37) in position over pin on end cover (38). Install end cover making sure that pin engages center hole in commutator. Align holes and install cap screws (39). Alternately and progressively tighten cap screws while rotating input shaft. Final tightening should be 18-22 ft.-lbs. torque.

Relocate the unit in vise so input shaft is up. Lubricate new seal (8) and carefully work seal over shaft and into bore with lip toward inside. Install new Teflon washer (7), brass washer (6) and

stepped washer (5) with flat side up. Install snap ring (4) with rounded edge inward. Place new felt seal (3) and water and dirt seal (2) over input shaft.

Remove unit from vise and remove fitting from port. Turn unit on its side with hose ports upward. Pour clean hydraulic fluid into inlet port, rotate input shaft until fluid appears at outlet port, then plug all ports.

FLOW DIVIDER

All Models

52. **R&R AND OVERHAUL.** The flow divider valve is located in rear side of multiple control valve assembly and is the second from bottom plug. See Fig. 55. Valve (spool) can be removed after removing plug and spring.

Service of flow divider valve consists of renewing parts. Carefully inspect spool and bore for scratches, grooves or nicks. Spool should fit bore snugly, yet be free enough to slide easily in bore with both spool and bore lubricated. Flow divider (6) spring has free length of 4.08 inches and should test 33.66 pounds when compressed to a length of 2.50 inches.

STEERING CYLINDER

Models with Internal Cylinder

53. **REMOVE AND REINSTALL.** On models so equipped, the power steering cylinder and piston are incorporated into the front support and removal of cylinder requires removal of the front support.

To remove front support, first remove hood, then remove radiator as outlined in paragraph 130. Support tractor and on tricycle models, remove wheels and pedestal from steering pivot shaft. On adjustable wide axle models, unbolt axle support from front support, center steering arm from pivot shaft and stay rod ball from stay rod support, then roll complete assembly away from tractor. On all models, disconnect the power steering lines from front support, then attach hoist to front support, unbolt from side rails and remove from tractor.

NOTE: If necessary, loosen engine front mounting bolts to provide additional removal clearance for front support.

Reinstall by reversing removal procedure and bleed power steering system as outlined in paragraph 38.

54. **OVERHAUL.** Before disassembling power steering cylinder, match mark the cylinder flange and front support. The cylinder sleeve (6 – Fig. 70) is bored off-center and the cam action resulting from rotating the cylinder will regulate the backlash between the teeth of pivot shaft (18) and teeth of piston (5). If the backlash prior to disassembly is satisfactory, the marks installed will provide the correct backlash during reassembly. If backlash is not correct, a point of reference will be established to simplify the backlash adjustment. Backlash should be adjusted to as near zero as possible without binding.

To remove the power steering cylinder and piston, first remove cap from top of pivot shaft, then straighten lock on nut, remove nut from top end of pivot shaft

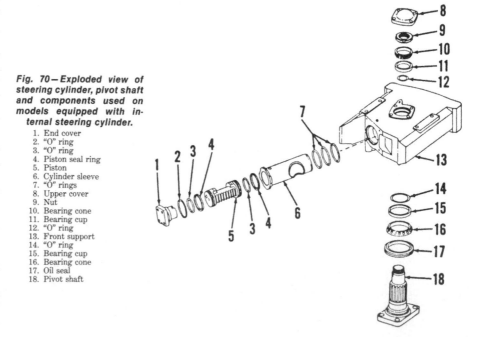

Fig. 70 – Exploded view of steering cylinder, pivot shaft and components used on models equipped with internal steering cylinder.

1. End cover
2. "O" ring
3. "O" ring
4. Piston seal ring
5. Piston
6. Cylinder sleeve
7. "O" rings
8. Upper cover
9. Nut
10. Bearing cone
11. Bearing cup
12. "O" ring
13. Front support
14. "O" ring
15. Bearing cup
16. Bearing cone
17. Oil seal
18. Pivot shaft

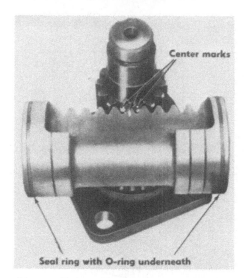

Fig. 71—Pivot shaft and piston are correctly meshed when timing marks align as shown.

and pull pivot shaft from bottom of front support. Remove cylinder end cover and pull cylinder and piston from upper bolster. Piston can now be removed from cylinder. Any further disassembly required is obvious.

Reassembly is the reverse of disassembly. Renew all "O" rings and seals. When installing pivot shaft be sure to position the marked center tooth of piston rack between the two punch marked teeth of the pivot shaft as shown in Fig.71. Tighten pivot shaft nut to provide a slight preload on pivot shaft and secure by bending lock flange on nut into notch in pivot shaft. Lobes are provided on flange of cylinder so mesh position of pivot shaft and piston can be adjusted to zero backlash (at straight ahead position) by using a small punch and hammer.

Models with External Cylinder

55. **R&R AND OVERHAUL.** To remove the external power steering cylinder, first disconnect lines from cylinder and plug lines to prevent oil drainage. Disconnect steering cylinder from center steering arm and cylinder rod anchor pin and remove from tractor.

56. With cylinder removed, move piston rod back and forth several times to clear oil from cylinder. Refer to Fig. 72 and proceed as follows: Place barrel of cylinder in a vise and clamp vise only enough to prevent cylinder from turning. Using a spanner wrench, turn cylinder head (3) until free end of retaining ring (7) appears in slot of barrel (9). Lift end of ring to outside of barrel and continue turning until all of ring is outside of barrel, lift nib from hole and remove the ring. Pull the piston rod (13) out of anchor end of cylinder. Loosen the set screw in the anchor (1) and remove anchor. Remove head from piston rod. Remove head from other end of barrel using the same procedure as before. All seals, "O" rings and backup washers are now available for inspection and/or renewal.

Clean all parts in a suitable solvent and inspect. Check cylinder for scoring, grooving and out-of-roundness. Light scoring can be polished out by using a fine emery cloth and oil providing a rotary motion is used during the polishing operation. A cylinder that is heavily scored or grooved, or that is out-of-round, should be renewed. Check piston rod and cylinder for scoring, grooving and straightness. Polish out very light scoring with fine emery cloth and oil, using a rotary motion. Renew rod and piston assembly if heavily scored or grooved, or if piston rod is bent. Inspect piston seal for frayed edges, wear and imbedded dirt or foreign particles. Renew seal if any of the above conditions are found.

NOTE: Do not remove the "O" ring (11 – Fig. 72) located under the piston seal unless renewal is indicated. Inspect balance of "O" rings, backup washers and seals and renew same if excessively worn.

Reassemble steering cylinder as follows: Lubricate piston seal and cylinder head "O" rings. Using a ring compressor, or a suitable hose clamp, install piston and rod assembly into cylinder. Install cylinder heads in cylinder so hole in cylinder heads will ac-

cept nib of retaining rings and pull same into its groove by rotating cylinder heads with a spanner wrench. Complete balance of reassembly by reversing the disassembly procedure.

Reinstall unit on tractor, then fill and bleed the power steering system as outlined in paragraph 38.

MULTIPLE CONTROL VALVE

All Models

57. **R&R AND OVERHAUL.** The multiple control valve is mounted on left front of clutch housing as shown in Fig. 73. The multiple control valve has a 12 gpm pump mounted on inner side which provides pressurized oil for operation of power steering, brakes and torque amplifier clutches as well as providing lubrication for the torque amplifier assembly and speed transmission assembly. The multiple control valve also contains the spools, valves and passages necessary to control these operations.

When servicing the multiple control valve, cleanliness is of the utmost importance as well as the avoidance of any nicks or burrs. When reinstalling the multiple control valve and pump assembly, be sure to use the same thickness gasket as original. Use a micrometer to measure gasket. Gaskets are available in light (0.011-0.019 inch) and heavy (0.016-0.024 inch) thicknesses.

To remove and reinstall the multiple control valve, refer to paragraph 47.

With unit removed, all spools, plungers and springs can be removed and the procedure for doing so is obvious.

However, note that before both by-pass valves (25–Figs. 74 and 74A) can

Fig. 73—View of multiple control valve (MCV) showing control linkage and hydraulic connections.

B. Brake pressure tube
C. Torque amplifier valve operating pivot
P. Transmission oil pressure switch
S. Supercharge line
T. Dump valve operating lever
O. Oil cooler lines

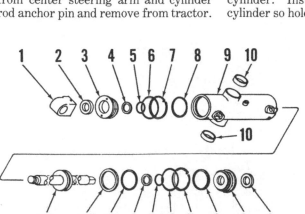

Fig. 72—Exploded view of external steering cylinder used on all models so equipped.

1. Rod end
2. Seal
3. Cylinder head
4. Back-up washer
5. "O" ring
6. "O" ring
7. Retaining ring
8. Back-up washer
9. Cylinder
10. Bushing
11. "O" ring
12. Piston seal
13. Piston rod

be removed, the small retaining pins must be removed. Pins are retained in position by the inner and outer plates.

Refer to the following table for spring test data and to Figs. 74 and 74A for spring location. Spring call-out numbers are in parentheses.

Brake Check Valve (17)
Free length—inches 0.500
Test length lbs. at in. ......0.5 at 0.300
Pressure Regulator Valve (19 & 20)
Outer spring free length—
    inches 3.508
Test length lbs. at in. .....64.5 at 2.170
Inner spring free length—
    inches 2.910
Test length lbs. at in. .....19.5 at 1.940
Lubrication Regulator Valve (24)
Free length—inches 1.072
Test length lbs. at in. ......5.2 at 0.787
Flow Divider (29)
Free length—inches 4.08
Test length lbs. at in. .....33.7 at 2.50
Oil Cooler By-Pass Valve (7)
Free length—inches 1.244
Test length lbs. at in. ......26 at 0.712
Drive Selector Valve (45 & 46)
Outer spring free length—
    inches 1.578
Test length lbs. at in. ......28 at 1.281
Inner spring free length—
    inches 1.880
Test length lbs. at in. ......18 at 1.254
Clutch Dump Valve (36)
Free length—inches 1.720
Test length lbs. at in. ....5 at 1.150

Reassembly is the reverse of disassembly. Use all new "O" rings and refer to paragraph 152 when adjusting "Torque-Amplifier" drive selector and to paragraph 144 when adjusting the "Torque-Amplifier" dump valve.

Safety relief valve (10—Figs. 74 and 74A) is heavily staked and cannot be disassembled. Valve should relieve at approximately 1900 to 2350 psi with tractor equipped with external steering cylinder; and approximately 1600 psi for tractors with internal steering cylinder. Faulty valves are renewed as a unit. Refer to paragraph 40 for information on checking valve.

OIL COOLER

All Models

58. **R&R AND OVERHAUL.** An oil cooler is incorporated into the power steering system. Of the 12 gpm supplied by the power steering pump, only five gpm is used to operate power steering and tractor controls. The remaining seven gpm is directed, via the flow divider valve, to the oil cooler where it is cooled and returned to lubricate the

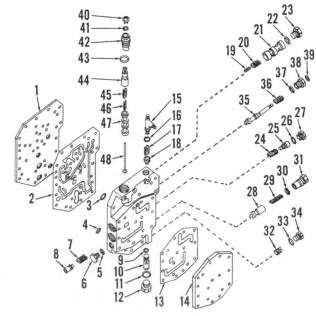

Fig. 74—Exploded view of multiple control valve for tractors not equipped with torque amplifier.
1. Inner plate
2. Inner gasket
3. "O" ring
4. "O" ring
5. "O" ring
6. Plug
7. Spring
8. By-pass valve
9. "O" ring
10. Safety relief valve
11. "O" ring
12. Plug
13. Cover gasket
14. Cover
15. Tee
16. "O" ring
17. Spring
18. Brake check valve
19. Spring (inner)
20. Spring (outer)
21. Pressure regulator valve
22. "O" ring
23. Plug
24. Spring
25. Lube valve
26. "O" ring
27. Plug
28. Flow divider
29. Spring
30. "O" ring
31. Plug
32. Flow control orifice
33. "O" ring
34. Plug

tractor differential and provide cooling for oil in the reservoir. Pressure regulation of the oil to oil cooler is controlled by oil cooler by-pass valve (8—Figs. 74 and 74A) in the multiple control valve.

Service of oil cooler involves only removal and reinstallation, or renewal of faulty units. Removal of oil cooler is obvious after removal of the radiator grill and hood and an examination of the unit. However, outlet and inlet hoses must be identified as they are removed from oil cooler pipes so oil circuits will be kept in the proper sequence.

TELESCOPING STEERING COLUMN

All Models

59. Telescoping steering column is standard on all models. To change steering wheel height, loosen locking collar (5—Fig. 64) by turning it counter-clockwise. Raise or lower steering wheel to desired height and tighten locking collar by turning it clockwise. Refer to paragraph 50 and Fig. 64 for removal and disassembly procedure.

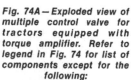

Fig. 74A—Exploded view of multiple control valve for tractors equipped with torque amplifier. Refer to legend in Fig. 74 for list of components except for the following:
35. Clutch dump valve
36. Spring
37. "O" ring
38. Plug
39. Seal
40. Retaining ring
41. "O" ring
42. Retaining screw
43. "O" ring
44. Stem
45. Spring (outer)
46. Spring (inner)
47. Spool (selector)
48. Guide

ENGINE AND COMPONENTS

Model 886 tractor (S.N. 14471 & below) is equipped with an engine having a bore and stroke of 3.875 x 5.085 inches and a displacement of 360 cubic inches. All Model 786 and Model 886 tractors (S.N. 14472 & above) are equipped with an engine having a bore and stroke of 3.875 x 5.060 inches and a displacement of 358 cubic inches.

Model 986 tractor is equipped with an engine having a bore and stroke of 4.300 x 5.004 inches and a displacement of 436 cubic inches.

Model 1086 tractor is equipped with an engine having a bore and stroke of 4.300 x 4.754 inches and a displacement of 414 cubic inches.

Model 1086 (S.N. 140023 & below) is equipped with DT-414; Model 1086 (S.N. 140024 & above) is equipped with DT-414B engine.

All engines are a six cylinder design. All engines are equipped with wet type sleeves. Dry type air filters are used in the air induction systems of all engines. Model 1086 engine is equipped with a turbocharger.

R&R ENGINE ASSEMBLY

All Models

60. To remove engine proceed as follows: Drain cooling system, disconnect battery cables; remove exhaust stack, radiator cap, front hood channel and both side hoods. Disconnect left and right turn hydraulic lines from steering hand pump and cap all lines and fittings. Disconnect oil cooler hoses at multiple control valve. Disconnect throttle cable from injection pump and support bracket. Disconnect oil pressure sending unit. If equipped, disconnect speedometer cable and transducer assembly. Disconnect power steering hand pump supply and return lines. Disconnect ether solenoid electrical lead. On models so equipped, shut off both suction and discharge valves on air conditioning compressor then disconnect suction line and move it back towards the control center. Cap the open compressor port. Partially remove left and right front lower panels and disconnect headlight wiring. Disconnect air conditioner quick connector and air conditioner wiring harness behind the left front panel. Shut off coolant valves and disconnect heater hoses. Remove subframe platform inside the control center, then disconnect the main electrical wiring harness, and all necessary remaining wires.

Remove the two mounting bolts on clutch housing ahead of the cranking motor. Install suitable jack stands to side rails and place a rolling floor jack under rear section of tractor. With tractor properly supported, unthread the remaining six mounting bolts from the speed transmission.

IMPORTANT: Do not completely remove the six mounting bolts. To do so will cause the engine to drop between side rails.

Drive wooden blocks in between lower bolster and front axle to prevent tilting of front section. Split tractor by rolling rear section away from front.

Remove air cleaner assembly and disconnect radiator hoses. Remove rear oil filter. Support engine with a sufficient hoist arrangement. Remove left side channel and bolts securing engine to right channel, then remove engine.

Installation of engine and tractor recoupling are the reverse of splitting and removal procedures. To avoid alignment problems when recoupling gear drive tractors, remove clutch assembly from flywheel and install it on the transmission input shaft. Recouple tractor, then secure clutch assembly to the flywheel through opening in bottom of clutch housing.

CYLINDER HEAD

All Models

61. To remove cylinder head, first drain cooling system and remove coolant temperature bulb or disconnect wire from sender. Unbolt and remove air cleaner assembly. On tractors so equipped, remove turbocharger as outlined in paragraph 125.

Disconnect and remove injection lines from fuel injectors. Cap all fuel connections immediately. Remove injector nozzles and place nozzle assemblies where they will not be damaged. Unbolt and remove exhaust and intake manifolds. Remove rocker arm cover, rocker arms, shaft assembly and push rods. Remove cylinder head retaining bolts and with a suitable hoist, lift off cylinder head using the lifting eyes.

Check cylinder head for warpage as follows: Place a straightedge across the machined (combustion) side of cylinder head and measure between cylinder head and straightedge. If a 0.003 inch feeler gage can be inserted between straightedge and cylinder head within any six-inch distance, the cylinder head should be refaced with a surface grinder, providing not more than 0.010 inch of material is removed.

The standard distance from rocker arm cover surface to combustion surface of cylinder head is as follows:

D-358 . . . . .3.890-3.910 inches
D-360, DT-414, DT-414B and
    D-436 . . . . .4.198-4.202 inches

NOTE: After cylinder head is resurfaced, check the valve head recession specifications outlined in paragraphs 63 and 64. Correct as necessary.

When reinstalling cylinder head, use new head gasket and make certain that gasket sealing surfaces are clean and dry. DO NOT use sealant or lubricants on head gasket, cylinder head or block.

CAUTION: Because of the minimum amount of clearance that exists between valves and piston tops, loosen rocker arm adjusting screws before installing the rocker arms and shaft assembly. Refer to paragraphs 63 and 64 for information concerning valve adjustment.

Fig. 75 — Tightening sequence for cylinder head retaining nuts on D-358 Diesel engine.

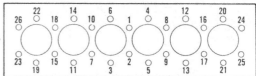

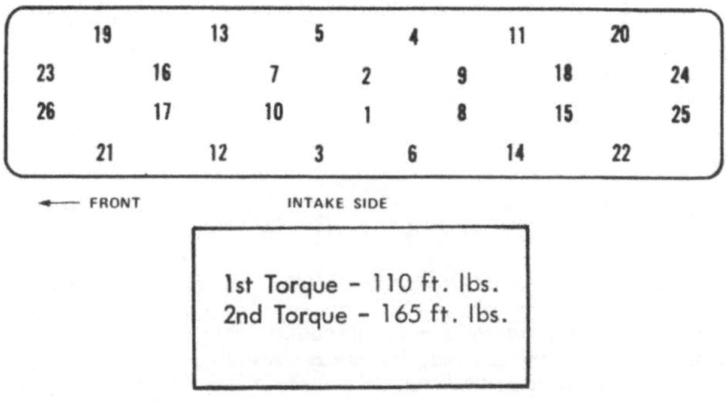

1st Torque   50 FT.-LBS.
2nd Torque   70 FT.-LBS.
3rd Torque   90 FT.-LBS.

Fig. 76 — Cylinder head cap screw tightening sequence for D-360, D-436 and DT-414 Diesel engines.

1st Torque – 110 ft. lbs.
2nd Torque – 165 ft. lbs.

On Models 786 and 886 equipped with D-358 engines, tighten cylinder head retaining nuts in three stages using the sequence shown in Fig. 75. Tighten nuts to a torque of 50 ft.-lbs. during the first step, 70 ft.-lbs. during the second step and 90 ft.-lbs. during the third step.

On Model 886 equipped with D-360 engine and Models 986 and 1086, tighten cylinder head cap screws in two steps using the sequence shown in Fig. 76. Tighten cap screws to a torque of 110 ft.-lbs. during the first step and 165 ft.-lbs. during the second step.

COOLING TUBES, WATER DIRECTORS AND NOZZLE SLEEVES

All Models

62. The cylinder head is fitted with brass injector nozzle sleeves that pass through the coolant passages. In addition, Models 786 and 886 equipped with D-358 engines use cooling jet tubes (one for each cylinder). Model 886 with D-360 engine and all Models 986 and 1086 use water directors (two for each cylinder) to direct a portion of the coolant to the valve seat and nozzle sleeve area. Cooling jet tubes, nozzle sleeves and water directors are available as service items.

To renew nozzle sleeves, remove injectors and cylinder head as outlined in paragraph 61. On D-358 engines, use I.H. tool (FES 112-4), turning it into the sleeve. Attach a slide hammer puller and remove sleeve. See Fig. 77. On D-360, D-436, DT-414 and DT-414B engines, use I.H. tool (FES 25-13), turning it into the sleeve. Attach a slide hammer puller and remove sleeve. See Fig. 78.

NOTE: Use caution during sleeve removal not to damage sealing areas in cylinder head. Under no circumstances should screwdrivers, chisels or other such tools be used in an attempt to remove injector nozzle sleeves.

When installing nozzle sleeves, be sure sealing areas are completely clean and free of scratches. Apply a light coat of "Grade B Loctite" on sealing surfaces of nozzle sleeves.

On D-358 engine use installing tool (FES 112-3) shown in Fig. 79 and drive

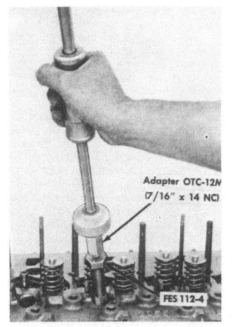
Fig. 77—On D-358 engines use I.H. tool No. FES 112-4 and slide hammer to remove injector nozzle sleeves.

Fig. 78—On D-360, D-436 and DT-414 engines use I.H. tool No. FES 25-13 (2) and slide hammer (1) to remove injector nozzle sleeves (3).

Fig. 79—Injector nozzle sleeve installing tool used on D-358 engine. Apply "Grade B Loctite" to sealing surface of nozzle sleeve.

Fig. 80—Drive nozzle sleeves into D-358 cylinder head until they bottom.

Fig. 81—On D-360, D-436 and DT-414 engines use installing tool I.H. No. FES 148-3 (1) and drive nozzle sleeves (2) in cylinder head until they bottom.

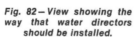
Fig. 82—View showing the way that water directors should be installed.
1. Water director
2. Exhaust valve
3. Water director
4. Intake valve

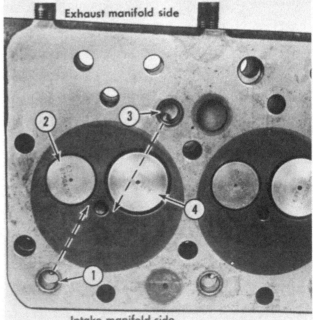

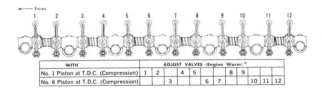

Fig. 83—Chart shows valve tappet gap adjusting procedure used on D-358 engines. Refer to text.*

WITH	ADJUST VALVES (Engine Warm)*											
No. 1 Piston at T.D.C. (Compression)	1	2		4	5			8	9			
No. 6 Piston at T.D.C. (Compression)			3			6	7			10	11	12

injector nozzle sleeves into their bores until they bottom as shown in Fig. 80. On D-360, D-436, DT-414 and DT-414B engines, use I.H. tool (FES 148-3) and drive injector nozzle sleeves into their bores until they bottom as shown in Fig. 81.

NOTE: Injector sleeves have an interference fit in their bores. When installing sleeves, be sure sleeve is driven straight with its bore and is completely bottomed.

To remove cooling jet tubes on D-358 engines, thread inside diameter and install a cap screw to assist in removal. Install new cooling jet tubes so coolant is directed between the valves of each cylinder. Tubes must be installed flush with cylinder head surfaces.

To remove water directors on D-360, D-436, DT-414 and DT-414B engines, use a slide hammer with a small jaw that will hook under water director. Install new water directors by tapping in place with small hammer. The directors must be recessed below cylinder head surface and the opening aimed as shown in Fig. 82.

VALVES AND SEATS

Models 786-886 (D-358)

63. Inlet and exhaust valves are not interchangeable. Inlet valves seat directly in cylinder head and exhaust valves seat on renewable seat inserts. Inserts are

available in oversizes of 0.0059 and 0.0157 inch. Valve rotators (Rotocaps) and valve stem seals are used on all valves.

When removing valve seat inserts, use the proper puller. DO NOT attempt to drive chisel under insert as counterbore will be damaged. Chill new insert with dry ice or liquid freon prior to installing. When new insert is properly bottomed it should be 0.008-0.030 inch below edge of counterbore. After installation, peen cylinder head material around complete outer circumference of valve seat insert.

Check valves and seats against the following specifications:

Inlet
Face and seat angle . . . . 45°
Stem diameter . . . . 0.3919-0.3923 in.
Stem to guide diametral clearance:
  Normal . . . . 0.0014-0.0026 in.
  Maximum allowable . . . . 0.006 in.
Seat width . . . . 0.076-0.080 in.
Valve run-out (max.) . . . . 0.001 in.
Valve tappet gap (warm) . . . . . 0.012 in.
Valve recession from face of cylinder head:
  Normal . . . . 0.039-0.055 in.
  Maximum allowable . . . . 0.120 in.
Exhaust
Face and seat angle . . . . 45°
Stem diameter . . . . 0.3911-0.3915 in.
Stem to guide diametral clearance:
  Normal . . . . 0.0022-0.0034 in.
  Maximum allowable . . . . 0.006 in.
Seat width . . . . 0.081-0.089 in.

Valve run-out (max.) . . . . 0.001 in.
Valve tappet gap (warm) . . . . . . 0.012 in.
Valve recession from face of cylinder head:
  Normal . . . . 0.047-0.063 in.
  Maximum allowable . . . . 0.120 in.

CAUTION: Due to close clearance between valves and pistons, severe damage can result from inserting feeler gage between valve stem and valve lever (rocker arm) with engine running. DO NOT attempt to adjust tappet gap with engine running.

To adjust valve tappet gap, crank engine to position number one piston at top dead center of compression stroke. Adjust six valves indicated on the chart shown in Fig. 83.

NOTE: The valve arrangement is exhaust-intake-exhaust-intake and so on, starting from the front of cylinder head.

Turn engine crankshaft one complete revolution to position number six piston at TDC (compression) and adjust the remaining six valves indicated on chart.

Models 886 (D-360) 986-1086

64. Inlet and exhaust valves are not interchangeable. On Model 886 equipped with D-360 engine, inlet valves seat directly in the cylinder head and exhaust valves seat on renewable seat inserts. Models 986 and 1086, inlet and exhaust valves seat on renewable seat inserts. Inserts are available in oversizes of 0.002 and 0.015 inch. Refer to Figs. 84, 85, 86 and 87 for dimensions to install oversize inserts. On all models valve rotators (Rotocoils) are used on all valves.

When removing valve seat inserts, use the proper puller. DO NOT attempt to drive chisel under insert as counterbore

VALVE SEAT INSERT CHART (OVERSIZE)

Engine	Oversize Insert	Diameter of Cylinder Head Counterbore (inches)	
		Intake	Exhaust
D-358	.0059" .0157"	None None	1.6594-1.6602 1.6693-1.6701
D-360	.002" .015"	None None	1.534-1.535 1.547-1.548
D and DT-414 436	.002" .015"	1.998-1.999 2.011-2.012	1.626-1.627 1.639-1.640

Fig. 84—Chart showing the diameter of cylinder head counterbore for oversize valve seat insert.

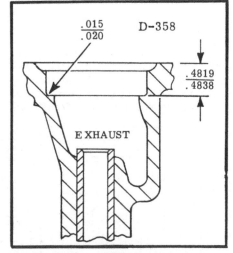

Fig. 85—Chart showing the correct radii and depth for valve seat installation of D-358 engines.

will be damaged. Chill new inserts with dry ice or liquid freon prior to installing. When new insert is properly bottomed, it should be 0.087-0.093 inch for exhaust valves and 0.120-0.127 inch for inlet valves below edge of counterbore.

Check valves and seats against the following specifications:

Inlet

Face and seat angle . . . . .30°
Stem diameter . . . . . . .0.3718-0.3725 in.
Stem to guide diametral clearance:
 Normal . . . . .0.0015-0.0032 in.
 Maximum allowable . . . . .0.006 in.
Seat width . . . . .0.075-0.085 in.
Valve run-out (max.) . . . . .0.0015 in.
Valve tappet gap (warm):
 D-360 .0.012 in.
 D-436 (S.N. 83591 & below) . .0.020 in.
 D-436 (S.N. 83592 & above) .0.025 in.*
 DT-414 (S.N. 149604 &
  below). . . . . .0.020 in.
 DT-414B (S.N. 149605 &
  above). . . . . .0.025 in.*
Valve recession from face of cylinder head:
 Normal . . . . .0.000-0.014 in.

Exhaust

Face and seat angle . . . . .45°
Stem diameter . . . . . . .0.3718-0.3725 in.
Stem to guide diametral clearance:
 Normal . . . . .0.0015-0.0032 in.
 Maximum allowable . . . . .0.006 in.
Seat width . . . . .0.075-0.085 in.
Valve run-out (max.) . . . . .0.0015 in.
Valve tappet gap (warm):
 D-360 .0.021 in.
 D-436 (S.N. 83591 & below) . .0.025 in.
 D-436 (S.N. 8359 & above) . . .0.025 in.
 DT-414 (S.N. 149604 &
  below). . . . . .0.025 in.
 DT-414B (S.N. 149605 &
  above). . . . . .0.025 in.
Valve recession from face of cylinder head:
 Normal . . . . .0.000-0.014 in.
*Engines equipped with camshaft assembly No. 1802023C91. Part number 691488C1 stamped on rear face of camshaft.

CAUTION: Due to close clearance between valves and pistons, severe damage can result from inserting feeler gage between valve stem and valve lever (rocker arm) with engine running. DO NOT at-

tempt to adjust valve tappet gap with engine running.

To adjust valve tappet gap, crank engine to position number one piston at top dead center of compression stroke. Adjust the six valves indicated on the chart shown in Fig. 88.

NOTE: The valve arrangement is intake-exhaust-intake-exhaust and so on starting from front of cylinder head.

Turn engine crankshaft one complete revolution to position number six piston at TDC (compression) and adjust the remaining six valves indicated on chart.

VALVE GUIDES AND SPRINGS

Models 786-886 (D-358)

65. The inlet and exhaust valve guides should be pressed into cylinder head until top of guides are 1.154-1.169 inches above spring recess in head. After installation, guides must be reamed to an inside diameter of 0.3940-0.3945 inch. Valve stem to valve guide diametral clearance should be 0.0014-0.0026 inch for inlet valves and 0.0022-0.0034 inch for exhaust valves. Maximum allowable stem clearance in all guides is 0.006 inch.

Inlet and exhaust valve springs are also interchangeable. Springs should test 145-159 pounds when compressed to a length of 1.346 inches. Renew any spring which is rusted, discolored or does not meet pressure test specifications.

Models 886 (D-360)-986-1086

66. The inlet and exhaust valve guides are interchangeable. Valve guides should be pressed into cylinder head un-

til the top of guides is above the spring recess in head as follows:
D-360
 Intake . . . . .1.002 in.
 Exhaust . . . . .1.282 in.
D-436 and DT-414
 Intake . . . . .1.217 in.
 Exhaust . . . . .1.297 in.

Guides are pre-sized; however, since they are a press fit in cylinder head, it is necessary to ream them to remove any burrs or slight distortion caused by the pressing operation. Inside diameter should be 0.374-0.375 inch. Valve stem to guide diametral clearance should be 0.0015-0.0032 inch with a maximum allowable clearance of 0.006 inch.

Inlet and exhaust valve springs are also interchangeable. Springs should have a free length of 2.340 inches and should test 156-164 pounds when compressed to a length of 1.562 inches. Renew any spring which is rusted, discolored or does not meet pressure test specifications.

VALVE TAPPETS (CAM FOLLOWERS)

Models 786-886 (D-358)

68. The 0.7862-0.7868 inch diameter mushroom type tappets operate directly in the unbushed crankcase bores. Tappets have an effective length of 1.8779-1.8937 inches. Clearance of tappets in the bores should be 0.0004-0.0023 inch. Tappets can be removed after removing oil pan and camshaft as outlined in paragraph 87. Oversize tappets are not available.

Models 886 (D-360)-986-1086

69. Valve tappets are barrel type and operate in unbushed bores in crankcase.

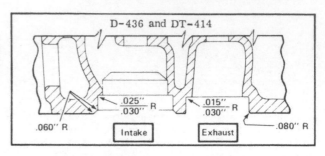

Fig. 87 — Chart showing the correct radii for valve seats installation for D-436 and DT-414 engines.

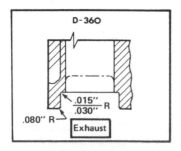

Fig. 86 — Chart showing correct radii for valve seat installation of D-360 engines.

Fig. 88 — Chart showing valve tappet gap adjusting procedure for D-360, D-436 and DT-414 engines. Refer to text.

WITH	ADJUST VALVES (Engine Warm)											
No. 1 Piston at T.D.C. (Compression)	1	2	3			6	7			10		
No. 6 Piston at T.D.C. (Compression)				4	5			8	9		11	12

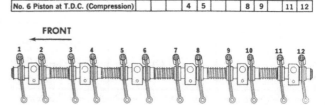

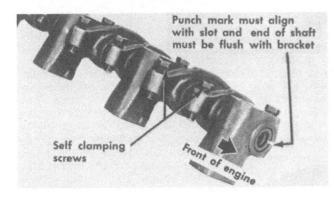

Punch mark must align with slot and end of shaft must be flush with bracket

Self clamping screws

Front of engine

Fig. 89 — On D-358 engines, align punch marks on front end of rocker shaft with slot in front mounting bracket. End of shaft must be flush with bracket.

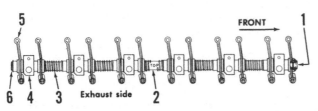

FRONT

5   6 4 3 Exhaust side 2   1

Fig. 90 — View showing the valve lever and shaft assembly.

1. Plug (2 used)
2. Shaft
3. Spring
4. Support bracket
5. Rocker arm
6. Snap ring (2 used)

Tappet diameter is 0.9965-0.9970 inch and length is 2.4370 inches and should operate with a clearance of 0.0025-0.0040 inch in the crankcase bore. Tappets are available in standard size only and removal requires removal of cylinder head as outlined in paragraph 61.

VALVE ROCKER ARM COVER

All Models

70. Removal of the rocker arm cover is obvious after removal of front hood. On Model 1086, remove turbocharger crossover inlet tube.

When reinstalling rocker arm cover, use new gasket to insure an oil tight seal. Torque rocker arm cover bolts to 25 in.-lbs.

VALVE TAPPET LEVERS (ROCKER ARMS)

Models 786-886 (D-358)

71. **REMOVE AND REINSTALL.** Removal of rocker arms and shaft assembly is obvious after removal of rocker arm cover. When reinstalling, tighten hold-down nuts on bracket to 47 ft.-lbs. torque.

72. **OVERHAUL.** To remove rocker arms from the one-piece shaft, remove bracket clamp bolts and slide all parts from shaft. Outside diameter of rocker shaft is 0.8491-0.8501 inch. The renewable bushings in rocker arms should have an operating clearance of 0.0009-0.0025 inch on rocker shaft with a maximum allowable clearance of 0.008 inch. Rocker arms may be refaced providing the original contour is maintained. Rocker arm adjusting screws are

self-locking. If they turn with less than 12 ft.-lbs. torque, renew adjusting screw and/or rocker arm.

Reassemble rocker arms on rocker shaft keeping the following points in mind: Thrust washers are used between each spring and rocker arms and spacer rings are used between rocker arms and brackets except between rear rocker arm and rear bracket. To insure that lubrication holes in rocker shaft are in correct position, align punch mark on front end of shaft with slot in front mounting bracket as shown in Fig. 89. End of shaft must also be flush with bracket. Rocker shaft clamp screws on brackets should be tightened to a torque of 10 ft.-lbs.

Models 886 (D-360)-986-1086

73. **REMOVE AND REINSTALL.** Removal of rocker arms and shaft

assembly is accomplished as follows: Remove all hoodsheets, heat shields, air cleaner assembly and (if equipped) turbocharger air outlet pipe. Rocker arm cover and the rocker arms and shaft assembly can now be removed. When reinstalling be sure to use new packing rings under the valve cover bolt washers. Torque valve cover bolts to 25 in.-lbs.

74. **OVERHAUL.** To remove rocker arms from the one piece shaft, remove snap rings from end of shaft and slide all parts from shaft. Refer to Fig. 90. Outside diameter of rocker shaft is 0.8491-0.8501 inch. The rocker arms should have an operating clearance of 0.0009-0.0039 inch on rocker shaft with a maximum allowable clearance of 0.007 inch. All rocker arms are interchangeable and may be refaced providing the original contour is maintained. If necessary to renew rocker shaft (2 – Fig. 90), press new plugs (1) in end of shafts.

When reassembling lay the shaft marked "TOP" up and the bolt grooves toward the assembler and refer to Fig. 90.

VALVE ROTATORS

All Models

75. The positive valve rotators used are called Rotocoils. For D-358 engine Rotocaps are used.

Normal servicing of valve rotators consists of renewing the units. It is important to observe valve action after engine is started. Valve rotator action can be considered satisfactory if valve rotates a slight amount of each time valve opens.

The Rotocoil rotates the valve at a slower speed than the Rotocap. Rotocaps should not be installed on engines where Rotocoils are specified.

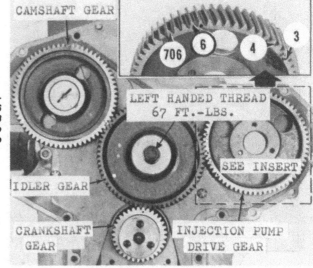

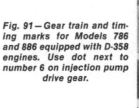

CAMSHAFT GEAR

706 6 4 3

LEFT HANDED THREAD 67 FT.-LBS.

SEE INSERT

IDLER GEAR

CRANKSHAFT GEAR

INJECTION PUMP DRIVE GEAR

Fig. 91 — Gear train and timing marks for Models 786 and 886 equipped with D-358 engines. Use dot next to number 6 on injection pump drive gear.

VALVE TIMING

Valve timing is correct when timing (punch) marks on timing gear train are properly aligned as shown in Figs. 91 and 92. To check valve timing on an assembled engine, follow the procedure outlined in the following paragraphs.

Models 786-886 (D-358)

76. To check valve timing, remove rocker arm cover and crank engine to position number one piston at TDC of compression stroke. Adjust number one cylinder intake valve tappet gap to 0.016 inch. Place a 0.004 inch feeler gage between valve lever and valve stem of number one intake valve. Slowly rotate crankshaft in normal direction until valve lever becomes tight on feeler gage. At this point, number one intake valve will start to open and timing pointer should be within the range of 23 to 29 degrees BTDC on the flywheel.

NOTE: One tooth "out of time" equals approximately 11 degrees.

Readjust number one intake valve tappet gap as outlined in paragraph 63.

Models 886 (D-360)-986-1086

77. To check valve timing, remove rocker arm cover and crank engine to position number one piston at TDC of compression stroke. Adjust number one cylinder intake tappet to specification shown in following chart:

D-360 (All engines) . . . . 0.016 in.
D-436 (S.N. 83591 & below) . . 0.024 in.
D-436 (S.N. 83592 & above) . 0.029 in.*
DT-414 (S.N. 149604 &
  below) . . . . 0.024 in.
DT-414B (S.N. 149605 &
  above) . . . . 0.029 in.*
* Engines equipped with camshaft assembly No. 1802023C91. Part number 691 488 C1 is stamped on rear face of camshaft.

Place a 0.004 inch feeler gage between valve lever and valve stem of number one intake valve. Slowly rotate crankshaft in normal direction until valve lever becomes tight on feeler gage. At this point number one intake valve will start to open and timing pointer on flywheel should be 27 to 33 degrees BTDC for D-360 engine and 21 to 27 degrees BTDC for all 400 series engines.

NOTE: One tooth "out of time" equals approximately 11 flywheel degrees.

Readjust number one intake valve tappet gap as outlined in paragraph 64.

TIMING GEAR COVER

All Models

78. To remove the timing gear cover, first remove hood, drain cooling system and disconnect battery cables. Disconnect air cleaner inlet hose and radiator hoses. Identify and disconnect power steering lines, then identify and disconnect hydraulic oil cooler lines. Plug or cap openings to prevent dirt or other foreign material from entering hydraulic system. Support tractor under clutch housing. On all models equipped with wide front axle, disconnect stay rod ball. Attach hoist to front support, unbolt front support from side rails and roll complete front assembly forward from tractor. Unbolt and remove fan, alternator and drive belts.

On Models 786 and 886 equipped with D-358 engine, unbolt and remove air cleaner assembly. Remove pump pulley, then unbolt and remove water pump and carrier assembly. Disconnect tachometer drive cable and remove tachometer drive unit from front cover. Do not lose the small driving tang when removing tachometer drive. Remove the three cap screws, flat washer and pressure ring from crankshaft. Tap crankshaft pulley with a plastic hammer to loosen pulley, then slide pulley off of the wedge rings. Remove cap screws re-

taining oil pan to timing cover and loosen remaining oil pan cap screws. Unbolt and remove timing gear cover. Reassemble by reversing the disassembly procedure. When installing crankshaft pulley, place one pressure ring on crankshaft with thick end towards engine. Install wedge rings in pulley bore so that slots in rings are 90 degrees apart. Slide pulley onto crankshaft and align with timing pin. Install pressure ring, flat pressure washer and three cap screws. Tighten cap screws evenly to a torque of 55 ft.-lbs.

On Models 886 equipped with D-360 engine, 986 and 1086, drain engine oil and remove oil pan. Remove three cap screws and retaining plate from end of crankshaft. Attach suitable puller (I.H. tool No. FES 10-17 or equivalent) and remove crankshaft pulley.

NOTE: Attach puller to tapped holes in pulley ONLY. The front pulley is also the vibration dampener and any other type puller can damage the elastic member.

Unbolt and remove oil pump from front of cover. The front oil seal is in the oil pump housing. Unbolt and remove timing gear cover. Reassemble by reversing the disassembly procedure. When reinstalling crankshaft pulley, heat in boiling water for a period not to exceed ONE hour. Then install on crankshaft and secure with retaining plate and cap screws torqued to 125 ft.-lbs.

NOTE: Do not install pulley cold, and when heating do not heat more than one hour as the elastic member can be damaged.

TIMING GEARS

Models 786-886 (D-358)

79. **CRANKSHAFT GEAR.** Crankshaft gear is a shrink fit on crankshaft. To renew the gear, it is recommended that crankshaft be removed from engine. Then, using a chisel and hammer, split gear at its timing slot.

The roll pin for indexing crankshaft gear on crankshaft must protrude approximately 0.078 inch. Heat new gear to 400 degrees F. and install it against bearing journal.

When reassembling, make certain all timing marks are aligned as shown in Fig. 91.

80. **CAMSHAFT GEAR.** Camshaft gear is a shrink fit on camshaft. To renew the gear, remove camshaft as outlined in paragraph 87. Gear can now be pressed off in conventional manner, using care not to damage the tachometer drive slot in end of camshaft. When reassembling, install thrust

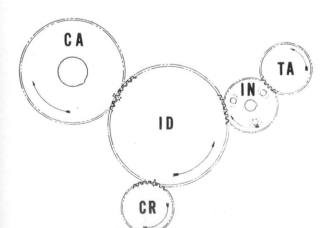

Fig. 92—Model 886 (D-360 engine), 986 and 1086 gear train and timing marks.
  CA. Camshaft gear
  CR. Crankshaft gear
  ID. Idler gear
  IN. Injection pump drive gear
  TA. Tachometer drive gear

plate and Woodruff key. Heat gear to 400 degrees F. and install it on camshaft.

NOTE: When sliding gear on camshaft, set thrust plate clearance at 0.004-0.017 inch.

Install camshaft assembly and make certain all timing marks are aligned as shown in Fig. 91.

81. **IDLER GEAR.** To remove the idler gear, first remove timing gear cover as outlined in paragraph 78. Idler gear shaft is attached to front of engine by a special (left hand thread) cap screw. Idler gear is equipped with two renewable needle bearings. A spacer is used between the bearings.

When installing idler gear, align all timing marks as shown in Fig. 91. Coat threads of special cap screw with "Grade B Loctite" and tighten cap screw to a torque of 67 ft.-lbs. End clearance of gear on shaft should be 0.008-0.013 inch.

Timing gear backlash should be as follows:

Idler to crankshaft gear,
 New gears . . . . 0.007-0.015 in.
 Used gears (max.) . . . . 0.0295 in.
Idler to camshaft gear,
 New gears . . . . 0.0035-0.0107 in.
 Used gears (max.) . . . . 0.0215 in.
Idler to injection pump gear,
 new gears . . . . 0.0021-0.012 in.
 Used gears (max.) . . . . 0.024 in.

82. **INJECTION PUMP DRIVE GEAR.** To remove the pump drive gear, first remove timing gear cover as outlined in paragraph 78. Remove pump drive shaft nut and washer and three hub cap screws. Attach puller (IH tool No. FES 111-2 or equivalent) to threaded holes in gear and pull gear and hub from shaft.

Fig. 93 — View showing crankshaft gear and oil pump drive spline.

1. Crankshaft
2. Drive spline
3. Crankshaft gear

When reassembling, make certain all timing marks are aligned a shown in Fig. 91. Use timing dot next to number 6 on injection pump drive gear. Refer to paragraph 114 and retime injection pump. Tighten pump drive shaft nut to a torque of 47 ft.-lbs. and the three hub cap screws to a torque of 17 ft.-lbs.

Models 886 (D-360)-986-1086

83. **CRANKSHAFT GEAR.** Crankshaft gear is a shrink fit on crankshaft. To renew the gear, it is recommended that crankshaft be removed from engine. When renewing crankshaft gear, oil pump drive spline (2 – Fig. 93) will also have to be renewed. With crankshaft removed, use a chisel and hammer and split oil pump drive spline and then split crankshaft gear at roll pin slot.

Install new roll pin in crankshaft if it was removed. Heat new gear and drive spline to 370-395 degrees F. Install crankshaft gear against shoulder and over roll pin. Then, slide drive spline against crankshaft gear.

When reassembling, make certain all timing marks are aligned as shown in Fig. 92.

84. **CAMSHAFT GEAR.** The camshaft gear is keyed and press fitted to camshaft. The camshaft gear can be removed using a suitable puller after camshaft is removed as outlined in paragraph 89.

Backlash between camshaft gear and idler gear should be 0.003-0.007 inch with a maximum allowable backlash of 0.016 inch.

When installing, heat gear to 370-395 degrees F. and press gear against shoulder of shaft. Make certain timing marks are aligned as shown in Fig. 92.

85. **IDLER GEAR.** To remove the idler gear, first remove timing gear cover as outlined in paragraph 78.

The idler gear rotates on two taper roller bearings and idler gear shaft is attached to front of cylinder block by a twelve point head cap screw. Removal of idler gear, bearings, bearing spacer and shaft is obvious.

The idler gear and idler gear shaft are available separately, but the two taper bearings, two bearing cups and bearing spacer must be renewed as an assembly.

When reinstalling, make certain timing marks are aligned as shown in Fig. 92 and tighten idler gear shaft cap screw to a torque of 85 ft.-lbs.

86. **INJECTION PUMP DRIVE GEAR.** To remove the injection pump drive gear, first remove pump access cover from front of timing cover. Turn engine until pointer (3 – Fig. 94) is

aligned with mark on pump hub, then remove three cap screws while holding pump shaft as shown.

Reassemble by reversing the disassembly procedure and making sure that timing marks are aligned.

CAMSHAFT AND BEARINGS

Models 786-886 (D-358)

87. **CAMSHAFT.** To remove the camshaft, first remove timing gear cover as outlined in paragraph 78. Remove rocker arm cover, rocker arms assembly and push rods. Remove engine side cover and secure cam followers (tappets) in raised position with clothes pins or rubber bands. Working through openings in camshaft gear, remove camshaft thrust plate retaining cap screws. Carefully withdraw camshaft assembly.

Recommended camshaft end play is 0.004-0.017 inch. Camshaft bearing journal diameter should be 2.2823-2.2835 inches for all journals.

Install camshaft by reversing the removal procedure. Make certain timing marks are aligned as shown in Fig. 91.

88. **CAMSHAFT BEARINGS.** To remove the camshaft bearings, first remove engine as outlined in paragraph 60 and camshaft as outlined in paragraph 87. Unbolt and remove clutch, flywheel and engine rear support plate. Remove expansion plug from behind camshaft rear bearing and remove the bearings.

NOTE: Camshaft bearings are furnished semi-finished and must be align bored after installation to an inside diameter of 2.2844-2.2856 inches.

Install new bearings so that oil holes in bearings are in register with oil holes in crankcase.

Normal operating clearance of camshaft journals in bearings is 0.0009-0.0033 inch. Maximum allowable clearance is 0.006 inch.

Fig. 94 — View showing timing marks for injection pump.

1. Idler gear
2. Pump drive gear
3. Timing pointer

When installing expansion plug at rear camshaft bearing, apply a light coat of sealing compound to edge of plug and bore.

Models 886 (D-360)-986-1086

89. **CAMSHAFT.** To remove the camshaft, first remove timing gear cover as outlined in paragraph 78. Remove idler gear as outlined in paragraph 85. Remove cylinder head as outlined in paragraph 61. Remove cam followers (tappets) from above. Working through openings in camshaft gear, remove camshaft thrust plate retaining cap screws. Carefully withdraw camshaft assembly.

Recommended camshaft end play is 0.005-0.013 inch; if excessive, renew thrust plate. Camshaft bearing journal diameter should be 2.2814-2.2825 inches for all journals.

Install camshaft by reversing the removal procedure. Make certain timing marks are aligned as shown in Fig. 92.

90. **CAMSHAFT BEARING.** To remove the camshaft bearings, first remove engine as outlined in paragraph 60 and camshaft as in paragraph 89. Unbolt and remove clutch, flywheel and engine rear support plate. Remove expansion plug from behind camshaft rear bearing and remove the bearings.

Install new bearings so that oil holes in bearings are in register with oil holes in crankcase.

Camshaft bearings are pre-sized and should need no final sizing if carefully installed. Camshaft journals should have a diametral clearance of 0.002-0.0066 inch in camshaft bearings.

When installing expansion plug at rear camshaft bearing, apply a light coat

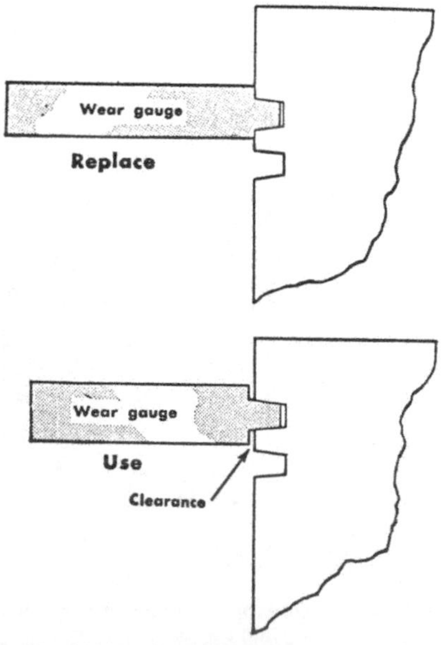

Fig. 95 — Perfect Circle Ring Gage No. 1 to check top ring groove of pistons used in D-358 engines.

of sealing compound to edge of plug and bore.

PISTON AND ROD UNITS

All Models

91. Connecting rod and piston assemblies can be removed from above after removing cylinder head as outlined in paragraph 61 and oil pan as outlined in paragraph 108.

Models 786 and 886 equipped with D-358 engines have cylinder numbers stamped on connecting rod and cap. Numbers on rod and cap should be in register and face towards camshaft side of engine. The arrow or FRONT marking stamped on the tops of pistons should be towards front of engine.

Models 886 (D-360), 986 and 1086 have numbers stamped on connecting rod and cap.

NOTE: Be sure to note and record the numbers on the rod and cylinder from which it came.

Be sure numbers on the rod face AWAY from camshaft, and the marking on top of the pistons is TOWARD camshaft side of engine.

On Models 786 and 886 equipped with D-358 engines, tighten connecting rod nuts to a torque of 45 ft.-lbs.

On Model 886 equipped with D-360 engine, 986 and 1086 equipped with DT-414, DT-414B and D-436 engines, tighten connecting rod cap screws in two steps. Tighten cap screw to a torque of 60 ft.-lbs. in first step and 130 ft.-lbs. in second step.

PISTONS, SLEEVES AND RINGS

Models 786-886 (D-358)

92. Pistons are not available as individual service parts, but only as matched units with wet type sleeves. New pistons have a diametral clearance in new sleeves of 0.0039-0.0047 inch when measured between piston skirt and sleeve at 90 degrees to piston pin.

The wet type cylinder sleeves should be renewed when out-of-round or taper exceeds 0.006 inch. Inside diameter of new sleeve is 3.8750-3.8754 inches. Cylinder sleeves can usually be removed by bumping them from the bottom, with a block of wood.

Before installing new sleeves, thoroughly clean counterbore at top of block and seal ring groove at bottom. All sleeves should enter crankcase bores full depth and should be free to rotate by hand when tried in bores without seal rings. After making a trial installation without seal rings, remove sleeves and install new seal rings, dry, in grooves in crankcase. Wet lower end of sleeve with

a soap solution or equivalent and install sleeve.

NOTE: The cut-outs in bottom of sleeve are for connecting rod clearance and must be installed toward each side of engine. Chisel marks are provided on top edge of cylinder sleeves to aid in correct installation. Align chisel marks from front to rear of engine.

If seal ring is in place and not pinched, very little hand pressure is required to press the sleeve completely in place. Sleeve flange should extend 0.003-0.005 inch above top surface of cylinder block. If sleeve stand-out is excessive, check for foreign material under the sleeve flange. The cylinder head gasket forms the upper cylinder sleeve seal and excessive sleeve stand-out could result in coolant leakage.

Pistons are fitted with three compression rings and one oil control ring. Prior to installing new rings, check top ring groove of piston by using Perfect Circle Piston Ring Gage No. 1 as shown in Fig. 95. If one or both, shoulders of gage touch ring land, renew the piston.

The top compression ring is a full keystone ring and it is not possible to measure ring side clearance in the groove with a feeler gage. Check fit of top ring as follows: Place ring in its groove and push ring into groove as far as possible. Measure the distance ring is below ring land. This distance should be 0.002-0.015 inch. Refer to Fig. 96 for view showing ring fit being checked using IH tools FES 68-3 and dial indicator FES 67.

The second compression ring is a taper face ring and is installed with largest outside diameter towards bottom of piston. Upper side of ring is marked TOP.

The oil control ring can be installed either side up, but make certain the coil

Fig. 96 — View showing ring fit being checked. Refer to Fig. 97 for the correct distance.

spring expander is completely in its groove.

Additional piston ring information is as follows:

Ring End Gap:
Compression rings .... 0.014-0.022 in.
Oil control ring .... 0.010-0.016 in.
Ring Side Clearance:
Top compression
(ring drop) .... 0.002-0.015 in.
Second compression . 0.0030-0.0042 in.
Third compression .. 0.0024-0.0034 in.
Oil control .... 0.0014-0.0024 in.

Models 886 (D-360)-986-1086

93. Pistons are not available as individual service parts, but only as matched units with wet type sleeves. The piston design for various engines differ in diameter, combustion dish, ring groove location and piston pin location. Be sure to identify proper piston part number when replacing pistons.

Model DT-414 S.N. 118713 and above utilize a new piston design having a revised top ring groove insert and a revised intermediate (second) ring groove to reduce piston crown failure and oil consumption. The new piston and ring can only be used in engines equipped with valve cover breathers. DO NOT use the new piston and ring in old crankcase breather engines. On engines prior to the new valve cover breather and pistons, it will be necessary to install the new breather and appropriate piston/sleeve assembly if the use of the new piston and ring is desired.

New pistons have a diametral clearance in new sleeves of 0.0045-0.0065 inch when measured between piston skirt and sleeve at 90 degrees to piston pin.

The wet type cylinder sleeves should be renewed when out-of-round or taper exceeds 0.004 inch.

Inside diameter of new sleeve is as follows:
D-360 .... 3.875-3.876 in.
DT-414 .... 4.300-4.301 in.
DT-414B .... 4.300-4.301 in.
D-436 .... 4.300-4.301 in.
Cylinder sleeves can usually be removed by bumping them from the bottom with a block of wood.

Before installing sleeves, thoroughly clean counterbore of block using a wire brush or steel wool. All sleeves should enter crankcase bores full depth and should be free to rotate by hand when tried in bores without "O" rings. With sleeves in bores without "O" rings, sleeve flange should extend 0.002-0.005 inch above top surface of cylinder block. If sleeves do not extend above surface of cylinder block, shims are available and can be placed under sleeve flange to obtain proper height. If sleeve stand-out is

excessive, check for foreign material under the sleeve flange.

NOTE: Sleeves can be held down with cap screws and large flat washers when checking sleeve flange for proper height.

After making the trial installation without "O" rings, remove sleeves and install the "O" rings in the following order: bottom, center and top. Lubricate "O" rings with petroleum jelly or equivalent and install sleeves. Be sure flange of sleeve is firmly seated in crankcase counterbore.

The top compression ring on D-360, DT-414 (S.N. 118712 & below) and D-436 engines is a full keystone ring. On DT-414 (S.N. 118713 & above) engine uses a modified top ring groove insert. DT-414B engine top compression ring is a full keystone ring.

The second (intermediate) ring on D-360 is a notched taper face ring. It is installed with the notch up. The second (intermediate) ring on D-436 and DT-414 (S.N. 118712 & below) can be either the half or full keystone design, depending on the type of piston used. It serves as both a compression and oil control ring.

The second (intermediate) ring used on DT-414 (S.N. 118713 & above) has a wider ring gap than the old design piston. Ring gap is 0.065-0.078 inch. It is not possible to measure ring side clearance in the groove with a feeler gage. Check fit of top ring on D-360 and the top two rings on DT-414, DT-414B and D-436 engines, as follows: Install ring in its groove and push ring into groove as far as possible. Measure the distance ring is below ring land. Refer to Fig. 97 for the correct distance. See Fig. 96 for view showing ring being checked using .000-.026 in.

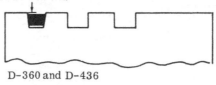

D-360 and D-436

.000-.028 .0045-.0215 in.

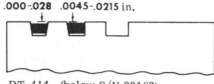

DT-414, (below S/N 39182)

.000-.028 in.

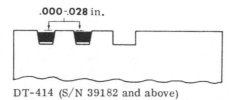

DT-414 (S/N 39182 and above)

Fig. 97 — Chart showing correct distance below ring land.

IH tools FS 68-3 and dial indicator FES 67.

The second compression is marked TOP; be sure that it is facing top of piston.

The oil control ring is a one-piece slotted chrome ring and can be installed either side up.

Additional piston ring information is as follows:

Ring End Gap
Top compression ring
D-360 .... 0.012-0.025 in.
D-436 .... 0.016-0.029 in.
DT-414 .... 0.016-0.029 in.
DT-414B .... 0.016-0.029 in.
Second compression ring
D-360 .... 0.020-0.033 in.
D-436 .... 0.020-0.033 in.
DT-414 (S.N. 140000 &
below) .... 0.020-0.033 in.
DT-414B (S.N. 140001 &
above) .... 0.065-0.078 in.
Oil control ring
D-360 .... 0.010-0.023 in.
D-436, DT-414,
DT-414B .... 0.010-0.028 in.
Ring Side Clearance
Top compression ring (ring drop)
D-360, D-436 .... 0.000-0.026 in.
DT-414 (S.N. 39181 &
below) .... 0.000-0.028 in.
DT-414 (S.N. 39182 &
above) .... 0.000-0.028 in.
Second compression ring (ring drop)
D-360, D-436 .... 0.003-0.005 in.
DT-414 (S.N. 39181 &
below) .... 0.005-0.022 in.
DT-414 (S.N. 39182 &
above) .... 0.000-0.028 in.
Oil control ring
D-360 .... 0.010-0.023 in.
D-436 .... 0.002-0.004 in.
DT-414, DT-414B ... 0.010-0.028 in.

PISTON PINS

All Models

94. The full floating type piston pins are retained in the piston bosses by snap rings. Specifications are as follows:
Piston pin diameter,
D-358 .... 1.4172-1.4173 in.
D-360, D-436, DT-414 and
DT-414B ...... 1.6248-1.6250 in.
Piston pin diametral clearance in piston,
D-358 .... 0.0000-0.0001T in.
D-360, D-436, DT-414 and
DT-414B ...... 0.0005-0.0010 in.
Piston pin diametral clearance in rod bushing,
D-358 .... 0.0005L-0.0010L in.
D-360, D-436, DT-414 and
DT-414B ...... 0.0006-0.0010 in.
Piston pin bushings are furnished semi-finished and must be reamed or honed for correct pin fit after they are pressed into connecting rods.

CONNECTING RODS AND BEARINGS

All Models

95. Connecting rod bearings are of the slip-in, precision type, renewable from below after removing oil pan and connecting rod cap. When installing new bearing inserts, make certain the projections on same engage slots in connecting rod and cap and that cylinder identifying numbers on rod and cap are in register and face TOWARDS camshaft side of engine on D-358 engine. On D-360, D-436, DT-414 and DT-414B, identifying numbers on rod and cap face AWAY from camshaft. Connecting rod bearings are available in standard size and undersizes of 0.010 and 0.020 inch for D-358 engines. Rod bearings are available for D-360, D-436, DT-414 and DT-414B engines in standard size and undersizes of 0.010 and 0.030 inch. Check crankshaft, crankpins and connecting rod bearings against the values which follow:

Crankpin diameter,
    D-358 . . . . 2.5185-2.5193 in.
    D-360, D-436, DT-414 and
      DT-414B . . . . . . 2.9977-2.9990 in.
Rod bearing diametral clearance,
    D-358 . . . . 0.0023-0.0048 in.
    D-360, D-436, DT-414 and
      DT-414B . . . . . . 0.0018-0.0051 in.
Rod side clearance,
    D-358 . . . . 0.006-0.010 in.
    D-360, D-436, DT-414 and
      DT-414B . . . . 0.007-0.015 in.
Rod bolt (or nut) torque,
    D-358 . . . . 45 ft.-lbs.
    D-360, D-436, DT-414 and
      DT-414B . . . . 130 ft.-lbs.

CRANKSHAFT AND MAIN BEARINGS

Models 786-886 (D-358)

96. Crankshaft is supported in seven main bearings. Main bearings are of the non-adjustable, slip-in, precision type, renewable from below after removing the oil pan and main bearing caps. Crankshaft end play is controlled by the flanged rear main bearing inserts. Removal of crankshaft requires R&R of engine. The crankshaft is counter-balanced by twelve weights bolted opposite to crankshaft throws. The weights are numbered consecutively from 1 to 12 which correspond to numbers stamped on crankshaft. The mounting holes in balance weights are offset. The wide edge goes toward connecting rod bearing. Balance weights are not serviced separately.

Check crankshaft and main bearings against the values which follow:

Crankpin diameter . . . 2.5185-2.5193 in.
Main journal diameter . 3.1484-3.1492 in.
Crankshaft end play . . . . 0.006-0.009 in.
Main bearing diametral
    clearance . . . . 0.0029-0.0055 in.
Main bearing bolt torque,
    Marked 10K or 10-9 . . . . 80 ft.-lbs.
    Marked 12K or 12-9 . . . . 97 ft.-lbs.
    Marked 10.9 . . . . 115 ft.-lbs.
    Marked 12.9 . . . . 140 ft.-lbs.
    Balance weight bolt torque . 57 ft.-lbs.
Main bearings are available in standard size and undersizes of 0.010 and 0.020 inch. Main bearing caps should be installed with numbered side toward camshaft side of engine.

Models 886 (D-360)-986-1086

97. Crankshaft is supported in seven main bearings and crankshaft end thrust is taken by the rear main bearing. Main bearings are of the non-adjustable, slip-in, precision type, renewable from below after removing oil pan and main bearing caps. Removal of crankshaft requires R & R of engine. Check crankshaft and main bearings against the values which follow:

Crankpin diameter . . . 2.9977-2.9990 in.
Main journal diameter . 3.3742-3.3755 in.
Crankshaft end play . . . . 0.006-0.012 in.
Main bearing diametral
    clearance . . . . 0.0018-0.0051 in.
Main bearing bolt torque . . . . 115 ft.-lbs.

Main bearings are available in standard size and undersize of 0.010 and 0.030 inch. Main bearing caps are numbered for correct location and numbered side of caps should be toward camshaft side of engine.

CRANKSHAFT SEALS

All Models

98. **FRONT.** To renew the crankshaft front oil seal, first remove left front hood and left side radiator support panel. Unbolt and move fan shroud to rear. Remove fan belt, fan and shroud.

On Models 786 and 886 equipped with D-358 engines, remove three cap screws, flat washer and pressure ring from crankshaft. Tap crankshaft pulley with a plastic hammer to loosen pulley, then slide pulley off the wedge rings. Remove wedge rings and oil seal wear ring from crankshaft. Remove and renew oil seal in conventional manner. Renew "O" ring on crankshaft. Inspect wear ring and timing pin for wear or other damage and renew as necessary. Use a non-hardening sealer on timing pin. When installing wear ring, timing pin must engage slot in crankshaft gear. Install one pressure ring on crankshaft with thick end against wear ring. Place wedge rings in pulley bore so slots in

rings are 90 degrees apart. Slide pulley on crankshaft, aligning slot in pulley with timing pin in wear ring. Install pressure ring, flat pressure washer and three cap screws. Tighten cap screws evenly to a torque of 57 ft.-lbs. Reassemble tractor by reversing disassembly procedure.

On Models 886 equipped with D-360 engine, 986 and 1086, remove the three cap screws and retaining plate from front of crankshaft pulley. Using a puller (IH tool No. FES 10-17) remove pulley. Unbolt and remove oil pump from front cover and renew seal. Remove wear sleeve from pulley only if it is worn, scratched or nicked.

NOTE: One side of wear sleeve is chamfered on the I.D. to aid in starting it on the crankshaft and opposite end is chamfered on the O.D. to aid in sliding oil seal over the sleeve.

When reinstalling crankshaft pulley, heat in boiling water for a period not to exceed one hour. Then install on crankshaft and secure with retaining plate and torque cap screws to 125 ft.-lbs.

Reassemble tractor by reversing disassembly procedure.

Fig. 98—*Splitting and removing rear oil seal. Refer to text.*

Fig. 99—*Use a chisel and split wear sleeve as shown.*

99. REAR. To renew the crankshaft rear oil seal, engines must be detached from clutch housing as follows: Drain cooling system, disconnect battery cables; remove exhaust stack, radiator cap, front hood channel and both side hoods. Disconnect left and right turn hydraulic lines from steering hand pump and cap all lines and fittings. Disconnect oil cooler hoses at multiple control valve. Disconnect throttle cable from injection pump and support bracket. Disconnect oil pressure sending unit. If equipped, disconnect speedometer cable and transducer assembly. Disconnect power steering hand pump supply and return lines. Disconnect ether solenoid electrical lead. On models so equipped, shut off both suction and discharge valves on air conditioning compressor, then disconnect suction line and move it back towards the control center. Cap the open compressor port. Partially remove left and right front lower panels and disconnect headlight wiring. Disconnect air conditioner quick connector and air conditioner wiring harness behind the left front panel. Shut off coolant valves and disconnect heater hoses. Remove subframe platform inside the control center, then disconnect main electrical wiring harness, and all necessary remaining wires.

Remove the two mounting bolts on clutch housing ahead of cranking motor. Install suitable jack stands to side rails and place a rolling floor jack under rear section of tractor. With tractor properly supported, unthread the remaining six mounting bolts from speed transmission.

IMPORTANT: Do not completely remove the six mounting bolts. To do so will cause the engine to drop between side rails.

Drive wooden blocks in between lower bolster and front axle to prevent tilting of front section. Split tractor by rolling rear section away from front.

Unbolt and remove clutch assembly, then unbolt and remove flywheel.

On Models 786 and 886 equipped with D-358 engines, unbolt and remove seal retainer. Check depth that old seal is installed in retainer, then remove seal. New oil seal may be installed in retainer in any of three locations; 1/16-inch above flush with retainer (new engine original position), flush with retainer or 1/16-inch below flush with retainer. Location of seal in retainer will depend on condition of the sealing surface on crankshaft. Use new gasket and install retainer and seal assembly. Tighten seal retainer cap screws evenly to a torque of 14 ft.-lbs. Install flywheel and tighten flywheel retaining cap screws to 85 ft.-lbs.

To avoid alignment problems when recoupling tractor, place clutch pressure plate and driven disc on shaft in clutch housing. Roll tractor together. Clutch can be bolted to flywheel after tractor is rejoined by working through opening at bottom of clutch housing. The balance of reassembly is the reverse of disassembly procedure.

On Models 886 equipped with D-360 engine, 986 and 1086, using a chisel, split and remove the oil seal (1 – Fig. 98).

NOTE: Do not damage the seal bore in support plate.

To remove oil seal wear sleeve from crankshaft, use a chisel to split the sleeve as shown in (1 – Fig. 99) being careful not to damage the crankshaft flange.

When installing wear sleeve and/or oil seal, International Harvester recommends the use of wear sleeve and oil seal installation tool set (IH tool No. FES 149-3). Press the wear sleeve on crankshaft flange.

NOTE: One side of wear sleeve is chamfered on the I.D. to aid in starting it on crankshaft and opposite end is chamfered on the O.D. to aid in installing oil seal.

Install new oil seal.

NOTE: When using IH tool to install sleeve and seal, pull them in until the installing tool bottoms against the centering plate.

Install flywheel, and tighten flywheel retaining cap screws to a torque of 123 ft.-lbs.

To avoid alignment problems when recoupling tractor, place clutch pressure plate and driven disc on shaft in clutch housing. Roll tractor together. Clutch can be bolted to flywheel after tractor is rejoined by working through opening at bottom of clutch housing. The balance of reassembly is the reverse of disassembly procedure on all models.

FLYWHEEL

All Models

100. To remove the flywheel, first split tractor as outlined in paragraph 99. Then, unbolt and remove clutch assembly. Remove cap screws and lift flywheel from crankshaft. When reinstalling, tighten flywheel retaining cap screws to the following torque values:

D-358 .85 ft.-lbs.
D-360, D-436, DT-414,
　DT414B . . . . .123 ft.-lbs.

To install a new flywheel ring gear, heat same to approximately 500 degrees F.

OIL PUMP

Models 786-886 (D-358)

101. The externally mounted gear type oil pump is located on right front side of engine and is driven by the camshaft gear. To remove the oil pump, disconnect the oil cooler lines, remove two cap screws securing oil filter base to crankcase, then unbolt pump assembly. Remove pump, filter and connecting pipes as an assembly. Slide pump off the pipes.

Remove plug (16 – Fig. 100) and withdraw pressure relief valve components (12 through 15). Remove rear cover (2); it may be necessary to tap rear cover with plastic hammer to free it from dowel pins (1). Lift out idler gear (4), then withdraw pumping drive gear (3). Remove two Woodruff keys (11), snap ring (7) and spacer (8) from drive gear shaft, then withdraw drive gear and shaft (10) from front cover (9). Separate front cover from pump body (6).

Inspect all parts for scoring, excessive wear or other damage. Pump covers (2 and 9) and pump body (6) are not serviced separately. Check pump against the following specifications:

Drive shaft end play
　(cover installed) . . . . . .0.000-0.002 in.
Drive shaft running
　clearance. . . . . .0.001-0.0032 in.
Idler gear to shaft
　clearance. . . . . .0.001-0.0032 in.
Pumping gears end
　clearance. . . . . .0.002-0.0038 in.

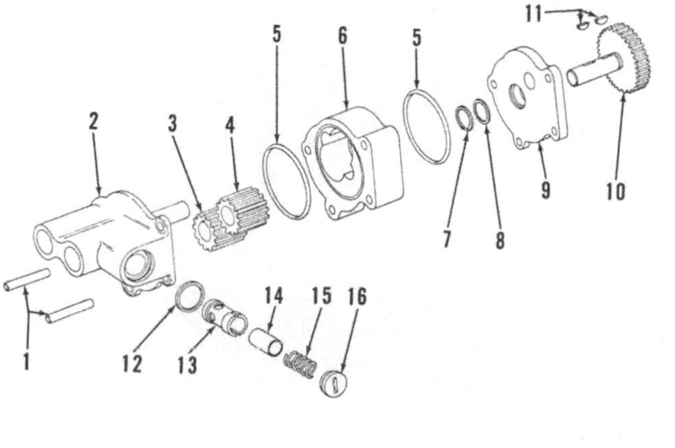

Fig. 100 – Exploded view of oil pump assembly used on D-358 engines.

1. Roll pin dowels
2. Rear cover
3. Pumping drive gear
4. Pumping idler gear
5. "O" rings
6. Pump body
7. Snap ring
8. Spacer
9. Front cover
10. Drive gear and shaft
11. Woodruff keys
12. Seal rings
13. Relief valve body
14. Relief valve piston
15. Spring
16. Plug

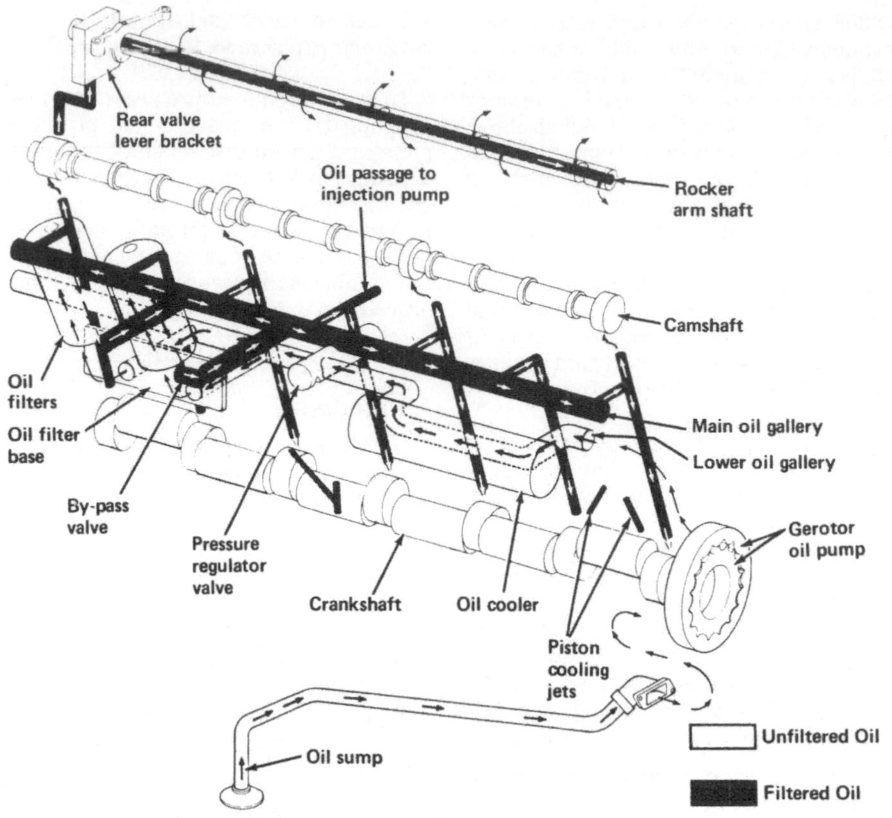

Fig. 105 — View of engine lubrication system.

Labels in figure: Rear valve lever bracket; Oil passage to injection pump; Rocker arm shaft; Camshaft; Oil filters; Oil filter base; Main oil gallery; Lower oil gallery; By-pass valve; Pressure regulator valve; Crankshaft; Oil cooler; Gerotor oil pump; Piston cooling jets; Oil sump; Unfiltered Oil; Filtered Oil

pulley. Using a puller (IH tool FES 10-17) remove pulley. Unbolt and remove oil pump from front cover.

Remove spacer plate (10 – Fig. 106), "O" ring (9) and inner and outer rotor (7 and 8).

Inspect all parts for scoring, excessive wear or other damage. Inner and outer rotor (7 and 8) and wear sleeve and seal (3 and 4) are serviced as matched sets.

Using a feeler gage and referring to Fig. 107, check radial clearance between housing (1) and outer rotor (2). Specified clearance is 0.0055 to 0.0085 inch. Refer to Fig. 108 and check side clearance using "Plastigage". Specified clearance for inner rotor (2) is 0.0018 to 0.0042 inch and 0.0014 to 0.0038 inch for outer rotor (1). Oil pressure at rated rpm should be 45-65 psi.

Renew the "O" rings and seal when reassembling and reinstalling oil pump.

NOTE: Be sure to install special cap screws with the nylon pellet insert.

OIL PRESSURE REGULATOR

Models 786-886 (D-358)

103. On D-358 engines, the oil pressure regulator valve is located in the externally mounted oil pump. See Fig. 100. Refer to paragraph 101 for regulator valve and spring specifications. Filter by-pass valve is located in the spin-on type oil filter. Oil pressure at 2300 rpm should be 35-50 psi.

Models 886 (D-360)-986-1086

104. The oil pressure regulator valve is located at rear of engine oil cooler and behind crankcase breather in main oil gallery. The by-pass valve is located in base of oil filters (Fig. 109). Removal of valves is obvious after an examination of assembly.

Oil pressure regulator valve (9) is available as a unit only. Check regulator valve, by-pass valve and springs against the following specifications:

Pumping gears to body
radial clearance ......0.007-0.012 in.
Pressure regulating spring,
Spring free length ....2.52 in.
Spring test and
length ....18-20 lbs. at 1.858 in.
Pressure regulating valve,
Piston diameter ......0.825-0.827 in.
Clearance in bore ....0.003-0.007 in.
Oil pressure at 2300 rpm......35-50 psi
Renew all "O" rings and gaskets when reassembling and reinstalling oil pump and filter assembly.

Models 886 (D-360)-986-1086

102. The oil pump is a GEROTOR type mounted on front cover and driven

directly by the crankshaft at engine speed. Refer to Fig. 105 showing the lube system. With the exception of the injection pump and turbocharger oil supply, there is no external piping to direct oil from one component to the next. Two types of pumps are used, one for turbocharged engines and one for naturally aspirated models. The difference being the width of the pump. The wide pump is used on turbocharged engines.

To remove the oil pump, first remove left front hood and left side radiator support panel. Unbolt and move fan shroud to rear. Remove fan belt, fan and shroud.

Remove the three cap screws and retaining plate from front of crankshaft

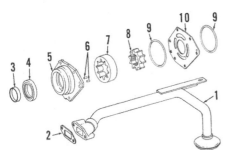

Fig. 106 — Exploded view of Gerotor type oil pump assembly used on D-360, D-436 and DT-414 engines.

1. Inlet tube	6. Dowels
2. Gasket	7. Rotor (outer)
3. Wear sleeve	8. Rotor (inner)
4. Oil seal	9. "O" rings
5. Housing	10. Plate

Fig. 107 — Checking radial clearance between housing and outer rotor. Refer to text.

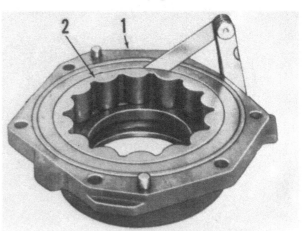

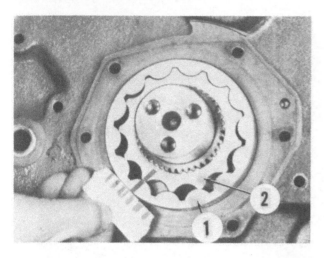

Fig. 108 — Using plastigage to check side clearance. Refer to text.
1. Outer rotor
2. Inner rotor

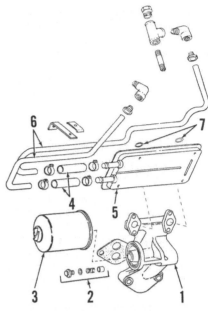

Pressure regulator valve:
Piston diameter . . . . . . 0.621-0.622 in.
Piston clearance in bore 0.002-0.004 in.
Spring free length . . . . 2-11/64 in.
Spring test and
length . . . . 14.5-16 lbs. at 1-15/64 in.
By-pass valve spring:
Spring free length . . . . 2-5/64 in.
Spring test and
length . . . . 6.12 lbs. at 59/64 in.
Oil pressure at 2600 rpm . . . . 45-65 psi

OIL JET TUBES

Models 886 (D-360)-986-1086

105. The D-360, D-436, DT-414 and DT-414B engines are equipped with 12 oil jet tubes located in main bearing bosses, which spray oil on sleeves for added lubrication and cooling of pistons.

When overhauling engine, make certain jet tubes are open and clean. Oil jet tubes need not be removed unless they are damaged.

OIL COOLER

Models 786-886 (D-358)

106. The engine oil cooler is mounted on oil filter base (1 – Fig. 110) which is

located on right side of engine. Removal and disassembly of the unit is obvious after an examination of the unit. Inspect oil cooler relief valve assembly (2) for freeness in its bore.

Normal cleaning of the oil cooler consists of blowing out water tubes with compressed air and flushing oil passages with a suitable cleaning solvent. Renew "O" rings and gasket when reassembling.

Models 886 (D-360)-986-1086

107. The engine oil cooler is horizontally mounted to the right side of crankcase.

To remove the oil cooler, first drain cooling system, then remove drain plug (2 – Fig. 111) from oil cooler and drain cooler. Unbolt cooler from crankcase and slide off of water tubes.

Normal cleaning of oil cooler consists of blowing out water tubes with compressed air or the use of a rotary brush or rod of proper diameter.

Fig. 110 — Exploded view of oil cooler and oil filter used on D-358 engines.
1. Oil filter and oil cooler base
2. Relief valve (oil cooler)
3. Oil filter
4. Hose connection
5. Oil cooler assembly
6. Water pipes to oil cooler
7. "O" rings

Using new gaskets and "O" rings, reinstall oil cooler by reversing the removal procedure.

OIL PAN

All Models

108. Removal of oil pan is conventional and on tricycle model tractors can be accomplished with no other disassembly. On models equipped with adjustable wide tread front axle, the stay rod must be disconnected from stay rod support and stay rod support from side

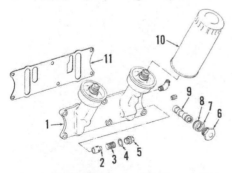

Fig. 109 — Exploded view of oil filter assembly used on D-360, D-436 and DT-414 engines.
1. Filter base
2. Valve (by-pass)
3. Spring
4. Gasket
5. Plug
6. Cap
7. "O" ring
8. Gasket
9. Valve (pressure regulating)
10. Filter
11. Gasket

Fig. 111 — Right side view of DT-414 engine; D-360 and D-436 are similar.
1. Oil cooler
2. Coolant drain plug
3. Coolant shut-off
4. Turbocharger
5. Oil filters
6. Coolant system filter
7. Exhaust temperature thermocouple

rails before removal of oil pan can be accomplished.

On International and "All Wheel Drive" models, disconnect stay rod (or bracket) from clutch housing and front axle from axle support, then raise front of tractor to provide clearance for oil pan removal.

DIESEL FUEL SYSTEM

The diesel fuel system of all Model 786 and Model 886 equipped with D-358 engines uses a Robert Bosch injection pump. Models 886 equipped with D-360 engine, 986 and 1086 use American Bosch injection pumps. All engines are of the direct injection type.

When servicing any unit of the diesel system, the maintenance of absolute cleanliness is of utmost importance. Of equal importance is the avoidance of nicks or burrs on any of the working parts.

Probably the most important precaution that service personnel can impart to owners of diesel powered tractors, is to urge them to use an approved fuel that is absolutely clean and free from foreign material. Extra precaution should be taken to make certain that no water enters the fuel storage tank.

FILTERS AND BLEEDING

All Models

109. Tractors are equipped with two spin on cartridge type fuel filters. The primary filter is the rear of the two filters. Filters should be serviced when engine shows signs of losing power.

110. To bleed the fuel system on Models 786 and 886 (D-358 engine) equipped with Robert Bosch injection pump, refer to Fig. 112 and proceed as follows: Open vent valves on top of primary and final filters (one valve on each). When outflowing fuel is free of air bubbles, close vent valves. Next, open vent screw on injection pump until fuel is flowing free of air bubbles, then retighten vent screw. Fuel system is now free of air.

111. To bleed the fuel system on Models 886 (D-360 engine), 986 and 1086 equipped with the American Bosch injection pump, refer to Fig. 113 and proceed as follows: Loosen vent valve on top of primary filter and allow fuel to run until a solid stream with no air bubbles appears; then, close vent valve. Open vent valve on final filter. Then, using the hand priming pump, pump fuel into final

Fig. 112—Left side view of D-358 engine used on Model 886. Model 786 is similar.

1. Oil filler and level gage
2. Final fuel filter
3. Primary fuel filter
4. Fuel injection pump
5. Compressor
6. Ether starting fluid
7. Engine coolant heater
8. Heater shut-off
9. Coolant filter and conditioner

filter until fuel coming out of final filter vent is free of air bubbles. Close vent on final filter and lock hand pump plunger in place. Fuel system is now free of air.

INJECTION PUMP

All Models

The Robert Bosch and American Bosch injection pumps are all of the rotary distributor type. Because of the special equipment needed, and skill required of servicing personnel, service of injection pumps is generally beyond the scope of that which should be attempted in the average shop. Therefore, this section will include only timing of pump to engine, removal and installation and the linkage

adjustments which control the engine speeds.

If additional service is required, the pump should be turned over to an International Harvester facility which is equipped for diesel service, or to some other authorized diesel service station. Inexperienced personnel should NEVER attempt to service diesel injection pumps.

REMOVE AND REINSTALL PUMP

Models 786-886 (D-358) (Robert Bosch)

112. To remove the Robert Bosch injection pump from the D-358 engine, first thoroughly clean injection pump,

Fig. 113—Left side view of D-436 engine. (D-360 and DT-414 similar).

1. Ether starting fluid
2. Fuel filter vent screw
3. Primary fuel filter
4. Final fuel filter
5. Oil filler and level gage
6. Crankcase water drain plug
7. Hand priming pump
8. Fuel injection pump
9. Compressor
10. Heater shut-off

fuel lines and side of engine. Shut off fuel at tank and remove timing hole cover screws from pump. Rotate cover 90 degrees and remove by prying evenly with two screwdrivers. Remove timing hole cover on right side of clutch housing. Using a bar on flywheel, turn engine in normal direction of rotation until timing line (TL – Fig. 114) on face cam is aligned with timing pointer (TP) in pump.

NOTE: The face cam has two timing lines. Near one of the lines, a letter "L" is etched on face cam. DO NOT use this timing line.

Disconnect throttle control rod and shut-off cable from pump. Disconnect fuel inlet and return lines from pump, then remove high pressure injection lines. Immediately cap or plug all openings. Unbolt and remove the rectangular cover from timing gear cover. Remove nut and washer from pump drive shaft, then remove three cap screws securing hub to drive gear. Remove nuts from pump mounting studs. Install a puller (IH tool No. FES 111-2 or equivalent) to tapped holes in drive gear and force drive shaft from hub. Remove injection pump.

CAUTION: Do not turn crankshaft while pump is removed.

When reinstalling pump, make certain that timing pointer (TP) and timing line (TL) are aligned, then reinstall pump by reversing removal procedure. Align scribe mark on pump mounting flange with punch mark on engine front plate. Tighten pump drive shaft nut to a torque of 47 ft.-lbs. and the three hub to drive gear cap screws to a torque of 17 ft.-lbs.

Bleed fuel system as in paragraph 110, then check and adjust injection pump timing as in paragraph 114.

Models 886 (D-360)-986-1086 (American Bosch)

113. To remove the American Bosch injection pump from D-360, D-436, DT-414 and DT-414B engines, first clean injection pump, fuel lines and side of engine. Remove timing hole cover on right side of clutch housing. Using a pry bar on flywheel teeth, turn crankshaft in normal direction of rotation until timing marks on flywheel are aligned at 18° BTDC for D-360, D-436 and DT-414 engines, and 16° BTDC for DT-414B engine; this is static timing. Disconnect control rod, fuel lines and oil supply line from injection pump. Cap or plug all openings immediately. Remove fuel filter to pump lines, left radiator support panel and pump access cover from front cover. Unbolt and remove pump and adapter plate as an assembly.

NOTE: If same pump is to be reinstalled, DO NOT rotate the pump drive shaft after pump has been removed.

When reinstalling pump with drive gear removed, secure with two cap screws (3 – Fig. 115) with timing mark on flywheel aligned with pointer at 18° BTDC for D-360, D-436 and DT-414 engines, 16° BTDC for DT-414B engine (it does not matter whether it is on No. 1 or No. 6 cylinder). Turn pump so that line mark (1 – Fig. 115 or 2 – Fig. 116) on pump hub are aligned with pointer.

Install drive gear, meshing it with the idler gear.

NOTE: Disregard any timing marks on the idler gear.

Hold pump shaft with socket and tighten cap screws in gear to a torque of 30 ft.-lbs.

Bleed fuel system as outlined in paragraph 111. If engine does not start after fifth or sixth revolution, the pump is out of time. Reposition engine crankshaft to

18° BTDC for D-360, D-436 and DT-414 engines, and 16° BTDC for DT-414B engine. Then, remove pump drive gear. Rotate pump drive shaft one revolution and reinstall gear. Pump is now in time.

Models 786-886 (D-358) (Robert Bosch)

114. **STATIC TIMING.** To check or adjust static timing, first shut off fuel at tank and remove pump timing hole screws. Rotate cover 90 degrees and remove cover by prying evenly with two screwdrivers. Remove timing hole cover from right side of clutch housing. Using a pry bar on flywheel teeth, turn crankshaft in normal direction of rotation until number one piston is coming up on compression stroke. Continue turning crankshaft until the 16° BTDC (for D-358 engine equipped with injection pump No. 3218 238 R91) or 18° BTDC (for D-358 engine equipped with injection pump No. 3228 168 R91 or No. 3228 169 R91), mark on flywheel is aligned with pointer on clutch housing. At this time, timing line (TL – Fig. 114) on face cam should be aligned with timing pointer (TP) on roller retainer ring. If not, first make certain that scribe line on pump mounting flange is aligned with punch mark on engine front plate. Then, unbolt and remove rectangular cover from front of timing gear cover. Loosen the three cap screws securing pump drive gear to pump shaft hub. Rotate hub as required to align line on face cam with timing pointer. Tighten hub retaining cap screws to a torque of 17 ft.-lbs. and install all removed covers.

Models 886 (D-360)-986-1086 (American Bosch)

115. **STATIC TIMING.** To check or adjust static timing, remove access cover Fig. 115 on early models or plug Fig. 116 on late models. Remove timing

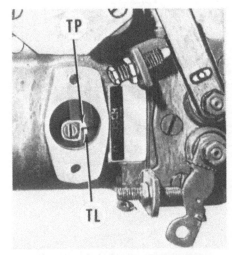

Fig. 114 – View of timing pointer (TP) and timing line (TL) on Robert Bosch injection pump.

Fig. 115 – View showing the timing pointer and timing mark on pump hub of D-360, D-436 and DT-414 engines.

Fig. 116 – View showing timing pointer on late model tractors. Refer to text.

hole cover from right side of clutch housing. Using a pry bar on flywheel teeth, turn crankshaft in normal direction of rotation until 18° BTDC (for D-360, D-436 and DT-414 engines) or 15° BTDC (for DT-414B engine) mark on flywheel is aligned with timing pointers on clutch housing.

At this time, timing mark on injection pump hub and pointer (Fig. 115 or 116) should be aligned. If timing marks are not aligned as shown, remove pump drive gear and align pump hub and pointer. Reinstall drive gear and torque to 30 ft.-lbs. Install access cover.

Models 786-886 (D-358) (Robert Bosch)

116. **HIGH IDLE, LOW IDLE & SHUT-OFF.** Start engine and bring to operating temperature. Disconnect control cable from throttle lever (T–Fig. 117). Move throttle lever rearward to high idle position, loosen jam nut and turn high idle screw (H) as required to obtain engine high idle speed of 2640 rpm. Tighten jam nut.

Move throttle lever forward until it contacts the low idle plunger (L). Loosen the locknut, then adjust the low idle and shut-off plunger capsule as required to obtain an engine low idle speed of 700 rpm. Tighten jam nut.

Place speed control lever at idle position on dash quadrant. Adjust control cable positioners (on cable mounting bracket directly behind injection pump) until the cable, when connected, will hold throttle lever (T) against low idle plunger. When speed control lever is moved to high idle position on dash quadrant, throttle lever (T) must contact high idle screw (H).

With engine operating at low idle speed, move speed control lever to the

right and downward from low idle position. At this time throttle lever (T) should push low idle plunger (L) into the shut-off capsule and engine should stop.

117. **MAXIMUM FUEL.** With engine high idle speed properly adjusted as outlined in paragraph 116, adjust rated load engine speed of 2400 rpm as follows: Attach a dynamometer to tractor and load engine to maintain rated rpm with throttle control lever in high idle position.

On Model 786, adjust maximum fuel stop screw (M–Fig. 117) to obtain 80 (pto) horsepower for D-358 engine equipped with injection pump No. 3228 168 R91 at 2400 rpm.

On Model 886, adjust maximum fuel stop screw (M) to obtain approximately 85 (pto) horsepower for D-358 engine equipped with injection pump No. 3218 238 R91 or 90 (pto) horsepower for D-358 engine equipped with injection pump No. 3228 169 R91 at rated load (2400) rpm.

CAUTION: Do not overfuel the engine or attempt to increase horsepower above the rated load.

Models 886 (D-360)-986-1086 (American Bosch)

118. **LOW & HIGH IDLE.** To adjust engine low and high idle speeds, start engine and bring to operating temperature. Disconnect throttle control rod from pump control lever. Hold pump rod lever in high idle position (rearward against stop) and check engine rpm. Engine high idle should be 2640 rpm for all models.

Adjust high idle stop screw (1–Fig. 118) on rear of pump as necessary to obtain correct high idle rpm.

Release pump control lever and check engine low idle rpm. Low idle speed

should be 700 rpm for all models. Adjust engine low idle rpm with screw (1–Fig. 119) on rear of pump.

With engine operating at low idle rpm, place throttle control lever in low idle position (in detent approximately ¾-inch from stop). Adjust length of throttle control rod to hold pump control lever in low idle position.

INJECTION NOZZLES

WARNING: Fuel leaves the injection nozzles with sufficient force to penetrate the skin. When testing, keep your person clear of the nozzle spray.

All Models

119. **TESTING AND LOCATING FAULTY NOZZLE.** If engine does not run properly and a faulty injection nozzle is suspected, or if one cylinder is misfiring, locate the faulty nozzle as follows: Loosen the high pressure line fitting on each nozzle holder in turn, thereby allowing fuel to escape at the union rather than enter the cylinder. As in checking spark plugs in a spark ignition engine, the faulty nozzle is the one that when its line is loosened least affects the running of the engine.

Remove the suspected nozzle as outlined in paragraph 120 or 121, place nozzle in a test stand and check the nozzle against the following specifications:

Models 786-886 (D-358)
Opening pressure, new . . . . 3000 psi
Opening pressure, used . . . . 2900 psi

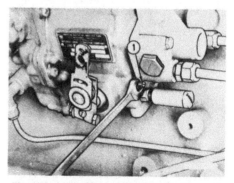

Fig. 118—View showing the high idle stop screw adjustment on American Bosch injection pump.

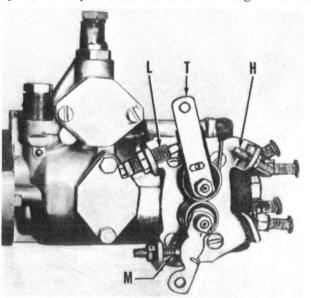

Fig. 117—View of Robert Bosch injection pump showing adjustment points.
H. High idle stop screw
L. Low idle & shut-off capsule
M. Maximum fuel stop screw
T. Throttle lever

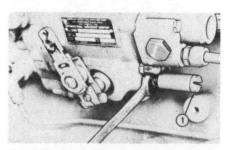

Fig. 119—View showing low idle adjusting screw on American Bosch injection pump.

Nozzle showing visible wetting on tip after 10 seconds at 2700 psi is permissible. Maximum leakage through return port is 10 drops in 1 minute at 2700 psi.

Models 886 (D-360) — 986 (D-436) — 1086 (DT-414 and DT-414B)

Opening pressure, new . . . 3600-3750 psi
Opening pressure, used . . . . 2900 psi

Nozzles showing visible wetting on tip after 5 seconds with pressure held at 100 psi below opening pressure is permissible.

Models 786-886 (D-358)

120. **R&R NOZZLES.** To remove any injection nozzle, first remove dirt from nozzle injection line, return line and cylinder head. Disconnect leakage return line and high pressure injection line from nozzle and immediately cap or plug all openings. Remove the injector clamp bolt and clamp and carefully withdraw the injector assembly from cylinder head.

NOTE: It is recommended that cooling system be drained before removing injectors. It is possible that injector nozzle sleeve may come out with injector and allow coolant to enter engine.

When reinstalling, tighten injector clamp bolt to a torque of 18 ft.-lbs.

Models 886-986-1086 (D-360, D-436, DT-414 and DT-414B)

121. **R&R NOZZLES.** To remove any injection nozzle, first remove left front hood. Then remove dirt from nozzle injection line, return line and cylinder head. Disconnect leakage return line and high pressure injection line from nozzle and immediately cap or plug all openings. Remove the injector cap screw and carefully withdraw the injector assembly from cylinder head.

NOTE: It is recommended that cooling system be drained before removing injectors. It is possible that injector nozzle sleeve may come out with injector and allow coolant to enter engine.

When reinstalling, tighten injector cap screw to a torque of 20 ft.-lbs.

Models 786-886 (D-358)

122. **OVERHAUL.** To disassemble the injector, clamp flats of nozzle body (3 – Fig. 120) in a vise with nozzle tip pointing upward. Remove nozzle holder nut (10). Remove nozzle tip (9) with valve (8) and spacer (7). Invert nozzle body (3) and remove spring seat (6), spring (5) and shims (4).

Thoroughly clean all parts in a suitable solvent. Clean inside the orifice end of

nozzle tip with a wooden cleaning stick. The 0.011 inch diameter orifice spray holes may be cleaned by inserting a cleaning wire of proper size. Cleaning wire should be slightly smaller than spray holes.

When reassembling, make certain all parts are perfectly clean and install parts while wet with clean diesel fuel. To check cleanliness and fit of valve (8) in nozzle tip (9), use a twisting motion and pull valve about ⅓ of its length out of nozzle tip. When released, valve should slide back to its seat by its own weight.

NOTE: Valve and nozzle tip are mated parts and under no circumstance should valves and nozzle tips be interchanged.

Install shims (4 – Fig. 120), spring (5) and spring seat (6) in nozzle body (3). Place spacer (7) and nozzle tip (9) with valve (8) in position and install holder nut (10). Tighten holder nut to a torque of 43 ft.-lbs.

Connect the assembled injector nozzle to a test pump and check opening pressure. Adjust opening pressure by varying number and thickness of shims (4). Shims for adjusting nozzle opening pressure are available in thicknesses of 0.039 to 0.0778 inch in increments of 0.001 inch. A shim difference of 0.001 inch will change opening pressure by 57 psi. Opening pressure should be adjusted to 3000 psi if old spring (5) is used or 3100-3300 psi if new spring is in-

stalled. Valve should not show leakage at orifice spray holes for 10 seconds at 2700 psi. Maximum allowable leakage through return port is 10 drops in 1 minute at 2700 psi.

Models 886-986-1086 (D-360, D-436, DT-414 and DT-414B)

123. **OVERHAUL.** Clamp the injector assembly upright in a vise using the injector retainer from the engine. Remove nozzle cap (9 – Fig. 121), then carefully remove nozzle (8). Remove body (1) from vise, invert the body and remove valve stop spacer (6), spring seat (5), spring (4), spacer shim (3), adjusting shims (2) and the other spacer shim (3). Thoroughly clean all parts in suitable solvent. Inspect mating surfaces of nozzles, valve stop spacer and nozzle cap for cracks, scratches or other defects. Light scratches can be removed by lapping; however, the sealing surfaces must remain perfectly flat. Orifice spray holes can be cleaned by inserting a cleaning wire of proper size. Cleaning wire should be slightly smaller than spray holes. Spray holes are 0.012 inch.

Carbon can be removed from exterior surfaces with a soft brass brush and clean diesel fuel.

When reassembling, be sure all parts are perfectly clean and install parts

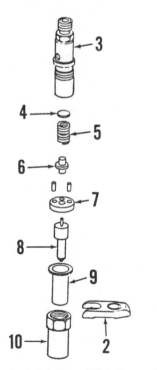

Fig. 120 — Exploded view of injector used on D-358 engines.

2. Clamp	
3. Body	7. Valve stop spacer
4. Adjusting shim	8. Valve
5. Spring	9. Nozzle
6. Spring seat	10. Nozzle cap

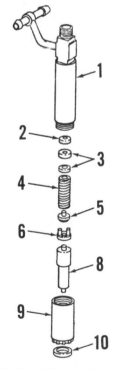

Fig. 121 — Exploded view of injector used on D-360, D-436 and DT-414 engines.

1. Body	5. Spring seat
2. Adjusting shims	6. Valve stop spacer
3. Spacer (2 used 0.060 inch)	8. Nozzle
	9. Nozzle cap
4. Spring	10. Washer

while wet with clean diesel fuel. To check cleanliness and fit of valve in nozzle (8), use a twisting motion and pull valve about ⅓ of its length out of nozzle. When released, valve should slide back to its seat by its own weight.

NOTE: Valve and nozzle (8) are mated parts and under no circumstance should valve and nozzles be interchanged. Lapping is not recommended on the valve and nozzle seat and faulty units must be renewed.

Be sure to install one spacer shim (3) on each side of adjusting shims (2). The spacer shims are 0.060 inch thick.

Reassemble injector by reversing the disassembly procedure. Tighten nozzle cap to a torque of 30 to 35 ft.-lbs. Adjust opening pressure to 3600-3750 psi. Injector should not show leakage at orifice spray holes for 5 seconds at 100 psi below opening pressure.

ETHER STARTING AID

All Models

124. On all diesel engines, it is necessary that ether be used as a starting aid at temperatures below freezing.

To test the ether spray pattern, disconnect ether line at spray nozzle and spray nozzle from manifold air inlet. Reconnect nozzle to ether line. Press ether injection button on dash and observe spray pattern. A good spray pattern is cone-shaped. Dribbling or no spray indicates a blocked spray nozzle or lack of ether pressure. Clean spray nozzle or install new can of ether as needed.

To change the ether fluid container, turn knurled adjusting screw clockwise until container can be removed. Install new container in the bail and tighten adjusting screw (counter-clockwise) while

guiding container head into position. Rotate container to make certain it is seated properly in injector body, then tighten adjusting screw to hold container firmly in position.

CAUTION: Ether must be in twelve ounce containers meeting ICC29 specifications.

NOTE: In warm temperatures, ether container can be removed and a protective plug installed in injector body. DO NOT operate tractor engine without either the ether container or protective plug in position.

DIESEL TURBOCHARGER

Model 1086 may be equipped with either an exhaust driven Schwitzer model 3LD turbocharger or an Airesearch model TO-4 turbocharger. The turbocharger consists of the following three main sections. The turbine, bearing housing and compressor.

Engine oil taken directly from the clean oil side of the engine oil filters, is circulated through the bearing housing. This oil lubricates the sleeve type bearings and also acts as a heat barrier between the hot turbine and the compressor. The oil seals used at each end of the shaft are of the piston ring type. When servicing the turbocharger, extreme care must be taken to avoid damaging any of the parts.

NOTE: When the engine has been idle for one month or more, after installation of a new or rebuilt turbocharger or after installing new oil filter elements, the turbocharger must be primed as outlined in paragraph 125.

REMOVE AND REINSTALL
Model 1086 (DT-414 and DT-414B)

125. To remove the turbocharger, first remove muffler, left and right front hood and air inlet and outlet pipes. Disconnect turbocharger oil inlet and oil drain tubes. Plug or cap all openings. Unbolt turbocharger from exhaust manifold, lift unit from engine and place it in a horizontal position on a bench.

To reinstall the turbocharger, reverse the removal procedure and prime the turbocharger as follows: With speed control lever in shut-off position, crank engine with starting motor approximately 30 seconds. Repeat this operation until engine oil pressure Tellite goes out. Start engine and operate at 1000 rpm for at least 2 minutes before going to higher speed.

126. OVERHAUL (SCHWITZER). Remove turbocharger as outlined in paragraph 125. Before disassembling, place a row of light punch marks across compressor cover, bearing housing and turbine housing to aid in reassembly. Clamp turbocharger mounting flange (exhaust inlet) in a vise and remove cap screws (3—Fig. 122), lockwashers and clamp plates (1). Remove compressor cover (17). Remove nut from clamp ring (4), expand clamp ring and remove bearing housing assembly from turbine housing (7).

CAUTION: Never allow the weight of bearing housing assembly to rest on either the turbine or compressor wheel vanes. Lay bearing housing assembly on a bench so that turbine shaft is horizontal.

Remove locknut (14—Fig. 122) and slip compressor wheel (13) from end of shaft. Withdraw turbine wheel and shaft (5) from bearing housing. Place bearing housing on bench with compressor side up. Remove snap ring (16), then using two screwdrivers, lift flinger plate insert (12) from bearing housing. Push spacer sleeve (11) from the insert. Remove oil deflector (10), thrust ring (20), thrust plate (9) and bearing (21). Remove "O" ring (18) from flinger plate insert (12) and remove seal rings (19) from spacer sleeve and turbine shaft.

Soak all parts in Bendix metal cleaner or equivalent and use a soft brush, plastic blade or compressed air to remove carbon deposits.

CAUTION: Do not use wire brush, steel scraper or caustic solution for cleaning, as this will damage turbocharger parts.

Inspect turbine wheel and compressor wheel for broken or distorted vanes. DO NOT attempt to straighten bent vanes. Check bearing bore in bearing housing, floating bearing (21) and turbine shaft

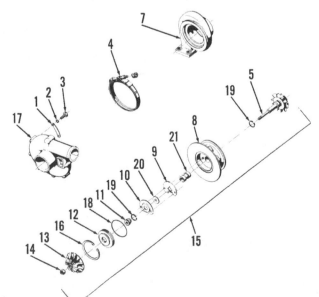

Fig. 122—Exploded view of Schwitzer Model 3LD turbocharger available for Model 1086 tractors equipped with DT-414 engines.

1. Clamp plate
2. Lockwasher
3. Cap screw
4. Clamp ring
5. Turbine wheel and shaft
7. Turbine housing
8. Bearing housing
9. Thrust plate
10. Oil deflector
11. Spacer sleeve
12. Flinger plate insert
13. Compressor wheel
14. Locknut
15. Turbine and compressor shaft assembly
16. Snap ring
17. Compressor cover
18. "O" ring
19. Seal rings
20. Thrust ring
21. Bearing

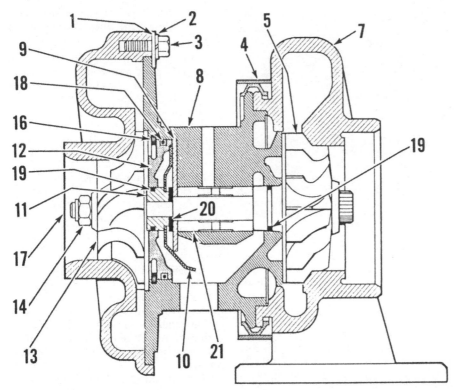

Fig. 123 – Cross-sectional view of Schwitzer turbocharger used on Model 1086 tractors. Refer to Fig. 122 for legend.

CAUTION: Do not rest weight of any parts on impeller or turbine blades. Weight of only the turbocharger unit is enough to damage the blades.

Remove clamp (1–Fig. 124), compressor housing (2) and diffuser (3). Remove cap screws (11) and lock plates (12) and clamp plates (13); then, remove turbine housing (8). Hold turbine shaft from turning using socket wrench at the center of turbine wheel (14) and remove locknut (22).

NOTE: Use a "T" handle to remove locknut in order to prevent bending turbine shaft.

Lift compressor impeller (21) off, then remove center housing from turbine shaft while holding shroud (7) onto center housing. Remove back plate retaining cap screws (10) and locks (9). Then, remove back plate (5), thrust bearing (15), thrust collar (16) and spring (4). Carefully remove bearing retainers (19) from ends and withdraw bearings (18) from center housing.

CAUTION: Be careful not to damage bearing or surface of center housing when removing retainers.

The center two retainers do not have to be removed unless damaged or unseated. Always renew bearing retainers if removed from groove in housing.

Soak all parts in Bendix metal cleaner or equivalent and use a soft brush, plastic blade or compressed air to remove carbon deposits.

CAUTION: Do not use wire brush, steel scraper or caustic solution for cleaning, as this will damage turbocharger parts.

Inspect turbine wheel and compressor wheel for broken or distorted vanes. DO NOT attempt to straighten bent vanes. Be sure all wheel blades are clean, as any deposits left on blades will affect balance. Inspect bearing bores in center housing (6–Fig. 124) for scored surfaces, out-of-round or excessive wear. Make certain bore in center housing is

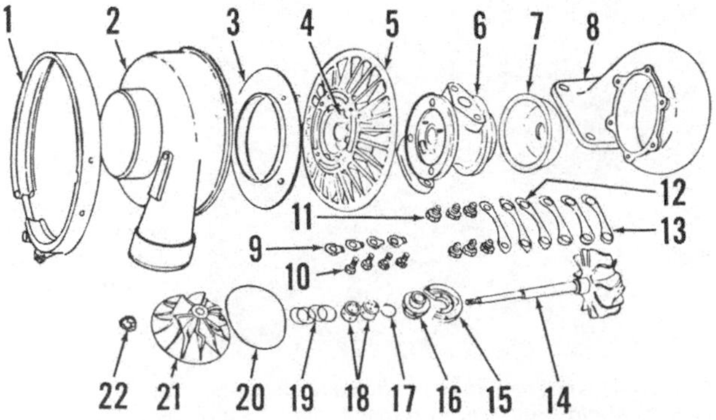

---

for excessive wear or scoring. Inspect flinger plate insert, spacer sleeve, oil deflector, thrust ring and thrust plate for excessive wear or other damage.

Renew all damaged parts and use new "O" ring (18) and seal rings (19) when reassembling. The seal ring used on turbine shaft is copper plated and is larger in diameter than the seal ring used on spacer sleeve. Refer to Figs. 122 and 123 as a guide when reassembling.

Install seal ring on turbine shaft, lubricate seal ring and install turbine wheel and shaft in bearing housing. Lubricate I.D. and O.D. of bearing (21), install bearing over end of turbine shaft and into bearing housing. Lubricate both sides of thrust plate (9) and install plate (bronze side out) on the aligning dowels. Install thrust ring (20) and oil deflector (10), making certain holes in deflector are positioned over dowel pins. Install new seal ring on spacer sleeve (11), lubricate seal ring and press spacer sleeve into flinger plate insert (12). Position new "O" ring (18) on insert, lubricate "O" ring and install insert and spacer sleeve assembly in bearing housing, then secure with snap ring (16). Place compressor wheel on turbine shaft, coat threads and back side of nut (14) with "Never-Seez" compound or equivalent, then install and tighten nut to a torque of 13 ft.-lbs. Assemble bearing housing to turbine housing and align punch marks. Install clamp ring, apply "Never-Seez" on threads and install nut

and torque to 10 ft.-lbs. Apply a light coat of "Never-Seez" around machined flange of compressor cover (17). Install compressor cover, align punch marks, and secure cover with cap screws, washers and clamp plates. Tighten cap screws evenly to a torque of 5 ft.-lbs.

Check rotating unit for free rotation within housings. Cover all openings until turbocharger is reinstalled.

Use a new gasket and install and prime turbocharger as outlined in paragraph 125.

127. OVERHAUL (AIRESEARCH) Remove turbocharger as outlined in paragraph 125. Mark across compressor housing, center housing and turbine housing to aid alignment when assembling.

Fig. 124 – Exploded view of Airesearch Model TO-4 turbocharger assembly used on some 1086 series tractors.

1. Clamp
2. Housing (compressor)
3. Diffuser
4. Spring
5. Backplate
6. Center housing
7. Turbine shroud
8. Turbine housing
9. Lock plates
10. Backplate cap screws
11. Turbine housing cap screws
12. Lock plates
13. Clamp plates
14. Turbine wheel and shaft
15. Thrust bearing
16. Thrust collar
17. Seal ring (2 used)
18. Bearing
19. Bearing retainers
20. Seal ring
21. Compressor impeller
22. Locknut

Fig. 125 — Checking end play of shaft after unit has been cleaned. Refer to text.

not grooved in area where seal (17) rides. Oil passage in thrust collar (16) must be clean and thrust faces must not be warped or scored. Ring groove shoulders must not have step wear.

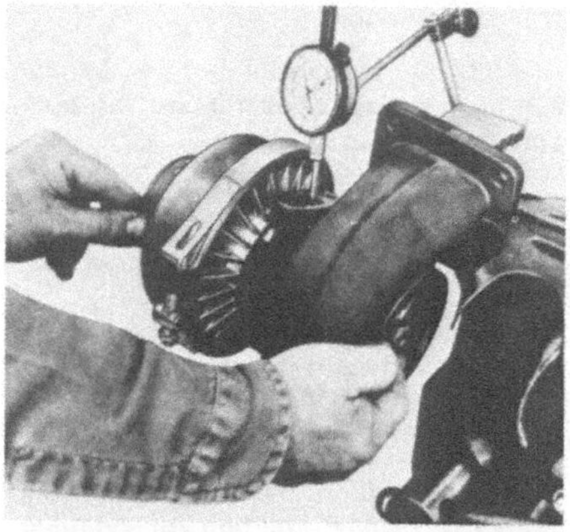

Fig. 126 — Checking radial clearance with dial indicator through oil outlet hole and touching shaft.

Fig. 127 — Install test gage as shown to check intake manifold pressure on turbocharged diesel engine.

1. Adapter
2. Bakelite handle   3. 30 psi gage

Fig. 128 — Test gage installed to check exhaust manifold pressure on turbocharged diesel engine.

1. ⅛-inch plug   3. Bakelite handle
2. Adapter   4. 30 psi gage

Check shaft end play and radial clearance when assembling.

If the bearing inner retainers (19) were removed, install new retainers using IH tool FES 57-3. Oil bearings (18) and install outer retainers using IH tool FES 57-3. Position the shroud (7) on turbine shaft (14) and install seal ring (17) in groove. Apply a light, even coat of engine oil to shaft journals, compress seal ring (17) and install center housing (6). Install new seal ring (17) in groove of thrust collar (16), then install thrust bearing so that smooth side of bearing (15) is next to the large diameter of collar. Install thrust bearing and collar assembly over shaft, making certain that pins in center housing engage holes in thrust bearing. Install new rubber seal ring (20), make certain that spring (4), if removed, is positioned in back plate (5), then install back plate making certain that seal ring is not damaged. Install lock plates (9) and cap screws (10), tightening screws to 40-60 inch-pounds torque. Install compressor impeller (21) and make certain that impeller is completely seated against thrust collar (16). Install locknut (22) and tighten to 18-20 inch-pounds torque, then use a "T" handle to turn locknut an additional 90°.

CAUTION: If "T" handle is not used, shaft may be bent when tightening nut (22).

Install turbine housing (8) with clamp plates (13) next to housing, tighten cap screws (11) to 100-130 inch-pounds torque, then bend lock plates (12) up around screw heads.

Check shaft end play and radial play at this point of assembly. If shaft end play (Fig. 125) exceeds 0.004 inch, thrust collar (16 – Fig. 124) and/or thrust bearing (15) is worn excessively. End play of less than 0.001 inch indicates incomplete cleaning (carbon not all removed) or dirty assembly and unit should be disassembled and cleaned. Refer to (Fig. 126) and check turbine shaft radial play. If it exceeds 0.007 inch, unit should be disassembled and bearings, shaft and/or center housing should be renewed.

Make certain that legs on diffuser (3 – Fig. 124) are aligned with spot faces on backplate (5) and install diffuser. Install compressor housing (2) and tighten nut of clamp (1) to 40-80 inch-pounds torque.

Check rotating unit for free rotation within the housing. Cover all openings until the turbocharger is reinstalled.

Use a new gasket and install and prime turbocharger as outlined in paragraph 125.

TESTING

Model 1086

128. Before testing the turbocharger, make certain the air filter system is clean and that the fuel injection pump is properly adjusted and delivering the correct amount of fuel to engine. A clogged air filter will cause intake manifold pressure to be low. Excessive fuel delivery will result in high exhaust manifold pressure and low fuel delivery will cause low exhaust manifold pressure.

Using the procedure as outlined below and referring to Fig. 129 for correct

Fig. 129 — Chart showing correct manifold pressure and rpm for testing turbocharger. Refer to text.

```
              Manifold Pressure Intake

1086(DT-414) Rated Load ....  8.35-11.30 psi @2400 rpm

1086(DT-414) Overload ......  4.42-7.36 psi @1800 rpm

1086(DT-414B) Rated Load ...  8.35-11.30 psi @2400 rpm

1086(DT-414B) Overload .....  6.63-9.58 psi @1800 rpm


              Manifold Pressure Exhaust

1086(DT-414) Rated Load ....  8.35-11.30 psi @2400 rpm

1086(DT-414) Overload ......  1.96-4.91 psi @1800 rpm

1086(DT-414B) Rated Load ...  7.36-10.31 psi @2400 rpm

1086(DT-414B) Overload .....  4.18-7.12 psi @1800 rpm
```

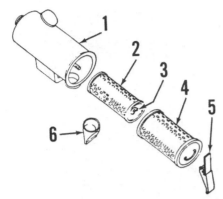

Fig. 130—Exploded view of typical air cleaner assembly used on D-358 equipped 786 and 886 tractors.

1. Housing
2. Safety element
3. Wing nut
4. Main element
5. Retainer
6. Dust unloader

rpm and psi reading. Connect tractor to a dynamometer or other loading device. Remove ¼-inch plug from intake manifold and install a 30 psi test gage as shown in Fig. 127. Start engine and set speed control lever at high idle position. Load engine to rated load rpm. Test gage will show intake manifold pressure. Continue to load engine until overload rpm is reached. Refer to Fig. 129 for correct psi reading. Stop engine, remove test gage and install plug. To check exhaust manifold pressure, remove either ⅛-inch plug from center of exhaust manifold. Connect a 30 psi test gage. See Fig. 128. Start engine and set speed control lever at high idle position. Load engine to rated load rpm. Test gage will show exhaust manifold pressure, refer to Fig. 129. Continue to load engine to overload rpm and reading should be same as chart Fig. 129. Stop engine, remove test gage and install plug.

AIR FILTER SYSTEM

All Models

129. All models are equipped with a dry type air cleaner with a safety filter

Fig. 131—Exploded view of air cleaner assembly used on D-360, D-436 and DT-414 equipped tractors.

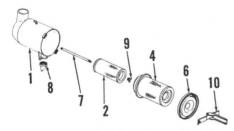

1. Housing
2. Safety element
4. Main element
6. End cover
7. Stud
8. Dust unloader
9. Wing nut
10. Retainer

Fig. 132—Exploded view of water pump assembly used on D-358 equipped 786 and 886 tractors.

1. Fan
2. Fan spacer
3. Pulley
4. Hub
5. Pump shaft and bearing
6. Alternator and water pump belt
7. Body
8. Gasket
9. "O" ring
10. Seal assembly
11. Impeller face ring
12. Impeller
13. Plastic screw

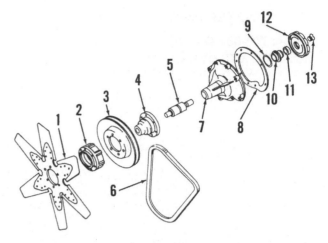

element (2–Figs. 130 and 131) which should be renewed at least once a year. DO NOT attempt to clean the safety element.

Large filter element (4–Figs. 130 and 131) can be cleaned by directing compressed air up and down the pleats on inside of element. Air pressure must not exceed 100 psi. An element cleaning tool (IH tool No. 407073R1) for use with compressed air, is available from International Harvester. Renew filter element after 10 cleanings or once a year, whichever comes first.

COOLING SYSTEM

RADIATOR

All Models

130. To remove radiator, first drain cooling system, then remove hood

skirts, hood and side panels. Remove the air cleaner inlet hose. Disconnect upper and lower radiator hoses from radiator and move them out of the way. Disconnect fan shroud from radiator and remove the radiator drain cock. Disconnect the oil cooler lines bracket. Remove the center cap screw from the four radiator mounts, then lift radiator straight up out of radiator support.

FAN

All Models

131. The fan is attached to the water pump and one belt drives the fan, water pump and alternator.

To remove the fan, first remove radiator as outlined in paragraph 130. Remove retaining cap screws and lift off the fan.

When reinstalling, adjust fan belt until a pressure of 25 pounds applied midway between water pump and crankshaft pulleys will deflect belt ⅞-inch.

Fig. 133—Sectional view of water pump and components for D-358 engine. Also dimension for installing fan hub and impeller.

4. Housing hub
5. Pump shaft and bearing
7. Body
9. "O" ring
10. Seal assembly
11. Impeller face ring
12. Impeller
13. Plastic screw

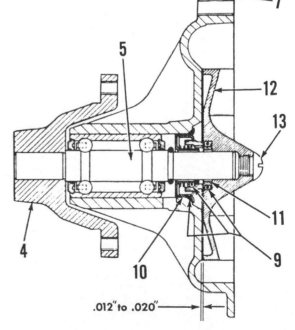

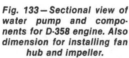

.012" to .020"

WATER PUMP

Models 786-886 (D-358)

132. R&R AND OVERHAUL. To remove the water pump, first remove radiator as outlined in paragraph 130. Loosen alternator mounting bolts and remove drive belt. Unbolt and remove fan (1–Fig. 132), spacer (2) and pulley (3). Remove cap screws securing pump body (7) to water pump carrier, then remove pump.

Disassemble water pump as follows: Remove plastic screw (13) and using a ½x2 inch NC cap screw for a jack screw in rear of impeller, force impeller (12) off rear end of shaft. Using two screwdrivers pry seal assembly (10) out of pump body. Support hub (4) and press out shaft. Press shaft and bearing assembly (5) out front of body. Make certain that body is supported as close to bearing as possible.

When reassembling, press shaft and bearing assembly into body using a piece of pipe so pressure is applied only to outer race of bearing. Bearing race should be flush with front end of body. Install new "O" ring (9) and seal (10). Press only on outer diameter of seal. Support shaft assembly and press hub on shaft until hub is flush with end of shaft. Install face ring (11) in impeller (12), then press impeller on shaft until there is a clearance of 0.012-0.020 inch between body and front of impeller (opposite fins). See Fig. 133. Install plastic screw.

Using a new gasket (8), reinstall pump by reversing the removal procedure. Install pulley, spacer and fan and adjust belt as outlined in paragraph 131.

Models 886 (D-360)-986-1086

133. R&R AND OVERHAUL. To remove the water pump, first remove fan as outlined in paragraph 131. Remove cap screws securing pump body (4–Fig. 134) to front cover, then remove pump.

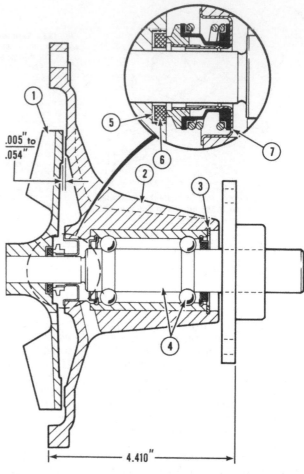

Fig. 135—Sectional view of water pump and components for D-360, D-436 and DT-414 engines. Also dimension for installing fan hub and impeller.

1. Impeller
2. Housing
3. Snap ring
4. Shaft and bearing assembly
5. Rubber bushing
6. Ceramic seal
7. Bearing housing seal

.005" to .054"

4.410"

Disassemble water pump as follows: Using a puller remove fan hub (3). Then, remove snap ring (5) and press shaft and bearing assembly (6) out of impeller (9) and housing (4). Make certain that body is supported as close to bearing as possible. Tap seal (7) out of housing (4) then, pry rubber bushing and ceramic seal (8) out of impeller (9).

When reassembling, press shaft and bearing assembly into body using a piece of pipe so pressure is applied only to outer race of bearing. Be sure bearing is bottomed in housing. Install snap ring. Press seal (7–Fig. 135) in housing, then

assemble ceramic seal (6) with identification mark toward rubber bushing (5). Moisten rubber bushing with water, and install rubber bushing and ceramic seal assembly in impeller. Be sure seal is in bottom of impeller bore. Support fan hub and press shaft through hub to dimension shown in Fig. 135. Support hub end of shaft and press impeller on shaft to a clearance of 0.005-0.054 inch, measured as shown in Fig. 135.

Using a new gasket (11–Fig. 134), reinstall pump by reversing the removal procedure. Install pulley and fan, then adjust belt as outlined in paragraph 131.

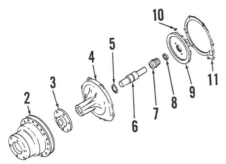

Fig. 134—Exploded view of water pump assembly for D-360, D-436 and DT-414 equipped tractors.

2. Pulley	7. Seal
3. Hub	8. Rubber bushing and
4. Housing	ceramic seal
5. Snap ring	9. Impeller
6. Bearing and shaft	10. Dowel (2 used)
assembly	11. Gasket

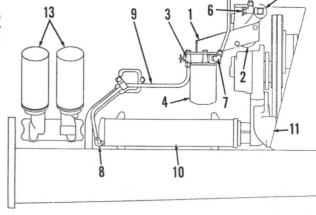

Fig. 136—View showing cooling filter-conditioner general layout for all D-360, D-436 and DT-414 engines.

1. Bracket
2. Spacer
3. Base
4. Element
5. Tee
6. Shut-off valve
7. Elbow
8. Elbow
9. Inlet tube
10. Oil cooler
11. Water pump
12. Outlet tube
13. Oil filter

COOLANT FILTER-CONDITIONER

All Models

135. The cooling system is equipped with a filtration system designed to prolong internal engine life and to increase cooling efficiency. Fig. 136 shows typical coolant filter-conditioner system for D-360, D-436 and DT-414 engines. Refer to (9 – Fig. 112) for view of coolant filter-conditioner system for D-358 engine.

Preventative maintenance is the best guard against cooling system problems. If a coolant filter-conditioner is added to a used tractor, flush cooling system with I.H. Cooling System Cleaner and then add I.H. Cooling System Conditioner. The water filter-conditioner element (4 – Fig. 136) or (9 – Fig. 112) should be changed after the first 100 hours and then every 400 hours.

WARNING: Use of stop leak additives in the cooling system will cause premature plugging of water filter.

ELECTRICAL

ALTERNATOR AND REGULATOR

All Models

136. Delco-Remy "DELCOTRON", Motorola and Niehoff alternators are used on all models. Some models may be equipped with Delco-Remy alternator No. 1103171, 1103183 or 1103184. Motorola alternator No. 107560C91 is available as an option on some 886, 986 and 1086 models. Niehoff alternator No. 146472C91 is used on late production 986 and 1086 tractors. Delco-Remy, Motorola and Niehoff alternators are equipped with internally mounted voltage regulators. "DELCOTRON" alternators use solid state regulator No. 1116387. Motorola alternators use solid state regulator No. 496686C1. Niehoff alternators use solid state regulator No. 77794C1. Neither regulator has any provision for adjustment.

CAUTION: Because certain components of the alternator can be damaged by procedures that will not affect a D.C. generator, the following precautions MUST be observed:

a. When installing batteries or connecting a booster battery, the negative post of battery must be grounded.

b. Do not attempt to polarize the Delco-Remy or Motorola alternators. Only Niehoff alternators should be polarized after reassembly.

c. Disconnect battery cables before removing or installing any electrical unit.

d. Do not operate alternator on an open circuit and be sure all leads are properly connected before starting engine.

Specification data for the alternators is as follows:

Alternator 1103171 (Delco-Remy)
Field current @ 80° F.:
  Amperes 4.0-5.0
  Volts 12
Cold output at specific voltage:
  Amperes @ rpm . . . . . .25.0 @ 2000
  Amperes @ rpm . . . . . .38.0 @ 5000
Rated output hot-Amperes 42
Alternator 1103183 (Delco-Remy)
Field current @ 80° F.:
  Amperes 4.0-5.0
  Volts 12
Cold output at specific voltage:
  Amperes @ rpm . . . . . .30.0 @ 2000
  Amperes @ rpm . . . . . .57.0 @ 5000
Rated output hot-Amperes 61
Alternator 1103184 (Delco-Remy)
Field current @ 80° F.:
  Amperes 4.0-5.0
  Volts 12
Cold output at specific voltage:
  Amperes @ rpm . . . . . .25.0 @ 2000
  Amperes @ rpm . . . . . .65.0 @ 5000
Rated output hot-Amperes 72
Alternator 107560C91 (Motorola)
Field current @ 80°F.:
  Amperes 2.7-3.2
  Volts 12.6
Cold output at specific voltage:
  Amperes @ rpm . . . . . .40.0 @ 2000
  Amperes @ rpm . . . . . .58.0 @ 5000
Rated output hot-Amperes 62
Alternator 146472C91 (Niehoff)
System voltage 12.0
Regulator setting, volts 14.5-14.6
Rated output hot,
  Volts 14.1-15.1
  Amperes 100.0

137. **ALTERNATORS (DELCO-REMY) 1103171, 1103183 OR 1103184; (MOTOROLA) 107560C91 TESTING AND OVERHAUL.** The only test which can be made without removal and disassembly of alternator is the regulator. If there is a problem with the battery not being charged, and battery and cable connectors have been checked and are good, check the regulator as follows: Operate engine at moderate speed and turn all accessories on and check the ammeter. If ammeter reading is within ten amperes of rated output as stamped on alternator frame (or refer to paragraph 136 for specifications) alternator is not defective. If ampere output is not within ten amperes of rated output, ground field winding by inserting a screwdriver into test hole (Fig. 137). If output is then within ten amperes of rated output, replace the regulator.

CAUTION: When inserting screwdriver in test hole, the tab is within ¾-inch of casting surface. Do not force screwdriver deeper than one inch into end frame.

If output is still not within ten amperes of rated output, alternator will have to be disassembled. Check field winding, diode trio, rectifier bridge and stator as follows:

To disassemble the alternator, first scribe match marks (M – Figs. 138 and 139) on the two frame halves (4 and 12), then remove the four through bolts. Pry frame apart with a screwdriver between stator frame (12) and drive end frame (4).

Stator assembly (8) must remain with slip ring end frame (12) when unit is separated.

NOTE: When frames are separated, brushes will contact rotor shaft at bearing

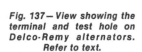

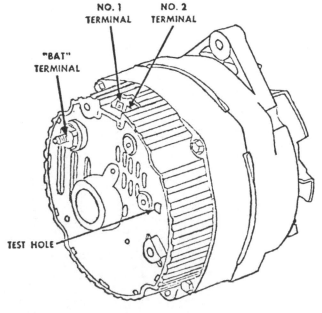

Fig. 137 – View showing the terminal and test hole on Delco-Remy alternators. Refer to text.

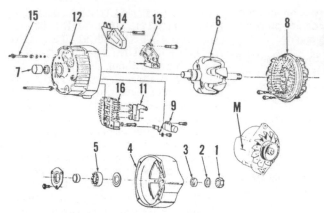

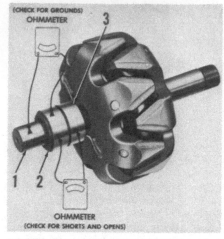

Fig. 138 – Exploded view of "DELCOTRON" alternator with internal mounted solid state regulator. Note match marks (M) on end frames.

1. Pulley nut
2. Washer
3. Spacer (outside drive end)
4. Drive end frame
5. Bearing
6. Rotor
7. Bearing
8. Stator
9. Capacitor
11. Diode-trio
12. Slip ring end frame
13. Brush holder
14. Solid state regulator
15. "Bat" terminal
16. Bridge rectifier

Fig. 140 – Removed rotor assembly showing test points when checking for grounds, shorts and opens.

area. Brushes MUST be cleaned of lubricant with a soft dry cloth if they are to be reused.

Clamp iron rotor (6) in a protected vise, only tight enough to permit loosening of pulley nut (1). Rotor and end frame can be separated after pulley and fan are removed. Check bearing surface of rotor shaft for visible wear or scoring. Examine slip ring surface for scoring or wear and rotor winding for overheating or other damage. Check rotor for grounded, shorted or open circuits using an ohmmeter as follows:

Refer to Fig. 140 and touch ohmmeter probes to points (1-2) and (1-3); a reading near zero will indicate a ground. Touch ohmmeter probes to slip ring (2-3); reading should be 5.3-5.9 ohms for Delco-Remy alternators and 3.8-4.6 ohms for Motorola alternators. A higher reading will indicate an open circuit and a lower reading will indicate a short. If windings are satisfactory, mount rotor in a lathe and check runout at slip rings using a dial indicator. Runout should not exceed 0.002 inch. Slip ring surfaces can be trued if runout is excessive or if surfaces are scored. Finish with 400 grit or finer polishing cloth until scratches or machine marks are removed.

Before removing stator, brushes or diode trio on Delco-Remy alternators,

refer to Fig. 141 and check for grounds between points A to C and B to C with an ohmmeter, using the lowest range scale. Then, reverse the lead connections. If both A to C readings or B to C readings are the same, brushes may be grounded because of defective insulating washer and sleeve at the two screws. If the screw assembly is not damaged or grounded, regulator is defective.

To test the brush assembly on Motorola alternators, first remove the integral voltage regulator (14 – Fig. 139) then remove the two screws securing brush assembly to the alternator.

Note use of flat washer under right side mounting screw.

Test brush assembly as shown on chart in Fig. 142. Renew brush assembly if readings do not correspond to specifications in chart. The original brush set may be reused if brushes are 3/16-inch or longer, and if brushes are not oil soaked, cracked or show evidence of grooves on the sides of brushes caused by vibration.

To test the diode trio on Delco-Remy alternators, first remove the stator. Then, remove the diode trio, noting insulator positions. Using an ohmmeter, refer to Fig. 143 and check between points A to D. Then, reverse ohmmeter lead connections. If diode trio is good, it

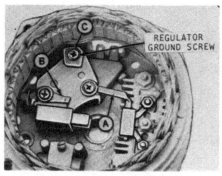

Fig. 141 – Test points for Delco-Remy brush holder. Refer to text.

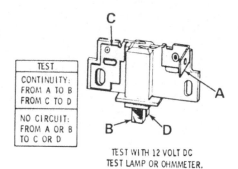

TEST
CONTINUITY: FROM A TO B FROM C TO D
NO CIRCUIT: FROM A OR B TO C OR D

TEST WITH 12 VOLT DC TEST LAMP OR OHMMETER.

Fig. 142 – Test points for Motorola brush holder. Refer to text.

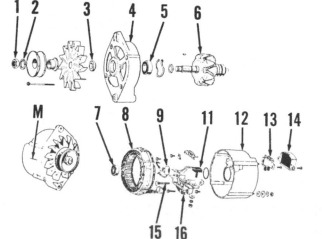

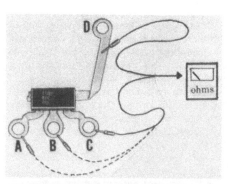

Fig. 139 – Exploded view of Motorola alternator with internal mounted solid state regulator. Note match marks (M) on end frames. Refer to legend in Fig. 138.

Fig. 143 – Delco-Remy diode trio test points. Refer to text.

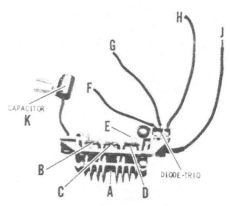

Fig. 144 — Rectifier bridge, capacitor and diode-trio assembly for Motorola alternators showing test points. Refer to text.

will give one high and one low reading. If both readings are the same, diode trio is defective. Repeat this test at points B to D and C to D.

To test the diode trio on Motorola alternators, first remove stator and rectifier/diode assembly as a unit. Unsolder leads from stator noting their relative positions. Refer to Fig. 144 and disconnect diode-trio from rectifier bridge assembly. Diode trio leads (F, G, H) must be disconnected to test unit. Using an ohmmeter, check between points F-J, G-J and H-J. If diode trio is good, a reading of 5-20 ohms should register. Reverse ohmmeter leads and test again. The meter should register infinite resistance. If both readings are the same, diode trio is defective.

The rectifier bridge (Figs. 144 and 145) has a grounded heat sink (A) and an insulated heat sink (E) that is connected to the output terminal. Connect ohmmeter to grounded heat sink (A) and flat metal strip (B). Then, reverse ohmmeter lead connections. If both readings are the same, rectifier bridge is defective. Repeat this test between points A-C, A-D, B-E, C-E, and D-E.

Test stator (8 – Figs. 138 and 139) windings for grounded or open circuits as follows: Connect ohmmeter leads successively between each pair of leads. A high reading would indicate an open circuit.

Fig. 145 — Rectifier bridge of Delco-Remy alternator showing test points. Refer to text.

NOTE: The three stator leads have a common connection in the center of the windings. Connect ohmmeter leads between each stator lead and stator frame. A very low reading would indicate a grounded circuit. A short circuit within stator windings cannot be readily determined by test because of the low resistance of windings.

Brushes and springs are available only as an assembly which includes brush holder (13 – Figs. 138 and 139). If brushes are reused, make sure all grease is removed from surface of brushes before unit is reassembled.

When reassembling Delco-Remy alternator, install regulator and then brush holder, springs and brushes. Push brushes up against spring pressure and insert a short piece of straight wire through the hole and through end frame to outside. Be sure that the two screws at points A and B (Fig. 141) have insulating washers and sleeves.

NOTE: A ground at these points will cause no output or controlled output. Withdraw wire only after alternator is assembled.

Capacitor (9 – Figs. 138 and 139) connects to the rectifier bridge and is grounded to the end frame. Capacitor protects the diodes from voltage surges.

Remove and inspect ball bearing (5 – Figs. 138 and 139). If bearing is in satisfactory condition, fill bearing ¼-full with suitable lubricant and reinstall. Inspect bearing (7) in slip ring end frame. This bearing should be renewed if its lubricant supply is exhausted; no attempt should be made to relubricate and reuse the bearing.

On Delco-Remy alternators press old bearing out towards inside and press new bearing in from outside until bearing is flush with outside of end frame. Saturate felt seal with SAE 20 oil and install seal.

To remove Motorola alternator rear bearing, first use suitable puller to remove bearing from rotor shaft.

NOTE: Do not damage slip rings. Make sure puller contacts inner race of bearing.

Renew bearing by pressing on rotor shaft until bearing contacts shoulder on shaft. When renewing rear bearing, also renew bearing retainer in rear housing.

Reassemble alternators by reversing the disassembly procedure. Tighten pulley nut to a torque of 50 ft.-lbs.

137A. ALTERNATOR (NIEHOFF) 146472C91 TESTING AND OVERHAUL. The only test which can be made without removal and disassembly of alternator is the regulator test. Make certain that batteries, cables and connections are in good condition. Connect

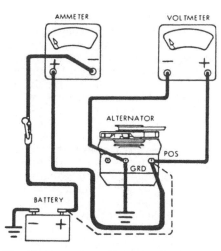

Fig. 146 — Connect ammeter and voltmeter in circuits on tractor to test Niehoff alternator charging system on late Model 986 and 1086 tractors.

an ammeter and voltmeter in circuits on tractor as shown in Fig. 146. Start engine and operate at high idle (2640) rpm. Initially, ammeter reading should be 90-100 amperes and voltmeter reading should be 14.1-15.1 volts. As the batteries are being recharged, voltage reading should return to regulator setting (14.5-14.6 volts). If voltage reading is too high (above 15.1 volts), alternator is OK; renew the regulator. If low or zero output current is observed, connect one end of an insulated jumper wire to the alternator ground as shown in Fig. 146A. Insert other end of wire into the hole in rear end cover to contact field terminal on regulator plate. If output is now obtained, alternator is OK and regulator must be renewed. If very low or no output is obtained, remove alternator and test stator winding, diode heat sink and field coil.

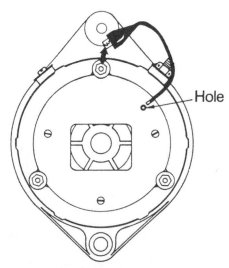

Hole

Fig. 146A — View showing jumper wire used in regulator test. Also used to polarize alternator after reassembly. Refer to text.

Remove pulley retaining nut and washer, then using a suitable puller, remove pulley assembly. Remove Woodruff key (11–Fig. 147), dust cover (1) and spacer (2). Refer to Fig. 148 and disconnect wires from diode heat sink as follows: Disconnect the three phase leads (black wires L) from three (S) terminals; output lead (red wire O) and field lead (red wire F) from (B+) terminal. Remove the three screws that secure regulator plate (14–Fig. 147) to rear cover (15). Remove the three nuts that retain rear cover, then remove rear cover (15). Refer to Fig. 148A and disconnect field lead (white wire F), "Y" lead (brown wire Y), ground wire (white wire G), battery lead (red wire P) and indicator light wire (black wire K) from regulator plate.

NOTE: During the following tests, do not hold ends of wires or ohmmeter leads with your fingers as a false resistance reading may be indicated.

Stator Winding Test. Check for open stator windings by connecting ohmmeter leads between each successive pair of stator phase leads (L–Fig. 148). Meter should read less than one ohm between each pair of stator phase windings. A higher reading would indicate a possible open winding and stator must be renewed.

Connect ohmmeter leads between alternator frame and each phase lead (L). Ohmmeter needle should not move. If any reading is shown on ohmmeter, stator winding is grounded and stator must be renewed.

Diode Heat Sink Test. Touch one ohmmeter lead to output terminal (B+–Fig. 148) and other lead to each of the stator phase terminals (S). Observe and record any readings. Reverse ohmmeter leads and test for reading at each phase terminal. There should be nearly equal reading for each terminal in one

test and no meter needle movement in second test with ohmmeter leads reversed. If no reading is observed in both tests or if a reading is observed in both tests, heat sink must be renewed.

Field Coil Test. Connect one ohmmeter lead to red field lead (F–Fig. 148) at heat sink end of alternator and other ohmmeter lead to white field lead (F–Fig. 148A) at regulator end of alternator. Ohmmeter reading should be 2.5-3.5 ohms. If not, renew main frame and field coil assembly (6–Fig. 147).

Connect one ohmmeter lead to either field lead (F–Fig. 148 or 148A) and the other lead to alternator frame. Ohmmeter needle should not move. If any reading is shown on ohmmeter, field coil is grounded and main frame and field coil must be renewed.

Disassembly. Balance of alternator disassembly is as follows: Remove the three screws and lift off heat sink (5–Fig. 147) from front of alternator. Remove regulator plate (14), spacers, washers and ground wire from alternator studs. Set alternator in a press and press rotor out rear of unit. Remove retaining cup (8) and ring (9) from rotor shaft and remove two rubber plugs from rear of rotor. Using a bearing removal tool, made up of two ¼ x 6 inch rods threaded into a piece of bar stock approximately ⅜ x 1¼ x 2¼ inches, insert rods through holes in rotor and press bearing (10) from rotor shaft. Remove front bearing retaining ring (3) and press bearing (4) from front of main frame (6). To separate stator (7) from main frame and field coil (6), remove three nuts from alternator studs. While holding stator, tap on mounting ears of main frame with a brass hammer. Note location of wires when separating parts.

Reassembly. "Loctite" should be used on all screws and nuts during reassem-

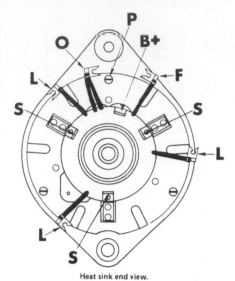

Heat sink end view.

Fig. 148–View showing wires disconnected from terminals on diode heat sink. Refer to text for color code and connections. Three pan head screws (P) secure heat sink to main frame.

bly. Place stator on alternator studs, making certain that threaded hole mounting ear of main frame is between relay terminal and battery terminal on stator. Feed the wires through the proper holes. Be sure that white field wire is between stator windings and behind stator wedge immediately next to the battery terminal on stator. Do not tighten down the stator at this time. Install front bearing (4–Fig. 147) in main frame (6) and secure with retaining ring (3). Install bearing (10) on rotor shaft until it seats. Install retaining ring (9) in groove on rotor shaft and install retaining cup (8) over the ring. Using special tool No. A10-104 as shown in Fig. 149,

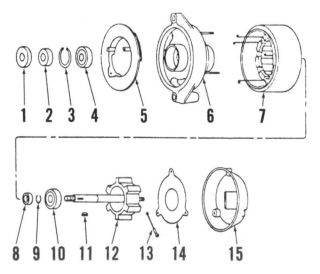

1   2   3   4   5   6   7

8   9   10   11   12   13   14   15

Fig. 147–Exploded view of Niehoff alternator used on late Model 986 and 1086.

1. Dust cover
2. Spacer
3. Retaining ring
4. Bearing
5. Diode heat sink
6. Main frame & field coil
7. Stator assy.
8. Retaining cup
9. Retaining ring
10. Bearing
11. Woodruff key
12. Rotor
13. Ground wire
14. Regulator plate
15. End cover

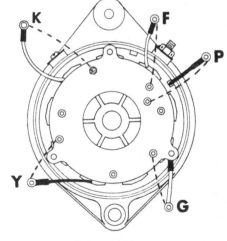

Regulator end view.

Fig. 148A–View showing regulator plate with wires disconnected. Broken lines show correct connection to terminals.

F. White field lead
P. Red battery lead
G. White ground wire
Y. Brown "Y" lead
K. Black indicator light wire

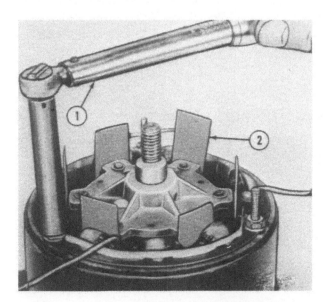

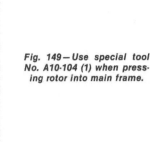

Fig. 149—Use special tool No. A10-104 (1) when pressing rotor into main frame.

press rotor into main frame until retainer cup (8 – Fig. 147) is against front bearing (4). Rotor must be free to turn at this time. Using six 1x3 inch pieces of 0.008 inch shim stock (Fig. 149A), place one shim between each rotor pole and adjacent stator poles. Move stator to provide the best equal clearance, then install and tighten stator nuts to a torque of 35 in.-lbs. Remove shims and check to be sure a clearance of 0.005-0.008 inch exists between all rotor poles and adjacent stator poles. Install spacers, washers and ground wire on alternator studs. Install regulator plate and connect all wires to terminals as shown in Fig. 148A. Install rear cover (15 – Fig. 147) and secure with nuts tightened to a torque of 26 in.-lbs. Install the three regulator support screws through rear cover.

Install heat sink (5) on front of alternator and tighten retaining screws to 20 in.-lbs. torque. Refer to Fig. 148 and connect the three phase leads (L) to terminals (S) and connect output lead (O) and field lead (F) to (B+) terminal.

Install spacer (2 – Fig. 147), dust cover (1), Woodruff key (11), fan and pulley. Secure with washer and locknut, tightened to a torque of 50 ft.-lbs.

NOTE: After alternator is reinstalled on tractor and connections are completed, polarize alternator as follows: Attach one end of insulated wire to alternator frame as shown in Fig. 146A. Insert other end of wire into hole in rear cover to contact field terminal on regulator plate. Make momentary contact and remove jumper wire. This will energize the field coil and re-establish the magnetic field of alternator.

Fig. 149A – Place 0.008 inch shim between each rotor pole and stator poles, then tighten stator retaining nuts.
1. Inch-pound torque wrench
2. Shims (0.008 inch)

STARTING MOTORS
All Models

138. Delco-Remy starting motors are used and specification data for these units is as follows:

Starter 1113439
Volts 12.0
Brush tension – Ounces 80
No-load test:
Volts 9.0
Amperes (min.) 75.0*
Amperes (max.) 105.0*
RPM (min.) 5000
RPM (max.) 8100
*Includes solenoid

STARTER SOLENOID
All Models

139. All starting motors are equipped with Delco-Remy solenoid switches. Specification data for these units is as follows:

Solenoid 1115530
Rated voltage 12.0
Current consumption,
Pull-in winding:
Volts 5.0
Amperes 26.0-29.0
Hold-in winding:
Volts 10.0
Amperes 18.0-20.0

CLUTCH
All Models

140. Models 786, 886 and 986 are fitted with a 12 inch dry disc clutch. Model 1086 is fitted with a 14 inch dry disc clutch. Spring loaded clutches are used. All models are equipped with hydraulic cylinder assist on clutch linkage.

Clutch wear is compensated for by adjusting clutch linkage and when engine clutch is adjusted, it will require that the transmission brake, torque amplifier dump valve and starter safety switch also be adjusted.

ADJUSTMENT
All Models

141. **ENGINE CLUTCH (INPUT BOOSTER SHAFT).** Remove link pin (2 – Fig. 150) and loosen jam nut (5) on booster pivot trunnion (1). Adjust trunnion until booster links (3) align with hole in clutch pedal lever arm (4). Insert link pin (2) and install the cotter pin. Tighten jam nut (5) against trunnion to lock input shaft linkage.

142. **ENGINE CLUTCH (OUTPUT BOOSTER SHAFT).** Remove clevis pin (8 – Fig. 150) and loosen jam nut (9).

Position bellcrank (10) so its bottom edge is 90 degrees to the front edge of the subframe bracket. With the bellcrank in this position, adjust the clevis (7) until its holes align with holes in rod (6). Insert pin (8) and install cotter pin. Tighten jam nut against clevis.

NOTE: Clevis (7) is used to adjust booster linkage only and must not be used to adjust pedal freeplay.

143. CLUTCH FREE PEDAL ADJUSTMENT. Disconnect transmission brake operating rod (11 – Fig. 150) by removing pins (12) and (13) from clevis (17) and lever (16). With clutch pedal against upper stop, measure the distance from raised stop on firewall to the arrow on clutch pedal. Depress clutch pedal until resistance is felt and measure the distance from stop on the firewall to the arrow on pedal. The difference should read 1 1/8 inch. If not, loosen adjusting rod lock bolt (14) and turn adjusting rod (15) until the 1 1/8 inch dimension is obtained, then tighten lock bolt.

144. DUMP VALVE ADJUSTMENT. The torque amplifier dump valve should be checked and adjusted each time the engine clutch is adjusted. With engine off, adjust dump valve operating rod by turning adjusting nut (19 – Fig. 150) until dump valve spool is 3/8-inch in the extended position as shown in Fig. 150. To be sure dump

valve is correctly adjusted start tractor and set rpm at 1000. Put in low gear and slowly engage clutch. The lube light on console should go off before any movement of tractor is noticed, if not, readjust.

145. TRANSMISSION BRAKE ADJUSTMENT. Transmission brake should be adjusted each time engine clutch is adjusted. Disconnect transmission brake operating rod (11 – Fig. 150) from lever (16) and loosen nut (18) from clevis (17). Depress clutch pedal until it contacts the stop. Holding pedal in this position, move lever (16) to full downward position. Adjust clevis (17) so pin (13) aligns with lever (16). Unscrew clevis 1/2-turn and install pin (13) and pin (12).

146. STARTER SAFETY SWITCH. The starter safety switch prevents tractor from being started except when engine clutch is disengaged. To adjust refer to (20 – Fig. 150) and loosen both adjustment nuts on switch, depress clutch pedal to stop, then adjust switch plunger until it is depressed 1/8-inch. Tighten adjustment nuts.

REMOVE AND REINSTALL

All Models

147. To remove the engine clutch, it is

first necessary to separate (split) engine from clutch housing as outlined in paragraph 148. With engine split from clutch housing, removal of clutch from flywheel is obvious. Refer to paragraph 150 for overhaul data.

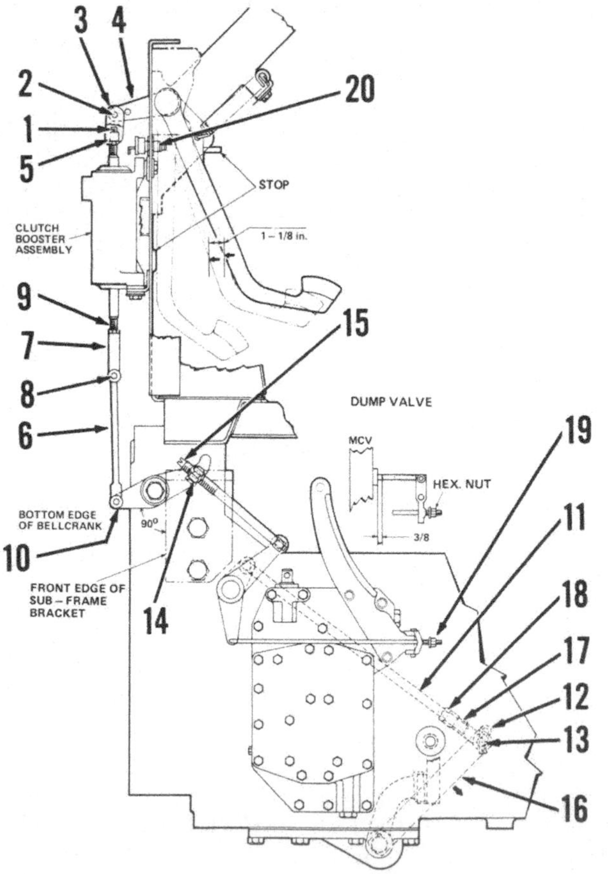

TRACTOR SPLIT

All Models

148. To split tractor for clutch service, disconnect battery cables and remove exhaust stack, radiator cap, front hood channel and left and right front side hood. On Models 986 and 1086, drain cooling system and on Models 786 and 886, disconnect temperature sending wire. On all models disconnect oil cooler hoses at multiple control valve and plug. Disconnect throttle cable at injection pump, support bracket and oil pressure sending unit. Disconnect power steering hand pump supply and return lines. If equipped, disconnect speedometer cable and transducer assembly. Disconnect ether solenoid. On models so equipped, shut off both suction and discharge valves on A/C compressor, then disconnect suction line and move it back towards control center. Cap open compressor port. Disconnect headlight wiring and remove left and right front lower panel at control center. Disconnect A/C wiring harness, heater hoses and plug the hoses. Remove subframe platform, then disconnect main wiring harness connectors. Remove two mounting bolts ahead of cranking motor. Install split stands to side rails and place a rolling floor jack under rear section of tractor. With tractor properly supported, begin to unthread the remaining six mounting bolts from clutch housing, then separate the tractor.

IMPORTANT: Do not completely remove the six mounting bolts. This will allow the engine to drop between side rails.

Rejoin tractor by reversing the splitting procedure; however in order to avoid any difficulty which might arise in trying to align splines of clutch assembly and transmission input shafts during mating of sections, most mechanics prefer to remove clutch from flywheel and place it over transmission input shaft. Clutch can be installed on flywheel after tractor sections are joined by working through the opening at bottom of clutch housing.

HYDRAULIC ASSIST CLUTCH CYLINDER

All Models

149. **R&R AND OVERHAUL.** Clutch assist cylinder is located on firewall in engine compartment. To remove the

Fig. 150 — View showing points of adjustment for hydraulic assist engine clutch linkage.

1. Trunnion
2. Pin
3. Booster link
4. Lever arm
5. Jam nut
6. Rod
7. Clevis
8. Pin
9. Jam nut
10. Bellcrank
11. Rod
12. Pin
13. Pin
14. Lock bolt
15. Rod
16. Lever
17. Clevis
18. Nut
19. Nut
20. Starter safety switch

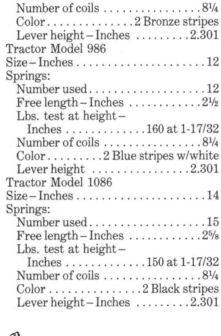

SERIES 786-886-986-1086

SERIES 786-886-986-1086

1. Body
2. Piston
3. Output rod
4. Spring (piston)

Fig. 151 — Exploded view of clutch hydraulic assist cylinder.

5. End cap
6. Retaining ring
7. Wiper
8. "O" ring
9. "O" ring
10. Teflon ring
11. "O" ring
12. Retaining ring
13. Input rod
14. "O" ring
15. "O" ring
16. Seal ring
17. Retaining ring
18. Washer
19. Spring
20. Cup
21. "O" ring
22. "O" ring
23. Wiper

CLUTCH SHAFT

All Models

151. The clutch shaft is part of the torque amplifier assembly and will be covered in the torque amplifier section.

TORQUE AMPLIFIER AND SPEED TRANSMISSION

The torque amplifier and speed (forward) transmission are both located in the clutch housing along with the hydraulic pump which supplies the power steering, brakes and torque amplifier. Any service on the torque amplifier requires that the entire speed transmission be disassembled before the torque amplifier can be removed. Therefore, this section will concern both units.

Power from the engine is applied directly to the torque amplifier and during operation, the torque amplifier is locked either in direct drive or torque amplifier (underdrive) position by hydraulic multiple disc clutches. There is no neutral position in either the torque amplifier or speed transmission as neutral position is provided for in the range (rear) transmission.

During operation in torque amplifier (underdrive) position, the clutch shaft is locked to the torque amplifier constant mesh output gear by a one-way clutch and an approximate 1/5-speed reduction occurs, resulting in about a 28 percent increase in torque.

All models have four speed transmissions. Adjustments and overhaul will be outlined in the following paragraphs.

clutch booster assembly disconnect linkage and hydraulic lines and remove the unit.

To disassemble the booster assembly use a press and compress output piston spring (4 – Fig. 151) and remove retainer ring (6). The balance of disassembly will be obvious after an examination of unit and reference to Fig. 151.

Reassembly is the reverse of disassembly procedure. Adjust the linkage as outlined in paragraph 141.

CLUTCH OVERHAUL

All Models

150. The disassembly and adjusting procedure for clutch will be obvious after an examination of the unit and reference to Fig. 152. Clutch (driven) disc is available as a unit only.

Specification data is as follows:

Tractor Models 786 and 886

Size – Inches	12
Springs:	
Number used	12
Free length – Inches	2-31/64
Lbs. test at height –	
Inches	140 at 1-17/32
Number of coils	8¼
Color	2 Bronze stripes
Lever height – Inches	2.301

Tractor Model 986

Size – Inches	12
Springs:	
Number used	12
Free length – Inches	2½
Lbs. test at height –	
Inches	160 at 1-17/32
Number of coils	8¼
Color	2 Blue stripes w/white
Lever height	2.301

Tractor Model 1086

Size – Inches	14
Springs:	
Number used	15
Free length – Inches	2⅝
Lbs. test at height –	
Inches	150 at 1-17/32
Number of coils	8¼
Color	2 Black stripes
Lever height – Inches	2.301

LINKAGE ADJUSTMENT

All Models

152. To adjust the torque amplifier (T.A.) linkage, loosen locknuts (1 – Fig. 153) and place control handle in the direct drive position. Adjust turnbuckle (2) so there is a slight gap between the T.A. spool snap ring (3) and the stop on the multiple control valve. Rotate the turnbuckle in direction of arrow until snap ring bottoms on the stop surface of multiple control valve. Rotate turnbuckle an additional ½-turn, then tighten locknuts. Loosen handle locking screw (4) and position handle so that an equal gap is obtained at each end of

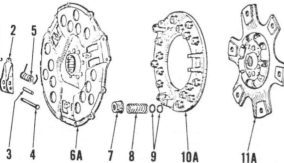

Fig. 152 — Exploded view of spring loaded clutches for all models.

1. Lever adjusting screw
2. Lever
3. Pivot pin
4. Lever pin
5. Lever spring
6. Back plate (786, 886 and 986 tractors)
6A. Back plate (1086 tractor)
7. Spring cup
8. Clutch spring
9. Washers
10. Pressure plate (786, 886 and 986 tractors)
10A. Pressure plate (1086 tractor)
11. Driven disc (786, 886 and 986 tractors)
11A. Driven disc (1086 tractor)

panel slot, then tighten locking screw. With handle in T.A. position, dimension between snap ring (3) and multiple control valve stop surface is approximately 1⅛ inches.

NOTE: The lube light should go off when T.A. is adjusted properly. If not, readjust.

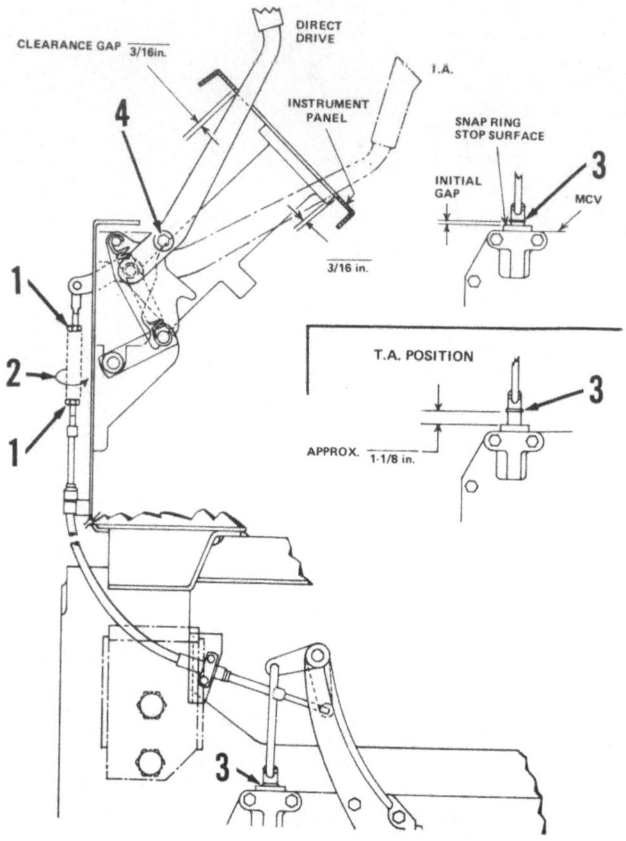

To adjust speed transmission linkage, remove floormats and subframe platform. Place speed control (7–Fig. 154) in first gear position after disconnecting it from speed shift lever (10). Place handle (9) at first speed mark on console.

Adjust ball joints (8) on rod (11) so that the mounting studs align with holes located in both the speed control lever (7) and speed lever (10). With levers adjusted properly there should be a ⅛-inch gap between handle (9) and end of the slot. Tighten locknuts on the ball joints and move speed selector handle through its ranges, checking for smooth operation.

REMOVE AND REINSTALL

All Models

NOTE: Model 786 is not available with cab; however, it is available with NON-ROPS frame with platform or with ROPS frame with canopy and platform. Reference to cab and air conditioning apply only to Models 886, 986 and 1086.

153. Removal of the torque amplifier or speed transmission requires removal of complete clutch housing from tractor. To remove clutch housing from tractor proceed as follows: Drain speed transmission and rear frame. Remove side step, batteries and battery trays. Partially remove both front lower panels on control center, then disconnect headlight wiring and remove both panels. Remove all hood sheets. Disconnect all hydraulic brake lines from brake valve. Disconnect all power steering lines, clutch assist booster lines and linkage. Disconnect torque amplifier cable at control center (cab). Disconnect throttle cable at injection pump and support bracket. Disconnect oil pressure sending unit wire, and if equipped, disconnect speedometer cable and the transducer electrical leads. Disconnect ether solenoid electrical lead. On models so equipped, disconnect air conditioning wiring and air conditioner quick-connect behind front panel. Shut off both air conditioner compressor valves, disconnect suction line and move it towards the rear. Disconnect heater hoses. On all models, remove subframe platform and disconnect all necessary wiring and brake lines. Block the park lock in disengaged position and disconnect speed transmission linkage. Disconnect

Fig. 153 – View showing points of adjustment of torque amplifier on all models.

1. Locknut
2. Turnbuckle
3. Snap ring
4. Locking screw

the park lock linkage, transmission brake linkage, hydraulic seat supply line and lube pressure warning switch. Install pto 89-526-8 gear locators. Support tractor under rear frame. Support control center (cab) or ROPS frame with suitable stands and place a rolling floor jack under front section. Remove both front Isomounts (platform supports) from the sides of speed transmission housing and raise control center just

enough to clear the starting motor. Disconnect the right supercharge tube assembly from the valve mounting bracket.

NOTE: Tractors with PFC (Pressure Flow Compensator) system, disconnect pump signal upper tube which takes the place of the right supercharge tube.

Remove mounting bolts securing speed transmission to rear frame and

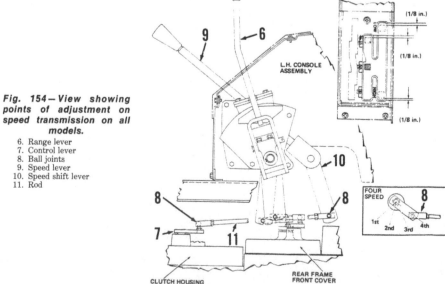

Fig. 154 – View showing points of adjustment on speed transmission on all models.

6. Range lever
7. Control lever
8. Ball joints
9. Speed lever
10. Speed shift lever
11. Rod

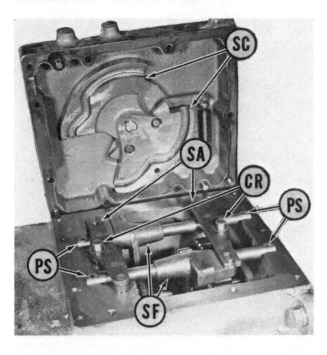

Fig. 155—View of speed transmission with cover removed.

CR. Cam rollers
PS. Phillips screws
SA. Shifter arms
SC. Shifter cam
SF. Shifter forks

roll front half of tractor forward, separating the tractor. Once the tractor has been split, remove speed transmission using a suitable hoist.

IMPORTANT: Leave side rail mounting bolts in side rails to hold up engine.

OVERHAUL

All Models

154. With clutch housing removed as outlined in paragraph 153, disassemble speed transmission and torque amplifier as follows: Disconnect and remove transmission brake operating rod, then remove snap ring, operating lever and Woodruff key from right end of clutch cross shaft. Loosen the two cap screws in clutch release fork, bump cross shaft toward left until the two Woodruff keys are exposed and remove Woodruff keys. Complete removal of cross shaft, release fork and throwout bearing assembly.

Unbolt and remove the multiple control valve and pump assembly. Complete removal of top cover retaining cap screws and lift off top cover and shifter cam assembly. Refer to Fig. 155 and lift off shifter arms (SA) and be careful not

to lose the two cam rollers (CR). Unstake and remove the four Phillips head screws (PS) which retain shifter rails and forks, then lift out the shifter rails and forks (SF).

Unbolt transmission main shaft bearing cage and as shaft and bearing cage

are pulled rearward, pull gears from top of clutch housing. See Fig. 156. Remove clutch housing bottom cover and brake assembly as shown in Fig. 157. Straighten lockwasher on rear end of countershaft and remove the nut. Remove the pto driven gear bearing retainer and pto driven gear and bearing assembly. Use a thread protector on rear end of countershaft, drive shaft forward and remove gears and spacers from bottom of clutch housing as shaft is moved forward. See Fig. 158.

Pull the three fluid supply tubes from housing as shown in Fig. 159.

At this time, the torque amplifier unit can be removed from clutch housing. However, in order to preclude any damage to sealing rings, or other parts, it is recommended that the torque amplifier unit be held together as follows: Install a cap screw and washer in rear end of clutch shaft to hold the direct drive gear assembly in place. Use a "U" bolt around clutch shaft to hold pto drive gear and carrier assembly in place. Install a lifting eye in front end of clutch shaft. Place clutch housing upright with bell end of housing upward. Attach a hoist to the previously installed lifting eye. Unbolt front carrier and the direct drive gear bearing cage (rear) from housing webs and carefully lift torque amplifier assembly from clutch housing.

Fig. 157—Speed transmission brake assembly is incorporated in clutch housing bottom cover. Brake pad operates against direct drive constant mesh gear.

Fig. 156—Mainshaft, mainshaft bearing assembly and gears shown removed from speed transmission on all models.

Fig. 158—View of speed transmission countershaft, gears and spacers.

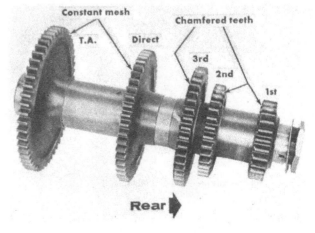

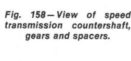

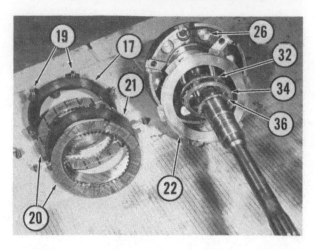

Fig. 162—Front (T.A. lock-up) clutch disassembled. Note the Teflon seal ring and bronze cup washer.
17. Backing plate
19. Springs (6 used)
20. Driven disc
21. Drive plate
22. Backing plate
26. Piston (6 used)
32. Teflon seal
34. Cup washer
36. Thrust washer

Fig. 159—Three fluid supply tubes to torque amplifier can be pulled straight out after removal of multiple control valve. Note "O" ring on each end of tube.

155. With the torque amplifier assembly removed from clutch housing, disassemble unit as follows: Remove the previously affixed holding fixtures and pull the pto drive gear and carrier assembly and the direct drive gear assembly from the unit. Unbolt and remove the pto drive shaft bearing cage from carrier assembly and remove bear-

Fig. 160—Pto drive shaft, bearing and bearing cage removed from T.A. and pto carrier.

ing cage and pto drive shaft as shown in Fig. 160. Pull torque amplifier drive gear and bearing assembly from T.A. carrier. Remove lubrication baffle and baffle springs from carrier. Remove seal ring, then remove the one-way clutch from drive gear and be sure to note how the unit is installed in the drive gear. See Figs. 161 and 166. It is possible to install the clutch with the wrong side forward and should this happen the torque amplifier would not operate. The quill shaft and output gear can be removed from bearing cage after removing the large internal snap ring. Drive gear and bearing removal from quill gear is obvious. Remove the large internal snap ring from front (lock-up) clutch and remove the clutch plates as shown in Fig. 162. Straighten tabs of lockwashers at rear of rear clutch, remove nuts and remove direct drive (rear) clutch. See Fig. 163. Remove the three cap screws and remove the front (lock-up) clutch carrier. Piston carrier is generally a press fit on clutch shaft and can be removed from clutch shaft after removing rear snap ring and thrust washer. Remove piston from piston carrier. See Fig. 164.

At this time, all parts of the torque amplifier assembly and speed transmission can be inspected and parts renewed as necessary. Procedure for removal of bearings is obvious. Refer to Figs. 165 through 168 for installation dimensions and information. Use new

"O" rings and lubricate all torque amplifier parts during assembly.

Reassemble components as follows: Install the six pistons in the front (T.A.) clutch with the widest edge toward clutch discs (front). Install the large piston in the piston carrier with smooth surface toward inside. Align the oil holes of piston carrier with oil holes of clutch shaft and press carrier on clutch shaft. Install thrust washers on each side of clutch carrier with grooved sides away from carrier and install snap rings. Use new gasket, mount front clutch carrier on piston carrier, tighten the three cap screws to 19-21 ft.-lbs. torque and bend

Fig. 163—Rear (direct drive) clutch disassembled. Note inner and outer springs.
37. Piston carrier   46. Inner spring
43. Driven plate   47. Outer spring
44. Drive plate   48. Backing plate

Fig. 161—The one-way clutch is positioned in inner bore of torque amplifier drive gear.
15. Ball bearing
29. T.A. Drive gear   33. One-way clutch

Fig. 164—View showing piston removed from piston carrier. Note seal rings on inside and outside diameter of piston. Also note the removed front (lock-up) clutch piston which is installed with widest edge toward front.
25. Clutch carrier
26. Lock-up piston
37. Piston carrier
39. Piston
40. Outer seal
41. Inner seal

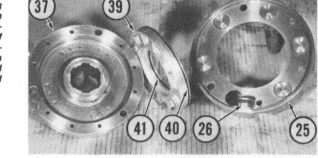

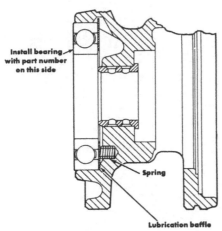

Fig. 165—Install lubrication baffle, baffle springs and torque amplifier drive gear bearing in T.A. carrier as shown.

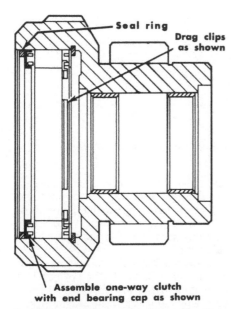

Fig. 166 — Assemble one-way clutch assembly in torque amplifier drive gear as shown. A snap ring and baffle are installed in bore ahead of one-way clutch. Note installation of bearing cap (bronze cup washer) and Teflon seal ring.

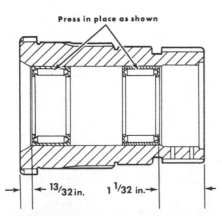

Fig. 167—Needle bearings are installed in direct drive output quill shaft to dimension shown.

down lockwashers. Start with back plate next to piston, then install a driven plate (internal spline) and alternate with driving plate (external spline). Place the nine aligning pins in rear side of piston carrier and place the double springs on pins.

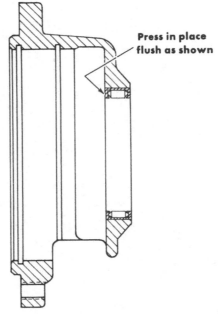

Fig. 168—Install needle bearing in direct drive output bearing cage as shown.

With clutch plates positioned, place the rear clutch backing plate over discs and install the three long bolts but do not tighten nuts until clutch splines have been aligned with the output gear. With clutch discs aligned and the backing plate positioned, tighten nuts to 75-85 ft.-lbs. torque on early production or tighten three bolts to 70-75 ft.-lbs. torque on late production units and bend down lockwashers. See Fig. 169 for cross sectional view of assembled clutches. Reassemble the output gear assembly, then install it in the rear clutch and secure it in position with the cap screw and washer used during removal.

Be sure snap ring and baffle is in one-way clutch bore in torque amplifier drive gear, then install the one-way clutch with drag clips toward front as shown in Fig. 166. Check one-way clutch operation before proceeding further. Slide assembly over hub of piston carrier. Turning clutch shaft same direction of engine rotation should lock the one-way clutch to rotate the drive gear. Opposite rotation should allow clutch to over-run.

Remove gear and one-way clutch and using heavy grease, position the bronze cup washer over end of one-way clutch, then install the Teflon seal ring. If the large ball bearing is installed on drive gear, place clutch discs over the drive

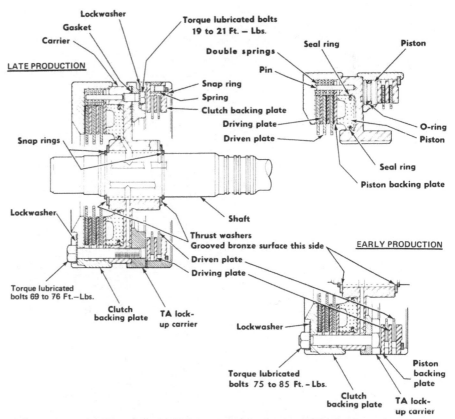

Fig. 169—Cross-sectional view of T.A. clutches after assembly.

1. Mainshaft
2. Bearing retainer
3. Bearing cup
4. Bearing cone
5. Shim
6. Bearing cage
7. Snap ring
8. 1st & 2nd sliding gear
9. 3rd & 4th sliding gear
10. Snap ring
11. Ball bearing
12. Bearing cage
13. Lock plate
14. Bearing
15. Transmission drive shaft
16. Constant mesh drive gear
17. Snap ring
18. Bushing
19. Oil seal
20. Pto drive shaft
21. Bearing retainer
22. Ball bearing
23. Snap ring
24. Oil seal
25. "O" ring
26. Bearing cage
27. Bearing (clutch pilot)
28. Bearing retainer
29. Shim (heavy)
30. Shim
31. Seal
32. Bearing cup
33. Bearing cone
34. Pto driven gear
35. Needle bearing
36. Countershaft
37. Snap ring
38. Spacer (long)
39. Constant mesh gear
40. Spacer
41. 3rd speed gear
42. 2nd speed gear
43. 1st speed gear
44. Ball bearing
45. Bearing retainer
46. Washer
47. Nut
48. Pto driven shaft

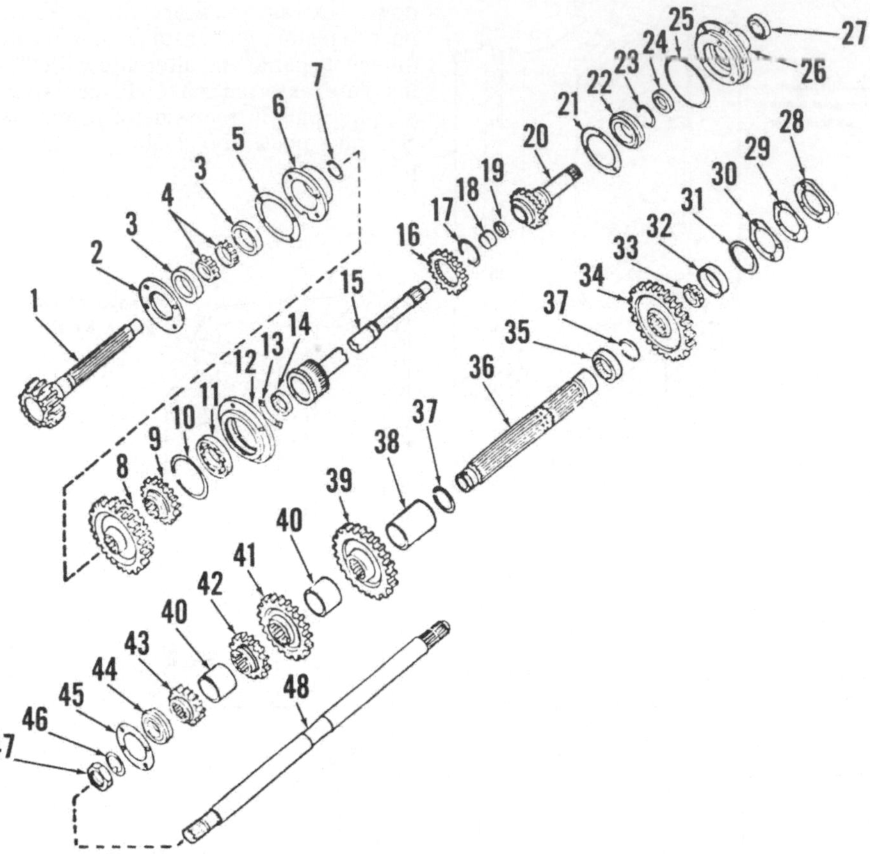

gear, install gear and position plate (discs) in clutch carrier. If the bearing is removed from drive gear, install the drive gear and feed clutch discs over the drive gear and into clutch carrier. Clutch discs are positioned as follows: Backing plate with no slots in external lugs next to pistons in carrier, driven disc (internal spline), driving disc (with slots in

Fig. 171 — Exploded view of speed transmission showing components used.

lugs) and driven disc. Place springs through lugs of center driving disc and rest them on lugs of backing plate. Position front backing plate with pins through springs and install the large snap ring.

Renew sealing rings on clutch shaft (5 – Fig. 170), install baffle springs and lubrication baffle (14) in T.A. carrier, with bent tab away from bearing, then install drive carrier over clutch shaft. If removed, install the pto drive shaft and bearing cage, then install the "U" bolt

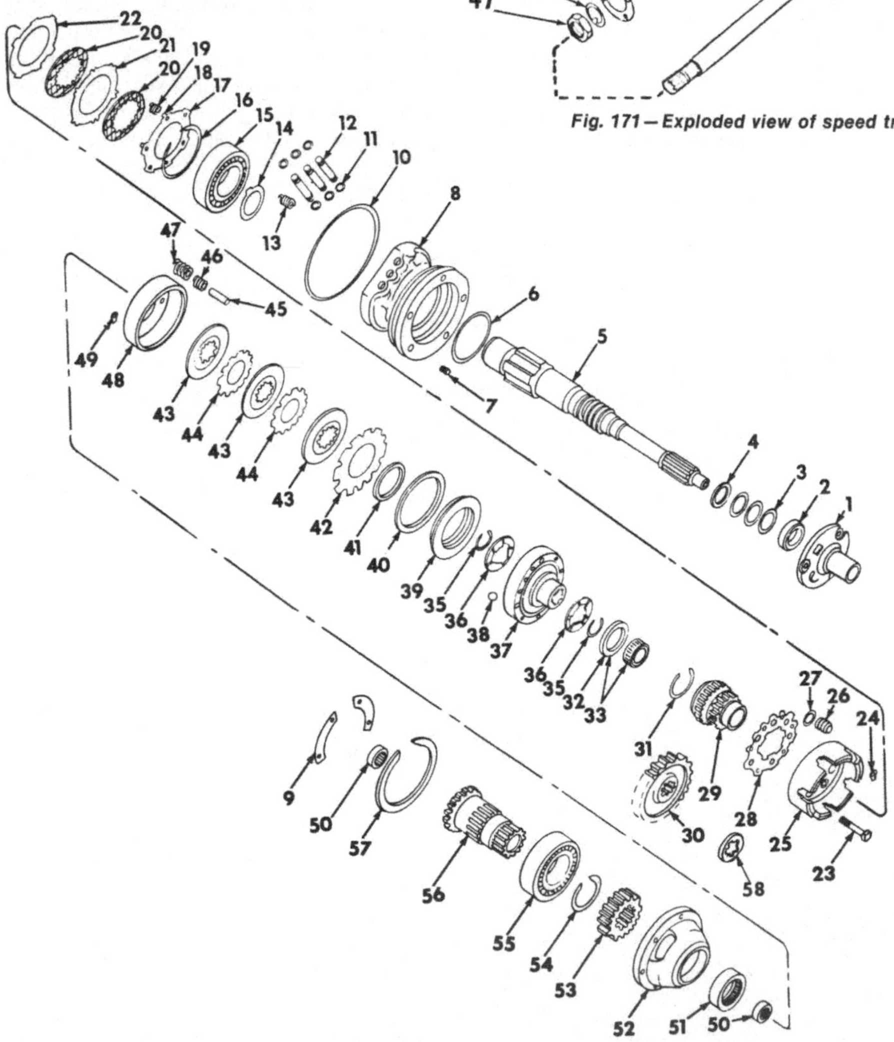

Fig. 170 — Exploded view of torque amplifier assembly showing component parts and their relative positions.

1. Bearing cage
2. Oil seal
3. Seal ring
4. Bearing retainer
5. Clutch shaft
6. "O" ring
7. Set screw
8. Carrier
9. Lock plate
10. "O" ring
11. "O" ring
12. Oil inlet tube
13. Spring (2 used)
14. Lubrication baffle
15. Ball bearing
16. Snap ring
17. Backing plate
18. Dowel pin
19. Clutch spring
20. Driven disc
21. Drive plate
22. Piston backing plate
23. Bolt
24. Lockwasher
25. Clutch carrier
26. Piston (T.A. lock-up)
27. "O" ring
28. Gasket
29. Drive gear
30. Drive gear
31. Snap ring
32. Teflon seal
33. One way clutch assembly
35. Snap ring
36. Thrust washer
37. Piston carrier
38. Steel ball
39. Piston
40. Seal, outer
41. Seal, inner
42. Backing plate
43. Driven disc
44. Drive disc
45. Guide pin
46. Clutch spring inner
47. Clutch spring outer
48. Backing plate
49. Lock washer
50. Needle bearing
51. Needle bearing
52. Bearing cage
53. Drive gear
54. Snap ring
55. Ball bearing
56. Quill shaft
57. Snap ring
58. Spacer

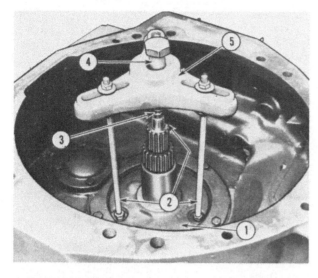

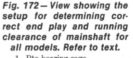

Fig. 172 — View showing the setup for determining correct end play and running clearance of mainshaft for all models. Refer to text.

1. Pto bearing cage
2. Rods and nuts (⅜ x 12 inch)
3. Spacer (ball bearing)
4. Puller bolt
5. Puller

Fig. 173 — Using inch-pound torque wrench to determine the correct end play. Refer to text.

Fig. 174 — Using feeler gages to check the clearance between the gears. Refer to text and to Fig. 175 for proper spacer.

1. Main shaft
2. Feeler gages
3. Spacer

Fig. 175 — Chart showing spacer size and part number. Refer to text.

Spacers	Part Number
.593 + .003 −	69053C1
.602 + .003 −	69054C1
.611 + .003 −	69055C1

clamp previously used to hold parts in position. Install lifting eye in front end of clutch shaft.

Attach torque amplifier unit to a hoist and lower assembly into clutch housing making sure the cut out portion in direct drive gear bearing cage is on bottom side. Install cap screws in both end bearing cages and tighten securely. Install the three oil inlet tubes. Start speed transmission lower shaft in front of clutch housing and install gears and spacers in the sequence shown in Fig. 158. Tighten nut on rear of shaft to 300 ft.-lbs. torque.

NOTE: On some early model tractors the bearing locknut is thinner than the nut used on late model tractors. So it may not be possible to torque the thin nut to 300 ft.-lbs. as the socket will slip off the nut. Shaft should rotate freely with no visible end play. If above conditions are not met, recheck installation.

To determine the proper spacer for correct end play and running clearance of the main shaft proceed as follows: Install a three jaw puller as shown in Fig. 172. Be sure to use a ball bearing spacer (3) between bolt and end of shaft. Using an inch-pound torque wrench (1 – Fig. 173) torque the puller bolt to 45 in.-lbs. which will give 200-250 pounds of force against the main shaft. Install the main shaft assembly with spacer that was removed and mounting bolts. Then using two feeler gages as shown in Fig. 174, check the clearance between the gears. If spacer is correct there should be 0.031-0.066 inch clearance. If not, use one of the spacers in Fig. 175 to obtain the correct clearance.

On all models, if mainshaft bearing assembly was disassembled, refer to Fig. 171 and reassemble as follows: Press forward bearing cup into bearing cage (6) with largest diameter rearward. Place bearing cones (4) in the cage and install rear bearing cup. Place end plate, original shims and bearing assembly over mainshaft (1). Press both bearings on shaft and install the snap ring. Start shaft in rear of clutch housing and place gears on shaft as shaft is moved forward. Install bearing retainer cap screws and rotate shaft while tightening cap screws. Shaft should turn freely with no visible end play. Vary shims as necessary.

Disassembly and reassembly of transmission top cover and shifter cams is obvious after an examination of the unit.

Balance of reassembly is the reverse of disassembly. Be sure to stake the shift rail retaining screws at the side of shaft (not front and rear) after installation.

On all models, reinstall the clutch housing assembly by reversing the removal procedure.

NOTE: When reinstalling the rear section on the clutch housing, it will be necessary to work the pto driven gear onto the pto shaft as the tractor is recoupled.

Adjust the clutch, T.A. dump valve and transmission brake linkage as outlined in paragraphs 141 through 146 and the torque amplifier linkage and speed transmission linkage as outlined in paragraph 152.

RANGE TRANSMISSION

The range transmission is located in the front portion of tractor rear frame. This transmission provides four positions: Hi (direct drive), Lo (underdrive), neutral and reverse. The Lo-range position provides a 3½ to one speed reduction and the reverse speed is about 25 percent faster than the same forward speed in Lo-range. The transmission park lock is also located in the range transmission.

To remove the range transmission gears and shafts, tractor rear main frame must be separated from clutch housing and the differential assembly removed. See Fig. 180 for an exploded view of range transmission shafts and

gears. Refer to paragraph 171 for linkage adjustment procedures.

R&R AND OVERHAUL

All Models

158. If proper split stands are available, rear frame can be separated from clutch housing before any disassembly is done. The following split allows service to be performed on the rear frame and axle housings.

159. To remove the range transmission, proceed as follows: Remove front hoods, side steps, batteries and battery trays. Shut off fuel supply. Disconnect necessary fuel lines and electrical connections on fuel tank sending unit. Remove fuel tank mounting bolts and with a suitable hoist, remove fuel tank. Drain hydraulic fluid from both sections of rear frame and both left and right axle housings. To prevent tractor front from tilting, block front axle bolster with wood blocks.

Support tractor under clutch housing with suitable jackstands. Place suitable support stands under sub-frame channel of cab or ROPS frame as shown in Fig. 181. Remove rear Isomount hex nut, clamp locking pliers on threaded plate and remove Isomount stud. See Fig. 182. Raise rear of cab or frame just

enough to clear the sub-frame, then adjust stands supporting the cab or ROPS frame.

IMPORTANT: Do not raise rear of cab or ROPS frame any higher than necessary to clear the sub-frame, otherwise damage to front Isomounts, steering and brake lines and clutch booster linkage may occur.

Disconnect brake lines and brake wear indicator switch, then remove brake housings, brake discs, plates and brake piston housing. Using a suitable hoist, remove axle housings.

NOTE: Be sure when splitting axle housing that the bull pinion shaft is kept from falling.

Disconnect actuating link from pto control valve, attach hoist to pto unit, then unbolt and remove pto unit and extension shaft.

NOTE: As pto unit is withdrawn, be sure to tip front of unit upward as shown in Fig. 183, otherwise damage to the oil inlet tube and screen will occur. Also leave the two short cap screws in place so unit will not separate during removal.

Fig. 181—View of supercharge tube and auxiliary valve assembly.

1. Right hand supercharge tube
2. Valve mounting bracket cover
3. Control center support

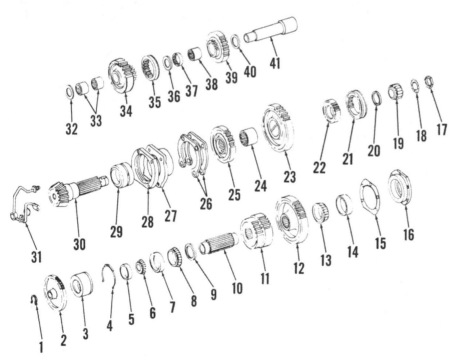

Fig. 180—Exploded view of range transmission gears and shafts used on all models.

1. Retainer
2. Hydraulic pump drive gear
3. Bearing cage
4. Retainer
5. Bearing cup
6. Bearing cone
7. Bearing cup
8. Bearing cone
9. Snap ring
10. Countershaft
11. Lo-Drive gear
12. Constant mesh gear
13. Bearing cone
14. Bearing cup
15. Shim
16. Bearing cage
17. Nut
18. Lockwasher
19. Roller bearing
20. Spacer
21. Hi-Lo shift collar
22. Shift collar carrier
23. Hi-Lo driven gear
24. Gear carrier
25. Reverse driven gear
26. Shims (0.004 & 0.007 in.)
27. Bearing cage
28. Retainer
29. Bearing assembly
30. Bevel pinion shaft
31. Lubrication tube
32. Thrust washer
33. Needle bearings
34. Reverse drive gear
35. Shift collar
36. Thrust washer
37. Needle bearing
38. Needle bearing
39. Reverse idler gear
40. Thrust washer
41. Reverse idler shaft

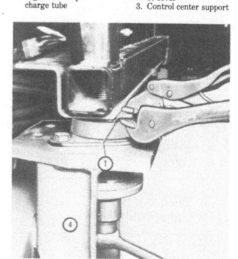

Fig. 182—View showing removal of Isomount stud.

1. Thread plate
2. Platform support

Fig. 183 — When removing pto unit, be sure to tip front end upward to prevent damage to oil inlet tube and screen.

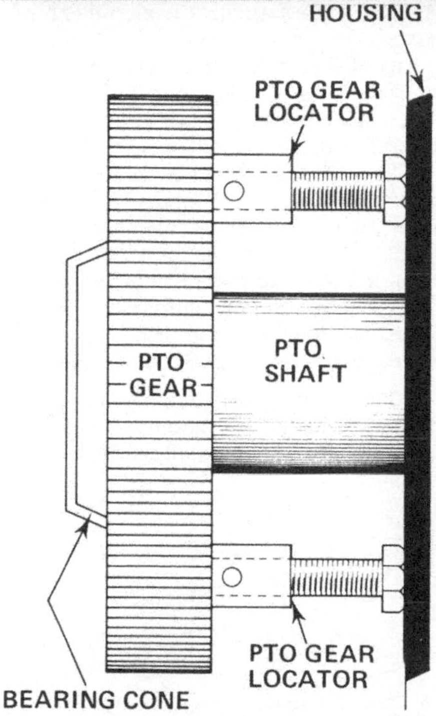

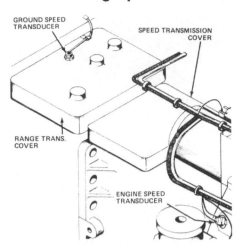

Fig. 187 — View of rear frame top cover showing transducer hookups.

Remove speed transmission bottom cover and install the 89-526-8 pto gear locators. See Fig. 185. Disconnect both manifold relief valve hydraulic lines (2 – Fig. 186) located at top of rear frame cover, also disconnect electrical connector (1 – Fig. 186).

NOTE: Tractors with PFC (Pressure Flow Compensator) system must have the signal tube disconnected at the motor priority valve (called manifold relief valve on tractors with open center system). The PFC system deals with the hydraulic hitch and draft control system and is outlined in the Hydraulic Lift section.

If equipped, disconnect engine speed and ground speed transducers located as shown in Fig. 187. Disconnect draft control and auxiliary valve linkages (1 and 2 – Fig. 188).

Tractors with PFC system, disconnect flow control cables. See Fig. 189. Disconnect right supercharge tube assembly (1 – Fig. 181) at auxiliary valve mounting bracket. Tractors with PFC system disconnect pump signal upper tube which takes the place of right supercharge tube on tractors equipped with open center system.

Disconnect auxiliary valve return tube (1 – Fig. 190) at right side of rear frame. Remove auxiliary valve bank mounting bolts (2) and lower auxiliary valve bank.

Disconnect hydraulic lines at auxiliary couplers. Remove coupler support bracket mounting bolts, loosen remaining bolt and lower coupler assembly for clearance. Disconnect draft and position control linkage. Remove hitch upper links and disconnect lift links from rockshaft arms. Unbolt hydraulic lift housing from rear frame, attach hoist and lift unit from rear frame.

NOTE: Before disconnecting any shift linkage, block park lock in disengaged position using a ¾-inch nut or equivalent.

Disconnect shift linkage at top of range transmission cover. Disconnect hydraulic seat dump tube. Remove mounting bolts securing speed transmission to rear frame and separate rear frame and speed transmission.

Remove range transmission cover. Remove dipstick, dipstick tube assembly from rear frame cover, then remove rear frame cover. Remove differential lube tube. Support differential with suitable hoist, remove bearing retainers, then lift out differential assembly. Remove range transmission shift rails and forks.

Fig. 185 — Cross-sectional view of pto driven gear showing installation of 89-526-8 pto gear locators.

Fig. 186 — View of manifold relief valve.

1. Electrical connector
2. Manifold relief valve hydraulic lines

Fig. 188 — View of draft control and auxiliary valve linkage.

1. Draft control linkage
2. Auxiliary control linkage

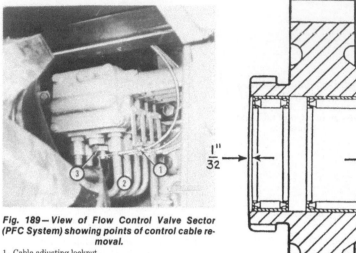

Fig. 189—View of Flow Control Valve Sector (PFC System) showing points of control cable removal.

1. Cable adjusting locknut
2. Cable
3. Valve

160. MAINSHAFT REMOVAL.

With range transmission removed and placed on a workbench or suitable stand, proceed as follows: Lock mainshaft and remove mainshaft end nut. Using a split collar and puller, remove and discard front mainshaft bearing (19—Fig. 180). Remove mainshaft bearing cage cap screws and lubrication assembly. Pull mainshaft out of rear and remove gears and collars from top of housing.

161. COUNTERSHAFT REMOVAL.

Remove horseshoe snap ring (1—Fig. 180) from pto driven shaft. Using a slide hammer, pull pto driven shaft forward out of housing. Remove countershaft front bearing cage (16—Fig. 180). On tractors equipped with open center system, remove retaining ring (4). Re-

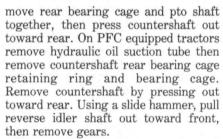

Fig. 200—Bearings in reverse idler gear are installed as shown.

move rear bearing cage and pto shaft together, then press countershaft out toward rear. On PFC equipped tractors remove hydraulic oil suction tube then remove countershaft rear bearing cage retaining ring and bearing cage. Remove countershaft by pressing out toward rear. Using a slide hammer, pull reverse idler shaft out toward front, then remove gears.

Clean and inspect all parts and renew any which show excessive wear or damage. Refer to Figs. 200 and 201 when renewing needle bearings in reverse idler and reverse drive gears. The mainshaft rear bearing assembly is

available only as a package of mated parts which will provide correct operating clearance. The package contains both bearing cups, both bearing cones and center spacer. If mainshaft rear bearing assembly is renewed, proceed as follows: Press both bearing cups into bearing retainer, with smallest diameters toward center, until they seat. Press rear bearing on mainshaft, with largest diameter toward gear, until it seats. Place bearing cage over mainshaft with flange toward pinion gear, then place spacer over shaft. Press front bearing on shaft with largest diameter toward front and rotate the cage as the bearing enters to insure alignment of parts.

Any disassembly or service required on transmission top cover will be obvious after an examination of the unit and reference to Fig. 202.

162. Reassembly sequence of range transmission is countershaft, reverse idler shaft, mainshaft and pto drive shaft.

To install countershaft, install snap ring (9—Fig. 180) on countershaft, then press on rear bearing cone (8) with largest diameter toward countershaft.

NOTE: Heat rear bearing cone in "HyTran" oil to 275°F. before installing on countershaft. Do not install bearing cold.

Fig. 190—View of right side rear frame showing auxiliary valve return tube (1) and auxiliary valve mounting bolt (2).

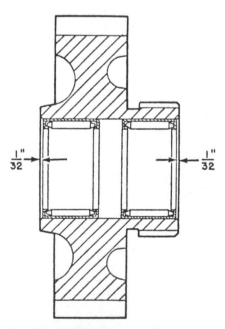

Fig. 201—Bearings in reverse drive gear are installed as shown.

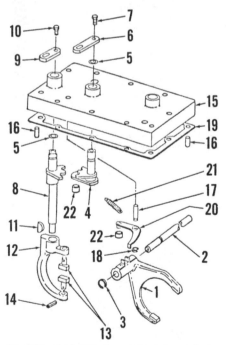

Fig. 202—Exploded view of range transmission top cover and shifting mechanism.

1. Hi-Lo shift fork
2. Shifter shaft
3. Retaining ring
4. Hi-Lo pivot arm and shaft
5. "O" ring
6. Hi-Lo shift lever
7. Bolt
8. Reverse fork shaft
9. Reverse shift lever
10. Bolt
11. Woodruff key
12. Reverse shift fork
13. Fork pads
14. Pin
15. Cover
16. Dowel
17. Pivot pin
18. Retaining ring
19. Gasket
20. Control arm
21. Spring
22. Detent roller

Note that rear bearing cone is narrower than front bearing cone. Install countershaft into housing from rear and at same time thread gears on shaft. Refer to Fig. 203 for proper gear placement. Using a long bolt and two large washers, insert bolt through the countershaft and with a washer at each end pull front bearing (13 – Fig. 180) onto shaft. Do not allow washers to contact bearing outer race. Install rear bearing cage, pto driven shaft bearing and spacer along with pto shaft and snap ring. On models equipped with PFC install pto shaft first and angle to left of rear frame. Install pto shaft bearing cage. Install hydraulic oil suction tube and then pto extension shaft.

163. COUNTERSHAFT BEARING SET-UP. On Models 786, 886 and 986, countershaft bearing set-up procedure is as follows: Using one 0.009 inch shim (15 – Fig. 180), assemble bearing cage (16). While rotating countershaft, tighten bearing cage bolts to 35 ft.-lbs. torque. If shaft binds before specified torque is obtained, disassemble bearing cage and add additional shims, one at a time, until countershaft end play measures 0.000 to 0.009 inch at specified torque. Countershaft bearing set-up shim is available in 0.009 inch only.

On Model 1086, countershaft bearing set-up procedure is as follows: Assemble

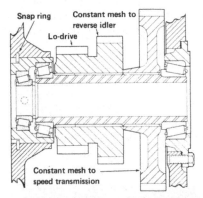

Fig. 203 – Cross-sectional view showing countershaft gear arrangement.

Snap ring
Constant mesh to reverse idler
Lo-drive
Constant mesh to speed transmission

Fig. 204 – When installing mainshaft be sure to align oil holes in mainshaft with holes in center.

countershaft front bearing cage without shims and while rotating shaft, tighten two opposite bolts to 150-in.-lbs. torque. Loosen bolts and retorque to 75 in.-lbs. while rotating shaft. Measure through two gage slots the distance from housing to outer surface of bearing cage, and average readings. If difference is greater than 0.010 inch, repeat procedure until desired range of 0.010 inch is obtained. Remove bearing cage and measure the flange next to the slots using a micrometer and average two readings. The desired shim pack is the difference of the two measurements plus 0.016 inch ± 0.001 inch. With the shim pack installed and two bolts torqued to 35 ft.-lbs., measure shaft end play while rotating shaft. If end play is not within 0.001 to 0.006 inch, add or remove shims until desired end play is obtained. Shims are available in thicknesses of 0.004, 0.005, 0.007, 0.012, 0.025, 0.029, and 0.035 inch.

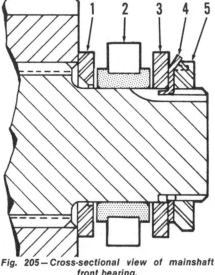

Fig. 205 – Cross-sectional view of mainshaft front bearing.

1. Spacer
2. Bearing
3. Spacer
4. Lock
5. Nut

Fig. 206 – Use IH tool No. 89-10-34 and feeler gage when checking pinion shaft setting.

164. REVERSE IDLER SHAFT. Reverse idler shaft installation is as follows: Install reverse idler shaft and gear assembly, largest gear to the rear.

NOTE: When installing forward and reverse shift collar, install beveled portion towards reverse drive gear. Three spacers are used, one at each end and one between gears.

Be sure to align shaft pin with slot in housing. After shaft is installed, place the rubber plug in slot with small radius against shaft. Install reverse shift shaft in reverse shift fork.

165. PTO DRIVEN SHAFT. Installation of the pto driven shaft is as follows: On tractors equipped with open center system, install pump drive gear and slide the pto shaft through the gear. Drive pto shaft into rear bearing assembly and install horseshoe snap ring (1 – Fig. 180). Install hitch pump assembly.

On tractors equipped with PFC hitch pump, remove hydraulic filter cover from rear housing and remove filters, baffle and jumper tube from the pump to the filter. Install pump assembly and a new gasket and "O" ring. With pump installed, insert two 3/8-inch dowels to center the pump. Install cap screws and secure flange to the housing. Install jumper tube, baffle and filter assemblies.

NOTE: When installing filters, be sure by-pass valve straps are 90 degrees from one another.

Install filter cover being careful not to damage the "O" ring.

167. MAINSHAFT ASSEMBLY. The cups, cones and spacers are provided as a unit eliminating the need of adjusting bearing preload for mainshaft rear bearing assembly. Mainshaft assembly is as follows: Assemble both

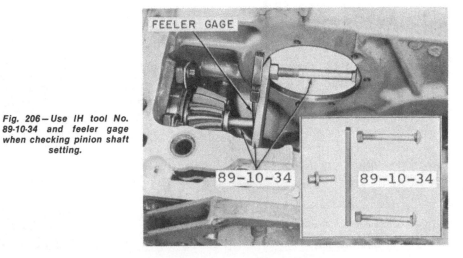

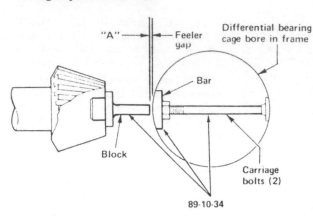

"A" — Feeler gap

Differential bearing cage bore in frame

Bar

Block

Carriage bolts (2)

89-10-34

Fig. 207 — Side view showing correct installation of pinion shaft locating tool.

rear of pinion shaft. They are marked with etched setting numbers 60 through 90. Use this setting number and refer to chart in Fig. 208 to determine correct feeler gap (A – Fig. 207). Add or remove shims at pinion shaft rear bearing to obtain determined gap (A) plus or minus 0.001 inch. For example: If a number 73 is etched on pinion shaft, feeler gap (A) would be 0.045 inch plus or minus 0.001 inch. Shims are available in thicknesses of 0.004 inch and 0.007 inch. After pinion shaft (mainshaft) has been adjusted, remove locating tool and install lubrication tube. Insert pto drive shaft through countershaft and bump into its rear bearing cone.

Install pads in reverse shifter fork, position shifter fork in reverse shift collar, then install fork shaft and Woodruff key.

169. DIFFERENTIAL BEARING SET-UP. The direct measuring method is fast and positive requiring the use of a depth micrometer. With differential in place, assemble both retainers, without shims, into rear frame. Tighten left side cap screws to 73 ft.-lbs. Tighten right side cap screws to 100 in.-lbs. while rotating differential. Loosen right side cap screws, retorque to 20 in.-lbs. while rotating differential, then without rotating differential, tighten to 50 in.-lbs. Tap retainers lightly. Measure through the three jackscrew holes, the distance from machined surface on main frame to outer surface of retainer and average readings. See Fig. 209. Remove right differential bearing retainer. Measure flange of retainer with a micrometer at three jackscrew holes and average the readings. The total shim pack is the difference of the two measurements ± 0.002 inch. Torque differential bearing retainer cap screws to 73 ft.-lbs.

Setting Numbers	GAP A Inches
60	.006
61	.009
62	.012
63	.015
64	.018
65	.021
66	.024
67	.027
68	.030
69	.033
70	.036
71	.039
72	.042
73	.045
74	.048
75	.051
76	.054
77	.057
78	.060
79	.063
80	.066
81	.069
82	.072
83	.075
84	.078
85	.081
86	.084
87	.087
88	.090
89	.093
90	.096

cups into cup carrier. Press rear cone assembly onto shaft against the shoulder. Press front cone on shaft while rotating carrier. Install mainshaft and gears into housing.

NOTE: Be sure to align oil holes in mainshaft and Hi-Lo driven gear carrier. See Fig. 204.

Install and tighten mainshaft rear bearing retaining cap screws to 85 ft.-lbs. Install a new mainshaft front bearing with the radius on the I.D. to rear. Drive bearing on until it clears threads of mainshaft enough to allow installation of the nut. Complete bearing installation by tightening nut. Once bearing is installed, remove nut and install spacer with chamfer towards rear and install lock tab and nut. Torque mainshaft nut to 100 ft.-lbs. and bend lock tab over. See Fig. 205.

NOTE: On tractors receiving a new pinion and ring gear, oil supply tube must be installed after pinion has been adjusted.

168. PINION SET-UP. Install pinion shaft (mainshaft) locating tool (IH tool No. 89-10-34 as shown in Figs. 206 and 207. A cone setting number is etched on

Fig. 208 — Chart used in determining feeler gap "A". Setting numbers (60 through 90) are etched on pinion shaft.

Fig. 209 — Direct measuring method of differential bearing set-up using a depth micrometer.
1. Retainer
2. Depth micrometer
3. Jackscrew holes

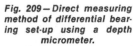

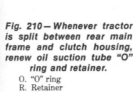

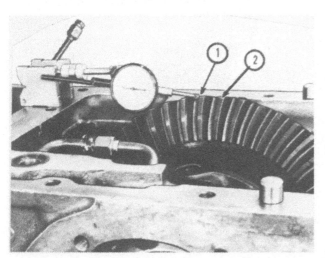

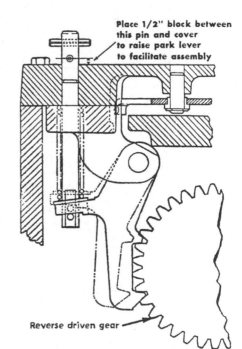

Fig. 210 — Whenever tractor is split between rear main frame and clutch housing, renew oil suction tube "O" ring and retainer.

O. "O" ring
R. Retainer
S. Bearing
25. Bearing
27. Nut
28. Reverse idler shaft
40. Pto shaft

170. RING AND PINION BACKLASH. After obtaining correct bearing adjustments, backlash between bevel drive gear and pinion must be checked.

NOTE: When adjusting backlash, rear frame must be bolted to the clutch housing, so that front of pinion shaft is supported.

Before bolting housings together, place the spacer on forward end of pto shaft, renew hydraulic pump suction tube "O" ring and retainer (Fig. 210) and join rear main frame to clutch housing.

The pinion shaft (mainshaft) must be forced forward and held forward when checking backlash. A bar may be used to pry the shaft forward.

The correct backlash is 0.005 to 0.015 inch. Take at least three readings at equally spaced intervals on ring gear using a dial indicator as shown in Fig. 211.

If backlash is at the low end on one side of gear, and outside the range on high end, look for dirt between ring gear and carrier. If no dirt is found, rotate ring gear 180 degrees on carrier and recheck backlash.

Moving shims from one side to the other of differential bearing retainers will provide proper backlash needed. Do not remove or add shims, or bearing preload will change. Moving a 0.010 inch shim from one side to the other will change backlash approximately 0.0075 inch.

NOTE: The oil supply tube for the mainshaft must be positioned to provide clearance between fitting that is centered in the pinion and the differential. The fitting should not be forced into the pinion, but the nose end of fitting should just enter hole in pinion and clear the differential.

Complete reassembly of tractor by reversing the disassembly procedure. Installation of range transmission top cover will be simplified if a ½-inch spacer is positioned between lower pin of park lock shaft and top cover as shown in Fig. 212.

LINKAGE ADJUSTMENT

All Models

171. To adjust range shift linkage and park lock linkage, just make sure all shift plate components are assembled and properly shimmed and that the range handle (A – Fig. 213) is in neutral position. Place the internal control levers reverse (B) and Hi-Lo (C) in their neutral position. Loosen locknuts on reverse shift and Hi-Lo joint links (E) and adjust the length so the mounting stud of the ball joints align with the holes in control levers (F) and (G) and shift levers (B) and (C). Assemble the nuts to the ball joints studs and tighten the locknuts. Move the range shift handle from neutral to the low position mark, then to high position and then reverse position mark. The handle must not touch the end of the selector slot in any of the shift positions and should clear by approximately ⅛-inch.

To adjust the park lock linkage, disconnect adjustable park lock link (P-Fig. 214) from pivot lever (W). Move range shift handle (A) to the park lock side of the console sector and let the handle rest in neutral detent position.

Fig. 211 — When installing the dial indicator, placement of the indicator against the ring gear is important. Put the indicator in a straight line with the tooth and at the very tip of the gear tooth. Do not install the indicator in the middle of the tooth or a false reading will result.

1. Dial indicator
2. Ring gear

Fig. 212 — When installing range transmission top cover, use a ½-inch spacer positioned as shown to hold park lock in disengaged position.

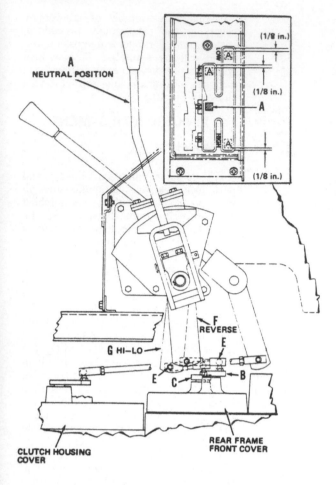

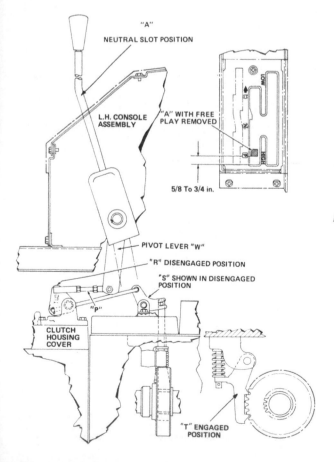

Fig. 213 — View showing adjustment of the range transmission shift linkage.

Fig. 214 — View showing adjustment of park lock.

Move park lock front bellcrank to full disengaged position. Adjust park lock link (P) as long as possible still being able to insert the mounting stud on the ball joint into the park pivot lever (W). Assemble the nut on the ball joint stud and tighten the locknut. Move range shift handle to park position and rock the rear wheels to ensure engagement of the pawl (T).

Now, move shift handle towards neutral until all looseness is removed and spring tension is felt. This should be about ¾-inch from end of the shift slot. When range shift lever is moved to neutral, the park lock should fully disengage. If shift handle is more than ¾-inch from end of slot adjust link (P) to remove excessive freeplay.

MAIN DRIVE BEVEL GEARS AND DIFFERENTIAL

The differential is carried on tapered roller bearings. The bearing on bevel gear side of differential is larger than that on opposite side and therefore, it is necessary to keep bearing cages in the proper relationship.

ADJUSTMENT

All Models

172. **CARRIER BEARING PRELOAD.** Differential carrier bearing preload can best be achieved by following tractor rear frame split procedure outlined in paragraph 159 with the exception to following note:

NOTE: Do not separate range transmission from speed transmission in split procedure. Rear frame must be attached to speed transmission to set differential and pinion backlash after carrier bearing preload is set.

With rear frame cover and hydraulic lift assembly removed set differential carrier bearing preload as outlined in paragraph 169.

173. **BACKLASH ADJUSTMENT.** After carrier bearing preload is obtained, refer to paragraph 170 for bevel gear and drive pinion backlash setting procedure.

R&R BEVEL GEARS

All Models

174. The main drive bevel pinion is also the range transmission mainshaft.

The procedure for removing, reinstalling and adjusting pinion setting is outlined in the range transmission sections (paragraphs 159 through 170). To remove the bevel ring gear, follow procedure outlined in paragraph 175 for R&R differential. The ring gear is secured by differential case bolts which should be tightened to a torque of 112 ft.-lbs.

R&R DIFFERENTIAL

All Models

175. All tractors are equipped with a four pinion type differential. To remove differential follow rear frame split procedure outlined in paragraph 159.

With differential removed from rear frame, use a puller and remove carrier bearings and note that bearing on bevel gear side of differential has a larger O.D. and contains more rollers than the opposite side bearing. Remove differential case bolts and separate differential assembly. Any further disassembly is obvious. Oil baffle and bearings can also be renewed at this time. Install baffle (3 – Fig. 215) with lip facing away from housing (19). Press bearing cone (1) onto housing after installing baffle.

When reassembling differential, tighten differential case bolts to 112 ft.-lbs. Refer to paragraphs 172 and 173 for carrier bearing preload and backlash adjustments.

DIFFERENTIAL LOCK

Models So Equipped

The differential lock is located on outside of right hand brake housing, and is hydraulically actuated. When operator holds the control valve in down position, this allows hydraulic oil pressure to actuate clutch assembly so that the two bull pinion shafts are locked together which causes the differential to act as a solid hub, Fig. 217. When control valve is released, the clutch assembly will disengage.

The differential lock is not intended for continuous use or when slippage is not a problem and therefore the unit is disengaged whenever the control valve is released.

176. **R&R AND OVERHAUL.** To remove the differential lock, disconnect necessary hydraulic lines. Loosen mounting bolts, then with a soft faced hammer tap cover to break the seal. Then, remove mounting bolts and cover. Any further disassembly will be obvious after referring to Fig. 218. Renew any parts showing excessive wear or other damage. Reinstall by reversing disassembly procedure.

NOTE: If both differential lock and brake housing seals are broken, reseal using "Loctite 515" gasket seal.

The differential lock pressure is 240-300 psi.

Models 886 S.N. 12997 & below, 986 S.N. 16735 & below and 1086 S.N. 28429 & below are equipped with manual differential lock actuating valve.

Models 886 S.N. 12998 & above, 986 S.N. 16736 & above, 1086 S.N. 28430 & above and all Model 786 are equipped with an electrically operated differential lock actuating valve.

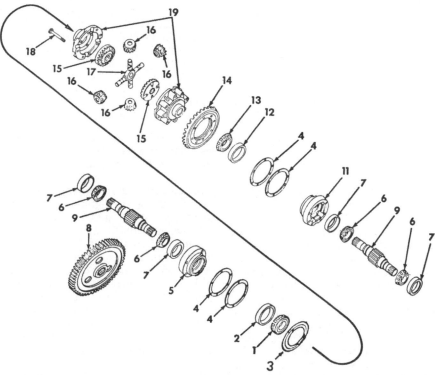

Fig. 215 — Exploded view of differential, bull pinions and associated parts.

1. Bearing cone (LH)
2. Bearing cup (LH)
3. Oil baffle
4. Bearing cage shims
5. Differential bearing cage (LH)
6. Bearing cone
7. Bearing cup
8. Bull gear
9. Bull pinion shaft
11. Differential bearing cage (RH)
12. Bearing cup (RH)
13. Bearing cone (RH)
14. Bevel ring gear
15. Bevel gears
16. Pinion gears
17. Spider
18. Case bolts (12 used)
19. Differential case

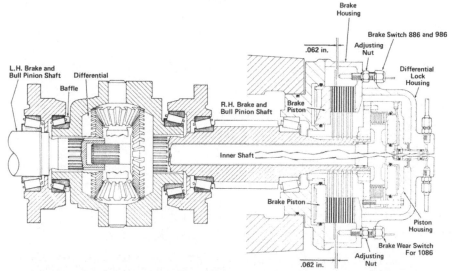

Fig. 217 — Cross-sectional view of differential lock on tractors so equipped.

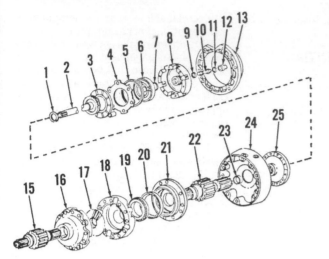

Fig. 218 — Exploded view of differential lock.

1. Snap ring
2. Inner shaft
3. Clutch assembly
4. Primary plate
5. "O" ring
6. Piston
7. "O" ring
8. Carrier
9. "O" ring
10. Spool
11. "O" ring
12. Thrust washer
13. Cover
15. Bull pinion shaft
16. Case
17. Bevel gear
18. Case
19. Bearing cone
20. Bearing cup
21. Retainer
22. Bull pinion shaft
23. Needle bearing
24. Housing
25. Piston housing assembly

FINAL DRIVE

The final drive assemblies on all models consist of the rear axle, bull gear and bull pinion. These assemblies can be removed from the tractor as a unit.

All Models Except High Clearance

180. **REMOVE AND REINSTALL.** Drain fluid from near frame and axle housing. Raise and support tractor under rear frame. Support front of the cab or ROPS frame and remove Isomount stud. Disconnect brake lines and brake wear switch. Remove rear wheel. Support axle housing with a suitable hoist. Remove axle housing bolts and split housing away from rear frame.

NOTE: Hold bull pinion shaft while making the split.

Axle installation is the reverse of removal procedure.

181. **OVERHAUL.** With final drive removed as outlined in paragraph 180, disassemble as follows: Remove brake piston housing and adjusting shims, then remove the bull pinion. Using a suitable puller, remove bearing cone (2 – Fig. 219) from inner end of axle shaft. Remove snap ring (3), then remove bull gear and snap ring (4). Unbolt axle outer bearing retainer and withdraw axle shaft assembly from carrier. Any further disassembly required is evident after an examination of the unit.

When reassembling, press rear axle outer bearing on axle using a pipe having a diameter which will contact inner race only. Press bearing on axle shaft until it bottoms against shoulder. If oil seal in outer bearing retainer is renewed, install with spring loaded lip toward inside and inner surface flush with inner surface of bore as shown in Fig. 220. Use shim stock or seal sleeve when installing seal retainer over axle. Install retainer cap screws but do not tighten them at this time. Install snap ring (4 – Fig. 219) bull gear, snap ring (3) and bearing on inner end of axle shaft. Place bull pinion in position on bull gear and install brake piston housing without shims or "O" ring. Use aligning dowels and install final drive assembly to the main frame. Torque axle carrier cap screws to 170 ft.-lbs.

NOTE: Hold bull pinion in position while installing the assembly so seals and bearings will not be damaged.

Install three cap screws through brake piston housing and tighten them evenly to a torque of 100 in.-lbs. while rotating bull pinion shaft to align and seat the bearings. Loosen cap screws, then retighten them evenly to a torque of 50 in.-lbs. while rotating the shaft. Next, without rotating bull pinion shaft, further tighten cap screws to a torque of 100 in.-lbs. Using a feeler gage, measure gap adjacent the cap screws. The correct shim pack is the average measured gap plus 0.015 inch. Shims are available in the following thicknesses: 0.003, 0.004, 0.007, 0.012, and 0.035 inch. Remove brake piston housing, install new "O" ring and oil seal, then reinstall brake piston housing with the shim pack.

To adjust axle shaft bearings, tighten three evenly spaced cap screws retaining axle outer cap (seal retainer) to axle carrier to a torque of 150 in.-lbs. while rotating axle. Then, without rotating axle, tighten evenly to 50 in.-lbs. Measure distance between outer cap and carrier next to the three tightened cap screws and average the three readings. Make a shim pack that will be within a plus or

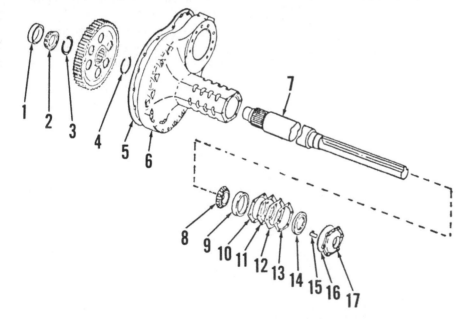

Fig. 219 — Exploded view of final drive assembly.

1. Bearing cup
2. Bearing cone
3. Snap ring
4. Snap ring
5. Gasket
6. Carrier
7. Rear axle shaft
8. Bearing cone
9. Bearing cup
10. Shim (light)
11. Shim (medium)
12. Shim (heavy)
13. Shim (ex. light)
14. Oil seal
15. Grease fitting
16. "O" ring
17. Cap

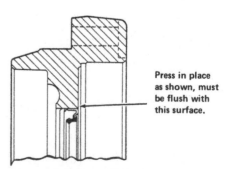

Press in place as shown, must be flush with this surface.

Fig. 220 — When installing rear axle cap (retainer) seal, position as shown with lip toward inside.

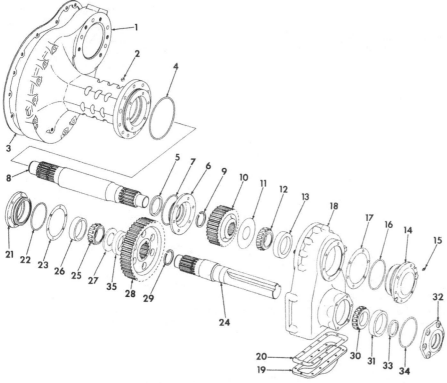

Fig. 221 — Exploded view of final drive assembly used on high clearance tractors.

1. Carrier	10. Drive gear	19. Pan
2. Plug	11. Grease retainer	20. Gasket
3. Gasket	12. Bearing cone	21. Bearing cap
4. "O" ring	13. Bearing cup	22. "O" ring
5. Oil seal	14. Bearing cap	23. Shim
6. Seal retainer	15. Grease fitting	24. Rear axle shaft
7. "O" ring	16. "O" ring	25. Inner bearing cone
8. Drive gear axle	17. Shim	26. Inner bearing cup
9. Snap ring	18. Housing	27. Axle spacer
		28. Driven gear
		29. Snap ring
		30. Outer bearing cone
		31. Outer bearing cup
		32. Bearing cap
		33. Oil seal
		34. "O" seal
		35. Gear spacer

Clean and inspect all parts and renew as necessary. Be sure to use all new "O" rings and gaskets and reassemble by reversing the disassembly procedure. However, delay final adjustment of shaft bearings until unit is installed on tractor. Refer to paragraph 183 for procedure.

183. To adjust drive gear axle bearings, remove all the shims (17–Fig. 221) from behind bearing cap (14). Install three cap screws (one every other hole) in bearing cap and tighten evenly to 150 in.-lbs. Bump bearing cap to insure bearings are seated, loosen the three cap screws, then retighten to 50 in.-lbs. Measure distance between bearing cap and drop housing next to the three cap screws, then select and install a shim pack within 0.002 inch of averaged readings. This will provide the shaft bearings with proper end play. Shims are available in 0.004, 0.005, 0.007, 0.012 and 0.0299 inch thicknesses.

To adjust bearings for rear axle shaft install bearing cap (21) without shims and tighten three evenly spaced cap screws to 150 in.-lbs. while rotating shaft. Bump bearing cap to insure bearings are seated, loosen three cap screws, then retighten to 75 in.-lbs. Measure distance between bearing cap and carrier next to three cap screws, then select and install a shim pack within 0.002 inch of averaged readings.

NOTE: Seals (5 and 33) are installed with lips facing toward inside. Use shim stock or a seal protector when installing retainer (6) or cap (32).

Complete reassembly of tractor and bleed brakes as outlined in paragraph 186.

minus 0.002 inch of the average measured distance and install between axle outer cap and carrier. After adjusting axle shaft bearings, tighten all cap screws securely and complete reassembly of tractor.

assembly. Remove bearing and snap ring from inner end of axle and bump axle out of bull gear. Any further disassembly required will be obvious after an examination of the unit and reference to Fig. 221.

High Clearance Models

182. **R&R AND OVERHAUL.** On high clearance model tractors, the complete final drive assembly is removed in the same manner as outlined in paragraph 180 although drop housing (18–Fig. 221) should also be drained.

With unit removed, proceed as follows: Remove pan (19) from housing (18). Remove bearing cap assembly (items 21, 22, 23 and 26), then remove inner bearing (25) and spacers (27 and 35) from end of axle (24). Remove outer bearing cap assembly (items 32, 34, 33 and 31), then bump axle outward and remove driven gear (28) from bottom of housing. Remove bearing cap assembly (items 13, 14, 16 and 17) from housing (18), then unbolt and remove housing (18) from carrier (1). Remove bearing (12), grease retainer (11) and drive gear oil seal retainer (6) and oil seal (5)

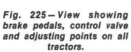

Fig. 225 — View showing brake pedals, control valve and adjusting points on all tractors.

A. Adjusting bracket
B. Valve spool
C. Cap screw
E. Pin
F. Locknut
G. Stop screw
K. Bolt
L. Stop
N. Clevis

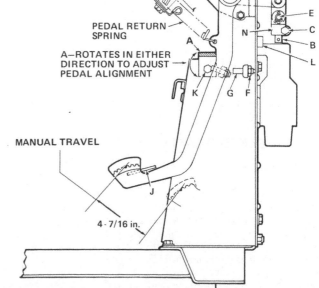

BRAKES AND CONTROL VALVE

Brakes on all models are power actuated hydraulically, self-adjusting and self-equalizing. All models use a wet type 8 inch diameter multi-disc brake system.

Brake operation can be accomplished with engine inoperative because of a one-way check valve located on the firewall of engine compartment. This check valve closes the hydraulic circuit when pressure ceases. Brakes will then operate with fluid that is trapped in the brake circuit. Brake hydraulic pressure for Models 786, 886 and 986 is 190 to 280 psi. Model 1086 brake hydraulic pressure is 135 to 255 psi.

Service (foot) brakes MUST NOT be used for parking or any other stationary job which requires tractor to be held in position. Even a small amount of fluid seepage would result in brakes loosening and severe damage to equipment or injury to personnel could result. USE PARK LOCK when parking tractor.

BRAKE ADJUSTMENT

All Models

185. Disconnect pedals from brake valve spools by removing pin (E – Fig. 225) from clevis (N). Adjust both brake pedals by loosening bolts (K) and rotate stop bracket (A) until pedal lock pawl (J) interlocks freely. Tighten bolts (K). With brake valve against stop (L) or the firewall, disconnect pedal return springs, locknut (F) and adjust stop screw (G) until full pedal travel is 4 7/16 inches. Tighten locknut (F), and reconnect pedals to brake valve spools. Loosen cap screws (C) on clevises (N) and rotate brake valve spools until equal amounts of slack is obtained in both pins (E). The brake spool valves should be adjusted equally. Tighten clevis cap screws. Both brake pedals should have:
1. Height adjustment
2. Full pedal travel
3. Equal braking

BLEED BRAKES

All Models

186. To bleed brakes, attach a length of hose to bleeder valve and place open end in a container. Start engine and run at low idle rpm. Depress brake pedal and while holding in this position open bleeder valve and when a solid flow of oil appears, close valve. Repeat operation on opposite brake. Check brake pedal

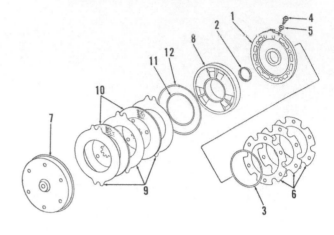

Fig. 226—Exploded view of brake assembly.

1. Brake piston housing
2. Oil baffle
3. "O" ring
4. Brake bleed screw
5. Bleeder valve
6. Shim
7. Brake housing
8. Brake piston
9. Separator plate
10. Disc
11. "O" ring (piston inner)
12. "O" ring (piston outer)

feel. If brake pedal operation feels spongy rather than having a solid feel, repeat bleeding operation.

NOTE: With engine not running, braking action should occur in last ⅓ of the pedal travel.

R&R AND OVERHAUL

All Models

187. Removal of either brake is accomplished by disconnecting the brake supply line and brake wear switch.

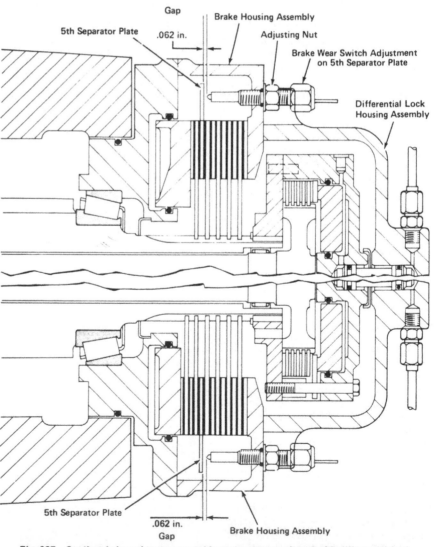

Fig. 227—Sectional view of brake assembly on tractors equipped with differential lock.

Remove brake housing mounting bolts, then remove brake assembly. Once removed, split piston housing from brake housing cover.

NOTE: The piston housing and brake housing are sealed together. The use of two four-inch long bolts may be helpful in splitting the housings.

Using compressed air under moderate control, force air into the brake supply fitting to remove brake piston. DO NOT stand in front of piston during removal. Remove brake discs and separator plates. Refer to Figs. 226 and 227. With brake disassembled, clean and examine parts. Renew any parts showing wear or damage.

Lubricate and install new "O" rings (11 and 12 – Fig. 226) into piston housing. Install two retaining pins into piston housing and install separator plates and brake discs (9 and 10).

NOTE: The fifth plate from outside serves as a brake wear indicator. See Fig. 227.

Seal brake housing cover and, if equipped, the differential lock housing cover using "Loctite 515" or equivalent. Install brake housing cover and, if equipped, the differential lock housing cover. Refer to paragraph 176 for differential lock information.

When assembly is completed, check external adjustments as outlined in paragraph 185, and bleed brakes as outlined in paragraph 186.

BRAKE CONTROL VALVE

All Models

188. **R&R AND OVERHAUL.** Remove brake control valve as follows: Remove left hood sheet. Disconnect brake pedal linkage and remove brake lines. Remove two mounting bolts then remove brake control valve.

189. With brake control valve removed, refer to Fig. 228 and disassemble as follows: Remove linkage yokes from spool ends. Remove body cap screws and withdraw valve spools and pistons from valve.

NOTE: Mark spool and piston assemblies for identification to aid in reassembling.

Install two 7/16-inch nuts on spool end, place nuts in a vise, then remove retaining locknut (7 – Fig. 228) and piston (6). Inspect valve body bores and spools for scratches or scoring and renew as necessary.

NOTE: Valve spools and body are matched set and cannot be serviced separately.

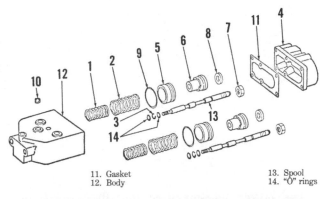

Fig. 228 – Exploded view of brake control valve. Gasket and "O" rings are available only in a kit. Valve body and spools are not serviced separately.

1. Return spring
2. Large piston spring
3. Snap ring
4. Body cap
5. Large piston
6. Small piston
7. Retaining nut
8. Piston retainer
9. "O" ring
10. Orifice plate
11. Gasket
12. Body
13. Spool
14. "O" rings

Fig. 230 – Installing test gages for pto operating pressure.

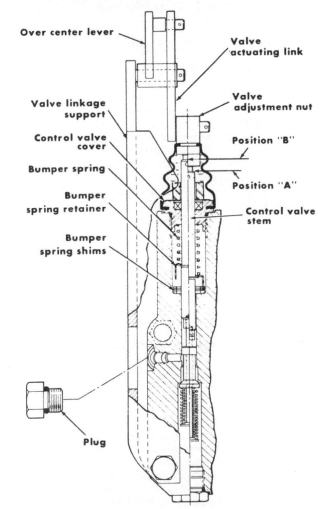

Fig. 231 – Cut-away view of the pto unit control valve used on early production models. Note positions "A" and "B". Refer to text for adjustment.

Over center lever

Valve actuating link

Valve linkage support

Valve adjustment nut

Control valve cover

Position "B"

Bumper spring

Position "A"

Bumper spring retainer

Control valve stem

Bumper spring shims

Plug

Inspect pistons for scoring or scratches. Renew all gaskets and "O" rings. Clean and inspect inlet orifice. Reassemble by reversing disassembly procedure.

After valve is installed, adjust linkage as outlined in paragraph 185 and bleed brakes as outlined in paragraph 186.

POWER TAKE-OFF

The power take-off used on all models is an independent type, having both 1000 and 540 rpm output shafts. The pto unit incorporates its own hydraulic pump which furnishes approximately three gpm of oil which is used to actuate the pto clutch and provide lubrication for the pto. Operation of the pto unit is controlled by a spool type valve located in a bore in left side of pto housing.

Models 886 (S.N. 16999 & below), 986 (S.N. 23999 & below) and 1086 (S.N. 47999 & below) are equipped with pto shown in Fig. 234. Models 886 (S.N. 17000 & above), 986 (S.N. 24000 & above), 1086 (S.N. 48000 & above) and all Model 786 are equipped with pto shown in Fig. 235.

OPERATING PRESSURE

Early Models 886-986-1086

190. To check operating pressure, run tractor until hydraulic fluid temperature is approximately 150°F. Refer to Fig. 230 and remove test port plug which is located at hole in linkage support (left side) and attach either an IH Flow-Rater, or a test gage capable of registering at least 600 psi. If necessary, loosen linkage support cap screws and push support downward as far as possible and retighten cap screws. Unscrew valve from adjusting nut and swing nut and actuating link up out of the way. Start engine and with pto engaged, operate engine at 2100 rpm (1000 pto rpm). Pull valve stem up until stem contacts bumper spring retainer (position "A" – Fig. 231) and hold in this position.

NOTE: This is partial engagement position.

If gage reads 41-46 psi, pressure is satisfactory at this point. If pressure is below 41 psi, remove control valve cover, bumper spring and bumper spring retainer and install bumper spring shims as required. Each shim will change pressure about five psi.

Reposition the actuating nut and screw valve stem into nut approximately the original position. Move control han-

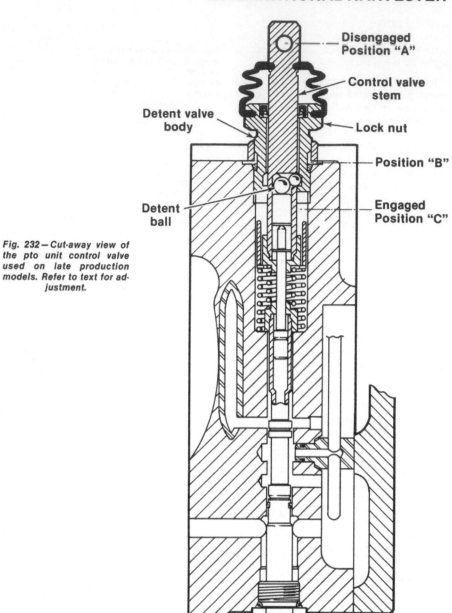

Fig. 232 – Cut-away view of the pto unit control valve used on late production models. Refer to text for adjustment.

dle and pull over-center link into up position (position "B" – Fig. 231) and check pressure. Gage should read 235 psi. If pressure is not as stated, turn valve

stem into adjusting nut to increase or out of nut to decrease pressure.

NOTE: This is the full engagement position.

Fig. 233 – Pto unit being removed. Note rear end being tipped downward to prevent damage to suction tube.

With these adjustments made, reduce engine speed to low idle rpm (pto engaged) at which time the pressure must not drop more than 40 psi from the 235 psi reading. Disengage pto and check operation of pto brake (anti-creep). Brake must stop pto rotation within a maximum of three seconds.

If above conditions cannot be met, remove and overhaul pto unit as outlined in paragraph 192.

Model 786 and Late Models 886-986-1086

191. To check operating pressure, run tractor until hydraulic fluid temperature is approximately 150°F. Refer to Fig. 230 and remove test plugs on left side of pto as shown. Install two gages capable of registering at least 600 psi and two orifices (IH 384328R1). Start engine with pto engaged, operate engine at 2100 rpm (1000 pto rpm). Push control valve to (position "B" – Fig. 232) and gage should read 41-46 psi at this point. If pressure is not in this range, loosen locknut and adjust detent valve body until a pressure of 41-46 psi is obtained. Tighten locknut. Push control valve to (position "C" – Fig. 232). Gages should read 210 psi and valve must remain in this position with no external force. Then, pull valve from "C" position to "A" position. The anti-creep brake pistons must stop output shaft in three seconds and pressure must be 210 psi at one gage and zero at the other. Reduce engine speed to low idle rpm and move valve to position "C". The pressure must not drop more than 40 psi from the 210 psi reading. Stop engine and control valve must automatically return to disengage position.

If above conditions cannot be met, remove and overhaul pto unit as outlined in paragraph 192.

R&R AND OVERHAUL

All Models

192. To remove pto unit, drain the rear main frame, then disconnect control lever rod and remove linkage support. Attach hoist to unit, complete removal of retaining cap screws, then pull unit rearward, tip rear end downward to prevent damage to suction tube assembly and remove unit from tractor main frame. See Fig. 233.

NOTE: If pto extension shaft remains with pto unit, it will be necessary to remove it from the pto unit input shaft before withdrawing unit from main frame.

Leave the two short cap screws which retain rear cover to housing in plate so the pto sections will be held together.

With the pto unit removed from tractor, use Fig. 234 for early production models and Figs. 235 and 235A for late production models as a reference and disassemble unit as follows: Remove the safety shield. Remove retainer (77), pump housing and suction tube assembly from carrier (67). See Fig. 236. Remove pump idler gear from housing and drive gear and Woodruff key from input shaft. If necessary, bearing (76) and suction tube (6) can be removed from pump housing. Note position of carrier (67 – Fig. 234 or 235) on housing (8), then unbolt and remove carrier. Remove large snap ring (60), retainer (59) and clutch discs (57). Place assembly in a press and using a straddle tool, depress clutch spring retainer (62). Remove snap ring (63), then lift out retainer and clutch release spring (61). Grasp hub of piston (54) and work it out of clutch cup (51). Place a bar across clutch cup and

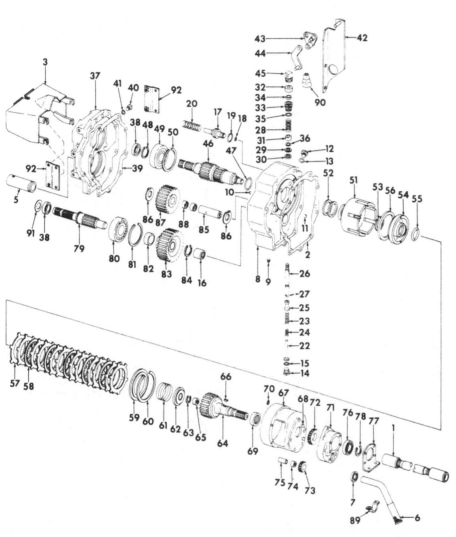

Fig. 234 – Exploded view of dual speed pto unit used on early production Models 886, 986 and 1086.

1. Extension shaft
2. Gasket
3. Safety shield
5. Protection tube
6. Suction tube
7. Seal
8. Housing
9. Plug
10. Steel ball
11. Steel ball
12. Plug
13. "O" ring
14. Plug
15. "O" ring
16. Bearing
17. Brake piston
18. "O" ring
19. "O" ring
20. Brake spring
22. Valve guide
23. Spring, light
24. Spring, heavy
25. Control valve
26. Guide stem
27. Dowel pin
28. Bumper spring
29. Shim
30. Sleeve spacer
31. Sleeve
32. Stop
33. Cover
34. Seal
35. "O" ring
36. Washer
37. Cover
38. Seal
39. Gasket
40. Plug
41. "O" ring
42. Linkage support
43. Over-center lever
44. Actuating link
45. Adjusting nut
46. Output shaft (1000 rpm)
47. "O" ring
48. Snap ring
49. Bearing
50. Snap ring
51. Clutch cup
52. Seal (Teflon)
53. Snap ring (Truarc)
54. Piston
55. "O" ring
56. Seal
57. Drive disc
58. Driven disc
59. Retainer
60. Snap ring
61. Release spring
62. Spring retainer
63. Snap ring
64. Drive shaft
65. Bearing
66. Woodruff key
67. Carrier
68. Dowel
69. Bearing
70. "O" ring
71. Pump housing
72. Drive gear
73. Idler gear
74. Bearing
75. Shaft
76. Bearing
77. Bearing retainer
78. Snap ring
79. Output shaft (540 rpm)
80. Bearing
81. Snap ring
82. Spacer
83. Driven gear (540 rpm)
84. Snap ring
85. Idler shaft
86. Thrust washer
87. Idler gear
88. Bearing
89. Support
90. Dust boot
91. Seal shield
92. Extension

apply a press to relieve pressure on Truarc snap ring (53). Remove snap ring and pull clutch cup from 1000 rpm shaft (46). Cover (37) can now be separated from housing (8) by removing the two short cap screws. See Fig. 237. Shafts and gears are now available for service. Brake (anti-creep) springs and pistons can be removed from rear side of housing as shown in Figs. 238 or 239. If control valve is to be disassembled, refer to Fig. 234 for early model tractors and to Fig. 235A for late model tractors. Disassembly will be obvious after an examination of the units and exploded views of valves. Inspect and renew any seals or "O" rings as necessary. Assemble by reversing disassembly procedure.

193. With unit disassembled, clean and inspect all parts and renew as necessary. Pay particular attention to clutch discs which should be free of scoring or warpage. Use all new "O" rings, seals and gaskets during reassembly.

While reassembly is the reverse of disassembly, the following points are to be considered during reassembly.

When installing rear cover (37 – Fig. 234 or 235) to housing (8), brake springs (20) will hold the sections apart so it will be necessary to use longer cap screws to pull the sections together. After sections are mated, original cap screws can be installed. Be sure step on end of idler shaft (85) is toward front of pto unit as shaft mates with edge of carrier (67).

If Teflon seal rings (52) used on hub of clutch cup are renewed, stretch seals on clutch cup. After the seals on clutch cup are installed, clamp a ring compressor around the seals to force them back to their original size.

When installing clutch piston (54) in clutch cup (51), be sure "O" ring is in inner bore of piston and outer seal (56) has lip facing away from piston hub. Use following methods to install piston in clutch cup. Obtain a piece of shim stock

four inches wide, 18 inches long and not more than 0.002 inch thick. Smooth all edges of shim stock to prevent any possibility of injuring piston outer seal, then roll shim stock and place it in clutch

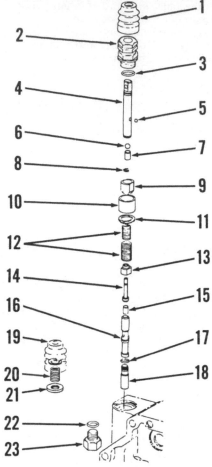

Fig. 235A – Exploded view of pto control valve.

1. Boot	13. Retainer
2. Body	14. Actuating rod
3. "O" ring	15. Pin
4. Stem	16. Spool
5. Ball (¼-inch)	17. "O" ring
6. Ball (11/32-inch)	18. Plug
7. Piston	19. Boot
8. Retaining ring	20. Spring
9. Retainer	21. Seal
10. Spacer	22. "O" ring
11. Washer	23. Plug
12. Springs	

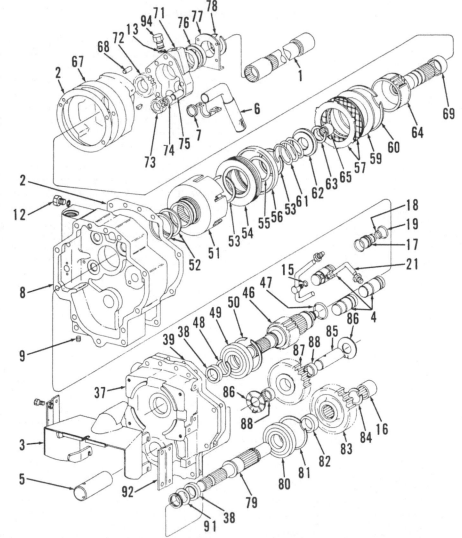

Fig. 236 – Oil pump assembly which is mounted on front end of carrier assembly.

6. Suction tube	72. Drive gear
67. Carrier	73. Idler gear
71. Pump housing	76. Bearing

Fig. 235 – Exploded view of dual speed pto unit used on all Model 786 and late production Models 886, 986 and 1086. Refer to legend in Fig. 234 except for following:

| 4. Anti-creep cylinders | 21. Tubes | 94. Regulator |

Fig. 237—View after cover is removed showing arrangement of gears and shafts for dual pto.

46. 1000 rpm shaft	85. Idler shaft
79. 540 rpm shaft	86. Thrust washer
83. 540 rpm gear	87. Idler gear

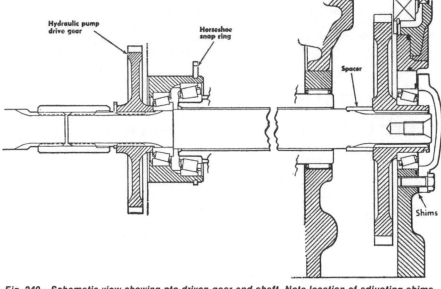

Fig. 240—Schematic view showing pto driven gear and shaft. Note location of adjusting shims.

Fig. 238—Rear side of housing showing brake pistons and springs for all models.

17. Piston	19. "O" ring
18. "O" ring	20. Spring

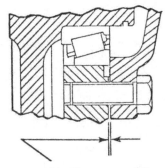

"Z" (Average of 3 readings)

Fig. 241—When adjusting bearing of pto driven gear, measure gap (Z) as shown. Refer to text for correct shim pack.

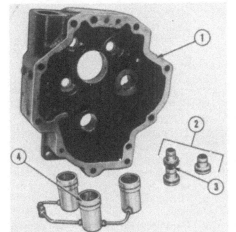

Fig. 239—Rear side of housing showing brake pistons and cylinders.

1. Housing	
2. Anti-creep brake pistons	3. Brake surface
	4. Cylinders

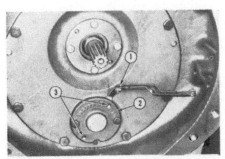

Fig. 242—Using a feeler gage to measure gap when adjusting pto shaft bearings. Refer to text.

1. Feeler gage   2. Pto bearing cover

cup and against bottom of cup. Lubricate piston and inside surface of shim stock liberally with oil, then start piston into clutch cup and as piston is moved into position, carefully maintain shim stock as nearly cylindrical as possible. Remove shim stock after piston is bottomed.

Clutch pack for early production models consists of 12 steel driven discs and five bronze drive discs. Late model pto clutch pack consists of nine steel driven discs and seven bronze drive discs. Install discs on early models as follows: Start with two steel discs, then add one bronze disc. Repeat this two-steel, one-bronze assembly until discs are installed. When properly assembled, clutch pack will start and end with two steel discs. Install discs on late models starting with one steel disc then one bronze disc. Repeat the one-steel, one-bronze assembly until discs are installed. When properly assembled, clutch pack will start and end with one steel disc. Install disc retainer (59) and snap ring (60). Use input shaft (64) to align discs.

Renew all bearings and seals as needed.

PTO DRIVEN GEAR

All Models

194. The pto driven gear, located at front of clutch housing, receives its drive from a hollow shaft which is splined to the backplate (cover) of the engine clutch. Removal of the pto driven gear requires that tractor be split between rear frame and clutch housing; at which time, gear and bearing assembly can be removed from bottom of clutch housing.

After tractor has been rejoined, bearing of pto driven gear should be checked, and if necessary, adjusted. Refer to Fig. 240 for a schematic view showing arrangement of pto shafts. To adjust bearing, use only two cap screws and install bearing retainer without shims or seal. Turn gear and shaft and tighten cap screws evenly to 10 in.-lbs. torque; then without turning gear and shaft, tighten cap screws to a torque of 20 in.-lbs. Now take three measurements around bearing retainer and average these readings.

See Figs. 241 and 242. Correct shim pack is gap (Z) plus 0.018-0.023 inch. Shims are available in thicknesses of 0.007 and 0.028 inch. Remove retainer, install shims and seal ring and tighten retainer cap screws.

HYDRAULIC LIFT SYSTEM

The hydraulic lift system provides load (draft) and position control in conjunction with a 3-point hitch. Load control is taken from the lower links and transferred through a torsion bar and sensing linkage to the main control valve located in the hydraulic lift housing. Torsion bar and sensing linkage are located in the rear main frame. The hydraulic lift housing, which also serves as the cover for the differential portion of the tractor rear main frame, contains the work cylinder, rockshaft, valving and the necessary linkage. Mounted on the top right side of the lift housing is the system relief valve and unloading valve along with control quadrant and levers. Also attached to the inside surface of the seat support, if tractor is equipped with an auxiliary system, are the auxiliary valves which control the hydraulic power to trailed or front mounted equipment. The pump which supplies the hydraulic system is attached to a plate which is mounted on left side of tractor rear main frame. The pump is driven by a gear located in the rear compartment of main frame on aft end of the pto driven shaft. All oil used by the hydraulic components of the tractor is drawn through a filter located in the rear main frame directly across from the hydraulic lift system pump.

All Model 786 and 886, Model 986 (S.N. 23999 & below and 28000 & above) and 1086 (S.N. 47999 & below and 55000 & above) are equipped with conventional open center hydraulic lift systems. Models 986 (S.N. 24000 through 27999) and 1086 (S.N. 48000 through 54999) are equipped with a hydraulic PFC (Pressure Flow Compensator) system.

The PFC system will provide flow and pressure required on demand by components in the hitch and/or auxiliary hydraulic system. This system will conserve fuel and horsepower until it is required to do work. The PFC system utilizes a five piston pressure and flow compensated axial piston pump located in the rear frame and driven by the pto shaft the same as all other models. Attached to the pump is a compensator block assembly which reads pressure signal messages from various components ordering variable flow rates to the pump.

All models are equipped with a supercharge circuit to prevent cavitation of the hydraulic pump during initial start up and high engine rpm in cool weather. This type of system routes all contaminants picked up from auxiliary couplers and remote cylinders or motors into the rear frame reservoir where they are picked up by the filter system. The supercharge system also prevents high volume hydraulic surges from entering the pump and filter area.

All Model 986 and 1086 tractors equipped with PFC systems, have a variable raise rate valve for draft control. It allows the operator to select a slow response setting for the draft control operation, while allowing a fast or slow raise when the position handle is actuated.

TROUBLESHOOTING

Models 786-886-986-1086 Equipped with Open Center Systems

195. The following are symptoms which may occur during the operation of the hydraulic lift system. By using this information in conjunction with the Check and Adjust information and the R&R and Overhaul information, no trouble should be encountered in servicing the hydraulic lift system.

1. Hitch will not lift. Could be caused by:
   a. Unloading valve orifice plugged or piston sticking.
   b. Unloading valve ball not seating or seat loose.
   c. Faulty main relief valve.
   d. Faulty cushion relief valve.
   e. Faulty or disconnected internal linkage.
2. Hitch lifts when auxiliary valve is actuated. Could be caused by:
   a. Unloading valve orifice plugged.
   b. Unloading valve piston sticking.
   c. Unloading valve ball not seating or valve seat loose.
3. Hitch lifts load too slowly. Could be caused by:
   a. Unloading valve seat leaking.
   b. Excessive load.
   c. Faulty main relief valve.
   d. Faulty cushion relief valve.
   e. Scored work cylinder or piston or piston "O" ring faulty.
4. Hitch will not lower. Could be caused by:
   a. Drop piston sticking or "O" ring damaged.
   b. Control valve spool sticking or spring faulty.
   c. Drop check valve piston sticking.

5. Hitch lowers too slowly. Could be caused by:
   a. Action control valve spool sticking.
   b. Action control valve linkage maladjusted.
   c. Drop check valve "O" ring damaged or pilot ball cage maladjusted.
6. Hitch lowers too fast with position control. Could be caused by:
   a. Action control valve malfunctioning.
7. Hitch will not maintain position. Could be caused by:
   a. Drop check valve in main control valve leaking.
   b. Work cylinder or piston scored or piston "O" ring damaged.
   c. Check valve pilot valve leaking or ball cage maladjusted.
   d. Cushion valve leaking or damaged.
8. Hydraulic system stays on high pressure. Could be caused by:
   a. Linkage maladjusted, broken or disconnected.
   b. Auxiliary valve not in neutral.
   c. Mechanical interference.
9. Hitch over-travels (depth variation). Could be caused by:
   a. Torsion bar linkage sticking and needs lubrication.
   b. Unloading valve orifice partially plugged.
10. Draft (load sensing too rapid in slow action position. Could be caused by:
    a. Action control linkage improperly adjusted.

Models 986-1086 Equipped with PFC systems

196. Problems which may occur during the operation of the hydraulic lift system and their possible causes are as follows:
1. Hitch leaks down – engine off. Could be caused by:
   a. Raise check poppet not seated.
   b. Drop poppet adjustment.
   c. Work cylinder piston "O" ring damaged or cracked draft control cylinder.
   d. Drop check poppet leaking past seat.
   e. Failed "O" ring under drop check valve locknut.
2. Hitch leaks down – engine on or off. Could be caused by:
   a. Damaged "O" ring on the drop poppet causing poppet to stick open.
   b. Damaged "O" ring on the drop cage.
   c. Damaged seal between valve and cylinder.
   d. Damaged "O" ring on work cylinder piston.
   e. Draft control cushion relief valve leaking.

4. Hitch will not lower. Could be caused by:
   a. Drop poppet misadjusted.
   b. Linkage misadjusted.
   c. Drop poppet pilot unseating rod broken.
   d. Drop rate screw turned in too far.
   e. Actuating rod or pin binding.
   f. Draft control valve spool binding in open position or spring damaged.

5. Hitch drops too fast. Could be caused by:
   a. Action valve misadjusted or piston sticking open.

6. Hitch chatters during lowering. Could be caused by:
   a. Excessive load on hitch.
   b. Damaged "O" ring between action valve and draft control valve.

7. Hitch raises slow and/or low pressure. Could be caused by:
   a. Variable raise rate control misadjusted.
   b. "O" ring in high pressure circuit damaged.
   c. Draft control cushion relief valve or cushion relief valve in motor priority valve leaking.
   d. Draft control cylinder scored or "O" ring on piston damaged.
   e. Motor priority valve spool sticking.
   f. Leak in signal circuit.
   g. Leak in pressure circuit.
   h. Motor valve signal check valve malfunction.

8. Draft response too fast. Could be caused by:
   a. Variable raise rate spool linkage misadjusted or return spring broken.
   b. Pump differential pressure setting incorrect.

9. Hitch raises when auxiliary circuit is used. Could be caused by:
   a. Draft control signal check valve stuck open or damaged.

10. Unable to raise load. Could be caused by:
    a. Hitch signal check valve stuck closed.

11. Hitch lowers very slowly. Could be caused by:
    a. Misadjusted position control linkage.
    b. Action control valve spool sticking.
    c. Drop poppet sticking.

12. Hitch raises or lowers but doesn't maintain position. Could be caused by:
    a. Leaking "O" ring in work cylinder piston.
    b. Scored work cylinder.
    c. Draft control cushion relief valve "O" ring damaged.

13. Hydraulic system stays on high pressure. Could be caused by:
    a. Lift linkage broken or misadjusted.
    b. Internal mechanical interference.

TEST AND ADJUSTMENT

Models 786-886-986-1086 Equipped With Open Center Systems

Before proceeding with any testing or adjusting, be sure the hydraulic pump is operating satisfactorily, hydraulic fluid level is correct and filter is in good condition. All tests should be conducted with hydraulic fluid at operating temperature which is normally 120-180 degrees F. Cycle system if necessary to insure that system is completely free of air.

197. **RELIEF VALVE.** On tractors equipped with an auxiliary hydraulic system, the relief valve can be tested as follows: Attach an IH Flo-Rater, or similar flow rating equipment, to any convenient outlet from an auxiliary control valve and be sure outlet hose from test unit is securely fastened in the hydraulic system reservoir. Start engine and run at rated speed. Manually hold auxiliary control valve in operating position, close valve of test unit and note the gage reading which should be 2000 to 2350 psi for Models 786, 886 and 986 or 2250 to 2600 psi for Model 1086. If pressure is not as stated, renew relief valve which is available only as a unit.

198. **PUMP FLOW RATE.** At this time, the hydraulic lift system pump delivery can be checked. At 2400 rpm the hydraulic lift pump should deliver 11.5 to 14 gpm at 2250 psi.

199. **AUXILIARY CONTROL VALVE.** To check the auxiliary control valve detent assembly, run engine at low idle rpm, pull control valve lever into operating position until it latches, then slowly close shut-off valve of test unit and observe the pressure gage. Valve control lever should unlatch and return to neutral at not less than 1650 psi nor more than 2000 psi for Models 786, 886 and 986 or not less than 1900 psi nor more than 2250 psi for Model 1086. If detent assembly does not operate properly, refer to paragraph 234.

Models 986-1086 Equipped With PFC Systems

200. **RELIEF VALVE.** Before proceeding with any testing or adjustments, be sure minimum pressure setting is above pump cut-off pressure and the pump meets all minimum performance standards. Refer to paragraph 204 for PFC Pump Flow Rate Test. Connect Flo-Rater inlet hose to top coupler port. Connect outlet hose from Flo-Rater to the bottom coupler port (Fig. 245) and open load restrictor on Flo-Rater. Lock motor control valve in run position to prevent unlatching and position flow control sector in fast or down position. See Fig. 246. Check proper hydraulic fluid level and allow fluid to reach an operating temperature of 100 to 150 degrees F. With engine operating at rated speed of 2400 rpm, close load valve on Flo-Rater and read pressure. Main system cushion relief valve should read 2450 to 2700 psi. If readings cannot be obtained, renew relief valve.

201. **AUXILIARY VALVE UNLATCHING PRESSURE.** With Flo-Rater connected as described in paragraph 200 and shown in Fig. 245, proceed with test as follows: Check for

Fig. 245 — View showing hook-up of Flo-Rater testing equipment into auxiliary or motor valve coupler.
1. Motor valve coupler
2. Flo-Rater
3. Lead (restrictor) valve

proper hydraulic fluid level and allow fluid to reach an operating temperature of 100 to 150 degrees F. Operate engine to rated speed of 2400 rpm. Shift valve control being tested to lower position and set flow control sector to fast or down position (Fig. 246). Restrict Flo-Rater and note unlatching pressure, which should occur between 2000 to 2450 psi.

IMPORTANT: Before checking unlatching pressure with control handle in the up position, reverse the Flo-Rater inlet and outlet hoses in coupler.

If specified pressure cannot be obtained, remove auxiliary valve dust cap which covers adjusting screw. Turn screw in for pressure increase, out for pressure decrease. Recheck pressure.

202. At this time the motor and auxiliary valve flow sector can be checked.

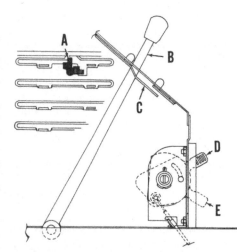

Fig. 246 — View showing motor or auxiliary valve control handle and flow control sector (PFC system) showing positions for checking flow rates and unlatching pressures. Refer to text.

A. Motor valve tab lock
B. Control valve handle
C. Neutral lock
D. Flow control sector (slow flow position)
E. Flow control sector (fast flow position)

Fig. 247 — View of Flow Control Valve Sector (PFC system) showing linkage and adjustment points. Refer to text.

1. Cable adjusting locknut
2. Cable
3. Valve

With flow sector control in the fast or down position, flow should be 16.5 gpm at rated engine rpm. With flow control sector in the slow or up position, reading should be 3 gpm for all PFC equipped models. If specified readings cannot be

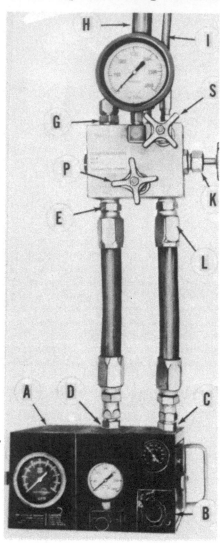

Fig. 248 — Component parts of 14-557 PFC Hydraulic System Analyzer and IH 14-51D Flo-Rater used for checking PFC systems on Models 986 and 1086 tractors so equipped.

A. 14-51D Flo-Rater
B. Load (restirctor) valve
C. Inlet
D. Outlet
E. Return in
G. Oil return
H. Pump pressure (flow-in)
I. Signal line
S. "S" signal pressure valve
K. Adjustable orifice
L. Flow out

obtained, adjust flow control sectors as follows: Loosen locking screw (1–Fig. 247) on cable of sector being tested. Observe Flo-Rater readings while sliding cable forward or backward until specified flow is obtained. Tighten locking screw. If flows are not attainable, check for bent linkage cable or a faulty sector valve. Inspect and renew if necessary.

203. **PFC HYDRAULIC ANALYZER.** To proceed with further testing of the Pressure Flow Compensator System, the 14-557 PFC Analyzer must be connected to hydraulic system as follows: Connect PFC Analyzer to Flo-Rater as shown in Fig. 248. Connect a return hose from the return (out) port of Analyzer (G–Fig. 248) to the top port of motor valve coupler. Remove service plug from pump flange assembly and install the 14-554-5 pump plug adapter. See Fig. 249. Install the 14-74-22 tube fitting (B–Fig. 250). Disconnect and plug signal line from flange (C and D–Fig. 250). Connect the "P" pump pressure hose to service port fitting and the "S" signal pressure hose to the signal port fitting. See Figs. 251 and 252 for completed hook-up.

204. **PUMP FLOW RATES.** With 14-557 PFC Hydraulic System Analyzer (Fig. 248) and Flo-Rater installed as described in paragraph 203, open all restrictors in PFC Analyzer and Flo-Rater. Lock motor control valve in run position and flow control sector in fast or down position. See Fig. 246. Check for proper hydraulic fluid level and allow fluid to reach an operating temperature of 100 to 150 degrees F. Operate engine at rated speed of 2600 rpm. Close signal (S–Fig. 248) valve on PFC Analyzer and restrict the Flo-Rater until a reading of 2200 psi is reached on Analyzer gage. The flow should be 16.5 gpm. Now close restrictor valve completely on Flo-Rater and read pump cut-off pressure. Readings should be 2450 to 2700 psi. If this reading is not obtained, run relief valve test. Now close adjustable orifice valve on PFC Analyzer and read pump standby pressure. Pressure should be 300 to 500 psi. If standby pressure is not obtained, renew pump and compensator block.

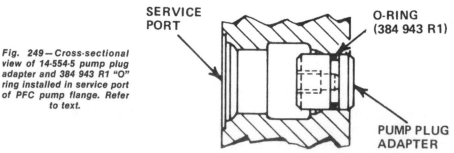

Fig. 249 — Cross-sectional view of 14-554-5 pump plug adapter and 384 943 R1 "O" ring installed in service port of PFC pump flange. Refer to text.

SERVICE PORT

O-RING (384 943 R1)

PUMP PLUG ADAPTER

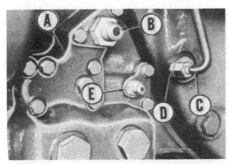

Fig. 250—View of PFC pump flange assembly showing installation of 14-74-22 test fitting. Refer to text.

A. Pump flange
B. 14-74-22 (1 1/16 inch 12 "O" ring boss x ⅝-inch tube JIC)
C. Signal line
D. Plug
E. Signal line fitting

Next open Analyzer "S" valve and adjustable orifice, then close valve "P." Adjust Flo-Rater load valve until a pressure reading of 780 to 820 psi is obtained on Analyzer gage. Close "S" valve and open "P" valve three turns. Adjust orifice valve on Analyzer until 1000 psi reading is obtained on Analyzer gage. After pressures for both "S" and "P" valves has been established, alternate opening and closing valves "S" and "P." Note that when "S" valve is open and "P" valve is closed a reading of 780 to 820 psi should resgister on Analyzer gage. When "S" valve is closed and "P" valve open the Analyzer gage should read 1000 psi. The pump differential pressures between "S" and "P" valves

should be 180 to 220 psi. Pump flow should be 16.5 gpm. If differential pressures cannot be obtained renew pump and rerun tests.

CONTROL LINKAGE ADJUSTMENT

All Models

205. **QUADRANT LEVERS.** The draft control lever should require five to ten pounds of force to pull from "LIGHT" to "HEAVY" setting. The position control lever should require four to eight pounds of force to pull from lower or forward position towards the rear. The draft and position control levers are adjusted by tightening or loosening cap screws in bearing support.

On Models 986 and 1086 equipped with PFC, a third lever (raise rate lever) requires one to three pounds of force to move from slow to fast position and is adjusted by tightening or loosening adjusting screw under the lever.

206. **DRAFT CONTROL OUTSIDE LEVER.** Place position control lever forward at offset, then back up ⅛-inch. Set draft control lever to full forward (HEAVY) position. There must be a ½ to 1 inch gap between front of draft control lever and front edge of slot in quad-

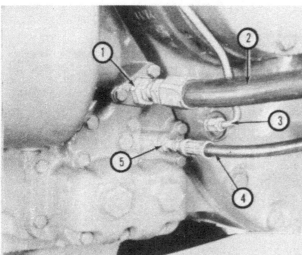

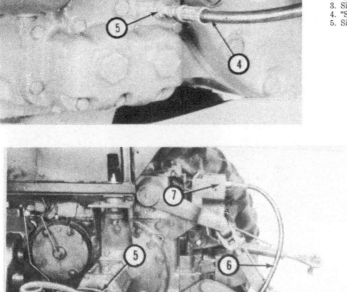

Fig. 251—View showing pump flange with PFC Analyzer connected to service and signal ports. Refer to text.

1. 14-74-22 tube fitting
2. "P" pump pressure line
3. Signal line plugged
4. "S" signal line from analyzer
5. Signal line adapter fitting

Fig. 252—View of tractor rear frame section with PFC Analyzer and Flo-Rater hook-up. Wheel is removed to show complete set-up.

1. 14-51D Flo-Rater
2. Load (restrictor) valve
3. 14-557 PFC Hydraulic Analyzer
4. "P" pump pressure line
5. "S" signal pressure line
6. Return hose to motor coupler top port
7. Motor coupler top port

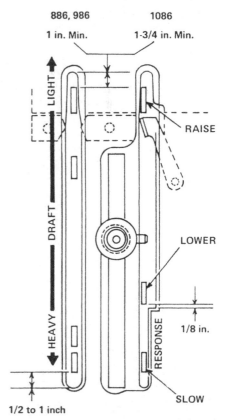

Fig. 253—Quadrant lever adjustment dimensions for draft and position controls. PFC models use a third lever (raise rate not shown) for position control response. Refer to text.

rant. Adjust linkage under platform as necessary.

With hitch unweighted, start engine and run at 1000 rpm. Move draft control lever slowly rearward until hitch just starts to raise.

NOTE: Hitch will raise all the way up.

Distance from rear edge of draft control lever to rear of slot in quadrant should be 1¾ to 2¼ inches. See Fig. 253. If dimension is incorrect, adjust internal sensing arm as follows:

With hitch unweighted, check adjustment of torsion bar stops (Figs. 254 or 255) and adjust if necessary until head of stop screws and left and right crank arms is as shown. Move position control lever forward to "LOWER" position (at offset in quadrant before action control sector). Remove upper link bracket. Loosen locknut on sensing arm and turn screw in until it bottoms. Place draft control lever two inches from rear of slot in quadrant. Start engine and turn sensing arm adjusting screw out (counter-clockwise) until hitch starts to raise. tighten locknut and reinstall upper link anchor bracket.

207. POSITION CONTROL LEVER. With hitch loaded with approximately 500 lbs. weight (do not use an implement), run engine at 1000 rpm.

NOTE: Implement cannot be used because it will not allow hitch to lower fully.

Set draft control lever in full forward position. Move position control lever midway in slot, then slowly move forward until hitch lowers completely. If

Fig. 255—Torsion bar adjustment for Models 786, 886 and 986.

position control lever is not at offset on console, loosen adjusting nut under console and position handle at offset. Retighten nut.

208. LIFT HEIGHT. To adjust lift height proceed as follows: Move position control lever to rear of slot and measure height "A" as shown in Fig. 256. "A" should be $39^3/_4$ inches for Models 786, 886 and 986 or 39 3/16 inches for Model 1086. If height correction is needed, loosen position adjusting screw nut and turn screw (S–Fig. 259) in or out until it contacts the internal lever with hitch at correct height.

209. DROP CHECK VALVE ADJUSTMENT. Connect a 3000 psi pressure gage to the service port of the pump flange. See Fig. 257. Start tractor and attach a weight or implement which will require 1000 psi of pump pressure to lift. Set draft control in "HEAVY" position. Move position control lever in the extreme lower position so that weight is off the hitch and record reading on gage. Set position control lever in raise position about halfway in the slot then move back toward the lower position slowly until hitch begins to lower. Note gage pressure reading. If gage reading is the same as the first reading then remove front plate from rear frame cover and adjust drop check valve as follows:

Loosen locknut (Fig. 258) and turn the drop valve clockwise in ¼-turns until the pressure gage reads 200 psi above the first reading, then turn valve counterclockwise ¼-turn and tighten locknut.

IMPORTANT: Always tighten drop valve locknut after each adjustment otherwise fluid leakage past the valve "O" ring will register on the gage and result in a false reading.

If gage pressure registers higher than the first test reading then adjust as follows: Loosen locknut and turn drop valve counter-clockwise in ¼-turns until gage reads same as first reading. Tighten locknut and repeat test and observe pressure readings.

Fig. 256— View of hydraulic hitch showing lift height dimension (A). Refer to text.

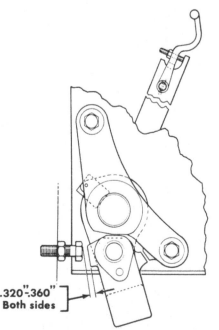

.320"-360" Both sides

Fig. 254—Torsion bar adjustment for Model 1086.

Fig. 257— View showing 3000 psi gage installed in PFC pump flange for testing drop check valve adjustment. Refer to text.
1. 14-74-22 tube fitting
2. 3000 psi gage
3. 14-513-4 male reducer
4. 14-513-3 female reducer
5. 05-70-16 female coupler

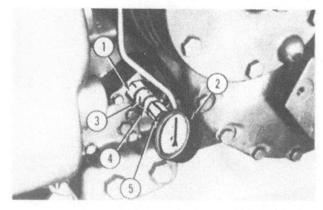

Models 786-886-986-1086 Equipped With Open Center Systems

210. POSITION CONTROL AND CYCLE TIME. With the above tests and adjustments made, hitch can be final checked for accuracy of position and the raise and lower times.

To check for position control accuracy, place a mark about midway of the quadrant, start engine and raise hitch by moving position control lever to rear of quadrant. Move position control lever forward to the affixed mark and measure distance from ends of lower links to ground. Repeat the operation and again measure the distance from ends of lower links to ground. These two measurements should not vary more than ⅛-inch. Now push position control lever forward to lower hitch, then move it rearward to the affixed mark. The measurement between ends of lower links and ground should not vary more than one inch from the measurements obtained in the first test. If differences are excessive, refer to paragraph 214.

211. To check hitch raise and lower times, be sure hydraulic fluid is at operating temperature and load hitch with an implement or weights. To check raise time, start engine and run at high idle rpm. Be sure hitch is in lowest position, then quickly move position control lever rearward to raise position. Hitch should reach full raise in three seconds or less.

212. To check the minimum lowering (drop) time, move the load control lever to the light load (rear) position and the position control lever forward to offset of quadrant. Now move the load control

lever forward to the heavy position. Hitch should completely lower in two seconds or less.

213. To check the maximum lowering (drop) time, repeat the minimum time operation given in paragraph 212 except the position control lever should be in the extreme forward slow action position. Hitch drop time should be six seconds or more.

214. If position control accuracy and

the hitch raise and lower times are not satisfactory, remove lift housing from rear main frame and inspect internal linkage. A visual inspection is sufficient to find any linkage defects. Renew any linkage or pins showing excessive wear or damage. Also check the main valve return spring and the main valve spool for binding. Spring can be renewed; however, if valve spool is defective, renew the spool and body assembly.

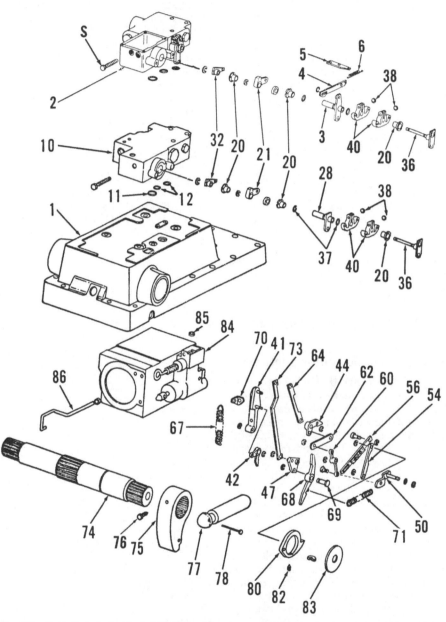

Fig. 258—With cover removed from front of lift housing, units shown are available for adjustment or service. (A) is Esna nut on forward end of action control valve actuating rod.

Fig. 259—Exploded view of hydraulic lift housing. Items 2 through 6 are for Models 986 and 1086 equipped with PFC system.

1. Lift housing	28. Position control shaft	73. Link
2. Motor priority valve (PFC)	32. Position control support	74. Rockshaft
3. Position control shaft (PFC)	36. Draft control shaft	75. Bellcrank
4. Walking beam (PFC)	37. "O" rings	76. Set screw
5. Spring retainer (PFC)	38. Washers	77. Connecting rod
6. Spring (PFC)	40. Control supports	78. Pin
10. Unloading valve body	41. Support	80. Cam
11. "O" ring	42. Lever	82. Set screw
12. "O" rings	44. Bellcrank	83. Retainer
20. Bearing	47. Draft control eccentric	84. Cylinder and valve assembly
21. Position control lever	50. Position control eccentric	85. Seal ring
	54. Link	86. Lubrication tube
	56. Link	
	60. Walking beam	
	62. Link	
	64. Lower rod	
	67. Balance spring	
	68. Sensing pickup arm	
	69. Pin	
	70. Spring support	
	71. Return spring	

Models 986-1086 Equipped With PFC Systems

215. **VARIABLE RAISE RATE CONTROL.** Place load on hitch allowing full travel of hitch from full lower to full upper positions. Disconnect the override linkage. With variable raise rate lever in slow position against the stop in the quadrant slot, run the engine at 2400 rpm. Make sure hydraulic fluid is at a temperature of 100 to 150 degrees F. With draft lever in off position, using the position lever, raise hitch to full upper position and then back to full lower. Record time of full cycle and average times of three full cycles. If average time is less than fifteen seconds, adjust slow speed control as follows: Loosen locknut on flow control spool and turn set screw on control spool **out** ¼-turn and repeat test. Turn screw out until a time of fifteen seconds or more is reached. Retighten locknut.

Place raise rate handle in fast position, and with draft control off, raise and lower hitch with position control lever 3 times. Average time should be not greater than 3 seconds. If time is more than 3 seconds, adjust fast control as follows:

Loosen locknut on flow control spool and turn screw **in** ¼-turn until a speed of three seconds or less is reached. Retighten locknut.

HYDRAULIC LIFT HOUSING

All Models (PFC and Open Center Systems)

216. **R&R AND OVERHAUL.** Disconnect battery and electrical connectors at rear frame cover. Remove fuel tank. Disconnect pto linkage, brake lines and hydraulic tubing from auxiliary couplers. Remove link pins from hitch arms and lower arms. Remove oil dipstick. For tractors equipped with open center systems disconnect hydraulic lines from draft control valve. For tractors with PFC, disconnect motor priority valve linkage and hydraulic tubing. Making sure entire housing cover area is clean, remove cover mounting bolts and with a suitable sling and hoist, remove lift housing cover. If available, mount housing cover on a suitable engine stand to avoid damage to control and sensing linkage.

217. Disconnect and remove sensing (return) spring (71 – Fig. 259) from pickup arm (68). Disconnect the draft control lever (73) lower linkages and remove control lever. Remove retainer and pin on valve spool linkage (56). Remove draft control sensing pick-up arm (68) from cylinder. Remove clip holding walking beam (60) to position control eccentric (50). Remove cap screw holding

stop support (41) to cylinder and remove linkage assembly from housing. Remove cap screws and lift out cylinder and valve assembly (84).

When reinstalling cylinder and valve assembly in housing, install the mounting cap screws loosely; then use a small pry bar to hold cylinder assembly against mounting boss in housing and tighten the mounting cap screws securely. Complete reassembly by reversing the disassembly procedure and adjust system as outlined in paragraphs 205 through 215.

218. **WORK CYLINDER AND PISTON.** To disassemble the cylinder assembly after lift housing is removed as outlined in paragraph 216, first remove the lubrication tube, then remove piston (8 – Fig. 260) by bumping open end of cylinder against a wood block. Cushion valve (11) and the check valve assembly (items 4, 5, 6 and 7) can also be removed. If necessary, lever pin (3) can also be renewed.

Inspect cylinder and piston for scoring or wear. Small defects may be removed using crocus cloth. Pay particular attention to the piston "O" ring and back-up ring as well as the check valve poppet. If

any doubt exists as to the condition of these items, be sure to renew them during assembly. Cushion valve (11) can be bench tested by using a hydraulic hand pump, gage, test body IH No. FES 64-7-1, adapter IH No. FES 64-7-2 and petcock IH No. FES 64-7-4. Valve should test 1600-1800 psi. Cushion valve adjusting screw is heavily staked in position and valve cannot be disassembled. If valve does not meet the above specifications renew the complete valve.

NOTE: When checking valve with hand pump bear in mind that the relief pressure obtained will be on the low side of the pressure range due to the low volume of oil being pumped.

Coat parts with IH "Hy-Tran" fluid before reassembly. Piston "O" ring is installed nearest closed end of piston.

219. **MAIN CONTROL VALVE.** To disassemble the main control valve after lift housing is removed as outlined in paragraph 216, first remove plug (30 – Fig. 260), spring (29) and ball (28) from end of check valve ball seat. Loosen locknut (35), then while counting the number of turns, remove the valve seat. Remove snap ring (20) and pull

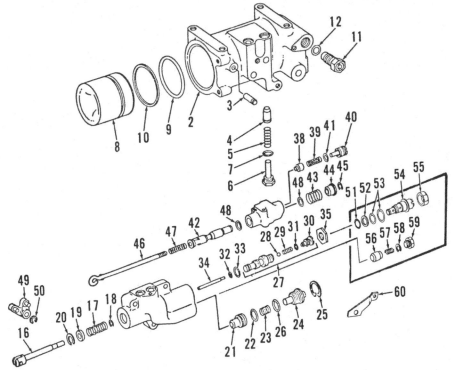

Fig. 260 — Exploded view of work cylinder, action control valve and main control valve assemblies. Items 51 through 56 are for Models 986 and 1086 equipped with PFC.

2. Cylinder	19. Spring retainer	32. "O" ring	47. Spring
3. Lever pin	20. Snap ring	33. "O" ring	48. "O" ring
4. Check valve poppet	21. Drop poppet valve	34. Actuating rod	49. Bellcrank
5. Spring	22. "O" ring	35. Locknut	50. Snap ring
6. Plug	23. Spring	38. Piston	51. "O" ring
7. "O" ring	24. Valve plug	39. Spring	52. Washer
8. Piston	25. Snap ring	40. Plug	53. "O" rings
9. "O" ring	26. "O" ring	41. "O" ring	54. Drop valve assy.
10. Back-up washer	27. Drop valve seat assy.	42. Variable orifice spool	55. Nut
11. Cushion valve	28. Ball	43. Return spring	56. Poppet
12. "O" ring	29. Spring	44. Bushing	57. Spring
16. Spool actuator	30. Plug	45. Snap ring	58. "O" ring
17. Spring	31. "O" ring	46. Actuating rod	59. Plug
18. Retaining ring			60. Spring anchor

spool assembly (16) from valve body. Compress return spring, remove snap ring (18) and disassemble spool assembly. Remove snap ring (25) and pull plug (24), spring (23) and poppet valve (21) from valve body.

Clean and inspect all parts. Spool return spring (17) has a free length of 2-21/64 inches and should test 18.4-21.6 pounds when compressed to a length of 1-29/32 inches. Check ball spring (29) has a free length of 59/64-inch and should test 3.5-4.1 pounds when compressed to a length of ¾-inch. Drop poppet valve spring (23) has a free length of 1-11/32 inches and should test five pounds when compressed to a length of 63/64-inch. If valve spool or spool bore show signs of scoring or excessive wear, renew complete valve assembly. All other parts are available for service.

Use all new "O" rings, dip all parts in IH "Hy-Tran" fluid and reassemble by reversing the disassembly procedure, keeping the following points in mind. Retainer (19) is placed on actuator (16) with relieved (chamfered) side toward spring (17). Drop valve seat is turned into valve body the same number of turns as were counted during removal. Tighten the locknut (35) to 45 ft.-lbs. torque and plug (30) to 10 ft.-lbs. torque.

After assembly, mount valve on cylinder and piston assembly.

220. ACTION CONTROL VALVE. To disassemble the action control valve

after lift housing is removed as outlined in paragraph 216, count the turns and remove the Esna nut from end of actuating rod (46–Fig. 260) and remove rod and spring (47). Remove retainer ring (45), then remove bushing (44), return spring (43) and variable orifice spool (42). The "O" ring (48) is located in I.D. of bushing bore. Remove plug (40), spring (39) and the drop control piston (38). Remove "O" ring from I.D. of orifice spool bore.

Clean and inspect all parts and renew any which show signs of excessive scoring or wear. Spool (42) and piston (38) should be a snug fit yet slide freely in their bores.

Use all new "O" rings, dip all parts in IH "Hy-Tran" fluid and reassemble by reversing the disassembly procedure. Count the turns when installing the

Esna nut on end of actuating rod so nut will be installed to original position.

After assembly, mount action control valve on the main control valve.

221. ROCKSHAFT. Renewal of rockshaft or rockshaft bushings requires removal of the hydraulic lift housing; however, if only the rockshaft seals are to be renewed, it is possible to pry seals out of their bores after removing the rockshaft lift arms. Seals are installed with lips toward inside and are driven into bores until bottomed. Dip seals in IH "Hy-Tran" fluid prior to installation.

With the lift housing removed, the rockshaft can be removed as follows: Remove set screws (76 and 82–Fig. 259) and slide rockshaft out right side of housing. Bushings can now be driven out of housing.

Rockshaft O.D. at bearing surface is larger on right side than on left. Rockshaft bearing surface O.D. is 3.020-3.022 inches for right side and 2.780-2.782 inches for left side. Rockshaft bushing I.D. is 3.023-3.028 inches for right side and 2.783-2.788 inches for left side. Rockshaft has 0.001-0.008 inch operating clearance in bushings. Use a piloted driver when installing bushings and install bushings with outer ends flush with bottom of seal counterbore.

When installing rockshaft, start rockshaft in right side of housing and align master splines of rockshaft, actuating hub (80) and bellcrank (75). Tighten hub and bellcrank set screws.

222. TORSION BAR ASSEMBLY. To remove the torsion bar assembly, it is necessary to drain rear main frame and remove pto unit, if so equipped. Refer to Figs. 265 and 266 and proceed as follows: Disconnect and remove hitch and drawbar support bracket assembly. Remove "U" bolt and sensing arm assembly. Remove cap screws in left end retainer of torsion bar tube and pull off left crank arm. Remove right hand torsion bar anchor bracket. Remove torsion bar and torsion bar tube from right side of rear frame.

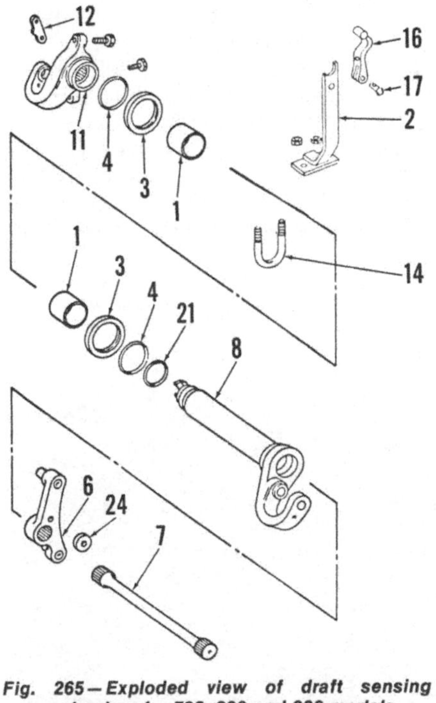

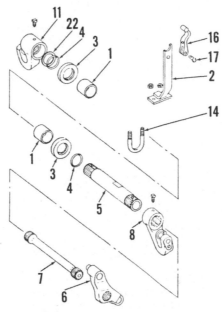

Fig. 266—Exploded view of draft sensing mechanism for 1086 models.

1. Bushing
2. Lower sensing arm
3. Retainer
4. "O" ring
5. Shaft
6. Anchor bracket
7. Torsion bar
8. Crank arm RH
11. Crank arm LH
14. "U" bolt
16. Upper sensing arm
17. Pin
22. Shim (0.054 in.)

Fig. 267—Exploded view of unloading valve assembly for all models equipped with open center system. Items 18 through 22 are for sequential lift.

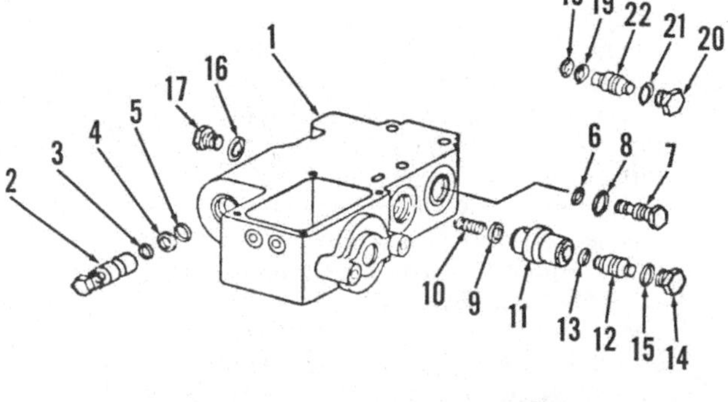

1. Valve body
2. Relief valve
3. "O" ring
4. Washer
5. "O" ring
6. "O" ring
7. Simultaneous lift plug
8. "O" ring
9. Washer
10. Spring
11. Unloading body
12. Unloading valve piston
13. "O" ring
14. Plug
15. "O" ring
16. "O" ring
17. Plug
18. "O" ring
19. "O" ring
20. Plug
21. "O" ring
22. Sequential raise valve

Fig. 265—Exploded view of draft sensing mechanism for 786, 886 and 986 models.

1. Bushing
2. Lower sensing arm
3. Retainer
4. "O" ring
6. Anchor bracket
7. Torsion bar
8. Crank arm and shaft (RH)
11. Crank arm (LH)
12. Retainer
14. "U" bolt
16. Upper sensing arm
17. Pin
21. "O" ring
24. Retainer

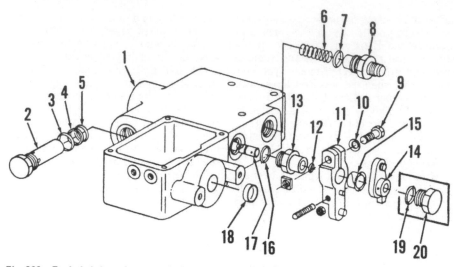

Fig. 268—Exploded view of motor priority valve for Model 986 and 1086 tractors equipped with PFC system. Items (19) and (20) in insert are for models without draft control.

1. Body	8. Motor spool plug	14. Lever (raise speed)
2. Relief valve	9. Bolt	15. Bearing (nylon)
3. "O" ring	10. Washer	16. "O" ring
4. Washer	11. Friction lever	17. Spool (variable raise)
5. "O" ring	12. "O" ring	18. Back-up plug
6. Spring	13. Retainer (variable	19. "O" ring
7. "O" ring	raise)	20. Plug

IMPORTANT: Do not chip, dent or scratch torsion bar or torsion bar failure may result.

On Models 786, 886 and 986, remove torsion bar anchor bracket and slide torsion bar out of tube. On Model 1086, remove torsion bar bracket, then remove roll pin and retainer plate from right crank arm. Remove torsion bar from tube. Bushing case and bushing, "O" rings and "O" ring retainers can now be renewed and procedure for doing so is obvious.

Reassemble by reversing the disassembly procedure.

Models 786-886-986-1086 Equipped With Open Center Systems

223. **UNLOADING VALVE ASSEMBLY.** Located at the right rear of lift housing cover is the unloading valve assembly. See Fig. 267. Contained within are the system relief valve (2), unloading valve (11 and 12), either a sequential raise valve (22) or simultaneous raise plug (7) and the draft and position lever control linkage. In most cases, these components can be removed for service without removing the entire unloading valve assembly. To remove the unloading valve, remove plug (14), unloading valve body and piston (11 and 12), washer (9) and spring (10). If renewal is required, valve must be renewed as an assembly only.

Relief valve (2) can be removed easily at any time. Relief valve adjustment plug is heavily staked and cannot be removed. Faulty relief valves must be renewed.

Check the unloading valve piston (12) as follows: Shake piston. Piston must not rattle nor should rod be loose or easy to turn. Place unit in a soft jawed vise and compress at least 0.045 inch. Blow through orifices in piston to insure that they are open and clean. If piston does not meet all of these conditions, renew the complete unit. If removal of entire unloading valve assembly is necessary, proceed as follows: Remove fuel tank. Disconnect draft and position control linkages and hydraulic lines at unloading valve assembly. Remove cover plate and disconnect the draft and control internal links from linkage going to inside of lift housing. Remove mounting bolts (one mounting bolt is inside unloading valve housing) and remove assembly from

rear frame. Clean and inspect linkage for wear or damage. Renew parts as necessary. Assembly is the reverse of removal procedure.

Models 986-1086 Equipped With PFC Systems

224. **MOTOR PRIORITY VALVE.** The motor priority valve assembly (Fig. 268) is located at the right rear of lift housing, same as the unloading valve assembly on Series 86 tractors equipped with open center systems. Contained within are the system relief valve, priority valve spool and linkage connecting the quadrant control levers to the internal position and draft links inside the lift housing. In addition to the draft and position control linkage on PFC equipped tractors is a variable raise rate control lever.

To service components in the motor priority valve assembly proceed as follows: Remove fuel tank. Disconnect control linkages (2—Fig. 269). Remove rear cover plate (5) and disconnect draft and position control links inside housing. Remove hydraulic lines (3). Remove mounting bolts and lift assembly from rear frame.

To remove variable raise rate spool, disconnect linkage to variable raise control arm. Loosen the cap screw and remove control arm from spool. Now remove variable raise rate spool. Clean and inspect spool for scratches or burrs.

The system relief valve can be removed and serviced easily without removal of priority valve assembly. Relief valve adjusting plug is heavily staked and cannot be removed. Faulty relief valves must be renewed.

Reassembly is the reverse of disassembly procedure. After assembly, adjust variable raise rate linkage as described in paragraph 215. Adjust draft

Fig. 269—View showing hydraulic line hook-ups and control linkage for motor priority valve assembly for Model 986 and 1086 tractors equipped with PFC system. Refer to text.

1. Motor priority valve
2. Control linkages
3. Hydraulic lines
4. 100 cu. in. attenuator
5. Cover plate

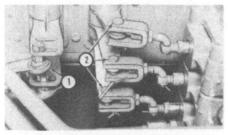

*Fig. 270 — View of right underside of control plat-
form showing draft, position and auxiliary valve
linkage.*

1. Draft and position
   linkage
2. Auxiliary valve linkage

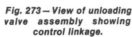

*Fig. 273 — View of unloading
valve assembly showing
control linkage.*

1. Cover plate
2. Position control
3. Draft control

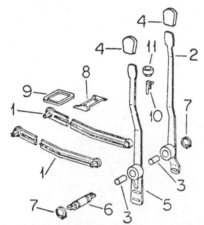

*Fig. 271 — Exploded view of draft and position
control linkage and quadrant lever assembly on
786, 886, 986 and 1086 models equipped with
open center systems.*

1. Draft and position
   rods
2. Position control
   handle
3. Bushing
4. Knob
5. Draft control handle
6. Pivot shaft
7. Snap ring
8. Support
9. Seal
10. Position handle stop
11. Knob

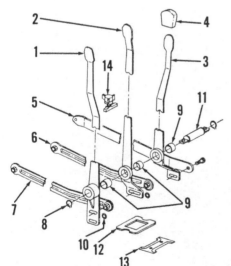

*Fig. 272 — Exploded view of draft, position and
raise rate control linkage and quadrant levers on
Models 986 and 1086 equipped with PFC.*

1. Draft control handle
2. Position control
   handle
3. Raise rate control
   handle
4. Knob
5. Raise rate rod
6. Position rod
7. Draft control rod
8. Retainer
9. Bushing
10. Washer
11. Pivot shaft
12. Seal
13. Support
14. Position handle stop

and position levers as described in paragraph 205.

Models 786-886-986-1086 Equipped With Open Center Systems

225. QUADRANT AND LEVER AS-
SEMBLY. The quadrant and lever assembly can be removed as follows: Remove knobs from control levers. Remove control lever support cover. On
underside of control platform, remove
cap screws securing lever bushings and
pivot shafts. See Fig. 270. Remove fuel
tank. Disconnect draft and position control linkage (1 – Fig. 271) from unloading
valve assembly (2 and 3 – Fig. 273). Remove cover plate from unloading valve
assembly and disconnect the draft and
position links from inside the unloading
valve housing. Drive roll pin from draft
control link and remove. Be very careful
not to drop the two "C" clips and roll pin
into lift housing. Remove necessary
mounting bolts and pull bearing support
along with the draft control inner shaft
out of unloading valve assembly. Remove position control bearing support
and shaft from unloading valve
assembly. Inspect shafts, bearings and
"O" rings and renew if necessary.

Reassemble by reversing disassembly
procedure. Be sure to align all marked
shafts and splines and adjust friction
tension on levers as outlined in
paragraph 205.

Models 986-1086 Equipped With PFC Systems

226. QUADRANT AND LEVER AS-
SEMBLY. Service of the quadrant and
lever assemblies on tractors equipped
with PFC is basically the same as open
center system models with the exception
of the variable raise rate lever and cor-
responding linkage to the variable raise
rate spool. See Fig. 269. An additional
lever is mounted on the control
quadrant. See (3 – Fig. 272). After dis-

assembly, inspect all shafts, bearings
and "O" rings and renew as necessary.

Reassemble by reversing disassembly
procedure. Be sure to align all marked
shafts and splines and adjust tension on
levers as outlined in paragraph 205. Adjust variable raise rate spool as outlined
in paragraph 215.

HYDRAULIC LIFT PUMP

Models 786-886-986-1086 Equipped With Open Center Systems

The pump for the hydraulic lift system
is located in the left forward end of the
differential portion of tractor rear main
frame. Pump is driven from a gear
mounted on aft end of the pto driven
shaft. See Fig. 274. Pump is 12 gpm
capacity.

227. **R&R AND OVERHAUL.** To remove the hydraulic pump, drain the

*Fig. 274 — Hydraulic lift system pump (8) is
mounted at left of differential compartment. Differential has been removed for illustrative purposes.*

10. Suction tube

38. Drive gear

tractor rear main frame, then unbolt and remove mounting plate, spacer and pump. Pump and spacer can now be separated from mounting plate by removing the attaching cap screws. See Fig. 275.

228. OVERHAUL CESSNA. With pump removed from spacer, proceed as follows: Remove pump drive gear and key, then unbolt and remove covers (2 and 14 – Fig. 276). Balance of disassembly will be obvious after an examination of the unit.

Pump specifications are as follows:

12 Gpm Pump

O.D. of shafts at bushings
(min.). . . . . 0.810 in.
I.D. of bushings in body and
cover (max.) . . . . 0.816 in.
Thickness (width) of gears
(min.). . . . . 0.572 in.
I.D. of gear pockets (max.) . . . . 2.002 in.
Max. allowable shaft to
bushing clearance . . . . 0.006 in.

When reassembling, use new diaphragms, gaskets, back-up washers, diaphragm seal and "O" rings. With open part of diaphragm seal (5) towards cover (2), work same into grooves of cover using a dull tool. Press protector gasket (6) and back-up gasket (7) into the relief of diaphragm seal. Install check ball (3) and spring (4) in cover, then install diaphragm (8) inside the raised lip of the diaphragm seal and be sure bronze face of diaphragm is toward pump gears. Dip gear and shaft assemblies in oil and install them in cover. Position wear plate (15) in pump body with the bronze side toward pump gears and cut-out portion toward inlet (suction) side of pump. In-

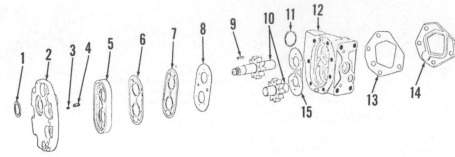

Fig. 276 — Exploded view of Cessna 12 gpm pump.

1. Oil seal
2. Cover
3. Ball
4. Spring
5. Diaphragm seal
6. Protector gasket
7. Back-up gasket
8. Pressure diaphragm
9. Key
10. Gears and shafts
11. "O" ring
12. Body
13. Gasket
14. Rear cover
15. Wear plate

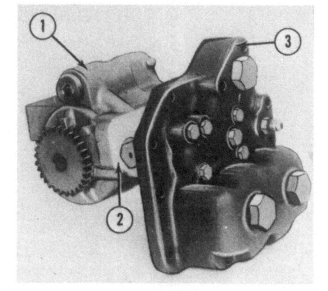

Fig. 277 — Hydraulic pump, compensator block and flange assembly for PFC equipped Model 986 and 1086 tractors.

1. Pump
2. Compensator block
3. Flange assembly

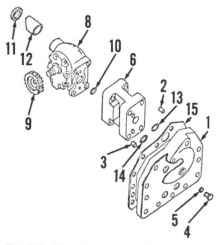

Fig. 275 — View of hydraulic lift pump, spacer and mounting flange on models equipped with open center system.

1. Mounting flange
2. Dowel
3. Dowel
4. Plug
5. "O" ring
6. Spacer
8. Pump
9. Drive gear
10. "O" ring
11. Seal
12. Suction tube
13. "O" ring
14. "O" ring
15. Gasket

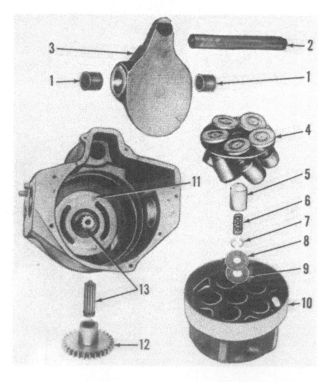

Fig. 278 — Exploded view of PFC axial piston pump.

1. Trunnion bearings
2. Pin
3. Swashplate
4. Pistons and retainer
5. Ball seat
6. Barrel spring
7. Split ring retainers
8. Keeper
9. Washer
10. Barrel
11. Port plate
12. Drive gear
13. Gear shaft (shown installed and removed)

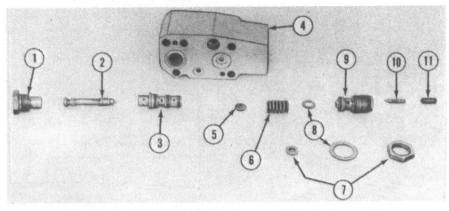

Fig. 279 — Exploded view of PFC system compensator block assembly.

1. Air bleed valve body
2. Compensator spool
3. Compensator spool body
4. Compensator block
5. Spring retainer
6. Bias spring
7. Locknuts
8. Washers
9. High pressure assembly
10. Dart
11. Adjustment screw

stall pump body over gears and shafts and install retaining cap screws. Torque cap screws to 20 ft.-lbs.

Check pump rotation. Pump will have a slight amount of drag but should turn evenly.

Models 986-1086 Equipped With PFC Systems

Located in the rear frame of the tractor is a five piston pressure and flow compensated (PFC) axial piston pump. The location and driv of the pump is the same as all other models. Maximum flow at rated engine speed is 17 gpm.

229. **REMOVE AND REINSTALL.** To remove the hydraulic pump, drain tractor rear frame, then unbolt and remove pump flange (3–Fig. 277), compensatorsator (2) and pump. Remove cap screws on flange and separate from

compensator block. Remove Allen bolt on compensator and separate from pump.

230. PFC PUMP. With pump removed from compensator block proceed as follows: Remove plug and spring from pump housing. Remove pump cover plate and then remove retaining rings from each side of trunnion caps. Remove trunnion bearings (1–Fig. 278), pin (2) and lift swashplate (3) out of pump housing. Remove pistons and retainer (4), ball seat (5) and barrel spring (6). Tap on drive gear and remove split rings (7), keeper and washer (8 & 9), then remove barrel (10). Lift our port plate (11), then push out drive gear (12) and gear shaft (13).

PFC pump and compensator are serviced as an assembly.

Clean and inspect all parts for scratches or wear. Service parts are not

available for the PFC pump. Reassemble pump by reversing disassembly procedure. Use RTV sealant on pump cover. Check pump rotation.

231. COMPENSATOR. Attached to the pump is a compensator block assembly. The compensator reads signal pressures from the various components of the hydraulic system and directs the pump to adjust the hydraulic flow required to do the work.

232. DISASSEMBLE COMPENSATOR. With pump, compensator and flange assembly removed as outlined in paragraph 229, proceed as follows: Remove compensator from pump dowel pins. Remove air bleed plug (1–Fig. 279), then remove complete compensator spool body.

Clean and inspect all components. Service parts are not available for the compensator. Reassemble by reversing disassembly procedure.

AUXILIARY CONTROL VALVE

All Models

233. Tractors may be equipped with up to three auxiliary valves and couplers and a motor valve and coupler. Auxiliary valves are located directly beneath the control levers mounted under the control platform.

Hydraulic flow for the motor valve, and each auxiliary valve is supplied by the hitch pump. Fluid returning from the supercharge line passes through the center ports of the auxiliary valve in neutral position. When the valve is actuated, the flow stops through the center ports and the system goes on pressur, directing fluid to the cylinder. The auxiliary valves have a "float" position which joins both sides of the cylinder, opening the center ports thus placing the system in a neutral position and avoiding automatic unlatching.

234. OVERHAUL AUXILIARY VALVE. To remove auxiliary valve disconnect control linkage, hydraulic lines and remove mounting bolts and then remove valves.

To disassemble, refer to Figs. 284 and 285 and proceed as follows: Remove end cap (26) and pull detent assembly and actuator from valve body (1). Remove sleeve (19) and pull remaining parts from the valve body (1). Check valve seat (7) and poppet (6) can be removed after plug (8) is removed. Detent (21, 22, 23 and 24) may be removed after rfemoving plug (25). Push out unlatching piston (18) out of actuator (13) using a long thin punch.

Inspect all parts for nicks, burrs, scoring and undue wear and renew parts as necessary. Use all new "O" rings and reassemble by reversing the disassembly procedure. Detent unlatching pressure

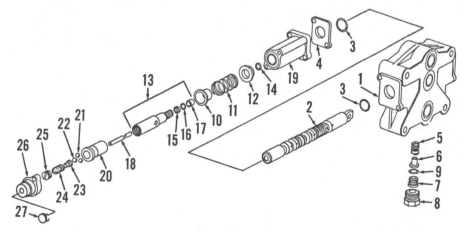

Fig. 284 — Exploded view of auxiliary valve for 786, 886, 986 and 1086 tractors equipped with open center systems.

1. Valve body
2. Spool
3. "O" ring
4. Retainer
5. Spring
6. Poppet
7. Seat
8. Plug
9. "O" ring
10. Spring cup
11. Spring
12. Spring retainer
13. Actuator assembly
14. "O" ring
15. Washer
16. "O" ring
17. "O" ring retainer
18. Unlatching piston
19. Sleeve housing
20. Sleeve
21. Ball
22. Ball
23. Washer
24. Detent spring
25. Adjusting screw
26. Cap
27. Plug

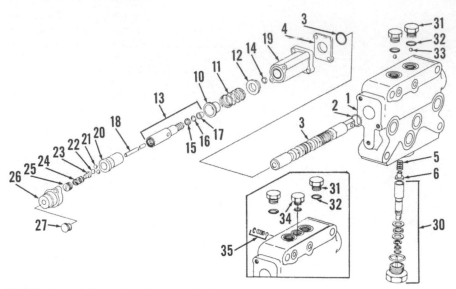

Fig. 285 — Exploded view of auxiliary valve for Models 986 and 1086 equipped with PFC. Insert shows motor valve with plug (34) and check valve and spring assembly (35).

1. Valve body
2. "O" ring
3. Spool
4. Retainer
5. Spring
6. Poppet
10. Spring cup
11. Spring
12. Spring retainer
13. Actuator assembly
14. "O" ring
15. Washer
16. "O" ring
17. "O" ring retainer
18. Unlatching piston
19. Sleeve housing
20. Sleeve
21. Ball
22. Ball
23. Washer
24. Detent spring
25. Adjusting screw
26. Cap
27. Plug
30. Flow Control valve assembly
31. Plug
32. "O" ring
33. Ball 3/16-in.
34. Plug
35. Check valve assembly

is adjusted by plug (25). Unit must unlatch at not less than 1650 psi or more than 2000 psi for 786, 886 and 986 models with open center systems or not less than 1900 psi or more than 2250 psi for Model 1086 with open center system. For Models 986 and 1086 with PFC systems, valve must unlatch at not less than 2000 psi or more than 2450 psi.

FLOW CONTROL VALVE (PFC SYSTEM)

235. Located on the auxiliary or motor valves, between the two lines going to each valve are the flow control valves. Fluid must pass through a cable operated valve spool before entering the

center ports. By placing the flow control sector in either the up or down position, flow to the auxiliary or motor valves can be regulated in slow or fast flow rate.

To service valve refer to Fig. 286 and proceed as follows: Disconnect cable (2) from valve, loosen locknut (3) and screw out valve.

Inspect valve (30 – Fig. 285) for scratches, burrs or undue wear and

renew if necessary. Use new "O" rings and reassemble by reversing disassembly procedure. Check and adjust flow control valve as outlined in paragraph 202.

CHECK VALVE

All Models

236. A double acting check valve is used with auxiliary system rear outlet which checks the flow of fluid in both directions and precludes the possibility of equipment dropping either during transport or while parked.

Removal and disassembly of the unit will be obvious upon examination of the unit and reference to Fig. 287.

HYDRAULIC SEAT

All Models So Equipped

237. A hydraulic controlled seat attachment is available. Fluid is supplied from a tee connection in the brake supply line. See Fig. 288. Fluid flows to the seat control valve only when seat is being raised. The speed of raise is controlled by a 0.054-0.058 inch drilled orifice in the control valve. Return oil from the single action cylinder is dumped back in reservoir. Seat ride control is adjustable and is controlled by rotating the ride control valve knob. This needle valve adjusts the variable orifice opening between the seat cylinder and a nitrogen filled accumulator.

238. **CONTROL VALVE.** To disassemble the control valve, remove pivot

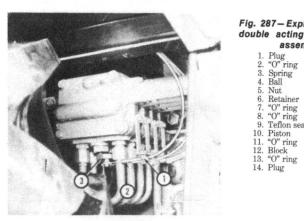

Fig. 286 — View of auxiliary valves and flow control valve cables for PFC equipped tractors.
1. Cable adjusting locknut
2. Cable
3. Valve

Fig. 287 — Exploded view of double acting check valve assembly.
1. Plug
2. "O" ring
3. Spring
4. Ball
5. Nut
6. Retainer
7. "O" ring
8. "O" ring
9. Teflon seal
10. Piston
11. "O" ring
12. Block
13. "O" ring
14. Plug

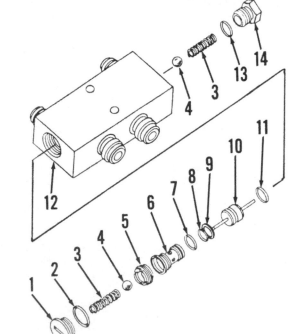

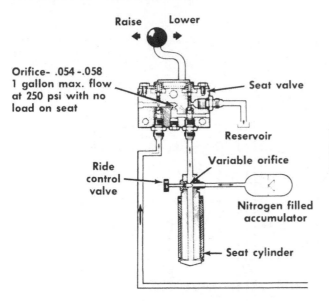

Fig. 288 — Schematic view of hydraulic circuit, control valve and cylinder of hydraulically controlled seat attachment.

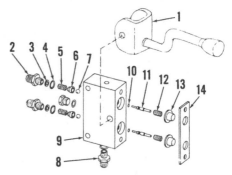

Fig. 289 — Exploded view of hydraulic seat control valve.

1. Control lever	8. Connector
2. Connector	9. Body
3. Washer	10. "O" ring
4. "O" ring	11. Piston
5. Spring	12. Spring
6. Retainer	13. Cap
7. Ball	14. Plate

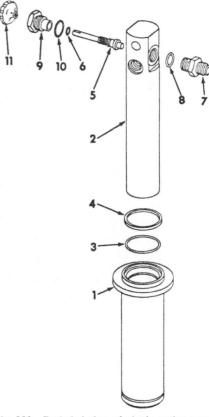

pin and control lever (1 – Fig. 289). Unbolt and remove plate (14), then withdraw caps (13), springs (12) and pistons (11) with "O" rings (10). Unscrew connectors (2) and remove washers (3), springs (5), retainers (6) and balls (7). Clean and inspect all parts for excessive wear or other damage.

Renew all "O" rings and reassemble by reversing the disassembly procedure.

239. CYLINDER. Disassembly of the single action seat cylinder is obvious after an examination of the unit and reference to Fig. 290. Clean and inspect all parts for excessive wear or other damage.

When reassembling, use new "O" rings (3, 6, 8 and 10) and new wiper ring (4).

CONTROL CENTER (CAB)

All Models So Equipped

240. REMOVE AND REINSTALL. Remove all front hood sheet metal. Shut off air conditioner suction side at compressor. Disconnect suction line and move it out of the engine compartment to the rear. After disconnecting and plugging the heater hoses move them back to the control center. Remove the left and right headlight panel covers. Disconnect the air conditioner quick connector and electrical connector. Remove main fuel tank, then disconnect the taillight wiring at the rear frame. Inside the control center remove the instrument panel cover and loosen the side bracket mounting screws. With the drivers seat positioned fully forward, remove the rear support bracket. Remove all knobs from all side control levers then remove the control console covers.

NOTE: Drivers seat must remain fully forward for clearance purposes.

If available, install and center the 17-259 Control Center Handler or equivalent hoist arrangement. Remove the right and left side steps. Remove the Isomount top hex nuts and lift off control center while observing any wiring, lines or linkage that might hang up on components. Installation of control center is reverse of removal procedure. On installation of Isomount hex nuts should be torqued to 68-77 ft.-lbs.

Fig. 290—Exploded view of single action seat cylinder.

1. Cylinder barrel	6. "O" ring
2. Cylinder ram	7. Connector
3. "O" ring	8. "O" ring
4. Wiper ring	9. Retainer
5. Ride control adjusting	10. "O" ring
screw	11. Knob

WIRING DIAGRAM (766 gas tractor, with external voltage)

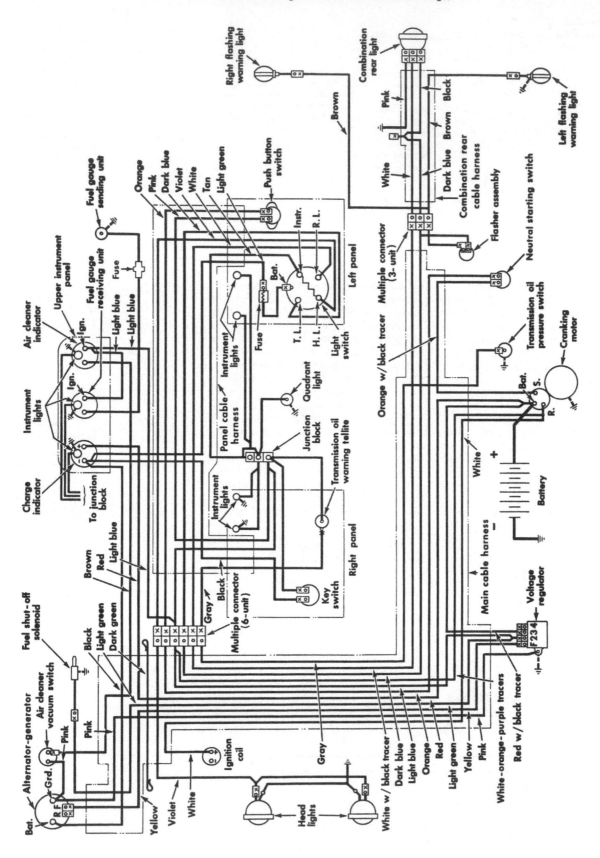

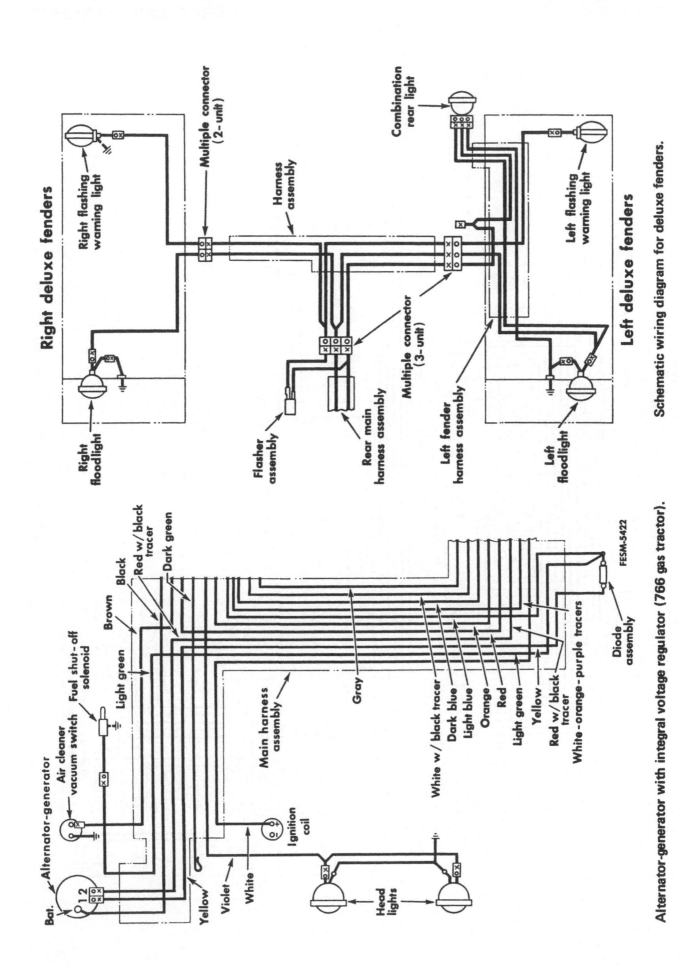

Right deluxe fenders

Right flashing warning light

Multiple connector (2- unit)

Harness assembly

Right floodlight

Flasher assembly

Rear main harness assembly

Multiple connector (3- unit)

Left fender harness assembly

Combination rear light

Left flashing warning light

Left deluxe fenders

Left floodlight

Schematic wiring diagram for deluxe fenders.

Black

Red w/ black tracer

Dark green

Brown

Light green

Air cleaner vacuum switch

Fuel shut- off solenoid

Alternator-generator

Main harness assembly

Gray

White w/ black tracer

Dark blue

Light blue

Orange

Red

Light green

Yellow

Red w/ black tracer

White – orange – purple tracers

Diode assembly

FESM-5422

Bat.

Ignition coil

Yellow

Violet

White

Head lights

Alternator-generator with integral voltage regulator (766 gas tractor).

93

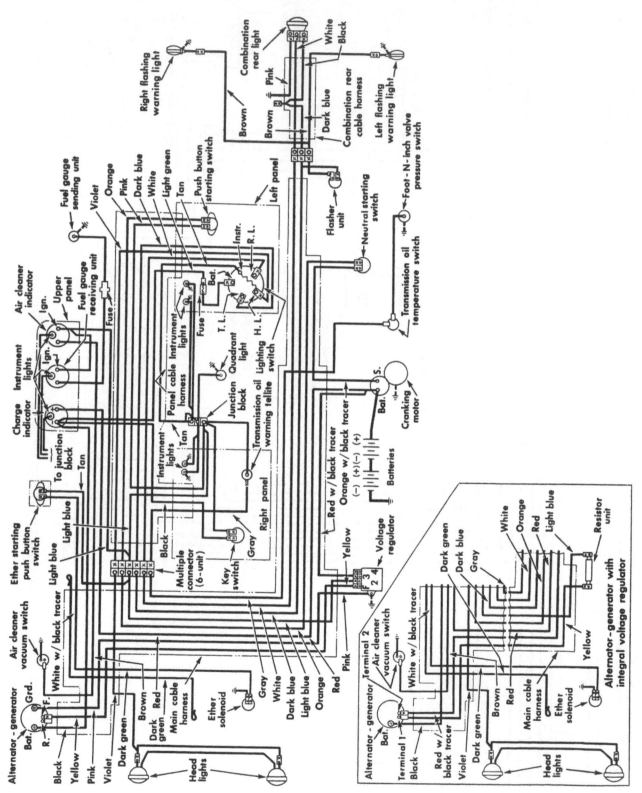

Hydrostatic Drive Tractors (966 and 1066) Schematic Wiring Diagram.

Gear drive tractors (766 diesel, 966 and 1066) schematic wiring diagram

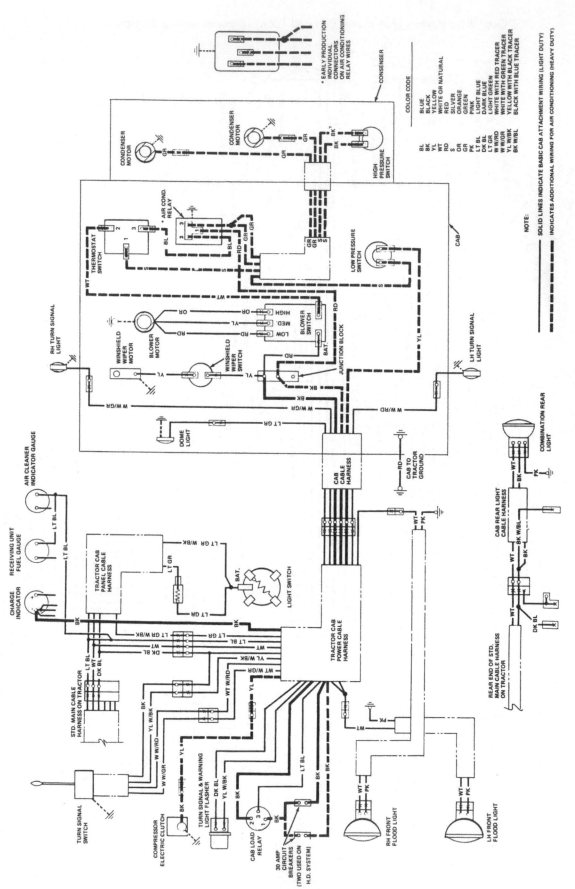

Schematic wiring diagram for cab.